AF690660

Metasurface Electromagnetics

The ACES Series on Computational and Numerical Modelling in Electrical Engineering

Andrew F. Peterson, PhD – Series Editor

The volumes in this series will encompass the development and application of numerical techniques to electrical and electronic systems, including the modelling of electromagnetic phenomena over all frequency ranges and closely related techniques for acoustic and optical analysis. The scope includes the use of computation for engineering design and optimization, as well as the application of commercial modelling tools to practical problems. The series will include titles for senior undergraduate and postgraduate education, research monographs for reference, and practitioner guides and handbooks.

Titles in the Series

K. Warnick, **"Numerical Methods for Engineering,"** 2010.
W. Yu, X. Yang and W. Li, **"VALU, AVX and GPU Acceleration Techniques for Parallel FDTD Methods,"** 2014.
A.Z. Elsherbeni, P. Nayeri and C.J. Reddy, **"Antenna Analysis and Design Using FEKO Electromagnetic Simulation Software,"** 2014.
A.Z. Elsherbeni and V. Demir, **"The Finite-Difference Time-Domain Method in Electromagnetics with MATLAB® Simulations, 2nd Edition,"** 2015.
M. Bakr, A.Z. Elsherbeni and V. Demir, **"Adjoint Sensitivity Analysis of High Frequency Structures with MATLAB®,"** 2017.
O. Ergul, **"New Trends in Computational Electromagnetics,"** 2019.
D. Werner, **"Nanoantennas and Plasmonics: Modelling, design and fabrication,"** 2020.
K. Kobayashi and P.D. Smith, **"Advances in Mathematical Methods for Electromagnetics,"** 2020.
V. Lancellotti, **"Advanced Theoretical and Numerical Electromagnetics, Volume 1: Static, stationary and time-varying fields,"** 2021.
V. Lancellotti, **"Advanced Theoretical and Numerical Electromagnetics, Volume 2: Field representations and the method of moments,"** 2021.
S. Roy, **"Uncertainty Quantification of Electromagnetic Devices, Circuits, and Systems,"** 2021.

Metasurface Electromagnetics

The Cagniard-DeHoop time-domain approach

Martin Štumpf

The Institution of Engineering and Technology

Published by SciTech Publishing, an imprint of The Institution of Engineering and Technology, London, United Kingdom

The Institution of Engineering and Technology is registered as a Charity in England & Wales (no. 211014) and Scotland (no. SC038698).

© The Institution of Engineering and Technology 2022

First published 2022

This publication is copyright under the Berne Convention and the Universal Copyright Convention. All rights reserved. Apart from any fair dealing for the purposes of research or private study, or criticism or review, as permitted under the Copyright, Designs and Patents Act 1988, this publication may be reproduced, stored or transmitted, in any form or by any means, only with the prior permission in writing of the publishers, or in the case of reprographic reproduction in accordance with the terms of licences issued by the Copyright Licensing Agency. Enquiries concerning reproduction outside those terms should be sent to the publisher at the undermentioned address:

The Institution of Engineering and Technology
Michael Faraday House
Six Hills Way, Stevenage
Herts, SG1 2AY, United Kingdom

www.theiet.org

While the author and publisher believe that the information and guidance given in this work are correct, all parties must rely upon their own skill and judgement when making use of them. Neither the author nor publisher assumes any liability to anyone for any loss or damage caused by any error or omission in the work, whether such an error or omission is the result of negligence or any other cause. Any and all such liability is disclaimed.

The moral rights of the author to be identified as author of this work have been asserted by him in accordance with the Copyright, Designs and Patents Act 1988.

British Library Cataloguing in Publication Data
A catalogue record for this product is available from the British Library

ISBN 978-1-83953-613-7 (hardback)
ISBN 978-1-83953-614-4 (PDF)

Typeset in India by MPS Ltd

If you are receptive and humble, mathematics will lead you by the hand.
(P.A.M. Dirac)

Contents

About the author

Martin Štumpf is an associate professor of theoretical electrical engineering with the Department of Radioelectronics, Brno University of Technology, Czech Republic. His research interests include modeling of electromagnetic wave phenomena with an emphasis on electromagnetic compatibility and antenna engineering. He is a senior member of the IEEE and the IEEE Antennas and Propagation and Electromagnetic Compatibility Societies.

Foreword

It is a privilege for me to write a Foreword to Dr Stumpf's book.

The title "*Metasurface Electromagnetics: The Cagniard–DeHoop Time-domain Approach*" takes my memory back to a number of decades ago.

After obtaining my M.Sc. degree in Electrical Engineering at Delft University of Technology, the Netherlands, in 1950, I got the opportunity to pursue research on frequency-domain electromagnetic scattering theory as a junior staff member at this university.

In 1955 I received an invitation from Professor Louis B. Schlichter, Head of the Department of Geophysics at the University of California at Los Angeles (UCLA) to come over for a year and participate in the activities of the "seismic scattering project", a project sponsored by a number of US oil companies and supervised by Dr Leon Knopoff.

After having completed the necessary administrative preliminaries, getting a leave of absence from Delft to be granted by the Minister of Education, and acquiring (in the midst of the Cold War) a US Exchange Visitors Visa, I embarked in Rotterdam on the Holland–America Line steamer "Rijndam" for an 11-day transatlantic voyage to New York via Southampton and Halifax. From New York onward, I had to start a 3-day journey by train, via Chicago and the impressive Rocky Mountains, to Los Angeles.

On my voyage, I had with me Louis Cagniard's book "*Réflexion et Réfraction des Ondes Séismiques Progressives*" (1939) that had been recommended to me by a Delft professor of applied mathematics. The book highly triggered me and has—in hindsight—proven to be decisive for the course of my scientific career.

Cagniard had modeled the seismic wave generated by a point source in a semi-infinite elastic solid. Instead of using the, by that time standard, Fourier transformation with respect to time and the horizontal space coordinates (by that time standard), he had used a time Laplace transformation with positive real transform parameter, relying for the transformation back to the time domain on the Czech scientist Lerch's uniqueness theorem. To accomplish this, Cagniard introduced in the spatial Fourier domain a complex-valued time parameter, for which he claimed that the physical solution was the calculated function expression at the positive real axis of the *complex-valued* time parameter. This step sounded strange to me: time is a *real-valued* parameter all the way through!

Upon arrival at L.A., Leon Knopoff made clear to me that a frequency-domain analysis of an earthquake of any sizable magnitude would imply that the interpreter's

infrastructure would have been demolished already infinitely long ago. Hence, only pulsed time behavior would be of any value in the mathematical modeling of the phenomenon. Leon's guideline was the start of my research in time-domain wave field analysis. I started in this direction by addressing some two-dimensional problems: Lamb's line source problems in elastodynamics and the plane-wave diffraction by a semi-infinite barrier which I solved with what I denoted as the "*modified Cagniard*" method, and I wrote several reports on the subject.

Leon greeted my solutions with enthusiasm and instructed me "to keep my mouth shut": he wanted them to serve as a surprise at a workshop he intended to organize at the California Institute of Technology, Pasadena, CA, the home of famous seismologists as Richter and Gutenberg, whom I was honored to meet there in person. Early 1957, this seminar took place and I presented my material. Right after finishing my presentation, a gentleman approached me and said: "*I had better stop my project of translating Cagniard's book into English.*" The gentleman proved to be the well-known geophysicist Charles Hewitt Dix! I told him that, in my opinion, his translation would be a most valuable addition to the geophysical literature in English.

The experience of my time in L.A. laid the basis for my Ph.D. thesis to be defended in Delft and was the start of my further researches in pulsed-wave problems. Other scientists followed with applying my theory to arrive at analytic solutions to a variety of problems. To refer to my "modified Cagniard method", professor Robert Burridge coined, in a paper published in *The Quarterly Journal of Mechanics and Applied Mathematics* (1971), the designation "*Cagniard–DeHoop*" transformation, a designation appearing in the literature ever since.

Up to now, the method has predominantly been applied to obtain analytic results. Dr. Stumpf has now taken the highly original step of letting the method be helpful and of great use in the parts of a computational code that handles the time behavior in the phenomenon under consideration. Examples of great interest can be found in detail in this, his book.

I feel most honored that Dr. Stumpf has proven to be a scientist who—with great competence—has embarked on giving a new direction to the "modified Cagniard" technique with which endeavor I wish him every success.

Adrianus T. de Hoop
Delft, the Netherlands
September 22, 2021

Preface

Thin layered structures are nowadays commonly encountered as composite materials in the aircraft, automotive and building industries in the form of purposefully engineered structures such as antenna radomes, radar absorbing materials, telecommunication sensors, optical coatings, solar cells and lots of others. Owing to the present-day ubiquity of pulsed electromagnetic field transfers effectuated via modern ultra-wide-band communication and radar systems, however, it is seems no longer adequate to limit electromagnetic scattering analyses of thin-layered structures by the steady-state assumption. Consequently, the present book is entirely dedicated to the exploration of *time-domain* electromagnetic fields in the presence of thin, high-contrast sheets, with an emphasis on metasurfaces combining magnetic and dielectric properties.

Since the number of problems that are amenable to exact solutions in terms of analytic functions is limited, our analysis is not restricted to analytical methods only, but attention is paid to the development of computational and approximate techniques too. All the solution methodologies presented in the book heavily rely on the Cagniard–DeHoop technique. Regrettably, perhaps because of its origins in seismology, it is felt that this powerful mathematical tool is still not fully appreciated in the electromagnetic and antenna community. Therefore, it is hoped that this book demonstrating the truly broad applicability of the Cagniard–DeHoop technique to achieving both analytical and numerical time-domain solutions will help, at least partly, to remedy the issue.

M. Štumpf
Brno, Czech Republic
September 2021

Acknowledgments

I am grateful to my wife and daughters for their forbearance they showed over years of writing the manuscript. Further, I wish to thank Lorentz Chair Emeritus Professor Adrianus T. de Hoop (Delft University of Technology) for his continued interest and advice. I also desire to thank Professors Giulio Antonini (University of L'Aquila), Ioan E. Lager (Delft University of Technology) and Guy A. E. Vandenbosch (KU Leuven) for their encouragement.

This research was financially supported by the Czech Science Foundation under Grant No. 20-01090S.

M. Š.

List of acronyms

CdH	Cagniard–DeHoop
CdH-MoM	Cagniard–DeHoop method of moments
EM	Electromagnetic
FD	Frequency domain
MoM	Method of moments
PEC	Perfect electric conductor
PMC	Perfect magnetic conductor
TEM	Transverse electromagnetic
TD	Time domain
WH	Wiener–Hopf

List of symbols

$\boldsymbol{E}$	Electric field strength $(\mathrm{V/m})$
$\boldsymbol{H}$	Magnetic field strength $(\mathrm{A/m})$
$\boldsymbol{J}$	Volume density of electric current $(\mathrm{A/m^2})$
$\boldsymbol{K}$	Volume density of magnetic current $(\mathrm{V/m^2})$
$\partial\boldsymbol{J}$	Surface density of electric current $(\mathrm{A/m})$
$\partial\boldsymbol{K}$	Surface density of magnetic current $(\mathrm{V/m})$
c_0	Electromagnetic wave speed in vacuum (= $299792458\,\mathrm{m/s}$)
μ_0	Magnetic permeability in vacuum (= $4\pi \times 10^{-7}\,\mathrm{H/m}$)
ϵ_0	Electric permittivity (= $1/c_0^2\mu_0$)
η_0	Electromagnetic wave admittance in vacuum (= $1/c_0\mu_0$)
ζ_0	Electromagnetic wave impedance in vacuum (= $c_0\mu_0$)
μ	Magnetic permeability $(\mathrm{H/m})$
ϵ	Electric permittivity $(\mathrm{F/m})$
σ	Electric conductivity $(\mathrm{S/m})$
G^{E}	Equivalent layer conductance (S)
C^{E}	Equivalent layer capacitance (F)
R^{M}	Equivalent layer resistance (Ω)
L^{M}	Equivalent layer inductance (H)
$\{x, y, z\}$	Cartesian coordinates with respect to $\{\boldsymbol{i}_x, \boldsymbol{i}_y, \boldsymbol{i}_z\}$ basis
$\{r, \phi, z\}$	Cylindrical coordinates with respect to $\{\boldsymbol{i}_r, \boldsymbol{i}_\phi, \boldsymbol{i}_z\}$ basis
$\{\partial_x, \partial_y, \partial_z\}$	Spatial differentiations with respect to $\{x, y, z\}$, respectively
$\{\partial_r, \partial_\phi, \partial_z\}$	Spatial differentiations with respect to $\{r, \phi, z\}$, respectively
t	Time coordinate (s)
$*_t$	Time-convolution operator (see (1.2))
∂_t	Time differentiation operator
∂_t^{-1}	Time integration operator (see (1.3))
$\mathrm{H}(t)$	Heaviside unit-step function
$\delta(t)$	Dirac-delta distribution
s	Time Laplace-transform parameter (s^{-1})
$\{\kappa, \sigma\}$	Slowness parameters along $\{\boldsymbol{i}_x, \boldsymbol{i}_y\}$, respectively $(\mathrm{s/m})$
$\{\gamma, p\}$	Slowness parameters along $\{\boldsymbol{i}_r, \boldsymbol{i}_z\}$, respectively $(\mathrm{s/m})$
$f(x) = O(\phi)$	The Landau order symbol big O (= $\lvert f(x)\rvert/\phi < A$)
$f(x) = o(\phi)$	The Landau order symbol little o (= $\lvert f(x)\rvert/\phi \to 0$),

where A is a real number and $\phi = \phi(x)$ is a positive function

Chapter 1
Introduction

The controlled synthesis of thin-film structures has given rise to major technological breakthroughs in electronics and semiconductor industry with diverse applications including micro- and nanoelectronic devices, data and energy storage technologies, integrated optical devices and optical coatings. Design methodologies of such structures are essentially based on computational tools for solving, in an approximate way, Maxwell's equations. Owing to the multiscale nature of thin-film problems, the standard, general-purpose numerical techniques lead to unacceptably high computational efforts and very limited insights into the physics of thin layers. Despite the still increasing number of applications relying entirely upon the pulsed electromagnetic (EM)-field operation, dedicated time-domain (TD) modeling techniques for analyzing the pulsed EM-field interaction with thin-layer structures are very rare in the relevant literature. With this gap in EM theory in mind, this book aims at developing new sophisticated modeling methods and closed-form solutions, thereby providing enablers for future developments of thin-layer-based technologies.

Significant research efforts have been recently invested in designing spatially periodic structures having exotic EM scattering properties. Such structures, widely known as *metamaterials*, are mostly designed by arranging a large number of electrically small scatterers into an ordered three-dimensional (3-D) array structure. Significant EM losses and design difficulties associated with bulky metamaterial structures can be avoided by organizing the scatterers in a two-dimensional (2-D) regular pattern of vanishing thickness (see Figure 1.1), thus yielding the notion of *metasurfaces*.

To ease the EM modeling of their crystal, highly inhomogeneous structure, the EM scattering by metasurfaces is commonly analyzed with the aid of bulk material parameters [1]. Indeed, under the influence of an external EM field, the EM fields associated with the scatterers' electric and magnetic dipole moments are averaged, which makes it possible to express the induced equivalent electric- and magnetic-current surface densities on a metasurface via the (bulk) electric and magnetic polarizabilities or, equivalently, via the equivalent sheet admittance and impedance [2]. The presence of such metasurfaces is then accounted for via a pair of cross-layer conditions, where the equivalent layer's admittance and impedance interrelate the (discontinuous) tangential electric and magnetic fields on its top and bottom sides. Mathematically, the model of a Huygens' metasurface bears strong similarities with the concept of *combined sheets* as introduced by Senior [3]. Hence, in accordance with a broad definition of metasurfaces [4], thin, highly contrasting

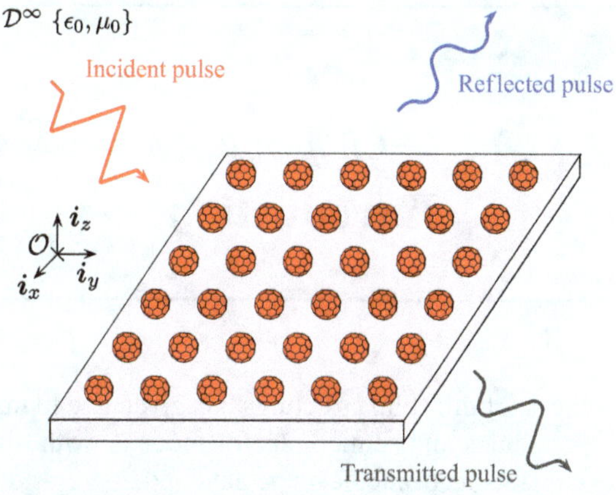

Figure 1.1 General metasurface consisting of small scatterers. The incident EM pulse is reflected against and transmitted through the metasurface

layers with combined magneto-dielectric properties can be viewed as a class of scalar metasurfaces. Since the TD analysis of such metamaterial structures is virtually *terra incognita* of EM-field theory, the purpose of this book is to explore the subject.

The opening chapter makes the reader acquainted with the structure of the book, its notational conventions and some mathematical concepts that are useful for analyzing the class of EM scattering problems under consideration.

1.1 Synopsis

First, after introducing notational conventions and some prerequisites including the definition of the Laplace transformation, in Chapter 2 we introduce the reader to joint transform methods and, in particular, to the Cagniard–DeHoop (CdH) technique that is the central mathematical tool of this book. Here, for the sake of illustration, the CdH inversion approach is first applied to analyzing the EM radiation from an impulsive point source in an unbounded, isotropic homogeneous medium. Subsequently, it is demonstrated that the CdH technique is suitable for solving wave propagation problems in horizontally layered media, thereby formulating the so-called generalized-ray theory.

Next, in Chapter 3, we derive thin-sheet high-contrast cross-layer conditions. Their various forms pertaining to layers supporting either electric or both electric and magnetic currents are discussed and frequently applied throughout the book.

In the following chapter, we provide a number of illustrative applications of the cross-layer conditions. For the sake of clarity, the analysis carried out in Chapter 4 is limited to the EM response of an *infinite* metasurface with combined magnetic and dielectric properties. Potentialities of metasurfaces in achieving the transparency for causal impulsive plane-wave, line-source and loop-antenna excitations are discussed in detail. Numerical examples demonstrating the intriguing TD reflection and transmission properties of metasurfaces are given.

Chapter 5 is concerned with the existence of EM waves propagating along the surface of a thin, highly contrasting layer. To that end, line-source excited pulsed EM surface effects are analytically studied with regard to the polarization of the exciting EM source, its (space–time) relation with respect to the field point of observation and the constitutive EM properties of the layer. It is shown that under certain circumstances, anomalous oscillatory EM transients on the surface of a thin slab can be excited. The existence of the EM surface phenomenon is further demonstrated by analyzing the pulsed EM signal transfer between two loops in a 3-D problem configuration.

In Chapter 6, we discuss the applications of the cross-layer conditions to modeling the TD response of *semi-infinite* screens. To that end, a powerful analytical methodology based on a combination of the CdH and Wiener–Hopf (WH) techniques is employed. Closed-form analytical TD solutions and their approximations are given for a variety of problem configurations including a penetrable semi-infinite metasurface with combined magneto-dielectric constitutive properties and their junctions. Accompanying TD edge-diffracted effects are illustrated via numerical examples.

Before formulating a general TD computational methodology capable of calculating the space–time induced current distribution over a bounded planar sheet, in Chapter 7 we introduce an approximate technique applying to the pulsed EM scattering by relatively *narrow* planar strips located in a homogeneous, isotropic and loss-free medium. Starting with the straightforward reciprocity-based formulation concerning the EM scattering by a narrow, perfectly electrically conducting (PEC) strip, it is shown that the proposed solution strategy can be readily generalized to account for complex material properties, thus enabling the efficient analysis of Drude-type plasmonic strips and metastrips.

Whenever analytical and approximate methods are no longer applicable, one has to resort to numerical solutions. Such a difficulty is handled in Chapter 8, where the pulsed EM-field interaction with a *bounded* planar screen embedded in a homogeneous, isotropic and loss-free medium is analyzed. It is shown that an efficient tool to tackle the TD EM scattering problem is the CdH method of moments (CdH-MoM), whose EM reciprocity-based formulation is thoroughly described. Again, the analysis is gradually extended from PEC screens to EM-penetrable metasurfaces exhibiting relatively complex (plasmonic and combined magneto-dielectric) material properties. Numerical results are finally validated with the aid of both Kirchhoff's and narrow-strip approximate solutions.

The CdH-MoM is well applicable to problem configurations consisting of a planar screen imposed on a multilayered background medium. This property is demonstrated in Chapter 9, where the pulsed EM scattering by a screen in the presence of a dielectric layer or/and a PEC ground plane is analyzed in detail. The CdH-MoM solutions of the analyzed problem configurations are supplemented with their approximations applying to a relatively narrow screen. The narrow-strip approximations are useful for gaining physical insights into the pulsed EM-field coupling phenomena as well as for demonstrating the validity of the numerical solution.

In order to further generalize the CdH-MoM, in Chapter 10, we propose a computational methodology for analyzing the pulsed EM scattering by a set of bounded, mutually coupled conductive screens located in both homogeneous and layered media. Partly for validation purposes, again, the numerical CdH-MoM solutions are accompanied with their narrow-strip approximations. Numerical examples clearly demonstrate the effect of mutual-coupling phenomena on the induced current response.

The pulsed EM-field coupling between two planar bounded metasurfaces with combined magneto-dielectric properties is the topic of Chapter 11. Relying, again, on the EM reciprocity-based formulation, it is shown that the CdH-MoM is capable of calculating induced electric and magnetic currents on the coupled metasurfaces. Furthermore, the problem of EM coupling between two metasurfaces is solved approximately under the assumption of their (relatively) small width. Again, the numerical CdH-MoM solution is validated with the aid of the pertaining asymptotic solution.

If the analyzed problem's dimensionality cannot be reduced, a fully (3+1)-D solution approach is needed. Therefore, in Chapter 12, we formulate a general, computational CdH-MoM methodology for analyzing the pulsed EM scattering by thin, highly contrasting planar screens of arbitrary shape. An emphasis is placed on the analysis of 3-D metasurfaces with combined magneto-dielectric properties. Numerical results obtained using the proposed CdH-MoM are finally compared with the corresponding Kirchhoff approximation, thus demonstrating the range of its applicability.

In Chapter 13, the EM reciprocity theorem is applied to analyze the pulsed EM scattering by various apertures in a planar screen. First, it is demonstrated that the 2-D problem of a slot in a PEC screen can be handled numerically with the help of the CdH-MoM, again. Consequently, in order to shed some light on the pertaining physics, closed-form solutions are derived under the narrow-slot approximation. In particular, we analyze the impact of a dielectric half-space of a homogeneous slot's filling as well the EM scattering by rectangular grooves and their mutual EM coupling. Finally, it is shown that the CdH-MoM is readily capable of analyzing the pulsed EM scattering by a bounded aperture of arbitrary shape.

The problem of interaction of pulsed EM waves with *time-varying* screens is touched upon in Chapter 14. Here, we demonstrate that the transient plane-wave scattering by a class of relatively simple metasurfaces can be analyzed analytically via a straightforward solution approach.

A dedicated application of the CdH-MoM is presented in Chapter 15, where the pulsed EM-field coupling to a straight transmission line is analyzed. Starting with a simple transmission line above the ideal PEC ground plane, an approximate approach to the inclusion of bounded electric and magnetic properties of the plane is proposed. Illustrative numerical results demonstrate the impact of both electric and magnetic constitutive parameters of the metasurface on the transient response of a transmission line.

In Chapter 16, we provide a few representative examples demonstrating the wide applicability of cross-layer conditions to solving diverse TD EM problems. First, a

general analytical CdH-based methodology for calculating short-range approximations is illustrated on the analysis of the pulsed EM-field loop-to-loop signal transfer across a thin conductive screen. The subsequent illustrative problem is concerned with the analytical description of the pulsed EM-field transfer via a thin layer with Drude-type plasmonic properties. Finally, the (modified) Kirchhoff approximation is applied to derive straightforward TD expressions describing the pulsed EM radiation from a slot-excited resonator antenna. Numerical examples clearly show the physical meaning of individual wave constituents forming the total wave motion.

1.2 Prerequisites

The mathematical description of EM wave field phenomena is effected via EM-field quantities that are, in general, functions of space and time. To register the position in the problem configuration under consideration, we employ a Cartesian reference frame with the origin $\mathcal{O}$ and its (standard) base vectors $\{\boldsymbol{i}_x, \boldsymbol{i}_y, \boldsymbol{i}_z\}$ that form, in the indicated order, a right-handed system. Hence, in accordance with common typographic conventions, a vector (or, more generally, a tensor of rank one) will be further represented by a bold-face italic symbol. For example, the position vector will be denoted by $\boldsymbol{r}$ and represented via the linear combination

$$\boldsymbol{r} = x\boldsymbol{i}_x + y\boldsymbol{i}_y + z\boldsymbol{i}_z, \tag{1.1}$$

where numbers $\{x, y, z\}$ are the Cartesian coordinates of the point of observation. The time coordinate is denoted by t. The partial differentiation with respect to a coordinate will be then denoted by ∂ that is supplied with the pertaining subscript. For example, the spatial differentiation with respect to z is denoted by ∂_z. The vectorial spatial differentiation (nabla) operator can be then written as $\boldsymbol{\nabla} = \partial_x\boldsymbol{i}_x + \partial_y\boldsymbol{i}_y + \partial_z\boldsymbol{i}_z$. The time convolution of two causal EM-field quantities is defined as

$$[f *_t g](\boldsymbol{r}, t) = \int_{\tau=0}^{t} f(\boldsymbol{r}, \tau)g(\boldsymbol{r}, t-\tau)\mathrm{d}\tau \quad \text{for all } t \geq 0. \tag{1.2}$$

If we let $g(\boldsymbol{r}, t) = \mathrm{H}(t)$, where $\mathrm{H}(t)$ denotes the Heaviside unit-step function, we get a special case of (1.2), that is,

$$\partial_t^{-1} f(\boldsymbol{r}, t) = \int_{\tau=0}^{t} f(\boldsymbol{r}, \tau)\mathrm{d}\tau \quad \text{for all } t \geq 0, \tag{1.3}$$

where ∂_t^{-1} denotes the time integration operator.

Throughout this book, EM wave field quantities will be frequently represented via their images in the Laplace-transform domain [5–7]. Owing to the universal property of causality, we are always interested in the behavior of an EM space–time quantity, say $f(\boldsymbol{r}, t)$, in the interval $\mathcal{T} = \{t \in \mathbb{R}; t > 0\}$ after its excitation at $t = 0$. Consequently, the one-sided Laplace transform of the causal EM quantity is defined by

$$\hat{f}(\boldsymbol{r}, s) = \int_{t=0}^{\infty} \exp(-st) f(\boldsymbol{r}, t)\mathrm{d}t, \tag{1.4}$$

where $\{s \in \mathbb{C}; s \geq s_0\}$ is the Laplace-transform parameter (complex frequency). A sufficient condition for the integral to exist is its absolute convergence along the line $\{s \in \mathbb{C}; \mathrm{Re}(s) = s_0\}$ that is parallel to the imaginary axis of the complex s-plane (see Figure 1.2). From practical reasons, we shall restrict ourselves to EM-field quantities that are bounded, in other words, that are of the zero exponential order as $t \to \infty$. As a consequence, $\exp(-st)f(\boldsymbol{r}, t) = o(1)$ as $t \to \infty$ to the right of the line of convergence. Throughout this region, the one-sided Laplace transform of a causal EM quantity does exist and is here, due to the analyticity of the transform kernel $\exp(-st)$, an analytic function of the complex frequency s.

Both analytical and computational methodologies developed in this book employ the CdH joint transform technique [8] that heavily relies on Lerch's uniqueness theorem [9, Appendix]. The uniqueness theorem states that if the Laplace transform defined by (1.4) is to hold for all s along a Lerch sequence $\mathcal{L} = \{s \in \mathbb{R}; s = s_0 + nh, s_0 > 0, h > 0, n = 0, 1, 2, \dots\}$ (see Figure 1.2), then a one-to-one mapping exists between the (causal) EM-field quantity $f(\boldsymbol{r}, t)$ and its Laplace-transform image $\hat{f}(\boldsymbol{r}, s)$. Hence, once we manage to cast the solution of an initial-boundary value EM problem in the form of Laplace transform (1.4) with $s \in \mathcal{L}$, its TD original immediately follows by inspection without the need for carrying out the actual inverse transform. More generally, the CdH approach eventually yields the s-domain EM-field quantity in the following form:

$$\hat{f}(\boldsymbol{r}, s) = \int_{\tau=T}^{\infty} \exp(-s\tau)\hat{F}(\boldsymbol{r}, \tau, s)\mathrm{d}\tau, \tag{1.5}$$

with $s \in \mathcal{L}$ and $T > 0$, where $\hat{F}(\boldsymbol{r}, \tau, s)$ is an analytic function in $\{s \in \mathbb{C}; \mathrm{Re}(s) > s_0\}$ that is bounded as $|s| \to \infty$. Owing to the s-dependency of $\hat{F}(\boldsymbol{r}, \tau, s)$, (1.5) does not have, in general, the form of the Laplace transform (1.4). Consequently, the desired TD original $f(\boldsymbol{r}, t)$ cannot be found directly by inspection [10, Section 6.6]. But, this difficulty can be readily circumvented with the aid of the Schouten–Van der Pol theorem [11, p. 1056]. To that end, the (causal) TD original $F(\boldsymbol{r}, \tau, t)$ of $\hat{F}(\boldsymbol{r}, \tau, s)$ is used in the Laplace transform (1.4) to write

$$\exp(-s\tau)\hat{F}(\boldsymbol{r}, \tau, s) = \int_{t=0}^{\infty} \exp(-st)[F(\boldsymbol{r}, \tau, t-\tau)\mathrm{H}(t-\tau)]\mathrm{d}t$$
$$\text{for } s \in \mathcal{L}. \tag{1.6}$$

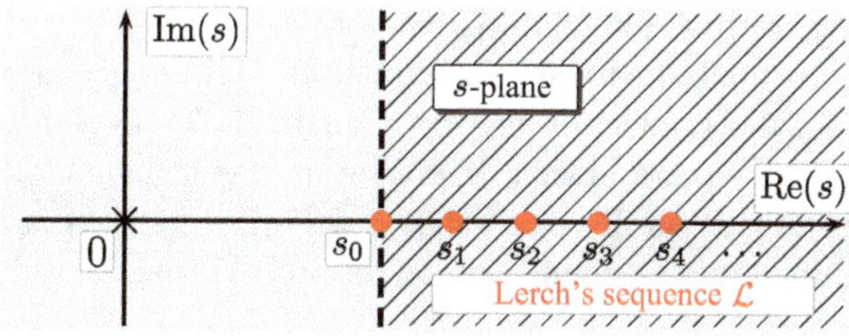

Figure 1.2 Complex s-plane with the line of convergence $\mathrm{Re}(s) = s_0$*, the region of regularity (the hatched area) and the Lerch sequence*

Substituting next (1.6) in (1.5) and changing the order of integration, we arrive at

$$\hat{f}(\boldsymbol{r}, s) = \int_{t=0}^{\infty} \exp(-st)\mathrm{d}t \int_{\tau=T}^{\infty} F(\boldsymbol{r}, \tau, t-\tau)\mathrm{H}(t-\tau)\mathrm{d}\tau \quad \text{for } s \in \mathcal{L}. \tag{1.7}$$

In the final step, Lerch's uniqueness theorem is employed to find the desired TD EM-field quantity by inspection

$$f(\boldsymbol{r}, t) = \int_{\tau=T}^{\infty} F(\boldsymbol{r}, \tau, t-\tau)\mathrm{H}(t-\tau)\mathrm{d}\tau. \tag{1.8}$$

Note that if $\hat{F}(\boldsymbol{r}, \tau, s)$ that occurs in (1.5) is s-independent, then $\hat{F}(\boldsymbol{r}, \tau, s) = F(\boldsymbol{r}, \tau)$ and hence $F(\boldsymbol{r}, \tau, t)\mathrm{H}(t) = F(\boldsymbol{r}, \tau)\delta(t)$, which via (1.8) yields $f(\boldsymbol{r}, t) = F(\boldsymbol{r}, t)\mathrm{H}(t - T)$. In this special case, however, the result $f(\boldsymbol{r}, t) = F(\boldsymbol{r}, t)\mathrm{H}(t - T)$ can be obtained directly from (1.5) upon relying on Lerch's uniqueness theorem, again. More details about the CdH technique and its alternatives are given in the following chapter.

References

[1] I. E. Lager, G. A. E. Vandenbosch and M. Štumpf. Electromagnetism in the electrical engineering classroom: dominant trends in teaching classical electromagnetic field theory and innovation vectors. *IEEE Antennas and Propagation Magazine*, 62(2):14–23, 2020.

[2] C. Pfeiffer and A. Grbic. Metamaterial Huygens' surfaces: tailoring wave fronts with reflectionless sheets. *Physical Review Letters*, 110(197401), 2013.

[3] T. B. A. Senior. Combined resistive and conductive sheets. *IEEE Transactions on Antennas and Propagation*, 33(5):577–579, 1985.

[4] C. Caloz and Z.-L. Deck-Léger. Spacetime metamaterials—part I: general concepts. *IEEE Transactions on Antennas and Propagation*, 68(3):1569–1582, 2019.

[5] D. V. Widder. *The Laplace Transform*. Princeton, NJ: Princeton University Press, 1946.

[6] G. Doetsch. *Introduction to the Theory and Application of the Laplace Transformation*. Berlin: Springer-Verlag, 1974.

[7] J. L. Schiff. *The Laplace Transform: Theory and Applications*. New York, NY: Springer-Verlag, 1999.

[8] A. T. de Hoop. A modification of Cagniard's method for solving seismic pulse problems. *Applied Scientific Research*, B(8):349–356, 1960.

[9] M. Štumpf. *Electromagnetic Reciprocity in Antenna Theory*. Hoboken, NJ: John Wiley & Sons, Inc., 2018.

[10] K. Aki and P. G. Richards. *Quantitative Seismology, 2nd Edition*. Sausalito, CA: University Science Books, 2009.

[11] A. T. de Hoop, *Handbook of Radiation and Scattering of Waves*. London, UK: Academic Press, 1995.

Chapter 2
Cagniard–DeHoop technique

The Cagniard–DeHoop (CdH) technique is a sophisticated mathematical tool for solving initial value problems in wave field physics that can be viewed as an ingenious simplification of the intricate joint transform procedure due to Cagniard [1, 2]. The technique has been initially developed for solving seismic pulse problems [3, 4], but thanks to its versatility, it has become popular in various research fields including acoustics [5], electromagnetics [6, 7, 8] and elastodynamics [9]. More recently, the CdH technique has been successfully applied to tackle several fundamental problems in antenna theory [10, 11] or electromagnetic (EM) compatibility [12, 13], for instance.

A wide class of initial value problems can be tackled with the aid of integral transformations applied with respect to temporal and spatial coordinates. In the CdH approach, a one-sided Laplace transformation is combined with the so-called slowness representation. The latter representation is, in fact, an inverse Fourier-type integral in which the Laplace-transform parameter, $s > 0$, is used as a scaling parameter. Thanks to this joint transformation procedure, the result expressed through the slowness integral can be cast into the form of the Laplace-transform integral (1.4) (see also (1.5)), from which the time-domain (TD) original immediately follows. Indeed, counting on Lerch's uniqueness theorem [14, Appendix] that ensures the existence of a one-to-one mapping between the original and its s-domain image, the inverse temporal transform can be carried out by inspection, thereby achieving the exact TD solution at once without actually computing the Laplace inversion.

The CdH technique yields closed-form analytical solutions directly in the TD, that is, in the original domain where any physical phenomena manifest themselves. Accordingly, such solutions provide valuable physical insights into the problem at hand, without obscuring the analysis by the concept of the time Fourier transformation with its real-valued frequency parameter. Furthermore, the CdH technique has proved to be useful in computational electromagnetics, namely, in developing TD integral-equation schemes termed as the Cagniard–DeHoop method of moments (CdH-MoM) [15] and in benchmarking and validation of purely numerical solution strategies.

2.1 A concise survey of joint transform methods

To illustrate the relation of the CdH technique to some alternative joint transform techniques, we shall analyze EM fields radiated from a short cylinder that carries uniformly distributed electric currents oriented along the z-direction (see Figure 2.1).

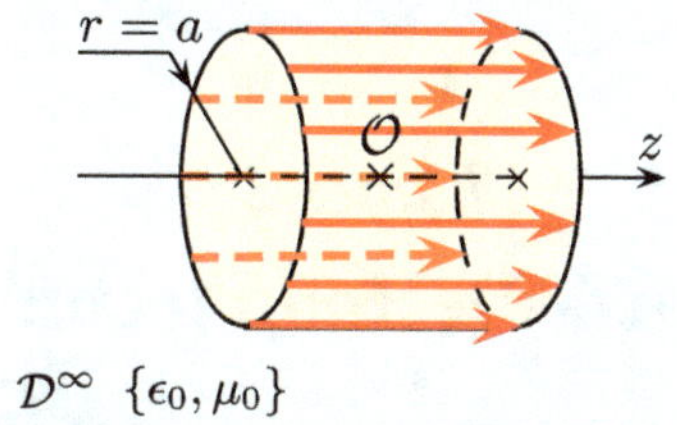

Figure 2.1 Fundamental electric-current source

The fundamental source is located in the homogeneous, isotropic and loss-free embedding that is described by its electric permittivity, ϵ_0, and magnetic permeability, μ_0. The cylinder emits EM fields that are governed by a circularly symmetric system of EM-field equations in vacuum whose form in the time Laplace-transform domain reads [16, Section 4.12]

$$-\partial_z \hat{H}_\phi = s\epsilon_0 \hat{E}_r, \tag{2.1a}$$

$$r^{-1}\partial_r (r\hat{H}_\phi) = s\epsilon_0 \hat{E}_z, \tag{2.1b}$$

$$\partial_r \hat{E}_z - \partial_z \hat{E}_r = s\mu_0 \hat{H}_\phi, \tag{2.1c}$$

where subscripts $\{r, z, \phi\}$ denote the radial, axial and azimuthal components of the radiated electric- and magnetic-field strengths and s is the (real-valued and positive) Laplace-transform parameter. Assuming the current element of moment $\hat{I}(s)\ell$, the EM-field equations are supplemented with the excitation condition expressing the jump of the magnetic-field strength across the current distribution, that is,

$$\lim_{r\downarrow a} \hat{H}_\phi(r, z, s) - \lim_{r\uparrow a} \hat{H}_\phi(r, z, s) = \hat{I}(s)[\mathrm{H}(z + \ell/2) - \mathrm{H}(z - \ell/2)]/2\pi a \quad \text{for all } z \in \mathbb{R}. \tag{2.2}$$

In (2.2), $\mathrm{H}(z)$ denotes the Heaviside unit-step function and $a > 0$ is the radius of the source cylinder. The problem formulated via (2.1a)–(2.2) will be solved under the wave-slowness representation taken in the axial z-direction. To show the notation, the representation is given for the axial electric-field strength

$$\hat{E}_z(r, z, s) = \frac{s}{2\pi\mathrm{i}} \int_{p=-\mathrm{i}\infty}^{\mathrm{i}\infty} \exp(-spz)\tilde{E}_z(r, p, s)\mathrm{d}p \quad \text{for } \{s \in \mathbb{R}; s > 0\}, \tag{2.3}$$

where p is the axial wave-slowness parameter. Under the representation, the governing EM-field equations (2.1a)–(2.1c) read

$$sp\tilde{H}_\phi = s\epsilon_0 \hat{E}_r, \tag{2.4a}$$

$$r^{-1}\partial_r (r\tilde{H}_\phi) = s\epsilon_0 \tilde{E}_z, \tag{2.4b}$$

$$\partial_r \tilde{E}_z + sp\tilde{E}_r = s\mu_0 \tilde{H}_\phi, \tag{2.4c}$$

while the excitation condition transforms to

$$\lim_{r\downarrow a} \tilde{H}_\phi(r,p,s) - \lim_{r\uparrow a} \tilde{H}_\phi(r,p,s) = \hat{I}(s)\ell\, \mathrm{i}_0(sp\ell/2)/2\pi a, \tag{2.5}$$

where $\mathrm{i}_0(x)$ represents the modified spherical Bessel function of the first kind and zero order. Combining (2.4a)–(2.4c) yields a modified Bessel differential equation for the transform-domain magnetic-field strength, i.e.

$$\partial_r \left[r^{-1}\partial_r(r\tilde{H}_\phi)\right] - s^2\gamma^2(p)\tilde{H}_\phi = 0. \tag{2.6}$$

Its general solution can be written in terms of the modified Bessel function of order one, that is, $\mathrm{I}_1(s\gamma r)$ for $\{0 \le r < a\}$ and $\mathrm{K}_1(s\gamma r)$ for $r > a$. Here, $\gamma(p)$ has the meaning of the radial slowness parameter that is defined by

$$\gamma(p) = (1/c_0^2 - p^2)^{1/2}, \quad \text{with} \quad \mathrm{Re}[\gamma(p)] \ge 0, \tag{2.7}$$

where $c_0 = (\epsilon_0\mu_0)^{-1/2} > 0$ denotes the wave speed in $\mathcal{D}^\infty$. The form of (2.6) suggests to represent the magnetic-field strength via the (z-component of the) electric-current potential [17, Eq. (26.3-2)]

$$\tilde{H}_\phi = -\partial_r \tilde{A}_z, \tag{2.8}$$

the transform-domain counterpart of which then solves the homogeneous modified Helmholtz equation, i.e.

$$r^{-1}\partial_r(r\partial_r\tilde{A}_z) - s^2\gamma^2(p)\tilde{A}_z = 0, \tag{2.9}$$

along with the corresponding excitation condition following from the substitution of (2.8) in (2.5), that is,

$$\lim_{r\downarrow a} \partial_r\tilde{A}_z(r,p,s) - \lim_{r\uparrow a} \partial_r\tilde{A}_z(r,p,s) = -\hat{I}(s)\ell\, \mathrm{i}_0(sp\ell/2)/2\pi a. \tag{2.10}$$

The solution of (2.9) can be written as

$$\tilde{A}_z(r,p,s) = \begin{cases} C\,\mathrm{I}_0(s\gamma r) & \text{for} \quad 0 \le r < a, \\ C\,\mathrm{K}_0(s\gamma r) & \text{for} \quad r > a, \end{cases} \tag{2.11}$$

thus representing, in virtue of the property of causality, a bounded transform-domain solution and C is the coefficient to be determined from the excitation condition (2.10). In this way, we readily obtain

$$C = \frac{\hat{I}(s)\ell}{2\pi} \frac{\mathrm{i}_0(sp\ell/2)}{s\gamma a[\mathrm{K}_1(s\gamma a) + \mathrm{I}_1(s\gamma a)]}. \tag{2.12}$$

Consequently, via the slowness representation (2.3), the electric-current potential function can be written as

$$\hat{A}_z(r,z,s) \simeq \frac{\hat{I}(s)\ell}{4\mathrm{i}\pi^2 a} \int_{p=-\mathrm{i}\infty}^{\mathrm{i}\infty} \exp(-spz)\frac{\mathrm{K}_0(s\gamma r)}{\mathrm{K}_1(s\gamma a)}\frac{\mathrm{d}p}{\gamma} \quad \text{for } r > a, \tag{2.13}$$

where we used $\mathrm{I}_1(s\gamma a) = O(a)$ as $a \downarrow 0$ and $\mathrm{i}_0(sp\ell/2) = 1 + O(\ell^2)$ as $\ell \downarrow 0$. It is now noted that (2.13) could also be found directly from (2.8) and from the relation for the axial electric-current distribution along a short current segment

$$\hat{I}(z,s) = 2\pi a \lim_{r\downarrow a} \hat{H}_\phi(r,z,s) = \hat{I}(s)\ell\,\delta(z), \tag{2.14}$$

where $\delta(z)$ is the one-dimensional (1-D) Dirac-delta distribution and

$$\hat{H}_\phi(r,z,s) \simeq \frac{s\hat{I}(s)\ell}{4\mathrm{i}\pi^2 a}\int_{p=-\mathrm{i}\infty}^{\mathrm{i}\infty} \exp(-spz)\frac{\mathrm{K}_1(s\gamma r)}{\mathrm{K}_1(s\gamma a)}\mathrm{d}p, \tag{2.15}$$

thus complying with (2.6) and with the integral representation for $\delta(z)$ in the limit $r \downarrow a$. Realizing next that (2.13) represents, in fact, the bounded point-source solution of the modified Helmholtz equation applying to the unbounded, isotropic and homogeneous medium $\mathcal{D}^\infty$ [17, Section 26.2], we immediately obtain [15, Eq. (A.4)]

$$\frac{\exp(-sR/c_0)}{4\pi R} = \frac{s}{4\mathrm{i}\pi^2}\int_{p=-\mathrm{i}\infty}^{\mathrm{i}\infty} \exp(-spz)\mathrm{K}_0(s\gamma r)\mathrm{d}p, \tag{2.16}$$

with $R = (r^2 + z^2)^{1/2}$ and we used $s\gamma a\mathrm{K}_1(s\gamma a) = 1 + O(a^2)$ as $a \downarrow 0$. Relation (2.16) is closely related to the identity frequently used by Schelkunoff [16, Eq. (15.6)]. Employing further an integral representation of the modified Bessel function

$$\mathrm{K}_0(s\gamma r) = \frac{1}{2\mathrm{i}}\int_{\sigma=-\mathrm{i}\infty}^{\mathrm{i}\infty} \exp[-s(\sigma y + \kappa(\sigma,p)|x|)]\frac{\mathrm{d}\sigma}{\kappa(\sigma,p)}, \tag{2.17}$$

where $\kappa(\sigma,p) = (1/c_0^2 - \sigma^2 - p^2)^{1/2}$ with $\mathrm{Re}[\kappa(\sigma,p)] \geq 0$, we get yet another slowness representation of the Green's function

$$\frac{\exp(-sR/c_0)}{4\pi R} = \frac{s}{(2\mathrm{i}\pi)^2}\int_{\sigma=-\mathrm{i}\infty}^{\mathrm{i}\infty}\mathrm{d}\sigma \int_{p=-\mathrm{i}\infty}^{\mathrm{i}\infty} \exp\{-s[\sigma y + pz + \kappa(\sigma,p)|x|]\}\frac{\mathrm{d}p}{2\kappa(\sigma,p)}. \tag{2.18}$$

Equation (2.18) is further taken as the point departure for introducing the classical inversion procedure due to de Hoop [3] and its alternative attributed to Strick (see [18, 19] and [20, Section 6.5]).

2.1.1 Classical CdH technique

Referring to the procedure introduced in the seminal paper [3, Eq. (3.10)], we replace the variables of integration in (2.18) through the following transformation:

$$\sigma = v\cos(\psi) - \mathrm{i}q\sin(\psi), \tag{2.19a}$$

$$p = v\sin(\psi) + \mathrm{i}q\cos(\psi), \tag{2.19b}$$

with $y = \varrho\cos(\psi)$ and $z = \varrho\sin(\psi)$ for $\varrho > 0$ and $\{0 \leq \psi < 2\pi\}$. Under this transformation, $\sigma y + pz = v\varrho$, $\sigma^2 + p^2 = v^2 - q^2$ and $\mathrm{d}\sigma\mathrm{d}p = \mathrm{i}\mathrm{d}v\mathrm{d}q$. Therefore, the right-hand side of (2.18) transforms to (cf. [3, Eq. (3.13)])

$$\frac{s}{4\pi^2\mathrm{i}}\int_{q=-\infty}^{\infty}\mathrm{d}q\int_{v=-\mathrm{i}\infty}^{\mathrm{i}\infty}\exp\{-s[v\varrho + \bar{\kappa}(v,q)|x|]\}\frac{\mathrm{d}v}{2\bar{\kappa}(v,q)}, \tag{2.20}$$

where we used $\bar{\kappa}(v,q) = [\Omega_0^2(q) - v^2]^{1/2}$ with $\Omega_0(q) = (1/c_0^2 + q^2)^{1/2} > 0$ and $\mathrm{Re}[\bar{\kappa}(v,q)] \geq 0$. The following step in the inversion procedure is the deformation of the path of integration in the complex v-plane to the so-called CdH path such that the propagation factor has the form of the Laplace-transform kernel. To that end, the integrand is first analytically continued away from the imaginary axis, while keeping $\mathrm{Re}[\bar{\kappa}(v,q)] \geq 0$, which implies the branch cuts along $\{\Omega_0(q) \leq |\mathrm{Re}(v)| < \infty, \mathrm{Im}(v) = 0\}$. By virtue of Cauchy's theorem and Jordan's lemma [17, p. 1054], the new CdH integration path is defined via

$$v\varrho + \bar{\kappa}(v,q)|x| = \tau, \tag{2.21}$$

where τ is a real-valued and positive parameter that will be associated with the time coordinate. Equation (2.21) is satisfied along a hyperbolic contour, say $\mathcal{C} \cup \mathcal{C}^*$ (here * denotes the complex conjugate), that is hence symmetrical with respect to $\mathrm{Im}(v) = 0$ (see Figure 2.2). Solving (2.21) for v, we obtain (cf. [3, Eq. (3.15)])

$$\mathcal{C} = \left\{v(\tau) = (\varrho/R^2)\tau + \mathrm{i}(|x|/R^2)\left[\tau^2 - R^2\Omega_0^2(q)\right]^{1/2}\right\}$$
$$\text{for } \{R\Omega_0(q) \leq \tau < \infty\}. \tag{2.22}$$

Along $\mathcal{C}$, we then have (cf. [3, Eq. (3.16)])

$$\bar{\kappa}[v(\tau),q] = (|x|/R^2)\tau - \mathrm{i}(\varrho/R^2)\left[\tau^2 - R^2\Omega_0^2(q)\right]^{1/2}, \tag{2.23}$$

and the Jacobian for mapping v to τ reads

$$\partial v/\partial\tau = \mathrm{i}\bar{\kappa}[v(\tau),q]\left[\tau^2 - R^2\Omega_0^2(q)\right]^{-1/2}. \tag{2.24}$$

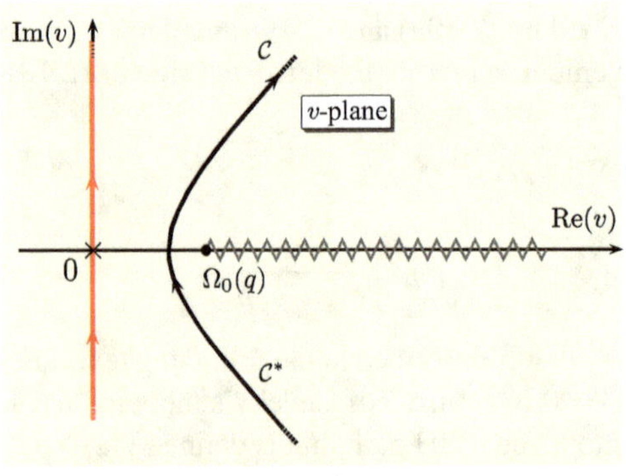

Figure 2.2 Complex v-plane with the CdH path $\mathcal{C} \cup \mathcal{C}^*$

Consequently, combining the contributions from C and C^* and introducing τ as the new variable of integration along C, we obtain (cf. [3, Eq. (3.18)])

$$\frac{s}{4\pi^2}\int_{q=-\infty}^{\infty}\mathrm{d}q\int_{\tau=R\Omega_0(q)}^{\infty}\exp(-s\tau)\left[\tau^2-R^2\Omega_0^2(q)\right]^{-1/2}\mathrm{d}\tau. \tag{2.25}$$

Next, we interchange the order of the integrations and find

$$\frac{s}{4\pi^2 R}\int_{\tau=R/c_0}^{\infty}\exp(-s\tau)\mathrm{d}\tau\int_{q=-Q(\tau)}^{Q(\tau)}\left[Q^2(\tau)-q^2\right]^{-1/2}\mathrm{d}q, \tag{2.26}$$

where we used $Q(\tau)=(\tau^2/R^2-1/c_0^2)^{1/2}$. The inner integral with respect q can be readily evaluated analytically and we end up with an expression having the form of the Laplace transform

$$\frac{s}{4\pi R}\int_{\tau=R/c_0}^{\infty}\exp(-s\tau)\mathrm{d}\tau. \tag{2.27}$$

The integration with respect to τ could be carried out analytically again, thereby getting the left-hand side of (2.18) that can be transformed to the TD at once. For later purposes, however, we shall next proceed in a more general way relying heavily on Lerch's uniqueness theorem of the one-sided Laplace transformation [14, Appendix]. In this manner, the inversion is achieved by inspection without actually carrying out the inverse Laplace transformation. Pursuing this line of reasoning, we note that the integral on the right-hand side of (2.27) is, as a matter of fact, the Laplace transform of a unit-step function. Therefore, employing the property $s\to\partial_t$, this way leads to

$$\frac{\partial_t \mathrm{H}(t-R/c_0)}{4\pi R}=\frac{\delta(t-R/c_0)}{4\pi R}, \tag{2.28}$$

which is a well-known result.

2.1.2 "Cartesian" CdH technique

For handling a number of problem configurations that are not rotationally symmetric (see [11, 13] and [15, Chapter 14], for example), the transformation of the slowness parameters as prescribed by (2.19a) and (2.19b) may not be purposeful anymore. In such cases, it is convenient to rewrite the left-hand side of (2.18) to

$$\frac{s}{(2\mathrm{i}\pi)^2}\int_{p=-\mathrm{i}\infty}^{\mathrm{i}\infty}\exp(-spz)\mathrm{d}p$$
$$\int_{\sigma=-\mathrm{i}\infty}^{\mathrm{i}\infty}\exp\left\{-s[\sigma y+\kappa(p,\sigma)|x|]\right\}\frac{\mathrm{d}\sigma}{2\kappa(\sigma,p)} \tag{2.29}$$

and proceed in the way that is described in the following text. The original integration contour along $\mathrm{Re}(\sigma)=0$ is in virtue of Cauchy's theorem and Jordan's lemma [17, p. 1054] deformed into a new CdH path that is defined via

$$\sigma y+\kappa(p,\sigma)|x|=ur\gamma(p)\quad \text{for all } \{1\le u<\infty\}, \tag{2.30}$$

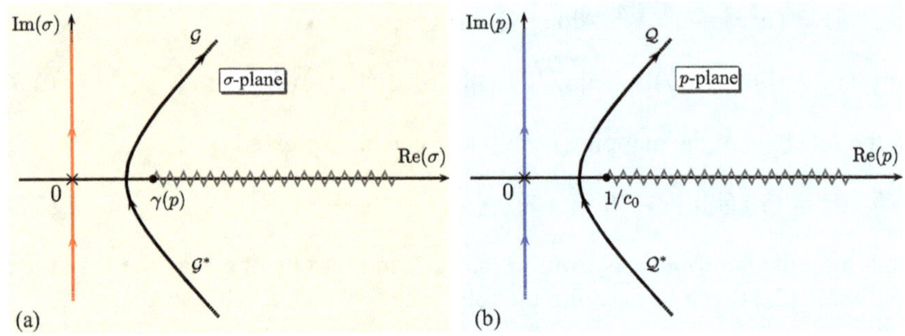

Figure 2.3 *(a) Complex σ-plane with the CdH path $\mathcal{G} \cup \mathcal{G}^*$ for $y > 0$. (b) Complex p-plane with the CdH path $\mathcal{Q} \cup \mathcal{Q}^*$ for $z > 0$*

where $r = (x^2 + y^2)^{1/2} > 0$ and $\gamma(p)$ is defined via (2.7). Upon solving (2.30) for σ, we find that the CdH path can be described via hyperbolic arcs $\mathcal{G} \cup \mathcal{G}^*$ (see Figure 2.3(a)), with

$$\mathcal{G} = \left\{\sigma(u) = \left[yu/r + \mathrm{i}|x|(u^2 - 1)^{1/2}/r\right]\gamma(p)\right\} \quad \text{for all } \{1 \le u < \infty\}. \tag{2.31}$$

Along $\mathcal{G}$, we then have

$$\kappa[p, \sigma(u)] = \left[|x|u/r - \mathrm{i}y(u^2 - 1)^{1/2}/r\right]\gamma(p), \tag{2.32}$$

and the relevant Jacobian reads

$$\partial\sigma/\partial u = \mathrm{i}\kappa[p, \sigma(u)](u^2 - 1)^{-1/2} \quad \text{for all } \{1 < u < \infty\}. \tag{2.33}$$

Combining next the contributions from $\mathcal{G}$ and $\mathcal{G}^*$, while introducing u as the new variable of integration, we arrive at

$$\frac{s}{4\mathrm{i}\pi^2} \int_{u=1}^{\infty} \frac{\mathrm{d}u}{(u^2 - 1)^{1/2}} \int_{p=-\mathrm{i}\infty}^{\mathrm{i}\infty} \exp\{-s[pz + \gamma(p)ur]\}\mathrm{d}p. \tag{2.34}$$

We next proceed in a similar fashion in the complex p-plane. Hence, relying on Cauchy's theorem and Jordan's lemma [17, p. 1054] again, the original integration contour is deformed into the CdH path, the parametrization of which is found from (cf. (2.21))

$$pz + \gamma(p)ur = \tau \quad \text{for } \{\tau \in \mathbb{R}; \tau > 0\}. \tag{2.35}$$

Thus, expressing p as a function of the time parameter τ, we obtain a hyperbolic CdH contour, say $\mathcal{Q} \cup \mathcal{Q}^*$ (see Figure 2.3(b)), where

$$\mathcal{Q} = \left\{p(\tau) = [z/R^2(u)]\tau + \mathrm{i}[ur/R^2(u)][\tau^2 - R^2(u)/c_0^2]^{1/2}\right\}$$
$$\text{for } \{R(u)/c_0 \le \tau < \infty\}, \tag{2.36}$$

and $R(u) = (u^2r^2 + z^2)^{1/2}$. Along $\mathcal{Q}$, we have

$$\gamma[p(\tau)] = [ur/R^2(u)]\tau - \mathrm{i}[z/R^2(u)][\tau^2 - R^2(u)/c_0^2]^{1/2}, \tag{2.37}$$

and the Jacobian of the mapping from p to τ can be expressed as

$$\partial p/\partial\tau = \mathrm{i}\gamma[p(\tau)][\tau^2 - R^2(u)/c_0^2]^{-1/2}. \tag{2.38}$$

Combining the contributions from $\mathcal{Q}$ and $\mathcal{Q}^*$ and introducing τ as the variable of integration, (2.34) can be cast into the following form:

$$\frac{sr}{2\pi^2}\int_{u=1}^{\infty}\frac{u\mathrm{d}u}{R^2(u)(u^2-1)^{1/2}}\int_{\tau=R(u)/c_0}^{\infty}\exp(-s\tau)\frac{\tau\mathrm{d}\tau}{[\tau^2 - R^2(u)/c_0^2]^{1/2}}. \tag{2.39}$$

Subsequently, we change the order of the integrations and get

$$\frac{sc_0}{2\pi^2}\int_{\tau=R/c_0}^{\infty}\exp(-s\tau)\tau\mathrm{d}\tau\int_{u=1}^{Q(\tau)}\frac{u\mathrm{d}u}{R^2(u)[Q^2(\tau)-u^2]^{1/2}(u^2-1)^{1/2}}, \tag{2.40}$$

where we introduced $Q(\tau) = (c_0^2\tau^2 - z^2)^{1/2}/r > 0$ and $R = R(1) = (r^2 + z^2)^{1/2}$. The inverse square-root singularities are handled via the substitution

$$u^2 = \cos^2(\varphi) + Q^2(\tau)\sin^2(\varphi) \quad \text{for } \{0 \le \varphi \le \pi/2\}, \tag{2.41}$$

which yields

$$\frac{sc_0}{2\pi^2}\int_{\tau=R/c_0}^{\infty}\exp(-s\tau)\tau\mathrm{d}\tau\int_{\varphi=0}^{\pi/2}\frac{\mathrm{d}\varphi}{R^2\cos^2(\varphi) + c_0^2\tau^2\sin^2(\varphi)}. \tag{2.42}$$

The integration with respect to φ can be found in closed form, that is,

$$\int_{\varphi=0}^{\pi/2}\frac{\mathrm{d}\varphi}{R^2\cos^2(\varphi) + c_0^2\tau^2\sin^2(\varphi)} = \frac{\pi}{2Rc_0\tau}, \tag{2.43}$$

and we end up with (2.27), again. The final transformation back to the TD can be carried out by inspection as in the previous subsection.

2.1.3 Strick's modification of Cagniard's technique

Another modification of Cagniard's inversion procedure that bears a similarity with Sommerfeld's representation [21, Chapter IV] is the one due to Strick (see [18, 19] and [20, Section 6.5]). A related inversion technique was also developed by Pekeris [22]. To demonstrate its relation to the procedure introduced by de Hoop (see Section 2.1.1), we replace the variables of integration in (2.18) via (cf. (2.19a) and (2.19b))

$$\sigma = w\cos(\psi'), \tag{2.44a}$$
$$p = w\sin(\psi'), \tag{2.44b}$$

where $\{0 \leq \psi' < 2\pi\}$, $\{\mathrm{Re}(w) = 0, 0 \leq \mathrm{Im}(w) < \infty\}$ and recall that $y = \varrho\cos(\psi)$ and $z = \varrho\sin(\psi)$ for $\varrho > 0$ and $\{0 \leq \psi < 2\pi\}$. Under this transformation, $\sigma y + pz = w\varrho\cos(\psi - \psi')$, $\sigma^2 + p^2 = w^2$ and $\mathrm{d}\sigma\mathrm{d}p = w\mathrm{d}w\mathrm{d}\psi'$, which allows to rewrite the right-hand side of (2.18) as

$$\frac{s}{(2\mathrm{i}\pi)^2}\int_{w=0}^{\mathrm{i}\infty}\exp[-s\bar{\kappa}(w)|x|]\frac{w\,\mathrm{d}w}{2\bar{\kappa}(w)}\int_{\psi'=0}^{2\pi}\exp[-sw\varrho\cos(\psi-\psi')]\mathrm{d}\psi', \quad (2.45)$$

where $\bar{\kappa}(w) = (1/c_0^2 - w^2)^{1/2}$ with $\mathrm{Re}[\bar{\kappa}(w)] \geq 0$ and $\{s \in \mathbb{R}; s > 0\}$. It is then straightforward to relate the integral with respect to ψ' to the modified Bessel function of the first kind and zero order [23, Section 3.71], which allows to rewrite (2.45) as

$$\frac{s}{2\mathrm{i}^2\pi}\int_{w=0}^{\mathrm{i}\infty}\mathrm{I}_0(sw\varrho)\exp[-s\bar{\kappa}(w)|x|]\frac{w\,\mathrm{d}w}{2\bar{\kappa}(w)}. \quad (2.46)$$

Expressing further $\mathrm{I}_0(sw\varrho)$ via the modified Bessel function of the second kind, that is, $\mathrm{I}_0(sw\varrho) = [\mathrm{K}_0(-sw\varrho) - \mathrm{K}_0(sw\varrho)]/\mathrm{i}\pi$ (see [23, p. 80]), and employing the reflection property $\mathrm{K}_0(z^*) = [\mathrm{K}_0(z)]^*$, (2.46) can be readily cast into its equivalent form

$$\frac{s}{2\pi^2}\mathrm{Im}\left\{\int_{w=0}^{\mathrm{i}\infty}\mathrm{K}_0(sw\varrho)\exp[-s\bar{\kappa}(w)|x|]\frac{w\,\mathrm{d}w}{\bar{\kappa}(w)}\right\}. \quad (2.47)$$

Following next the line of reasoning pursued in [19], (2.47) is transformed to the TD, which yields

$$\frac{\partial_t}{2\pi^2}\mathrm{Im}\left\{\int_{w=0}^{\mathrm{i}\infty}\frac{\mathrm{H}(t - w\varrho - \bar{\kappa}(w)|x|)}{[(t-\bar{\kappa}(w)|x|)^2 - w^2\varrho^2]^{1/2}}\frac{w\,\mathrm{d}w}{\bar{\kappa}(w)}\right\}. \quad (2.48)$$

The argument of the Heaviside unit-step function suggests to deform the integration contour into the CdH path defined via

$$w\varrho + \bar{\kappa}(w)|x| = \tau \quad \text{for } \{\tau \in \mathbb{R}; \tau > 0\}, \quad (2.49)$$

which is permissible thanks to Cauchy's theorem and Jordan's lemma. The hyperbolic arc in the complex w-plane, say $\mathcal{H}$ (see Figure 2.4), satisfying (2.49) can be described by

$$\mathcal{H} = \left\{w(\tau) = (\varrho/R^2)\tau + \mathrm{i}(|x|/R^2)(\tau^2 - R^2/c_0^2)^{1/2}\right\}$$
$$\text{for } \{R/c_0 \leq \tau < \infty\} \quad (2.50)$$

and $R = (x^2 + \varrho^2)^{1/2}$. Consequently, upon introducing the time parameter τ as the variable of integration, it is straightforward to rewrite (2.48) to an equivalent form bearing a similarity with the relevant line-source solution, that is,

$$\frac{\partial_t}{2\pi^2}\mathrm{Re}\left\{\int_{\tau=R/c_0}^{t}\frac{w(\tau)}{(t-\tau)^{1/2}[t-\tau+2w(\tau)\varrho]^{1/2}}\frac{\mathrm{d}\tau}{(\tau^2 - R^2/c_0^2)^{1/2}}\right\}. \quad (2.51)$$

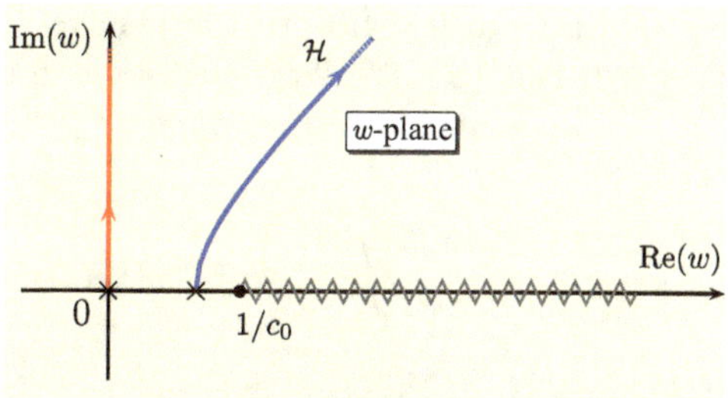

Figure 2.4 Complex w-plane with the CdH path $\mathcal{H}$

Representation (2.51) is frequently used in tackling seismic pulse problems (see [19] and [20, Section 6.5], for instance). It can be verified that the result of integration is $(\pi/2R)\mathrm{H}(t - R/c_0)$, thus leading to (2.28), again. Perhaps a more straightforward way to demonstrate the equivalence is to represent the modified Bessel function in (2.47) via (cf. (2.17))

$$\mathrm{K}_0(sw\varrho) = \int_{u=1}^{\infty} \exp(-sw\varrho u)\frac{\mathrm{d}u}{(u^2-1)^{1/2}}, \tag{2.52}$$

which upon changing the order of integration leads to

$$\frac{s}{2\pi^2}\int_{u=1}^{\infty} \frac{\mathrm{d}u}{(u^2-1)^{1/2}}\mathrm{Im}\left\{\int_{w=0}^{\mathrm{i}\infty} \exp\{-s[w\varrho u + \bar{\kappa}(w)|x|]\}\frac{w\,\mathrm{d}w}{\bar{\kappa}(w)}\right\}. \tag{2.53}$$

Subsequently, the relevant CdH path is defined by (cf. (2.49))

$$w\varrho u + \bar{\kappa}(w)|x| = \tau, \tag{2.54}$$

with the real-valued and positive time parameter τ, again. Employing the corresponding values of $w(\tau)$ and $\bar{\kappa}[w(\tau)]$ along the new path, (2.53) can be readily cast into a form that is similar to (2.39), that is,

$$\frac{s\varrho}{2\pi^2}\int_{u=1}^{\infty} \frac{u\mathrm{d}u}{D^2(u)(u^2-1)^{1/2}}\int_{\tau=D(u)/c_0}^{\infty} \exp(-s\tau)\frac{\tau\mathrm{d}\tau}{[\tau^2 - D^2(u)/c_0^2]^{1/2}}, \tag{2.55}$$

with $D(u) = (x^2 + u^2\varrho^2)^{1/2}$. As a matter of fact, (2.55) follows from (2.39) upon replacing x with z and vice versa. Owing to the spherical symmetry of the point-source solution, it can be hence concluded the TD counterpart of (2.55) is equal to (2.28), again.

2.2 EM radiation in a layered medium

The CdH technique has been originally developed for analyzing the seismic wave propagation in the horizontally layered approximation to Earth's crustal structure.

Accordingly, the CdH technique is ideally suited for (and thus mostly applied to) studying the pulsed wave propagation in planarly layered structures (e.g. [5–7]). We hence devote this section to an exact analytical representation of the EM wave motion in a horizontally stratified medium that is also known as the generalized-ray theory [8,9,24].

The problem configuration under consideration is depicted in Figure 2.5. Here, the layered medium consists of N (homogeneous) domains, each of which being characterized by its (scalar, real-valued and positive) electric permittivity, $\epsilon_n > 0$, magnetic permeability, $\mu_n > 0$, and by its thickness $d_n = z_{n+1} - z_n > 0$ for $n = \{2, \ldots, N-1\}$. The corresponding EM wave speed is $c_n = (\epsilon_n \mu_n)^{-1/2} > 0$. The outer half-spaces, $\mathcal{D}_1$ and $\mathcal{D}_N$, extend over $\{-\infty < x < \infty, -\infty < y < \infty, -\infty < z < z_2\}$ and $\{-\infty < x < \infty, -\infty < y < \infty, z_N < z < \infty\}$, respectively. The constitutive parameters, ϵ_n and μ_n, can be hence regarded as real and positive piecewise constant functions of the z-coordinate. Since these material parameters are independent of the horizontal coordinates, x and y, as well as of the time coordinate, t, the analyzed structure is shift invariant with respect to the horizontal xy-plane, and time invariant.

For the sake of clarity, we shall further limit our analysis to the EM wave fields radiated from a *vertical electrical dipole*. Such a source can be described by (the non-vanishing component of) the electric-current volume density $J_z(\boldsymbol{r}, t)$. Assuming that the dipole source of moment $I(t)\ell$ is located at $(0, 0, z_S)$, the current density in the s-domain can be represented by

$$\hat{J}_z(\boldsymbol{r}, s) = \hat{I}(s)\ell\, \delta(x)\delta(y)\delta(z - z_S), \tag{2.56}$$

where $\hat{I}(s)$ is the s-domain counterpart of the (causal) source signature and $z = z_S$ denotes the source level. In contrast to the EM fields radiated from a horizontal dipole that in general consists of TE (= Transverse Electric, with respect to the vertical z-direction) and TM (= Transverse Magnetic) field components, the EM fields due to a vertical electric dipole are purely of the TM-type, i.e. there is no z-component of the magnetic-field strength. Consequently, such EM fields can be represented by a single scalar potential function only [21, Chapter VI]. For a more general description accounting for the combination of both TE- and TM-type fields, we refer the reader to [8].

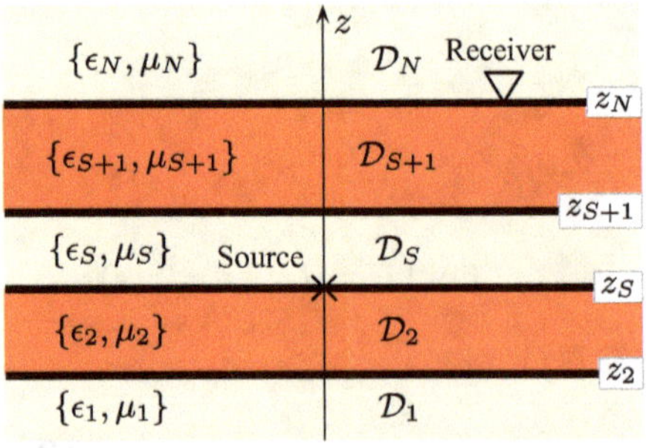

Figure 2.5 Point-source radiation in a 3-D horizontally layered medium

We next take advantage of the shift invariance in the xy-plane and represent the radiated EM fields through the wave-slowness representation over the horizontal plane. To show the notation, we give the integral expression for the vertical component of the electric-field strength (cf. (2.18))

$$\hat{E}_z(x,y,z,s) = (s/2\mathrm{i}\pi)^2 \int_{\kappa=-\mathrm{i}\infty}^{\mathrm{i}\infty} \mathrm{d}\kappa \int_{\sigma=-\mathrm{i}\infty}^{\mathrm{i}\infty} \exp[-s(\kappa x + \sigma y)]\tilde{E}_z(\kappa,\sigma,z,s)\mathrm{d}\sigma, \tag{2.57}$$

where the slowness parameters, κ and σ, are purely imaginary and s (= Laplace-transform parameter) is kept real-valued and positive. The slowness representation entails the property $\partial_x \to \tilde{\partial}_x = -s\kappa$ and $\partial_y \to \tilde{\partial}_y = -s\sigma$. As a matter of fact, expression (2.57) can be viewed as a special form of the standard inverse Fourier spatial transformation whose (real-valued) parameters along the x- and y-directions are replaced with the (scaled) ones $-\mathrm{i}s\kappa$ and $-\mathrm{i}s\sigma$, respectively.

Under the wave-slowness representation (2.57), we postulate that the tangential EM-field components are related to their (vertical electric) dipole source via

$$\tilde{E}_x(\kappa,\sigma,z,s) = \hat{I}(s)\ell\,\tilde{\partial}_x\partial_z\tilde{G}(\kappa,\sigma,z|z_S,s)/s\epsilon(z), \tag{2.58a}$$
$$\tilde{E}_y(\kappa,\sigma,z,s) = \hat{I}(s)\ell\,\tilde{\partial}_y\partial_z\tilde{G}(\kappa,\sigma,z|z_S,s)/s\epsilon(z), \tag{2.58b}$$
$$\tilde{H}_x(\kappa,\sigma,z,s) = \hat{I}(s)\ell\,\tilde{\partial}_y\tilde{G}(\kappa,\sigma,z|z_S,s), \tag{2.58c}$$
$$\tilde{H}_y(\kappa,\sigma,z,s) = -\hat{I}(s)\ell\,\tilde{\partial}_x\tilde{G}(\kappa,\sigma,z|z_S,s), \tag{2.58d}$$

where we introduced the (transform-domain) Green's function, $\tilde{G} = \tilde{G}(\kappa,\sigma,z|z_S,s)$, in which $z|z_S$ indicates the vertical coordinate of the field point, z, and of the source point, z_S (see Figure 2.5). It is noted that for the excitation under consideration, the transform-domain expressions (2.58a)–(2.58d) bear a similarity to the ones pertaining to an unbounded, homogeneous and isotropic space [17, Eqs. (26.1-34) and (26.1-35)]. Across any *source-free* interface of the layered medium, the tangential components of the radiated EM fields are continuous, which can be expressed through the continuity-type boundary conditions in the transform domain as

$$\lim_{z\downarrow z_n}\{\tilde{E}_x,\tilde{E}_y\}(\kappa,\sigma,z,s) = \lim_{z\uparrow z_n}\{\tilde{E}_x,\tilde{E}_y\}(\kappa,\sigma,z,s)$$
$$\text{for all } n = \{2,\ldots,N\}\setminus S, \tag{2.59a}$$

that is, $z_n \neq z_S$, and

$$\lim_{z\downarrow z_n}\{\tilde{H}_x,\tilde{H}_y\}(\kappa,\sigma,z,s) = \lim_{z\uparrow z_n}\{\tilde{H}_x,\tilde{H}_y\}(\kappa,\sigma,z,s)$$
$$\text{for all } n = \{2,\ldots,N\}. \tag{2.59b}$$

The use of expressions (2.58a)–(2.58d) in (2.59a) and (2.59b) yields the boundary conditions to be satisfied by the transform-domain Green's function across any

source-free interface. Hence, denoting the Green's function pertaining to the nth layer by $\tilde{G}_n$, the boundary conditions can be written as

$$\lim_{z \downarrow z_n} \tilde{G}_n(\kappa, \sigma, z|z_S, s) = \lim_{z \uparrow z_n} \tilde{G}_{n-1}(\kappa, \sigma, z|z_S, s)$$
$$\text{for all } n = \{2, \ldots, N\}, \tag{2.60a}$$

and

$$\lim_{z \downarrow z_n} \partial_z \tilde{G}_n(\kappa, \sigma, z|z_S, s)/\epsilon_n = \lim_{z \uparrow z_n} \partial_z \tilde{G}_{n-1}(\kappa, \sigma, z|z_S, s)/\epsilon_{n-1}$$
$$\text{for all } n = \{2, \ldots, N\} \setminus S, \tag{2.60b}$$

that is, $z_n \neq z_S$. At the interface where the vertical dipole is located, the tangential components of the electric-field strength are no longer continuous. Their jump across the source interface at $z = z_S$ can be associated with the transform-domain excitation condition

$$\lim_{z \downarrow z_S} \partial_z \tilde{G}_S(\kappa, \sigma, z|z_S, s) - \lim_{z \uparrow z_S} \partial_z \tilde{G}_{S-1}(\kappa, \sigma, z|z_S, s) = -1. \tag{2.60c}$$

The transform-domain source-type representations (2.58a)–(2.58d) can also be used to express the boundary conditions applying on EM-impenetrable interfaces. Indeed, (2.58a) and (2.58b) imply that $\partial_z \tilde{G} = 0$ when approaching a perfectly electrically conducting (PEC) interface, while (2.58c) and (2.58c) yield $\tilde{G} = 0$ to be met on a perfectly magnetically conducting (PMC) interface.

With the boundary conditions at our disposal, we can next determine the transform-domain Green's functions $\tilde{G}_n$ in the subdomains of the problem configuration. The EM-field quantities pertaining to the nth layer then directly follow from (2.58a)–(2.58d), where we replace $\tilde{G}$ by $\tilde{G}_n$. Hence, we next assume that the (transform-domain) Green's function in any source-free homogeneous layer, $\mathcal{D}_n$, can be generally expressed as the superposition of two 1-D waves

$$\tilde{G}_n(\kappa, \sigma, z|z_S, s) = W_n^- \exp[-s p_n(\kappa, \sigma)(z_{n+1} - z)]$$
$$+ W_n^+ \exp[-s p_n(\kappa, \sigma)(z - z_n)]$$
$$\text{for all } n = \{1, \ldots, N\}, \tag{2.61}$$

where

$$p_n(\kappa, \sigma) = (1/c_n^2 - \kappa^2 - \sigma^2)^{1/2}, \quad \text{with} \quad \text{Re}[p_n(\kappa, \sigma)] \geq 0. \tag{2.62}$$

The first term on the right-hand side of (2.61) represents a downgoing wave propagating in the negative z-direction away from the upper interface at $z = z_{n+1}$, while the second term, likewise, can be associated with an upgoing wave propagating in the positive z-direction away from the lower interface at $z = z_n$. The amplitudes of these 1-D transform-domain waves, W_n^+ and W_n^-, will be determined with the aid of the boundary conditions. In view of the property of causality, we take $W_1^+ = 0$ and $W_N^- = 0$, which eliminates the (unphysical) upgoing and downgoing waves in

the outer half-spaces $\mathcal{D}_1$ and $\mathcal{D}_N$, respectively. The amplitudes pertaining to the nth interface (see Figure 2.6) can be mutually interrelated via scattering relations

$$W_{n-1}^- = \bar{S}_n^{--} W_n^- + \bar{S}_n^{-+} W_{n-1}^+ + X_{n-1}^-, \tag{2.63a}$$
$$W_n^+ = \bar{S}_n^{+-} W_n^- + \bar{S}_n^{++} W_{n-1}^+ + X_n^+, \tag{2.63b}$$

where parameters X_n^+ and X_{n-1}^- are introduced to incorporate the effect of the source at $z = z_n$, and the modified scattering parameters are defined via

$$\{\bar{S}_n^{+-}, \bar{S}_n^{--}\} = \{S_n^{+-}, S_n^{--}\} \exp(-s p_n d_n), \tag{2.64a}$$
$$\{\bar{S}_n^{++}, \bar{S}_n^{-+}\} = \{S_n^{++}, S_n^{-+}\} \exp(-s p_{n-1} d_{n-1}),$$
$$\text{for all } n = \{2, \ldots, N\}. \tag{2.64b}$$

With reference to Figure 2.6, it is seen that relations (2.63a) and (2.63b) interrelate the amplitudes of waves propagating away from the interfaces with the amplitudes of waves propagating towards the interface. Upon enforcing the continuity-type boundary conditions (2.60a) and (2.60b), we obtain the scattering parameters pertaining to the nth interface in the following form:

$$S_n^{+-} = \frac{p_n/\epsilon_n - p_{n-1}/\epsilon_{n-1}}{p_n/\epsilon_n + p_{n-1}/\epsilon_{n-1}}, \tag{2.65a}$$
$$S_n^{--} = \frac{2\,p_n/\epsilon_n}{p_n/\epsilon_n + p_{n-1}/\epsilon_{n-1}}, \tag{2.65b}$$
$$S_n^{-+} = \frac{p_{n-1}/\epsilon_{n-1} - p_n/\epsilon_n}{p_{n-1}/\epsilon_{n-1} + p_n/\epsilon_n}, \tag{2.65c}$$
$$S_n^{++} = \frac{2\,p_{n-1}/\epsilon_{n-1}}{p_{n-1}/\epsilon_{n-1} + p_n/\epsilon_n}, \quad \text{for all } n = \{2, \ldots, N\}. \tag{2.65d}$$

The excitation condition (2.60c) applying across the source interface then leads to the source-coupling coefficients

$$X_S^+ = X_{S-1}^- = \frac{1}{s(p_S + p_{S-1})}. \tag{2.66}$$

Finally, if an impenetrable interface is present, then we can find

$$S_2^{+-} = -1 \quad \text{or/and} \quad S_N^{-+} = -1 \quad \text{on PMC}, \tag{2.67a}$$
$$S_2^{+-} = +1 \quad \text{or/and} \quad S_N^{-+} = +1 \quad \text{on PEC}. \tag{2.67b}$$

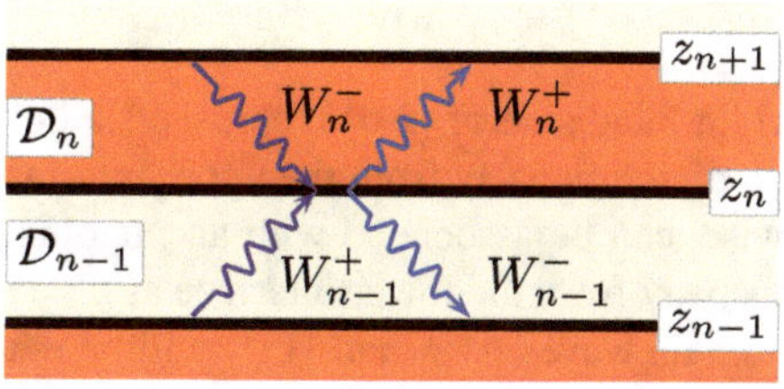

Figure 2.6 Wave amplitudes at a source-free interface

The wave amplitudes of the entire N-layered structure with its $N-1$ interfaces can be arranged in a $[(2N-2)\times 1]$ array, say $\boldsymbol{W}$, and interrelated using the $(2N-2)\times(2N-2)$ modified scattering array, denoted by $\underline{\bar{\boldsymbol{S}}}$, as

$$\boldsymbol{W} = \underline{\bar{\boldsymbol{S}}} \cdot \boldsymbol{W} + \boldsymbol{X}. \tag{2.68}$$

The only nonzero elements of the excitation $[(2N-2)\times 1]$ array $\boldsymbol{X}$ on the right-hand side of (2.68) are X_S^+ and X_{S-1}^- (see (2.66)). The detailed structure of the system of (2.68) is shown in Figure 2.7. The system of equations is next solved using a Neumann iteration as

$$\boldsymbol{W} = \boldsymbol{X} + \underline{\bar{\boldsymbol{S}}} \cdot \boldsymbol{X} + \underline{\bar{\boldsymbol{S}}}^2 \cdot \boldsymbol{X} + \underline{\bar{\boldsymbol{S}}}^3 \cdot \boldsymbol{X} + \cdots, \tag{2.69a}$$

which can be expressed in a more precise form, viz

$$\boldsymbol{W} = \sum_{q=0}^{Q} \underline{\bar{\boldsymbol{S}}}^q \cdot \boldsymbol{X} + \underline{\bar{\boldsymbol{S}}}^{Q+1} \cdot \boldsymbol{W} \quad \text{for } Q \geq 0. \tag{2.69b}$$

The second term on the right-hand side of (2.69b) is, in fact, the remainder of the geometric series. Clearly, the iterative solution is convergent if $|\underline{\bar{\boldsymbol{S}}}| < 1$, which is met for any passive layered structure. The procedure starts with $\boldsymbol{X}$, representing the excitation term at the source level, $z = z_S$, and then proceeds with higher-order (generalized-ray) constituents $\underline{\bar{\boldsymbol{S}}}^q \cdot \boldsymbol{X}$ for $q = \{1, 2, \ldots\}$ each containing q interactions with interfaces. The scattering array depends on the Laplace transform parameter s through (exponential) propagation factors only (see (2.64a) and (2.64b)). Since these factors represent a time delay in their original domain, the TD counterpart of each higher generalized ray arises at the field point at a later time than the previous one. Consequently, the number of wave constituents is in any bounded time window of observation always finite. Moreover, the amplitude of a generalized ray gets weaker and weaker due to successive reflections or/and transmissions when it propagates through the layered structure. Therefore, the number of generalized-ray

$$\begin{bmatrix} W_1^- \\ W_2^+ \\ \cdots \\ W_{S-1}^- \\ W_S^+ \\ \cdots \\ W_{N-1}^- \\ W_N^+ \end{bmatrix} = \begin{bmatrix} 0 & 0 & \bar{S}_2^{--} & 0 & \cdots & \cdots & \cdots \\ 0 & 0 & \bar{S}_2^{+-} & 0 & \cdots & \cdots & \cdots \\ \cdots & \cdots & \cdots & \cdots & \cdots & \cdots & \cdots \\ & \bar{S}_S^{-+} & 0 & 0 & \bar{S}_S^{--} & 0 & \cdots \\ & \bar{S}_S^{++} & 0 & 0 & \bar{S}_S^{+-} & 0 & \cdots \\ \cdots & \cdots & \cdots & \cdots & \cdots & \cdots & \cdots \\ 0 & \cdots & \cdots & \bar{S}_N^{-+} & 0 & \cdots & \cdots \\ 0 & \cdots & \cdots & \bar{S}_N^{++} & 0 & \cdots & \cdots \end{bmatrix} \cdot \begin{bmatrix} W_1^- \\ W_2^+ \\ \cdots \\ W_{S-1}^- \\ W_S^+ \\ \cdots \\ W_{N-1}^- \\ W_N^+ \end{bmatrix} + \begin{bmatrix} 0 \\ 0 \\ \cdots \\ X_{S-1}^- \\ X_S^+ \\ \cdots \\ 0 \\ 0 \end{bmatrix}$$

Figure 2.7 Scattering-array description

constituents involved in any computational procedure is further limited by an appropriate energy-based criterion that simply does away with wave constituents whose amplitude is below a detection threshold.

A generic transform-domain generalized-ray constituent, further denoted by $\tilde{R}$, can be described by an s-dependent factor representing the source signature, by the propagation factor representing its propagation along the total trajectory passed in the vertical direction, and, finally, by an algebraic factor accounting for its coupling to the source and its (possibly multiple) wave transmissions through and reflections against the interfaces. Accordingly, we may write

$$\tilde{R}(\kappa, \sigma, z, s) = \hat{Q}(s)\tilde{\Pi}(\kappa, \sigma) \exp\left[-s \sum_k p_k(\kappa, \sigma) Z_k\right], \tag{2.70}$$

where $\hat{Q}(s)$ can be associated with the source signature, $\tilde{\Pi}(\kappa, \sigma)$ incorporates the wave interaction with the interfaces and the exponential factor can be associated with the propagation factor. Here, $Z_k \geq 0$ denotes the *total* vertical propagation path the generalized ray has traversed in the kth layer $\mathcal{D}_k$. The generic generalized-ray coefficient (2.70) is amenable to inversion by the CdH technique, which is closely described in the next subsection.

2.2.1 Transformation to the space–time domain

The inversion of the transform-domain generalized-ray wave constituent (2.70) to the TD can be accomplished by the classical CdH technique, an application of which to the inversion of the scalar Green's function is given in Section 2.1.1. In the inversion procedure, we first represent (2.70) through the wave-slowness integral (2.57), viz

$$\begin{aligned}\hat{R}(x, y, z, s) = \hat{Q}(s)\,(s/2\mathrm{i}\pi)^2 \int_{\kappa=-\mathrm{i}\infty}^{\mathrm{i}\infty} \mathrm{d}\kappa \int_{\sigma=-\mathrm{i}\infty}^{\mathrm{i}\infty} \tilde{\Pi}(\kappa, \sigma)\\ \times \exp\left[-s\left(\kappa x + \sigma y + \sum_k p_k(\kappa, \sigma) Z_k\right)\right] \mathrm{d}\sigma\\ \text{for all } x \in \mathbb{R}, y \in \mathbb{R}, Z_k \geq 0 \text{ and } \{s \in \mathbb{R}; s > 0\}\end{aligned} \tag{2.71}$$

and recall that the vertical slowness parameter, p_k, is defined by (2.62). The s-domain generic wave constituent is further written in the following form:

$$\hat{R}(x, y, z, s) = s^2 \hat{Q}(s) \hat{G}(x, y, z, s), \tag{2.72}$$

where $\hat{G}(x, y, z, s)$ is the generalized-ray Green's function, viz

$$\begin{aligned}\hat{G}(x, y, z, s) = (1/2\mathrm{i}\pi)^2 \int_{\kappa=-\mathrm{i}\infty}^{\mathrm{i}\infty} \mathrm{d}\kappa \int_{\sigma=-\mathrm{i}\infty}^{\mathrm{i}\infty} \tilde{\Pi}(\kappa, \sigma)\\ \times \exp\left[-s\left(\kappa x + \sigma y + \sum_k p_k(\kappa, \sigma) Z_k\right)\right] \mathrm{d}\sigma.\end{aligned} \tag{2.73}$$

We shall next demonstrate that the CdH technique enables to cast the integral representation (2.71) into the form of the Laplace-transform integral (cf. (1.4))

$$\hat{G}(x, y, z, s) = \int_{\tau=T_1}^{T_2} \exp(-s\tau)K(x, y, z, \tau)\mathrm{d}\tau, \tag{2.74}$$

where $\{s \in \mathbb{R}; s > 0\}$ and $0 < T_1 < T_2$. Consequently, on account of Lerch's uniqueness theorem [14, Appendix], the TD original can be uniquely determined directly from (2.74) as

$$G(x, y, z, t) = K(x, y, z, t)[\mathrm{H}(t - T_1) - \mathrm{H}(t - T_2)] \quad \text{for } t > 0. \tag{2.75}$$

Once the TD Green's function is available, the TD generalized-ray constituent follows from the inverse of (2.72) as

$$R(x, y, z, t) = \partial_t^2 Q(t) *_t G(x, y, z, t), \tag{2.76}$$

which finally yields

$$R(x, y, z, t) = \begin{cases} 0 & \text{for } t \leq T_1, \\ \partial_t^2 \int_{\tau=T_1}^{\min(t,T_2)} Q(t-\tau)K(x, y, z, \tau)\mathrm{d}\tau & \text{for } t > T_1. \end{cases} \tag{2.77}$$

The TD expression (2.77) implies that T_1 and T_2 can be associated with the arrival times of wave fronts.

The CdH transformation that leads to the generalized ray TD Green's function (2.75) follows essentially the lines of reasoning described in Section 2.1.1. Accordingly, we first replace in (2.73) the variables of integration according to

$$\kappa = \gamma\cos(\phi) - \mathrm{i}q\sin(\phi), \tag{2.78a}$$
$$\sigma = \gamma\sin(\phi) + \mathrm{i}q\cos(\phi), \tag{2.78b}$$

with $x = r\cos(\phi)$ and $y = r\sin(\phi)$ for $r = (x^2 + y^2)^{1/2} > 0$ and $\{0 \leq \phi < 2\pi\}$. Under this mapping, $\kappa x + \sigma y = \gamma r$, $\kappa^2 + \sigma^2 = \gamma^2 - q^2$ and $\mathrm{d}\kappa\mathrm{d}\sigma = \mathrm{i}\mathrm{d}\gamma\mathrm{d}q$ and, consequently, (2.73) is transformed to

$$\hat{G}(x, y, z, s) = (1/4\pi^2\mathrm{i}) \int_{q=-\infty}^{\infty} \mathrm{d}q \int_{\gamma=-\mathrm{i}\infty}^{\mathrm{i}\infty} \bar{\Pi}(\gamma, q) \times \exp\left[-s\left(\gamma r + \sum_k \bar{p}_k(\gamma, q)Z_k\right)\right]\mathrm{d}\gamma, \tag{2.79}$$

where $\bar{\Pi}(\gamma, q)$ and $\bar{p}_k(\gamma, q)$ correspond to $\tilde{\Pi}(\kappa, \sigma)$ and $p_k(\kappa, \sigma)$, respectively, via the variable transformations (2.78a) and (2.78b). Hence, using (2.62), we write

$$\begin{aligned} \bar{p}_k(\gamma, q) &= (1/c_k^2 + q^2 - \gamma^2)^{1/2} \\ &= [\Omega_k^2(q) - \gamma^2]^{1/2}, \quad \text{with} \quad \mathrm{Re}[\bar{p}_k(\gamma, q)] \geq 0, \end{aligned} \tag{2.80}$$

where $\Omega_k(q) = (1/c_k^2 + q^2)^{1/2} > 0$. Next, the integrand in the integral with respect to γ is analytically continued into the complex γ-plane away from its imaginary axis. At the same time, we keep $\text{Re}(\bar{p}_k) \geq 0$ to have the integrand single-valued. This implies the branch cuts along the real axis $\{\text{Im}(\gamma) = 0, \Omega_k(q) \leq |\text{Re}(\gamma)| < \infty\}$ with branch points at $\gamma = \pm\Omega_k(q)$. Since the denominator of $\bar{\Pi}(\gamma, q)$ never goes to zero in a finite part of the γ-plane, no poles can occur. Apart from the branch points, the algebraic term $\bar{\Pi}(\gamma, q)$ has no other singularities and is of $o(1)$ as $|\gamma| \to \infty$. Consequently, under the application of Cauchy's theorem and Jordan's lemma, the original integration can be replaced with a new one along which

$$\gamma r + \sum_k \bar{p}_k(\gamma, q) Z_k = \tau \tag{2.81}$$

is met for a real-valued and positive (time) parameter τ. Equation (2.81) can be satisfied along a section of the real axis $\{\text{Im}(\gamma) = 0, 0 \leq \text{Re}(\gamma) < \min_k[\Omega_k(q)]\}$ and along a hyperbolic path that is symmetrical with respect to $\text{Im}(\gamma) = 0$ and goes to infinity. This hyperbolic path, known as the *body-wave path* $\mathcal{C}_\text{B} \cup \mathcal{C}_\text{B}^*$ (see Figure 2.8), can be symbolically represented by

$$\mathcal{C}_\text{B} = \left\{\gamma_\text{B} = \gamma_\text{B}(q, \tau); T_\text{B}(q) \leq \tau < \infty\right\}, \tag{2.82}$$

where the lower limit $T_\text{B}(q)$ corresponds to the point of intersection of the body-wave path with the real γ-axis, further denoted by $\gamma = \gamma_0(q)$. If $Z_k > 0$ for all k, the intersection point is located between the origin, $\gamma = 0$ and $\min_k[\Omega_k(q)]$. At this point, τ attains its minimum value, which through the differentiation of (2.81), leads to

$$r - \gamma \sum_k Z_k / \bar{p}_k(\gamma, q) = 0, \tag{2.83}$$

at $\gamma = \gamma_0(q)$. Since the point of intersection lies in $\{0 < \gamma_0(q) < \Omega_k(q)\}$, we can let

$$\gamma_0(q) = \Omega_k(q) \sin[\theta_k(q)] \quad \text{for all } k, \tag{2.84}$$

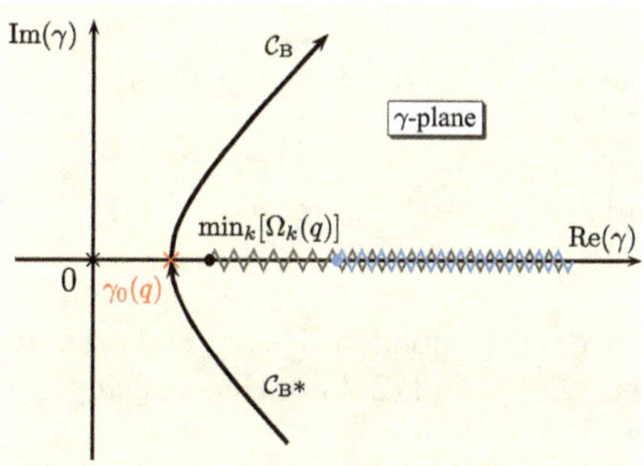

Figure 2.8 Complex γ-plane with the body-wave path $\mathcal{C}_\text{B} \cup \mathcal{C}_\text{B}^*$

with $\{0 \leq \theta_k(q) \leq \pi/2\}$. This implies that

$$\bar{p}_k(\gamma, q) = \Omega_k(q)\cos[\theta_k(q)] \quad \text{at } \gamma = \gamma_0(q). \tag{2.85}$$

Consequently, through (2.83), we find that

$$r = \sum_k Z_k \tan[\theta_k(q)] \quad \text{at } \gamma = \gamma_0(q). \tag{2.86}$$

Upon substituting (2.84)–(2.86) in (2.81), we arrive at the expression for the lower limit in the parametric representation of the body-wave path (2.82), viz

$$T_{\mathrm{B}}(q) = \sum_k \Omega_k(q) \frac{Z_k}{\cos[\theta_k(q)]}. \tag{2.87}$$

In accordance with Fermat's principle, T_{B} can be interpreted as the least time for a disturbance to traverse the distance from the source to the point of observation (see Figure 2.9). In this context, (2.84) can be interpreted as Snell's law of refraction. Finally, the body-wave integration path for late times looks as a straight line as follows from its asymptotic late-time behavior:

$$\gamma_{\mathrm{B}} = \frac{\tau}{r - \mathrm{i}\sum_k Z_k}\left[1 + O\left(\tau^{-2}\right)\right] \quad \text{as } \tau \to \infty. \tag{2.88}$$

Depending on the relative location of the observation point with respect to the source, it can happen that due to the reflection against a layer (characterized by a wave speed higher than the wave speeds of layers where the generalized-ray wave constituent propagates from) through which the generalized-ray constituent does not propagate, head waves occur. In such a case, the body-wave path must be supplemented by a loop along the branch cut denoted as the *head-wave path* $\mathcal{C}_{\mathrm{H}} \cup \mathcal{C}_{\mathrm{H}}^{*}$ (see Figure 2.10), where

$$\mathcal{C}_{\mathrm{H}} = \left\{\gamma_{\mathrm{H}} = \gamma_{\mathrm{H}}(q, \tau); T_{\mathrm{H}}(q) \leq \tau \leq T_{\mathrm{B}}(q)\right\}. \tag{2.89}$$

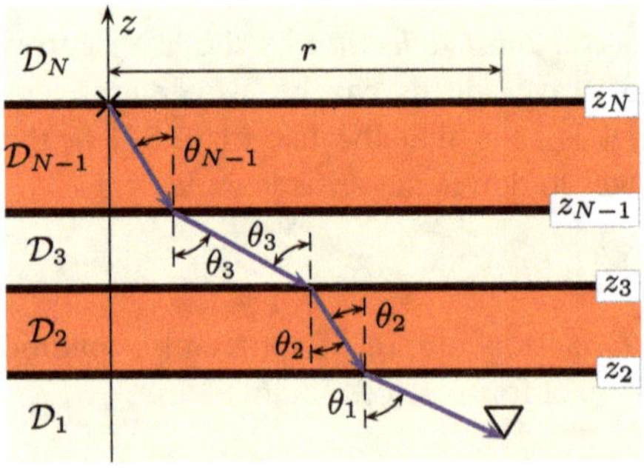

Figure 2.9 Least-time ray trajectory in a layered medium

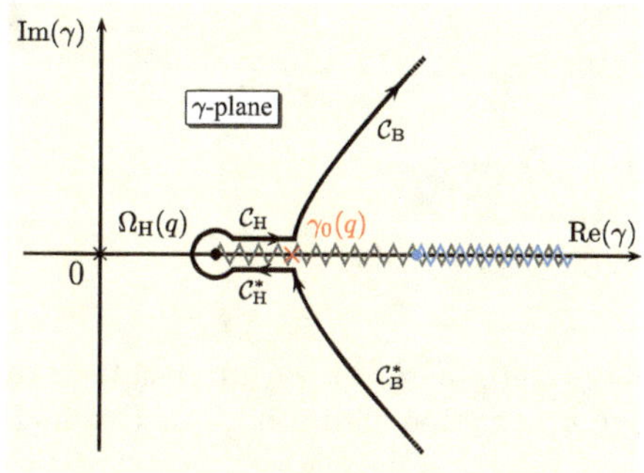

Figure 2.10 Complex γ-plane with both body-wave path $C_B \cup C_B^*$ and head-wave path $C_H \cup C_H^*$

Recall that $T_B(q)$ is given by (2.87) and the lower limit, $T_H(q)$, corresponds to the branch point $\Omega_H = \Omega_H(q)$ that is located at $\{\mathrm{Im}(\gamma) = 0; 0 \le \Omega_H(q) < \gamma_0(q)\}$. Consequently, it can be found via (2.81) as

$$T_H(q) = \Omega_H(q)r + \sum_k [\Omega_k^2(q) - \Omega_H^2(q)]^{1/2} Z_k. \tag{2.90}$$

In fact, $T_H(q)$ at $q = 0$ can be associated with the arrival time of the head-wave constituent. It is worth noting that the (existence) condition $\Omega_H(q) < \gamma_0(q)$ is equivalent to

$$r > \sum_k Z_k \frac{\Omega_H/\Omega_k}{[1 - (\Omega_H/\Omega_k)^2]^{1/2}}, \tag{2.91}$$

which tells that the head-wave constituent is present only if the horizontal source/receiver offset and corresponding contrast in EM properties are large enough. As a matter of fact, (2.91) is for $q = 0$, viz

$$r > \sum_k Z_k \frac{c_k/c_H}{[1 - (c_k/c_H)^2]^{1/2}}, \tag{2.92}$$

known as the condition for *total reflection* by the layer corresponding to the wave speed c_H. The ratio of wave speeds can be also interpreted in terms of "critical angles". To that end, it is seen that the use of (2.84) in the existence condition $\Omega_H(q) < \gamma_0(q)$ motivates to define the critical angles via

$$\sin[\theta_k^c(q)] = \Omega_H(q)/\Omega_k(q). \tag{2.93}$$

This implies that $\sin[\theta_k(q)] < \sin[\theta_k^c(q)]$. Apparently, condition (2.91) can be then rewritten in a more compact form as (cf. (2.86))

$$r > \sum_k Z_k \tan[\theta_k^c(q)]. \tag{2.94}$$

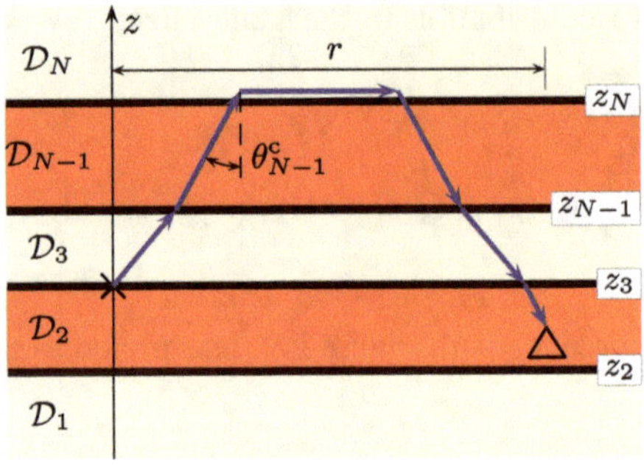

Figure 2.11 Head-wave ray trajectory in a layered medium

This geometric relation for $q = 0$ is depicted in Figure 2.11, where the critical refraction occurs at the interface $z = z_N$. In the following sections, the body- and head-wave contributions to the generalized-ray constituent are discussed separately.

2.2.1.1 Body-wave contribution

The body-wave contribution to the generalized-ray Green's function can be obtained from (2.79) by deforming the original integration path in the complex γ-plane to the body-wave path $\mathcal{C}_{\rm B} \cup \mathcal{C}^*_{\rm B}$ (see (2.82)). Then, upon introducing the time parameter τ as the new variable of integration and using Schwartz's reflection principle, we arrive at

$$\hat{G}_{\rm B}(x,y,z,s) = (1/2\pi^2)\int_{q=-\infty}^{\infty} {\rm d}q \int_{\tau=T_{\rm B}(q)}^{\infty} \exp(-s\tau){\rm Im}\left[\bar{\Pi}\left(\gamma_{\rm B},q\right)\frac{\partial\gamma_{\rm B}}{\partial\tau}\right]{\rm d}\tau, \tag{2.95}$$

where the Jacobian of mapping from $\gamma_{\rm B}$ to τ is given by (cf. (2.83))

$$\frac{\partial\gamma_{\rm B}}{\partial\tau} = \left[r - \gamma_{\rm B}\sum_k Z_k/\bar{p}_k(\gamma_{\rm B},q)\right]^{-1}. \tag{2.96}$$

Changing the order of the integrations leads to

$$\hat{G}_{\rm B}(x,y,z,s) = \int_{\tau=T_{\rm B}(0)}^{\infty} \exp(-s\tau){\rm d}\tau\frac{1}{2\pi^2}\int_{q=-Q_{\rm B}(\tau)}^{Q_{\rm B}(\tau)} {\rm Im}\left[\bar{\Pi}\left(\gamma_{\rm B},q\right)\frac{\partial\gamma_{\rm B}}{\partial\tau}\right]{\rm d}q, \tag{2.97}$$

where $q = Q_{\rm B}(\tau)$ denotes the inverse mapping of $\tau = T_{\rm B}(q)$. It is observed now that (2.97) has the form of (2.74). Consequently, relying on Lerch's uniqueness theorem, we get

$$G_{\rm B}(x,y,z,t) = K_{\rm B}(x,y,z,t){\rm H}[t - T_{\rm B}(0)], \tag{2.98}$$

with (cf. (2.75))

$$K_{\rm B}(x,y,z,t) = \frac{1}{2\pi^2}\int_{q=-Q_{\rm B}(t)}^{Q_{\rm B}(t)} {\rm Im}\left[\bar{\Pi}\left(\gamma_{\rm B},q\right)\frac{\partial\gamma_{\rm B}}{\partial t}\right]{\rm d}q. \tag{2.99}$$

Finally, the body-wave contribution to the generalized-ray constituent follows as (cf. (2.77))

$$R_{\mathrm{B}}(x,y,z,t) = \begin{cases} 0 & \text{for } 0 \le t \le T_{\mathrm{B}}(0), \\ \partial_t^2 \int_{\tau=T_{\mathrm{B}}(0)}^{t} Q(t-\tau) K_{\mathrm{B}}(x,y,z,\tau)\mathrm{d}\tau & \text{for } t > T_{\mathrm{B}}(0), \end{cases} \tag{2.100}$$

where $T_{\mathrm{B}}(0)$ is, in accordance with Figure 2.9, interpreted as the body-wave arrival time.

2.2.1.2 Head-wave contribution

For a head wave to exist, the condition $\Omega_{\mathrm{H}}(q) < \gamma_0(q)$ must be satisfied. It is seen, however, that $\Omega_{\mathrm{H}}(q) = |q| + O(q^{-1})$ and, by virtue of (2.84), that $\gamma_0(q) = |q|\sin(\theta) + O(q^{-1})$ with $\sin(\theta) < 1$ as $|q| \to \infty$. Therefore, the existence condition can be satisfied only for a bounded interval $|q| < Q_0$. In fact, $T_{\mathrm{H}}(q) = T_{\mathrm{B}}(q)$ holds true just at the limiting value $q = Q_0$.

Hence, starting from (2.79), where we replace the integration contour in the complex γ-plane with the head-wave path $\mathcal{C}_{\mathrm{H}} \cup \mathcal{C}_{\mathrm{H}}^*$ (see (2.89)), the head-wave contribution to the generalized-ray Green's function can be expressed via

$$\hat{G}_{\mathrm{H}}(x,y,z,s) = (1/2\pi^2) \int_{q=-Q_0}^{Q_0} \mathrm{d}q \int_{\tau=T_{\mathrm{H}}(q)}^{T_{\mathrm{B}}(q)} \exp(-s\tau) \mathrm{Im}\left[\bar{\Pi}\left(\gamma_{\mathrm{H}}, q\right) \frac{\partial \gamma_{\mathrm{H}}}{\partial \tau}\right] \mathrm{d}\tau, \tag{2.101}$$

where the Jacobian of mapping from γ_{H} to τ is given by (cf. (2.96))

$$\frac{\partial \gamma_{\mathrm{H}}}{\partial \tau} = \left[r - \gamma_{\mathrm{H}} \sum_k Z_k / \bar{p}_k(\gamma_{\mathrm{H}}, q) \right]^{-1}. \tag{2.102}$$

Upon interchanging the order of the integrations, we find

$$\begin{aligned} \hat{G}_{\mathrm{H}}(x,y,z,s) &= \int_{\tau=T_{\mathrm{H}}(0)}^{T_{\mathrm{B}}(0)} \exp(-s\tau)\mathrm{d}\tau \frac{1}{2\pi^2} \int_{q=-Q_{\mathrm{H}}(\tau)}^{Q_{\mathrm{H}}(\tau)} \mathrm{Im}\left[\bar{\Pi}\left(\gamma_{\mathrm{H}}, q\right) \frac{\partial \gamma_{\mathrm{H}}}{\partial \tau}\right] \mathrm{d}q \\ &+ \int_{\tau=T_{\mathrm{B}}(0)}^{T_0} \exp(-s\tau)\mathrm{d}\tau \frac{1}{2\pi^2} \left[\int_{q=-Q_{\mathrm{H}}(\tau)}^{-Q_{\mathrm{B}}(\tau)} + \int_{q=Q_{\mathrm{B}}(\tau)}^{Q_{\mathrm{H}}(\tau)} \right] \mathrm{Im}\left[\bar{\Pi}\left(\gamma_{\mathrm{H}}, q\right) \frac{\partial \gamma_{\mathrm{H}}}{\partial \tau}\right] \mathrm{d}q, \end{aligned} \tag{2.103}$$

where $q = Q_{\mathrm{H}}(\tau)$ denotes the inverse of the mapping $\tau = T_{\mathrm{H}}(q)$ for $q \ge 0$ and the upper limit $T_0 > 0$ in the second part of (2.103) is the common value of $T_{\mathrm{H}}(q)$ and $T_{\mathrm{B}}(q)$ at $q = Q_0$. Consequently, on account of Lerch's uniqueness theorem, the head-wave contribution to the TD generalized-ray Green's function can be expressed as

$$\begin{aligned} G_{\mathrm{H}}(x,y,z,t) &= K_{\mathrm{H};1}(x,y,z,t)\{\mathrm{H}[t - T_{\mathrm{H}}(0)] - \mathrm{H}[t - T_{\mathrm{B}}(0)]\} \\ &+ K_{\mathrm{H};2}(x,y,z,t)\{\mathrm{H}[t - T_{\mathrm{B}}(0)] - \mathrm{H}(t - T_0)\}, \end{aligned} \tag{2.104}$$

with (cf. (2.75))

$$K_{\mathrm{H};1}(x,y,z,t) = \frac{1}{2\pi^2}\int_{q=-Q_{\mathrm{H}}(t)}^{Q_{\mathrm{H}}(t)} \mathrm{Im}\left[\bar{\Pi}\left(\gamma_{\mathrm{H}},q\right)\frac{\partial\gamma_{\mathrm{H}}}{\partial t}\right]\mathrm{d}q, \tag{2.105}$$

$$K_{\mathrm{H};2}(x,y,z,t) = \frac{1}{2\pi^2}\left[\int_{q=-Q_{\mathrm{H}}(t)}^{-Q_{\mathrm{B}}(t)} + \int_{q=Q_{\mathrm{B}}(t)}^{Q_{\mathrm{H}}(t)}\right]\mathrm{Im}\left[\bar{\Pi}\left(\gamma_{\mathrm{H}},q\right)\frac{\partial\gamma_{\mathrm{H}}}{\partial t}\right]\mathrm{d}q. \tag{2.106}$$

Finally, the head-wave contribution to the generalized-ray constituent follows as (cf. (2.77))

$$R_{\mathrm{H}}(x,y,z,t) = \begin{cases} 0, & \text{for } t \le T_{\mathrm{H}}(0) \\ \partial_t^2 \int_{\tau=T_{\mathrm{H}}(0)}^{t} Q(t-\tau)K_{\mathrm{H};1}(x,y,z,\tau)\mathrm{d}\tau, & \text{for } T_{\mathrm{H}}(0) < t < T_{\mathrm{B}}(0) \\ \partial_t^2 \displaystyle\int_{\tau=T_{\mathrm{H}}(0)}^{T_{\mathrm{B}}(0)} Q(t-\tau)K_{\mathrm{H};1}(x,y,z,\tau)\mathrm{d}\tau & \\ \quad + \partial_t^2 \displaystyle\int_{\tau=T_{\mathrm{B}}(0)}^{\min(t,T_0)} Q(t-\tau)K_{\mathrm{H};2}(x,y,z,\tau)\mathrm{d}\tau, & \text{for } t > T_{\mathrm{B}}(0) \end{cases} \tag{2.107}$$

where $T_{\mathrm{H}}(0)$ and $T_{\mathrm{B}}(0)$ denote the head-wave and body-wave arrival times, respectively.

2.2.2 *Numerical implementation*

Apart from some special cases to which straightforward analytical expressions apply (see, e.g. (2.22) and [7, 10]), the parametric representations of CdH paths (2.82) and (2.89) must be obtained from (2.81) with the aid of numerical methods. Examples of such numerical procedures are briefly described in the following sections.

2.2.2.1 Body-wave contribution

In the first step, the arrival time of the body-wave wave contribution, $T_{\mathrm{B}}(0)$, is determined. This can be done by solving (2.83) numerically using the *regula-falsi* method for the intersection of the body-wave path, $\mathcal{C}_{\mathrm{B}} \cup \mathcal{C}_{\mathrm{B}}^*$, with $\mathrm{Im}(\gamma) = 0$ for $q = 0$, previously denoted by $\gamma_0(0)$. The desired arrival time is subsequently found

by substituting $\gamma = \gamma_0(0)$ in (2.81). Pseudocode of this procedure leading to $\gamma_0(q)$ (and hence to $T_{\mathrm{B}}(q)$) can be written as

INIT

$\gamma_u = \gamma_u(q); \gamma_w = \gamma_w(q);$

$n := 0;$

$\varepsilon :=$ prescribed precision;

$\gamma^{[0]} := \gamma_u$

WHILE $\left|\dfrac{\gamma_w - \gamma^{[n]}}{\partial_\gamma \tau(\gamma_w) - \partial_\gamma \tau(\gamma^{[n]})}\partial_\gamma \tau(\gamma^{[n]})\right| > \varepsilon$

$n := n + 1;$

$\gamma^{[n]} := \gamma^{[n-1]} - \dfrac{\gamma_w - \gamma^{[n-1]}}{\partial_\gamma \tau(\gamma_w) - \partial_\gamma \tau(\gamma^{[n-1]})}\partial_\gamma \tau(\gamma^{[n-1]});$

END

$T_{\mathrm{B}}(q) := \gamma^{[n]} r + \sum_k \bar{p}_k(\gamma^{[n]}, q) Z_k;$

where $\partial_\gamma \tau$ is a shorthand notation for $\partial \tau / \partial \gamma$, the expression for which is given on the left-hand side of (2.83). In the first block of the code, we calculate two initial values, viz

$$\gamma_u(q) = \Omega_m(q)\frac{r}{\left[r^2 + \left(\sum_k Z_k\right)^2\right]^{1/2}}, \tag{2.108a}$$

$$\gamma_w(q) = \Omega_m(q)\frac{r}{\left(r^2 + Z_m^2\right)^{1/2}}, \tag{2.108b}$$

where γ_u is then taken as the initial value of the iterative procedure, $\gamma^{[0]}$, and γ_w is the second point of its initial line segment. In (2.108a) and (2.108b), $\Omega_m(q) = \min_k[\Omega_k(q)]$ and Z_m is the vertical propagation path corresponding to Ω_m. With these starting values, the iterative procedure as indicated in the while loop of the code

is rapidly convergent. Finally, once $\gamma_0 = \gamma^{[n]}$ is achieved with prescribed precision $\varepsilon > 0$, the corresponding arrival time follows from (2.81).

In the second step, the inverse mapping $q = Q_\mathrm{B}(\tau)$ with respect to $\tau = T_\mathrm{B}(q)$ is determined for any $\tau > T_\mathrm{B}(0)$ using *Newton's method*. The numerical procedure can follow these lines

INIT

regula falsi: $\gamma_0 := \gamma_0(q^{[0]})$;

$n := 0$;

$\varepsilon :=$ prescribed precision;

$\mathcal{F}(\tau, \gamma_0, q^{[0]}); \partial_q \mathcal{F}(\gamma_0, q^{[0]})$;

WHILE $\left|(\mathcal{F}/\partial_q\mathcal{F})(\tau, \gamma_0, q^{[n]})\right| > \varepsilon$

$n := n + 1$;

$q^{[n]} := q^{[n-1]} + \mathcal{F}(\tau, \gamma_0, q^{[n-1]}) / [\partial_q \mathcal{F}(\gamma_0, q^{[n-1]})]$;

regula falsi (update): $\gamma_0 := \gamma_0(q^{[n]})$;

END

$Q_\mathrm{B}(\tau) := q^{[n]}$.

The first part of the code starts by calling the regula-falsi procedure from the previous step to find the intersection point in the complex γ-plane, $\gamma_0(q)$, for an initial value of q. For the latter, we may use

$$q^{[0]} = \left[\frac{\tau^2}{r^2 + \left(\sum_k Z_k\right)^2} - \frac{1}{c_M^2}\right]^{1/2}, \tag{2.109}$$

where $c_M = \max_k(c_k)$. Next, we use the standard Newton's method to solve $\mathcal{F} = 0$ for q with (cf. (2.81))

$$\mathcal{F}(\tau, \gamma, q) = \tau - \gamma r - \sum_k \bar{p}_k(\gamma, q) Z_k, \tag{2.110a}$$

$$\partial_q \mathcal{F}(\gamma, q) = -q \sum_k Z_k / \bar{p}_k(\gamma, q) \quad \text{for any } \tau > T_{\mathrm{B}}(0), \tag{2.110b}$$

and the (regularly updated) intersection point γ_0. Once the iterative procedure is terminated, the resulting $Q_{\mathrm{B}}(\tau) = q^{[n]}$ is readily obtained with any prescribed precision $\varepsilon > 0$.

Finally, we can apply Newton's method, again, to find a parametric representation of the CdH path for any $\tau > T_{\mathrm{B}}(0)$ with $\{-Q_{\mathrm{B}}(\tau) < q < Q_{\mathrm{B}}(\tau)\}$ (see (2.97)). The pertaining iterative procedure may run along the following lines:

INIT

$\gamma^{[0]} := \tau / (r - \mathrm{i} \sum_k Z_k);$

$n := 0;$

$\varepsilon :=$ prescribed precision;

WHILE $\left| (\mathcal{F} / \partial_\gamma \mathcal{F}) (\tau, \gamma^{[n]}, q) \right| > \varepsilon$

$n := n + 1;$

$\gamma^{[n]} := \gamma^{[n-1]} + \mathcal{F}(\tau, \gamma^{[n-1]}, q) / [\partial_\gamma \mathcal{F}(\gamma^{[n-1]}, q)];$

END

$\gamma(\tau, q) := \gamma^{[n]}.$

In the initial part of the code, we use the asymptotic expression (2.88) as the point of departure for the iterative procedure. Consequently, in the while loop, we solve $\mathcal{F} = 0$ (see (2.110a)) iteratively, again, but now until the value of γ on the

CdH path is achieved with the prescribed precision $\varepsilon > 0$. In the iterative procedure, we use the differentiation of (2.110a) with respect to γ, i.e.

$$\partial_\gamma \mathcal{F}(\gamma, q) = \gamma \sum_k Z_k / \bar{p}_k(\gamma, q) - r. \tag{2.111}$$

Once the iterative procedure is terminated, the calculated values of γ_{B} along the CdH path can be used to carry out the integral (2.99) using the Gauss–Legendre numerical integration method, for instance. Since the integrand shows the inverse square root singularity at $q = Q_{\mathrm{B}}(t)$, we stretch the variable of integration according to

$$q = Q_{\mathrm{B}}(t)\sin(\vartheta) \quad \text{for } \{-\pi/2 \le \vartheta \le \pi/2\}, \tag{2.112}$$

with the Jacobian $\mathrm{d}q/\mathrm{d}\vartheta = Q_{\mathrm{B}}(t)\cos(\vartheta)$ for $t > T_{\mathrm{B}}(0)$. Finally, the time convolution with the source waveform in (2.100) does not present any numerical difficulties and can be readily evaluated numerically via the trapezoidal integration rule, for example.

2.2.2.2 Head-wave contribution

In the first step, the arrival time of the head-wave wave contribution $T_{\mathrm{H}}(0)$ is determined. Since the corresponding γ-value is known, the arrival time directly follows from (2.90) for $q = 0$ as

$$T_{\mathrm{H}}(0) = r/c_{\mathrm{H}} + \sum_k (1/c_k^2 - 1/c_{\mathrm{H}}^2)^{1/2} Z_k. \tag{2.113}$$

In the second step, the inverse mapping $q = Q_{\mathrm{H}}(\tau)$ can be found from (2.81) with $\gamma = \Omega_{\mathrm{H}}(q)$ as

$$Q_{\mathrm{H}}(\tau) = \left\{ \left[\frac{\tau}{r} - \sum_k \left(\frac{1}{c_k^2} - \frac{1}{c_{\mathrm{H}}^2} \right)^{1/2} \frac{Z_k}{r} \right]^2 - \frac{1}{c_{\mathrm{H}}^2} \right\}^{1/2}. \tag{2.114}$$

Further, the upper time limit $T_0 > T_{\mathrm{B}}(0)$ in the second term on the right-hand side of (2.103) can be determined for $q = \pm Q_0$, where $T_{\mathrm{H}}(q) = T_{\mathrm{B}}(q)$ is met. This can be solved through Newton's iteration, again, along these lines:

INIT

$q^{[0]} := q_u$;

$n := 0$;

$\varepsilon :=$ prescribed precision;

WHILE $\left|(\mathcal{G}/\partial_q\mathcal{G})(\gamma_0, q^{[n]})\right| > \varepsilon$

$n := n + 1;$

$q^{[n]} := q^{[n-1]} + \mathcal{G}(\gamma_0, q^{[n-1]})/[\partial_q\mathcal{G}(\gamma_0, q^{[n-1]})];$

regula falsi (update): $\gamma_0 := \gamma_0(q^{[n]});$

END

$Q_0 := q^{[n]};$

$T_0 := \Omega_{\mathrm{H}}(Q_0)r + \sum_k (1/c_k^2 - 1/c_{\mathrm{H}}^2)^{1/2} Z_k.$

In the first part of the code, we calculate the initial value of q for the iterative procedure. For that purpose, we may use

$$q_u = \left[\frac{r^2}{Z_M^2}\left(\frac{1}{c_M^2} - \frac{1}{c_{\mathrm{H}}^2}\right) - \frac{1}{c_{\mathrm{H}}^2}\right]^{1/2}, \tag{2.115}$$

where $c_M = \max_k(c_k) < c_{\mathrm{H}}$ and Z_M is the vertical propagation path corresponding to c_M. In the subsequent while loop, we apply Newton's technique to solve $\mathcal{G} = 0$, with (cf. (2.81) and (2.90))

$$\mathcal{G}(\gamma, q) = \gamma r + \sum_k \bar{p}(\gamma, q)Z_k - r\Omega_{\mathrm{H}}(q) - \sum_k (1/c_k^2 - 1/c_{\mathrm{H}}^2)^{1/2} Z_k, \tag{2.116a}$$

$$\partial_q\mathcal{G}(\gamma, q) = q\sum_k Z_k/\bar{p}(\gamma, q) - qr/\Omega_{\mathrm{H}}(q). \tag{2.116b}$$

As an intermediate step, we apply the regular-falsi procedure, again, to find the intersection point, $\gamma_0(q)$, that corresponds to $T_{\mathrm{B}}(q)$. Once the iterative procedure is carried out, we end up with $Q_0 = q^{[n]}$ and thus with T_0 via (2.90). On account of the fact that also $\mathrm{d}T_{\mathrm{H}}(q)/\mathrm{d}q = \mathrm{d}T_{\mathrm{B}}(q)/\mathrm{d}q$ at $q = \pm Q_0$, the rate of convergence of Newton's method is not quadratic in this case.

Finally, Newton's method is employed to find a parametric representation of the CdH path for any $\{T_{\mathrm{H}}(0) < \tau < T_{\mathrm{B}}(0)\}$ with $\{-Q_{\mathrm{H}}(\tau) < q < Q_{\mathrm{H}}(\tau)\}$ and $\{T_{\mathrm{B}}(0) < \tau < T_0\}$ with $\{-Q_{\mathrm{H}}(\tau) < q < -Q_{\mathrm{B}}(\tau)\}$ and $\{Q_{\mathrm{B}}(\tau) < q < Q_{\mathrm{H}}(\tau)\}$ (see (2.103)). To that end, we may apply the same procedure as for the

body-wave path but with $\Omega_{\rm H}$ as the initial value. Therefore, the iterative procedure can be implemented along the following lines:

INIT

$\gamma^{[0]} := \Omega_{\rm H}(q)$;

$n := 0$;

$\varepsilon :=$ prescribed precision;

WHILE $\left|(\mathcal{F}/\partial_\gamma \mathcal{F})(\tau, \gamma^{[n]}, q)\right| > \varepsilon$

$n := n + 1$;

$\gamma^{[n]} := \gamma^{[n-1]} + \mathcal{F}(\tau, \gamma^{[n-1]}, q)/[\partial_\gamma \mathcal{F}(\gamma^{[n-1]}, q)]$;

END

$\gamma(\tau, q) := \gamma^{[n]}$;

where $\mathcal{F}$ and its derivative with respect to γ are given by (2.110a) and (2.111), respectively. The iterative procedure can be carried out for given values of τ and q, thereby achieving the head-wave CdH path within any prescribed precision $\varepsilon > 0$. With the CdH path at our disposal, we can evaluate the integrations in both (2.105) and (2.106). To handle numerical difficulties associated with inverse square root singularities at the integration end points, we may stretch the variable of integration according to

$$q = Q_{\rm H}(t)\sin(\vartheta) \quad \text{for } \{-\pi/2 \le \vartheta \le \pi/2\}, \tag{2.117}$$

with the Jacobian $\mathrm{d}q/\mathrm{d}\vartheta = Q_{\rm H}(t)\cos(\vartheta)$ for $\{T_{\rm H}(0) < t < T_{\rm B}(0)\}$. Likewise, we may use

$$q^2 = Q_{\rm B}^2(t)\cos^2(\vartheta) + Q_{\rm H}^2(t)\sin^2(\vartheta) \quad \text{for } \{0 \le \vartheta \le \pi/2\}, \tag{2.118}$$

with the Jacobian

$$\frac{\mathrm{d}q}{\mathrm{d}\vartheta} = \frac{\left[Q_{\rm H}^2(t) - Q_{\rm B}^2(t)\right]\sin(\vartheta)\cos(\vartheta)}{\left[Q_{\rm B}^2(t)\cos^2(\vartheta) + Q_{\rm H}^2(t)\sin^2(\vartheta)\right]^{1/2}} \quad \text{for } \{T_{\rm B}(0) < t < T_0\}. \tag{2.119}$$

To evaluate the integral over $\{-Q_{\rm H}(\tau) < q < -Q_{\rm B}(\tau)\}$, we may use the property that the integrand is an even function of q. Finally, the time convolution with the source waveform in (2.107) does not introduce any difficulties and can be hence readily computed numerically via the trapezoidal integration rule, for instance.

2.3 Concluding remarks

The scattering-array formalism can be elucidated on a relatively simple EM canonical problem. For this purpose, we shall analyze the EM radiation from a vertical electric dipole located over a dielectric half-space. Hence, with reference to the general problem configuration shown in Figure 2.5, the problem configuration under consideration consists of three domains, $\mathcal{D}_1$, $\mathcal{D}_2$ and $\mathcal{D}_3$, and two interfaces extending along $z = z_2$ and $z = z_3$ (see Figure 2.12). The EM properties of the lower half-space $\mathcal{D}_1$ are described by $\{\epsilon_1, \mu_0\}$, and both $\mathcal{D}_2$ and $\mathcal{D}_3$ are characterized by $\{\epsilon_2, \mu_0\}$. The corresponding EM wave speeds are $c_1 = (\mu_0\epsilon_1)^{-1/2}$ and $c_2 = (\mu_0\epsilon_2)^{-1/2}$. The impulsive vertical electric dipole is located at $z = z_3$, that is, at the height $d_2 = z_3 - z_2 > 0$ above the lower half-space.

With the help of (2.65a)–(2.65d), the scattering parameters corresponding to the interface at $z = z_2$ can be expressed as

$$S_2^{+-} = \frac{p_2/\epsilon_2 - p_1/\epsilon_1}{p_2/\epsilon_2 + p_1/\epsilon_1} \quad \text{and} \quad S_2^{--} = \frac{2\,p_2/\epsilon_2}{p_2/\epsilon_2 + p_1/\epsilon_1}, \tag{2.120a}$$

while the zero contrast in the EM properties between $\mathcal{D}_2$ and $\mathcal{D}_3$ implies that $p_3/\epsilon_3 = p_2/\epsilon_2$, and hence

$$S_3^{-+} = \frac{p_2/\epsilon_2 - p_3/\epsilon_3}{p_2/\epsilon_2 + p_3/\epsilon_3} = 0 \quad \text{and} \quad S_3^{++} = \frac{2\,p_2/\epsilon_2}{p_2/\epsilon_2 + p_3/\epsilon_3} = 1. \tag{2.120b}$$

Moreover, since the dipole source is located at $z = z_3$, we use $S = 3$ and $p_3 = p_2$ in (2.66) and get

$$X_3^+ = X_2^- = 1/2sp_2. \tag{2.121}$$

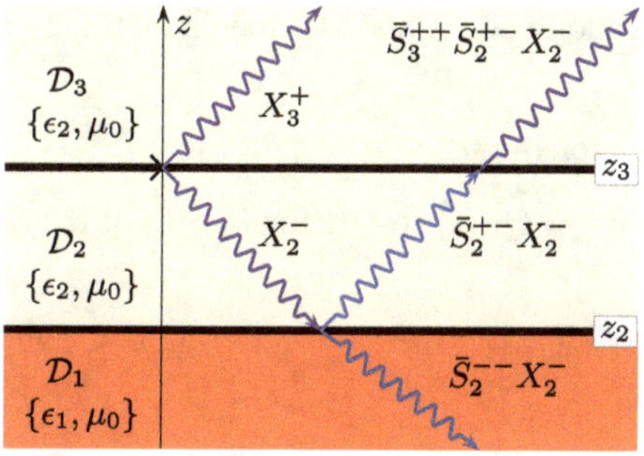

Figure 2.12 Wave amplitudes associated with a vertical electric dipole above a dielectric half-space

Recall that in line with (2.62), p_1 and p_2 represent the vertical slowness parameters in $\mathcal{D}_1$ and $\mathcal{D}_2$ that are given by

$$p_{1,2}(\kappa,\sigma) = (1/c_{1,2}^2 - \kappa^2 - \sigma^2)^{1/2}, \quad \text{with} \quad \mathrm{Re}[p_{1,2}(\kappa,\sigma)] \geq 0, \tag{2.122}$$

respectively.

The wave amplitudes in the analyzed three-layered structure can be interrelated via the $[4 \times 4]$ scattering array. With reference to Figure 2.7 showing the generic scattering-array description, the system of equations pertaining to the three-layered structure has the following form:

$$\begin{bmatrix} W_1^- \\ W_2^+ \\ W_2^- \\ W_3^+ \end{bmatrix} = \begin{bmatrix} 0 & 0 & \bar{S}_2^{--} & 0 \\ 0 & 0 & \bar{S}_2^{+-} & 0 \\ 0 & \bar{S}_3^{-+} & 0 & 0 \\ 0 & \bar{S}_3^{++} & 0 & 0 \end{bmatrix} \cdot \begin{bmatrix} W_1^- \\ W_2^+ \\ W_2^- \\ W_3^+ \end{bmatrix} + \begin{bmatrix} 0 \\ 0 \\ X_2^- \\ X_3^+ \end{bmatrix}. \tag{2.123}$$

Recall that the modified scattering parameters in the scattering array follow from (2.120a) and (2.120b) via (2.64a) for $n = 2$ and via (2.64b) for $n = 3$. Furthermore, it is noted that the upgoing wave in $\mathcal{D}_1$ and downgoing wave in $\mathcal{D}_3$ are identically zero, that is, we let $W_1^+ = W_3^- = 0$. The system of equations (2.123) can be solved iteratively, as indicated in (2.69a) and (2.69b). Adopting this approach, it is readily found that

$$\begin{bmatrix} W_1^- \\ W_2^+ \\ W_2^- \\ W_3^+ \end{bmatrix} = \begin{bmatrix} 0 \\ 0 \\ X_2^- \\ X_3^+ \end{bmatrix} + \begin{bmatrix} \bar{S}_2^{--} X_2^- \\ \bar{S}_2^{+-} X_2^- \\ 0 \\ 0 \end{bmatrix} + \begin{bmatrix} 0 \\ 0 \\ 0 \\ \bar{S}_3^{++} \bar{S}_2^{+-} X_2^- \end{bmatrix}. \tag{2.124}$$

Apparently, on the right-hand side of (2.124), we may identify the excitation array, $\underline{\bar{S}}$, and two subsequent generalized-ray constituents, $\underline{\bar{S}} \cdot \boldsymbol{X}$ and $\underline{\bar{S}}^2 \cdot \boldsymbol{X}$, representing one and two interactions with the interfaces, respectively (see Figure 2.12). All higher generalized-ray constituents are identically zero, that is, $\underline{\bar{S}}^q \cdot \boldsymbol{X} = \boldsymbol{0}$ for all $q = \{3, 4, \ldots\}$. Upon using (2.120a)–(2.121) with (2.64a) and (2.64b) in the solution (2.124), we may write

$$W_1^- = \bar{S}_2^{--} X_2^- = S_2^{--} X_2^- \exp(-sp_2 d_2) = \frac{2\, p_2/\epsilon_2}{p_2/\epsilon_2 + p_1/\epsilon_1} \frac{\exp(-sp_2 d_2)}{2sp_2} \tag{2.125a}$$

for the downgoing wave amplitude in $\mathcal{D}_1$, and,

$$W_2^+ = \bar{S}_2^{+-} X_2^- = S_2^{+-} X_2^- \exp(-sp_2 d_2) = \frac{p_2/\epsilon_2 - p_1/\epsilon_1}{p_2/\epsilon_2 + p_1/\epsilon_1} \frac{\exp(-sp_2 d_2)}{2sp_2} \tag{2.125b}$$

for the upgoing wave amplitude in $\mathcal{D}_2$, and

$$W_2^- = X_2^- = \frac{1}{2sp_2} \tag{2.125c}$$

for the downgoing wave amplitude in $\mathcal{D}_2$, and, finally,

$$\begin{aligned} W_3^+ &= X_3^+ + \bar{S}_3^{++}\bar{S}_2^{+-}X_2^- = X_3^+ + S_3^{++}S_2^{+-}X_2^- \exp(-2sp_2d_2) \\ &= \frac{1}{2sp_2} + \frac{p_2/\epsilon_2 - p_1/\epsilon_1}{p_2/\epsilon_2 + p_1/\epsilon_1}\frac{\exp(-2sp_2d_2)}{2sp_2} \end{aligned} \tag{2.125d}$$

for the upgoing wave amplitude in $\mathcal{D}_3$. In accordance with Figure 2.12, the wave motion in $\mathcal{D}_3$ consists of the primary wave radiated from the source and the secondary wave reflected by the lower dielectric half-space.

With the wave amplitudes at our disposal, we may specify the transform-domain Green's function in all subdomains of the problem configuration. Hence, using (2.125a) with $W_1^+ = 0$ in (2.61) for $n = 1$, we obtain

$$\begin{aligned} \tilde{G}_1(\kappa, \sigma, z|z_3, s) &= W_1^- \exp[-sp_1(z_2 - z)] \\ &= S_2^{--}X_2^- \exp\{-s[p_1(z_2 - z) + p_2d_2]\} \quad \text{in } \mathcal{D}_1. \end{aligned} \tag{2.126a}$$

Furthermore, using (2.125b) and (2.125c), we get

$$\begin{aligned} \tilde{G}_2(\kappa, \sigma, z|z_3, s) &= W_2^- \exp[-sp_2(z_3 - z)] + W_2^+ \exp[-sp_2(z - z_2)] \\ &= X_2^- \exp[-sp_2(z_3 - z)] + S_2^{+-}X_2^- \exp[-sp_2(z + d_2 - z_2)] \quad \text{in } \mathcal{D}_2. \end{aligned} \tag{2.126b}$$

Finally, using $W_3^- = 0$ and $p_3 = p_2$, we find

$$\begin{aligned} \tilde{G}_3(\kappa, \sigma, z|z_3, s) &= W_3^+ \exp[-sp_2(z - z_3)] \\ &= X_3^+ \exp[-sp_2(z - z_3)] + S_3^{++}S_2^{+-}X_2^- \exp[-sp_2(z + 2d_2 - z_3)] \\ &\quad \text{in } \mathcal{D}_3. \end{aligned} \tag{2.126c}$$

Once the Green's functions are determined, they can be used to specify the (transform-domain) radiated EM fields in any subdomain of the problem configuration. The tangential EM-field components, for instance, directly follow from (2.58a) to (2.58d). Their final inversion back to the TD can be essentially carried out using the methodology described in Section 2.2.1.

References

[1] L. Cagniard. *Réflexion et réfraction des ondes séismiques progressives*. Paris: Gauthier-Villars, 1939.

[2] L. Cagniard. *Reflection and Refraction of Progressive Seismic Waves*. New York, NY: McGraw-Hill, 1962. (Translation by Flinn, E., Dix, C. H. of Réflexion et réfraction des ondes séismiques progressives, Paris, Gauthier-Villars, 1939).

[3] A. T. de Hoop. A modification of Cagniard's method for solving seismic pulse problems. *Applied Scientific Research*, B(8):349–356, 1960.

[4] A. T. de Hoop. *Representation Theorems for the Displacement in an Elastic Solid and Their Application to Elastodynamic Diffraction Theory*. PhD thesis, Delft University of Technology, the Netherlands, 1958.

[5] A. T. de Hoop and J. H. M. T. Van der Hijden. Generation of acoustic waves by an impulsive point source in a fluid/solid configuration with a plane boundary. *The Journal of the Acoustical Society of America*, 75(6):1709–1715, 1984.

[6] A. T. de Hoop. Pulsed electromagnetic radiation from a line source in a two-media configuration. *Radio Science*, 14(2):253–268, 1979.

[7] B. J. Kooij. The electromagnetic field emitted by a pulsed current point source above the interface of a nonperfectly conducting earth. *Radio Science*, 31(6):1345–1360, 1996.

[8] M. Štumpf, A. T. de Hoop and G. A. E. Vandenbosch. Generalized ray theory for time-domain electromagnetic fields in horizontally layered media. *IEEE Transactions on Antennas and Propagation*, 61(5):2676–2687, 2013.

[9] J. H. M. T. Van der Hijden. *Propagation of Transient Elastic Waves in Stratified Anisotropic Media*. PhD thesis, Delft University of Technology, the Netherlands, 1987.

[10] A. T. de Hoop, M. Štumpf and I. E. Lager. Pulsed electromagnetic field radiation from a wide slot antenna with a dielectric layer. *IEEE Transactions on Antennas and Propagation*, 59(8):2789–2798, 2011.

[11] M. Štumpf. Pulsed vertical-electric-dipole excited voltages on transmission lines over a perfect ground—a closed-form analytical description. *IEEE Antennas and Wireless Propagation Letters*, 17(9):1656–1658, 2018.

[12] M. Štumpf and A. T. de Hoop. Loop-to-loop pulsed electromagnetic signal transfer across a thin metal screen with Drude-type dispersive behavior. *IEEE Transactions on Electromagnetic Compatibility*, 60(4):885–889, 2017.

[13] M. Štumpf and G. Antonini. Lightning-induced voltages on transmission lines over a lossy ground—an analytical coupling model based on the Cooray-Rubinstein formula. *IEEE Transactions on Electromagnetic Compatibility*, 62(1):155–162, 2020.

[14] M. Štumpf. *Electromagnetic Reciprocity in Antenna Theory*. Hoboken, NJ: John Wiley & Sons, Inc., 2018.

[15] M. Štumpf. *Time-Domain Electromagnetic Reciprocity in Antenna Modeling*. Hoboken, NJ: John Wiley & Sons, Inc., 2019.

[16] S. A. Schelkunoff. *Electromagnetic Waves*. New York, NY: D. Van Nostrand Company, Inc., 9th printing, 1957.

[17] A. T. de Hoop. *Handbook of Radiation and Scattering of Waves*. London, UK: Academic Press, 1995.

[18] E. Strick. Propagation of elastic wave motion from an impulsive source along a fluid/solid interface, II. Theoretical pressure pulse. *Philosophical Transactions of the Royal Society of London. Series A, Mathematical and Physical Sciences*, 251(1000): 465–523, 1959.

[19] D. V. Helmberger. The crust-mantle transition in the Bering Sea. *Bulletin of the Seismological Society of America*, 58(1):179–214, 1968.

[20] K. Aki and P. G. Richards. *Quantitative Seismology, 2nd Edition*. Sausalito, CA: University Science Books, 2009.
[21] A. Sommerfeld. *Partial Differential Equations in Physics*. New York, NY: Academic Press, Inc., 1949.
[22] C. L. Pekeris. The seismic buried pulse. *Proceedings of the National Academy of Sciences of the United States of America*, 41(9):629–639, 1955.
[23] G. N. Watson. *A Treatise on the Theory of Bessel Functions*. London, UK: Cambridge University Press, 1922.
[24] A. T. de Hoop. Acoustic radiation from impulsive sources in a layered fluid. *Nieuw Archief voor Wiskunde*, 6(4):111–129, 1988.

Chapter 3

Thin-sheet high-contrast saltus-type conditions

The widespread use of composite materials in the aircraft, automotive and building industries raised the need for efficient computational methodologies capable of providing their electromagnetic (EM) scattering characteristics. Owing to the multiscale nature of structures involving composites, however, both standard direct-discretization techniques (e.g. the finite difference time-domain (TD) method) and volumetric integral-equation-based approaches (e.g. the method of moments) suffer from the heavy computational burden or/and meshing issues. A way out of the difficulty has been offered by thin-sheet transition conditions (e.g. [1, Section 2.4]) that considerably simplify calculations by reducing the actual volumetric solution domain to the corresponding surface one. Accordingly, thin-sheet saltus-type conditions suitable for analyzing pulsed EM fields in the presence of thin-sheet layers are the main subject of this chapter. Their TD formulation concerning thin-film structures with dielectric and conductive properties has been introduced [2], while the idea behind the thin-sheet cross-layer conditions can be traced back to the works of Levi–Civita from the beginning of the twentieth century [3, p. 19].

3.1 Problem formulation

The problem configuration under consideration is shown in Figure 3.1. It consists of a layer of thickness $\delta > 0$ that is located in a homogeneous, isotropic and loss-free embedding $\mathcal{D}^\infty$. The EM properties of the sheet are described by its (scalar, real-valued and positive) electric permittivity ϵ, electric conductivity σ, magnetic permeability μ and by parameter Γ representing linear magnetic hysteresis losses. The latter parameter has been largely introduced for the sake of symmetry. In the most of practical situations magnetic hysteresis losses are negligible, so that we can let $\Gamma = 0$. The background medium is described by its electric permittivity ϵ_0 and by magnetic permeability μ_0.

The planar structure is supposed to be irradiated by an external EM source, whose action is represented by the incident electric- and magnetic-field strengths $\{\hat{\boldsymbol{E}}^{\mathrm{i}}, \hat{\boldsymbol{H}}^{\mathrm{i}}\} = \{\hat{\boldsymbol{E}}^{\mathrm{i}}, \hat{\boldsymbol{H}}^{\mathrm{i}}\}(\boldsymbol{r}, s)$, respectively. Consequently, the EM field scattered by

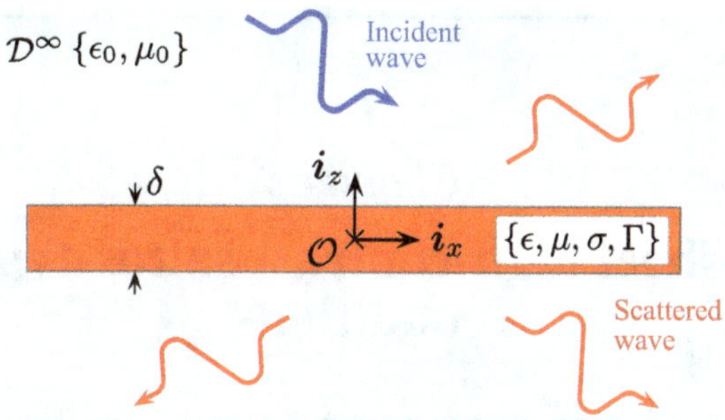

Figure 3.1 *Thin, high-contrast planar sheet embedded in a homogeneous, isotropic and loss-free background medium*

the sheet is defined as the difference between the total EM field in the configuration and the incident EM field. Accordingly, we write

$$\{\hat{\boldsymbol{E}}^{\mathrm{s}}, \hat{\boldsymbol{H}}^{\mathrm{s}}\}(\boldsymbol{r}, s) = \{\hat{\boldsymbol{E}}, \hat{\boldsymbol{H}}\}(\boldsymbol{r}, s) - \{\hat{\boldsymbol{E}}^{\mathrm{i}}, \hat{\boldsymbol{H}}^{\mathrm{i}}\}(\boldsymbol{r}, s)$$
$$\text{for all } \boldsymbol{r} \in \mathbb{R}^3 \text{ and } \{s \in \mathbb{C}; s \geq s_0\}. \tag{3.1}$$

Within the domain occupied by the layer, say

$$\mathcal{V} = \{(x, y) \in \Sigma \subset \mathbb{R}^2, -\delta/2 < z < \delta/2\} \subset \mathbb{R}^3, \tag{3.2}$$

the EM fields are governed by source-free Maxwell's equations [4, Section 18.3], whose volume-integrated forms read [2, Eqs. (4) and (5)]

$$\int_{\boldsymbol{r}\in\partial\mathcal{D}} \boldsymbol{\nu} \times \hat{\boldsymbol{E}}\,\mathrm{d}A = -\int_{\boldsymbol{r}\in\mathcal{D}} (s\mu\hat{\boldsymbol{H}} + \Gamma\hat{\boldsymbol{H}})\mathrm{d}V, \tag{3.3a}$$

$$\int_{\boldsymbol{r}\in\partial\mathcal{D}} \boldsymbol{\nu} \times \hat{\boldsymbol{H}}\,\mathrm{d}A = \int_{\boldsymbol{r}\in\mathcal{D}} (s\epsilon\hat{\boldsymbol{E}} + \sigma\hat{\boldsymbol{E}})\mathrm{d}V, \tag{3.3b}$$

where $\mathcal{D}$ denotes a subdomain of $\mathcal{V}$ and $\partial\mathcal{D}$ is its closed boundary surface with the piecewise continuous unit vector $\boldsymbol{\nu}$ oriented along its outward normal.

3.2 Transition conditions for a thin high-contrast layer with dielectric and conductive properties

In this section, it is assumed that EM properties of the layer show a relatively high contrast with respect to the surrounding medium in the electric permittivity and conductivity only, that is, $\mu = \mu_0$ and $\Gamma = 0$. Consequently, associating $\mathcal{D}$ in (3.3a) and (3.3b) with a pillbox perpendicularly crossing the layer of thickness $\delta > 0$, we readily arrive at the following cross-layer conditions:

$$\boldsymbol{i}_z \times \hat{\boldsymbol{E}}^+ - \boldsymbol{i}_z \times \hat{\boldsymbol{E}}^- = o(\delta), \tag{3.4a}$$

$$\boldsymbol{i}_z \times \hat{\boldsymbol{H}}^+ - \boldsymbol{i}_z \times \hat{\boldsymbol{H}}^- = (G^{\mathrm{E}} + sC^{\mathrm{E}})\hat{\boldsymbol{E}}^\pm + o(\delta), \tag{3.4b}$$

as $\delta \downarrow 0$ for all $(x, y) \in \Sigma$, where $\hat{\boldsymbol{E}}^{\pm} = \hat{\boldsymbol{E}}(x, y, \pm\delta/2, s)$ denotes the electric-field strength distribution on the top and bottom sides of the layer and a similar definition applies to the magnetic-field strength. Furthermore, the parameters G^{E} and C^{E} are given by [2, Eqs. (2) and (3)]

$$G^{\mathrm{E}} = \int_{\zeta=-\delta/2}^{\delta/2} \sigma(\zeta)\mathrm{d}\zeta, \tag{3.5a}$$

$$C^{\mathrm{E}} = \int_{\zeta=-\delta/2}^{\delta/2} \epsilon(\zeta)\mathrm{d}\zeta, \tag{3.5b}$$

and have the meaning of the *equivalent layer conductance and capacitance*, respectively. Owing to the high-contrast with respect to the embedding, these parameters are of order $O(1)$, i.e. they are bounded constants. As a matter of fact, (3.4a) tells us that the tangential component of the electric-field strength is essentially continuous across the layer, while (3.4b) represents the jump of (the tangential component of) the magnetic-field strength that is proportional to the contrast sheet electric-current density

$$\partial\hat{\boldsymbol{J}}^{\mathrm{s}}(x, y, s) = (G^{\mathrm{E}} + sC^{\mathrm{E}})\hat{\boldsymbol{E}}(x, y, 0, s). \tag{3.6}$$

After multiplying (3.6) by $\boldsymbol{i}_z \cdot$ and using the jump condition (3.4b), it is seen that $\boldsymbol{i}_z \cdot \partial\hat{\boldsymbol{J}}^{\mathrm{s}} = 0$, which implies that the contrast sheet electric-current density has only components tangential to the plane of the layer. Since the incident EM fields are continuous across the sheet, the transition conditions (3.4a) and (3.4b) can also be written in terms of the scattered fields, that is,

$$\boldsymbol{i}_z \times \hat{\boldsymbol{E}}^{\mathrm{s};+} - \boldsymbol{i}_z \times \hat{\boldsymbol{E}}^{\mathrm{s};-} = o(\delta), \tag{3.7a}$$

$$\boldsymbol{i}_z \times \hat{\boldsymbol{H}}^{\mathrm{s};+} - \boldsymbol{i}_z \times \hat{\boldsymbol{H}}^{\mathrm{s};-} = \partial\hat{\boldsymbol{J}}^{\mathrm{s}}(x, y, s) + o(\delta), \tag{3.7b}$$

as $\delta \downarrow 0$ for all $(x, y) \in \Sigma$. Upon introducing the basis vectors $\{\boldsymbol{i}_r, \boldsymbol{i}_\phi, \boldsymbol{i}_z\}$ pertaining to the circular cylindrical coordinate system with the radial, azimuthal and z-coordinates denoted by $\{r, \phi, z\}$, respectively, we may take $\boldsymbol{i}_\phi \cdot$ of (3.7a) and $\boldsymbol{i}_r \cdot$ of (3.7b) and get

$$\hat{E}_r^{\mathrm{s};+} - \hat{E}_r^{\mathrm{s};-} = o(\delta), \tag{3.8a}$$

$$\hat{H}_\phi^{\mathrm{s};+} - \hat{H}_\phi^{\mathrm{s};-} = -\partial\hat{J}_r^{\mathrm{s}}(r, \phi, s) + o(\delta), \tag{3.8b}$$

as $\delta \downarrow 0$ for all $(r, \phi) \in \Sigma$. In a similar fashion, taking $\boldsymbol{i}_r \cdot$ of (3.7a) and $\boldsymbol{i}_\phi \cdot$ of (3.7b), we readily obtain

$$\hat{E}_\phi^{\mathrm{s};+} - \hat{E}_\phi^{\mathrm{s};-} = o(\delta), \tag{3.9a}$$

$$\hat{H}_r^{\mathrm{s};+} - \hat{H}_r^{\mathrm{s};-} = \partial\hat{J}_\phi^{\mathrm{s}}(r, \phi, s) + o(\delta), \tag{3.9b}$$

as $\delta \downarrow 0$ for all $(r, \phi) \in \Sigma$. The form of saltus-type conditions represented by (3.9a) and (3.9b) has been previously used to describe the loop-to-loop pulsed EM-field signal transfer through a thin conductive sheet [5].

3.3 Transition conditions for a thin high-contrast layer with combined magnetic and dielectric properties

The cross-layer conditions derived in the previous section can be generalized by assuming that the thin sheet shows a high contrast with respect to the surrounding medium also in its magnetic properties. This assumption leads to the concept of scalar *metasurfaces* supporting both electric and magnetic currents that has been originated by Senior [6] in the real-frequency domain (FD). More recently, the TD transition conditions across a metasurface have been put forward and applied in [7]. A similar reasoning has also been pursued in [8] to model the effect of a localized fault on a transmission line.

Starting, again, with the space-integrated Maxwell's equations (3.3a) and (3.3b) in which $\mathcal{D}$ is associated with a pillbox of vanishing height, we find

$$\boldsymbol{i}_z \times \hat{\boldsymbol{E}}^+ - \boldsymbol{i}_z \times \hat{\boldsymbol{E}}^- = -(R^{\mathrm{M}} + sL^{\mathrm{M}})\langle \hat{\boldsymbol{H}} \rangle_-^+ + o(\delta), \tag{3.10a}$$

$$\boldsymbol{i}_z \times \hat{\boldsymbol{H}}^+ - \boldsymbol{i}_z \times \hat{\boldsymbol{H}}^- = (G^{\mathrm{E}} + sC^{\mathrm{E}})\langle \hat{\boldsymbol{E}} \rangle_-^+ + o(\delta), \tag{3.10b}$$

as $\delta \downarrow 0$ for all $(x, y) \in \Sigma$, where we applied the trapezoidal rule to the integration normal to the boundary. Furthermore, in (3.10a) and (3.10b), we used the shorthand notation

$$\langle f \rangle_-^+ = (f^+ + f^-)/2, \tag{3.11}$$

and introduced the *equivalent layer resistance and inductance* (see (3.5a) and (3.5b))

$$R^{\mathrm{M}} = \int_{\zeta=-\delta/2}^{\delta/2} \Gamma(\zeta)\mathrm{d}\zeta, \tag{3.12a}$$

$$L^{\mathrm{M}} = \int_{\zeta=-\delta/2}^{\delta/2} \mu(\zeta)\mathrm{d}\zeta, \tag{3.12b}$$

respectively, that are, again, supposed to be of order $O(1)$. In contrast to (3.4a), (3.10a) indicates that the tangential component of the electric-field strength is no longer continuous across the sheet, but jumps there by virtue of equivalent magnetic currents flowing on the sheet. Apparently, the latter currents vanish whenever $L^{\mathrm{M}} \downarrow 0$ and $R^{\mathrm{M}} \downarrow 0$, a consequence of which (3.10a) and (3.10b) boil down to (3.4a) and (3.4b), respectively.

In the ensuing step, we shall make use of the transition conditions (3.10a) and (3.10b) to interrelate the radial and azimuthal components of the EM fields on the two sides of the sheet. To that end, we first take $\boldsymbol{i}_\phi\ \cdot$ of (3.10a) and $\boldsymbol{i}_r\ \cdot$ of (3.10b), which leads to

$$\hat{E}_r^+ - \hat{E}_r^- = -(R^{\mathrm{M}} + sL^{\mathrm{M}})\langle \hat{H}_\phi \rangle_-^+ + o(\delta), \tag{3.13a}$$

$$\hat{H}_\phi^+ - \hat{H}_\phi^- = -(G^{\mathrm{E}} + sC^{\mathrm{E}})\langle \hat{E}_r \rangle_-^+ + o(\delta), \tag{3.13b}$$

while taking $\boldsymbol{i}_r\ \cdot$ of (3.10a) and $\boldsymbol{i}_\phi\ \cdot$ of (3.10b) yields

$$\hat{E}_\phi^+ - \hat{E}_\phi^- = (R^{\mathrm{M}} + sL^{\mathrm{M}})\langle \hat{H}_r \rangle_-^+ + o(\delta), \tag{3.14a}$$

$$\hat{H}_r^+ - \hat{H}_r^- = (G^{\mathrm{E}} + sC^{\mathrm{E}})\langle \hat{E}_\phi \rangle_-^+ + o(\delta), \tag{3.14b}$$

as $\delta \downarrow 0$ for all $(r, \phi) \in \Sigma$. The transition conditions (3.13a)–(3.14b) are next combined to express the tangential electric field one side of the sheet in terms of the tangential magnetic-field strengths on its two sides (see [6, Eq. (10)]). Hence, we define the equivalent layer impedances and admittances, that is,

$$\hat{Z}^{\mathrm{M}} = R^{\mathrm{M}} + sL^{\mathrm{M}} = 1/\hat{Y}^{\mathrm{M}}, \tag{3.15a}$$

$$\hat{Y}^{\mathrm{E}} = G^{\mathrm{E}} + sC^{\mathrm{E}} = 1/\hat{Z}^{\mathrm{E}}, \tag{3.15b}$$

and combine (3.13a) and (3.13b) to get

$$\hat{E}_r^{\pm} = -\hat{Z}^{\mathrm{E}}\left(1 \pm \frac{\hat{Z}^{\mathrm{M}}}{4\hat{Z}^{\mathrm{E}}}\right)\hat{H}_\phi^{+} + \hat{Z}^{\mathrm{E}}\left(1 \mp \frac{\hat{Z}^{\mathrm{M}}}{4\hat{Z}^{\mathrm{E}}}\right)\hat{H}_\phi^{-}. \tag{3.16}$$

In a similar way, a combination of (3.14a) and (3.14b) results in

$$\hat{E}_\phi^{\pm} = \hat{Z}^{\mathrm{E}}\left(1 \pm \frac{\hat{Z}^{\mathrm{M}}}{4\hat{Z}^{\mathrm{E}}}\right)\hat{H}_r^{+} - \hat{Z}^{\mathrm{E}}\left(1 \mp \frac{\hat{Z}^{\mathrm{M}}}{4\hat{Z}^{\mathrm{E}}}\right)\hat{H}_r^{-}, \tag{3.17}$$

assuming $\delta \downarrow 0$, again. According to [6, Eq. (10)], if the property $\hat{Z}^{\mathrm{M}} = 4\hat{Z}^{\mathrm{E}}$ applies, then the metasurface is "opaque" and relations (3.16) and (3.17) condense to surface-impedance (Leontovich) boundary conditions on the two sides of the layer. It is noted, however, that the pertaining transmission coefficient, say $\hat{T}(s)$, can be viewed as the response of a passive system. Consequently, $\hat{T}(s)$ is an analytic function of the complex frequency s in the entire right half of the complex-frequency plane. Upon recalling the property that if such an analytic function vanishes along a segment of a curve in $\mathrm{Re}(s) > 0$ it must vanish identically throughout the region of analyticity, it is concluded that condition $\hat{Z}^{\mathrm{M}} = 4\hat{Z}^{\mathrm{E}}$ can hardly be, in general, met with passive media in the layer.

Finally, the equivalent layer Kirchhoff-circuit parameters $\{C^{\mathrm{E}}, G^{\mathrm{E}}, L^{\mathrm{M}}, R^{\mathrm{M}}\}$ (see (3.5a) with (3.5b) and (3.12a) with (3.12b)) describing the constitutive EM properties of a thin, highly contrasting combined layer can be used to define the *layer conductance ratio* [2, Eq. (39)]

$$\eta_{\mathrm{E}} = G^{\mathrm{E}}/\eta_0, \tag{3.18}$$

and the *layer admittance time constant* [2, Eq. (40)]

$$\tau_{\mathrm{E}} = C^{\mathrm{E}}/G^{\mathrm{E}} = 1/\alpha^{\mathrm{E}}. \tag{3.19}$$

In a similar fashion, we may define the *layer resistance ratio*

$$\eta_{\mathrm{M}} = R^{\mathrm{M}}/\zeta_0, \tag{3.20}$$

where $\zeta_0 = \eta_0^{-1}$, and

$$\tau_{\mathrm{M}} = L^{\mathrm{M}}/R^{\mathrm{M}} = 1/\alpha^{\mathrm{M}}, \tag{3.21}$$

denoting the *layer impedance time constant*.

References

[1] T. B. A. Senior and J. L. Volakis. *Approximate Boundary Conditions in Electromagnetics*. London, UK: IEE, 1995.

[2] A. T. de Hoop and L. Jiang. Pulsed EM field response of a thin, high-contrast, finely layered structure with dielectric and conductive properties. *IEEE Transactions on Antennas and Propagation*, 57(8):2260–2269, 2009.

[3] H. Bateman. *Electrical and Optical Wave-motion*. Cambridge, UK: Cambridge University Press, 1915.

[4] A. T. de Hoop. *Handbook of Radiation and Scattering of Waves*, London, UK: Academic Press, 1995.

[5] M. Štumpf and A. T. de Hoop. Loop-to-loop pulsed electromagnetic signal transfer across a thin metal screen with Drude-type dispersive behavior. *IEEE Transactions on Electromagnetic Compatibility*, 60(4):885–889, 2018.

[6] T. B. A. Senior. Combined resistive and conductive sheets. *IEEE Transactions on Antennas and Propagation*, 33(5):577–579, 1985.

[7] M. Štumpf. Time-domain modeling of thin high-contrast layers with combined dielectric and magnetic properties. *IEEE Antennas and Wireless Propagation Letters*, 19(6):969–971, 2020.

[8] A. T. de Hoop and I. E. Lager. Transient waves along electrical transmission lines. Waves in (1+1)-spacetime. In *13th European Conference on Antennas and Propagation*, pp. 1–5, 2019.

Chapter 4
Pulsed EM-field response of an infinite metasurface

In this chapter, we shall provide illustrative applications of the thin-sheet, high-contrast boundary conditions introduced in Chapter 3. In particular, we shall explore reflection and transmission properties of a metasurface excited by impulsive electromagnetic (EM) sources. Starting from a simple, plane-wave-excited two-dimensional (2-D) configuration, the problem is gradually extended by analyzing the time-domain (TD) response to a localized line-source excitation and the loop-to-loop pulsed EM-field signal transfer across the sheet. In all the analyzed problems, a discussion on achieving the EM transparency is given.

4.1 Pulsed EM plane-wave induced response of a metasurface

We shall next analyze the pulsed EM-plane wave response of a thin, highly contrasting sheet. The infinite sheet occupies an unbounded domain $\{-\infty < x < \infty, -\infty < y < \infty, -\delta/2 < z < \delta/2\}$, where $\delta > 0$ denotes its relatively small thickness (see Figure 4.1). The EM properties of the sheet are described by $\{\epsilon, \mu, \sigma, \Gamma\}$, again (see Section 3.1), while the surrounding (homogeneous, isotropic and loss-free) medium is described by $\{\epsilon_0, \mu_0\}$, thus implying the EM wave speed $c_0 = (\epsilon_0\mu_0)^{-1/2} > 0$ and the wave admittance $\eta_0 = 1/c_0\mu_0$.

For the sake of simplicity, we assume that the planar structure under consideration as well as its plane-wave excitation is y-independent. The pulsed EM scattering by such a sheet has been previously studied in [1].

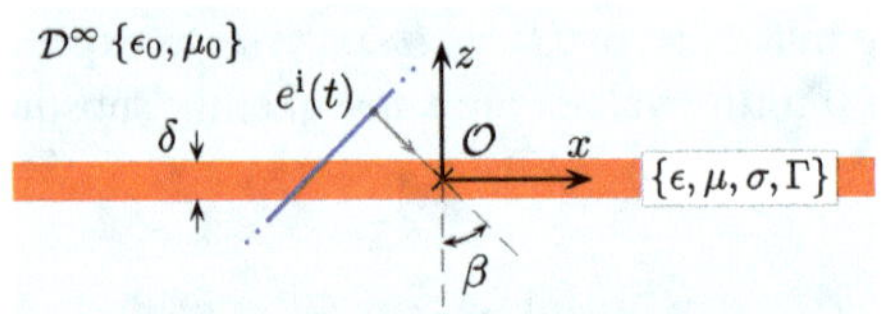

Figure 4.1 Pulsed EM plane wave impinging upon a metasurface

4.1.1 E-polarized waves

The electric-field strength of the incident E-polarized EM plane wave propagating along the positive x-direction and the negative z-direction can be in the s-domain described by

$$\hat{E}_y^{\mathrm{i}}(x,z,s) = \hat{e}^{\mathrm{i}}(s)\exp\{-s[\kappa_0 x - p(\kappa_0)z]\}, \tag{4.1}$$

where $\hat{e}^{\mathrm{i}}(s)$ is the plane-wave amplitude, κ_0 is the slowness parameter in the x-direction and $p(\kappa) = (1/c_0^2 - \kappa^2)^{1/2}$ with $\mathrm{Re}(p) \geq 0$ has the meaning of the slowness parameter in the z-direction. It is noted that the reference of the plane wave is chosen such that it hits the sheet at the origin $\mathcal{O}$ at $t = 0$. For the angle of incidence β as depicted in Figure 4.1, the slowness parameters are $\kappa_0 = \sin(\beta)/c_0$ and $p(\kappa_0) = \cos(\beta)/c_0$ with $\{0 \leq \beta \leq \pi/2\}$.

The point of departure for the following analysis are the saltus-type conditions (3.10a) and (3.10b) for the non-vanishing tangential EM-field components, that is,

$$\hat{E}_y^+ - \hat{E}_y^- = \hat{Z}^{\mathrm{M}}(s)\langle \hat{H}_x\rangle_-^+, \tag{4.2a}$$
$$\hat{H}_x^+ - \hat{H}_x^- = \hat{Y}^{\mathrm{E}}(s)\langle \hat{E}_y\rangle_-^+, \tag{4.2b}$$

where superscript $^+$ (or superscript $^-$) denotes the limit of the corresponding EM-field component as approaching $z \downarrow \delta/2$ (or $z \uparrow -\delta/2$) with $\delta \downarrow 0$, respectively, and the meaning of $\langle\rangle_-^+$ is specified by (3.11). The total electric-field strength above the sheet consists of the incident and reflected plane waves, while the field below the metasurface can be described by the transmitted plane wave. Accordingly, we write

$$\begin{aligned}\hat{E}_y(x,z,s) = {} & \hat{e}^{\mathrm{i}}(s)\exp\{-s[\kappa_0 x - p(\kappa_0)z]\} \\ & + \hat{R}_\perp \hat{e}^{\mathrm{i}}(s)\exp\{-s[\kappa_0 x + p(\kappa_0)z]\} \quad \text{for all } x \in \mathbb{R} \text{ and } z > 0,\end{aligned} \tag{4.3}$$

and

$$\hat{E}_y(x,z,s) = \hat{T}_\perp \hat{e}^{\mathrm{i}}(s)\exp\{-s[\kappa_0 x - p(\kappa_0)z]\} \quad \text{for all } x \in \mathbb{R} \text{ and } z < 0. \tag{4.4}$$

Yet unknown coefficients, $\hat{R}_\perp$ and $\hat{T}_\perp$, have the meaning of reflection and transmission coefficients, respectively, that will be next determined with the aid of (4.2a) and (4.2b). To that end, we make use of $\hat{H}_x = \partial_z \hat{E}_y/s\mu_0$ to express the solution for the magnetic field, whose limiting values are subsequently substituted in the transition conditions. In this way, we obtain

$$Y_\perp(\kappa_0)(1 - \hat{R}_\perp - \hat{T}_\perp) = \hat{Y}^{\mathrm{E}}(s)(1 + \hat{R}_\perp + \hat{T}_\perp)/2, \tag{4.5a}$$
$$1 + \hat{R}_\perp - \hat{T}_\perp = \hat{Z}^{\mathrm{M}}(s)Y_\perp(\kappa_0)(1 - \hat{R}_\perp + \hat{T}_\perp)/2, \tag{4.5b}$$

where $Y_\perp(\kappa_0) = p(\kappa_0)/\mu_0 = \eta_0 \cos(\beta)$. Solving this system of equations for $\hat{T}_\perp$ and $\hat{R}_\perp$, we find

$$\hat{T}_\perp = \frac{1}{2}\left[\frac{1-\hat{Y}^{\mathrm{E}}(s)/2Y_\perp(\kappa_0)}{1+\hat{Y}^{\mathrm{E}}(s)/2Y_\perp(\kappa_0)} + \frac{1-\hat{Z}^{\mathrm{M}}(s)Y_\perp(\kappa_0)/2}{1+\hat{Z}^{\mathrm{M}}(s)Y_\perp(\kappa_0)/2}\right], \tag{4.6}$$

$$\hat{R}_\perp = \frac{1}{2}\left[\frac{1-\hat{Y}^{\mathrm{E}}(s)/2Y_\perp(\kappa_0)}{1+\hat{Y}^{\mathrm{E}}(s)/2Y_\perp(\kappa_0)} - \frac{1-\hat{Z}^{\mathrm{M}}(s)Y_\perp(\kappa_0)/2}{1+\hat{Z}^{\mathrm{M}}(s)Y_\perp(\kappa_0)/2}\right]. \tag{4.7}$$

Apparently, the incident EM plane wave is not reflected if $\hat{R}_\perp = 0$, which occurs when

$$\hat{Y}^{\mathrm{E}}(s)\hat{Y}^{\mathrm{M}}(s) = Y_\perp^2(\kappa_0) = \eta_0^2 \cos^2(\beta). \tag{4.8}$$

Under this condition, the incident EM plane wave passes through the metasurface with the transmission coefficient

$$\hat{T}_\perp|_{\hat{R}_\perp=0} = \frac{1-\hat{Y}^{\mathrm{E}}(s)/2Y_\perp(\kappa_0)}{1+\hat{Y}^{\mathrm{E}}(s)/2Y_\perp(\kappa_0)} = \frac{1-\hat{Z}^{\mathrm{M}}(s)Y_\perp(\kappa_0)/2}{1+\hat{Z}^{\mathrm{M}}(s)Y_\perp(\kappa_0)/2}. \tag{4.9}$$

Notice that (4.7) for $\hat{Z}^{\mathrm{M}}(s) = 0$ yields the "scattering coefficient" of E-polarized waves pertaining to a thin sheet with dielectric and conductive properties [2, eq. (31)].

With the transmission and reflection coefficients at our disposal, we may find the TD counterpart of (4.3) and (4.4). For the sake of simplicity, we shall consider a loss-free metasurface whose equivalent layer impedance and admittance (see (3.15a) and (3.15b)) are given by $\hat{Z}^{\mathrm{M}} = sL^{\mathrm{M}}$ and $\hat{Y}^{\mathrm{E}} = sC^{\mathrm{E}}$. Upon defining

$$\Psi_\perp = 2Y_\perp(\kappa_0)/C^{\mathrm{E}}, \tag{4.10}$$

$$\Omega_\perp = 2/L^{\mathrm{M}}Y_\perp(\kappa_0), \tag{4.11}$$

the transmission and reflection coefficients can be cast into the following form:

$$\hat{T}_\perp = -\frac{1}{2}\left(\frac{s-\Psi_\perp}{s+\Psi_\perp} + \frac{s-\Omega_\perp}{s+\Omega_\perp}\right), \tag{4.12}$$

$$\hat{R}_\perp = -\frac{1}{2}\left(\frac{s-\Psi_\perp}{s+\Psi_\perp} - \frac{s-\Omega_\perp}{s+\Omega_\perp}\right), \tag{4.13}$$

whose transformation to TD is straightforward. Consequently, we readily arrive at the following TD transmission and reflection coefficients:

$$T_\perp(t) = -\delta(t) + [\Psi_\perp \exp(-\Psi_\perp t) + \Omega_\perp \exp(-\Omega_\perp t)]\mathrm{H}(t), \tag{4.14}$$

$$R_\perp(t) = [\Psi_\perp \exp(-\Psi_\perp t) - \Omega_\perp \exp(-\Omega_\perp t)]\mathrm{H}(t), \tag{4.15}$$

respectively, where $\delta(t)$ is the one-dimensional Dirac-delta distribution. Finally, the total TD electric-field strength is found from (4.3) as

$$E_y(x,z,t) = e^{\mathrm{i}}[t - \kappa_0 x + p(\kappa_0)z] + R_\perp(t) *_t e^{\mathrm{i}}[t - \kappa_0 x - p(\kappa_0)z] \quad \text{for all } x \in \mathbb{R}, z > 0 \text{ and } t > 0, \tag{4.16}$$

and the TD field transmitted through the metasurface follows from (4.4) as

$$E_y(x,z,t) = T_\perp(t) *_t e^{\rm i}[t - \kappa_0 x + p(\kappa_0) z] \quad \text{for all } x \in \mathbb{R}, z < 0 \text{ and } t > 0. \tag{4.17}$$

4.1.2 H-polarized waves

The analysis of H-polarized plane waves in the presence of a metasurface is carried out along the lines followed in Section 4.1.1. Thus, the corresponding incident EM plane-wave in the s-domain is first specified via (cf. (4.1))

$$\hat{H}_y^{\rm i}(x,z,s) = \eta_0 \hat{e}^{\rm i}(s) \exp\{-s[\kappa_0 x - p(\kappa_0) z]\}. \tag{4.18}$$

Across the thin layer, the total EM fields satisfy the following transition conditions (cf. (4.2a) and (4.2b)):

$$\hat{H}_y^+ - \hat{H}_y^- = -\hat{Y}^{\rm E}(s)\langle \hat{E}_x \rangle_-^+, \tag{4.19a}$$

$$\hat{E}_x^+ - \hat{E}_x^- = -\hat{Z}^{\rm M}(s)\langle \hat{H}_y \rangle_-^+ \quad \text{for all } x \in \mathbb{R}. \tag{4.19b}$$

Repeating the approach as pursued in the previous subsection, the transition conditions are next applied to determine the corresponding transmission and reflection coefficients. After a few steps of straightforward algebra, we obtain (cf. (4.6) and (4.7))

$$\hat{T}_\parallel = \frac{1}{2}\left[\frac{1 - \hat{Z}^{\rm M}(s) Y_\parallel(\kappa_0)/2}{1 + \hat{Z}^{\rm M}(s) Y_\parallel(\kappa_0)/2} + \frac{1 - \hat{Y}^{\rm E}(s)/2Y_\parallel(\kappa_0)}{1 + \hat{Y}^{\rm E}(s)/2Y_\parallel(\kappa_0)}\right], \tag{4.20}$$

$$\hat{R}_\parallel = \frac{1}{2}\left[\frac{1 - \hat{Z}^{\rm M}(s) Y_\parallel(\kappa_0)/2}{1 + \hat{Z}^{\rm M}(s) Y_\parallel(\kappa_0)/2} - \frac{1 - \hat{Y}^{\rm E}(s)/2Y_\parallel(\kappa_0)}{1 + \hat{Y}^{\rm E}(s)/2Y_\parallel(\kappa_0)}\right], \tag{4.21}$$

where $Y_\parallel(\kappa_0) = \epsilon_0/p(\kappa_0) = \eta_0/\cos(\beta)$. Considering a loss-free metasurface, again, the coefficients can be expressed in the following form (cf. (4.12) and (4.13)):

$$\hat{T}_\parallel = -\frac{1}{2}\left(\frac{s - \Omega_\parallel}{s + \Omega_\parallel} + \frac{s - \Psi_\parallel}{s + \Psi_\parallel}\right), \tag{4.22}$$

$$\hat{R}_\parallel = -\frac{1}{2}\left(\frac{s - \Omega_\parallel}{s + \Omega_\parallel} - \frac{s - \Psi_\parallel}{s + \Psi_\parallel}\right), \tag{4.23}$$

where we used (cf. (4.10) and (4.11))

$$\Psi_\parallel = 2Y_\parallel(\kappa_0)/C^{\rm E}, \tag{4.24}$$

$$\Omega_\parallel = 2/L^{\rm M} Y_\parallel(\kappa_0). \tag{4.25}$$

The s-domain coefficients given by (4.22) and (4.23) can be readily transformed to TD (cf. (4.14) and (4.15))

$$T_\parallel(t) = -\delta(t) + [\Omega_\parallel \exp(-\Omega_\parallel t) + \Psi_\parallel \exp(-\Psi_\parallel t)]{\rm H}(t), \tag{4.26}$$

$$R_\parallel(t) = [\Omega_\parallel \exp(-\Omega_\parallel t) - \Psi_\parallel \exp(-\Psi_\parallel t)]{\rm H}(t), \tag{4.27}$$

and further substituted in the final expressions for the total magnetic-field strength. This way leads to (cf. (4.16) and (4.17))

$$H_y(x,z,t) = \eta_0 e^{\mathrm{i}}[t - \kappa_0 x + p(\kappa_0)z] + \eta_0 R_{\|}(t) *_t e^{\mathrm{i}}[t - \kappa_0 x - p(\kappa_0)z]$$
$$\text{for all } x \in \mathbb{R},\, z > 0 \text{ and } t > 0, \tag{4.28}$$

and

$$H_y(x,z,t) = \eta_0 T_{\|}(t) *_t e^{\mathrm{i}}[t - \kappa_0 x + p(\kappa_0)z]$$
$$\text{for all } x \in \mathbb{R},\, z < 0 \text{ and } t > 0. \tag{4.29}$$

As can be deduced from (4.21), the zero reflection occurs when (cf. (4.8))

$$\hat{Y}^{\mathrm{E}}(s)\hat{Y}^{\mathrm{M}}(s) = Y_{\|}^2(\kappa_0), \tag{4.30}$$

which takes the form $(C^{\mathrm{E}}/L^{\mathrm{M}})^{1/2} = \eta_0/\cos(\beta)$ for a loss-free combined layer. Finally, it is noted that the zero plane-wave reflection can also be encountered when the H-polarized plane-wave hits the interface of two dielectric half-spaces at the so-called *Brewster's angle* [3, Section 9.6].

4.1.3 *Illustrative example*

As an example, we shall analyze the electric-field distribution in the presence of a metasurface that is irradiated by the E-polarized plane wave, whose signature is described by the unipolar power-exponential pulse

$$e^{\mathrm{i}}(t) = e_{\mathrm{m}}(t/t_{\mathrm{r}})^n \exp[-n(t/t_{\mathrm{r}} - 1)]\mathrm{H}(t), \tag{4.31}$$

where we take the unit amplitude $e_{\mathrm{m}} = 1.0\,\mathrm{V/m}$ and $n = 2$. Furthermore, the pulse-time width is related to a distance $h > 0$ as $c_0 t_{\mathrm{w}} = h$ and the pulse-rise time follows from $t_{\mathrm{r}} = t_{\mathrm{w}} n^{n+1}\exp(-n)/\Gamma(n+1) \simeq 0.5413\,t_{\mathrm{w}}$, where $\Gamma(x)$ is the Euler gamma function. For a loss-free metasurface, the transparency condition (4.8) is equivalent to $(C^{\mathrm{E}}/L^{\mathrm{M}})^{1/2} = \eta_0\cos(\beta)$. Figure 4.2(a) demonstrates the field spatial distribution at $c_0 t/h = 0.10$ as calculated for $\beta = \pi/4$ with $C^{\mathrm{E}}/\eta_0 t_{\mathrm{w}} = 1.0$ and $L^{\mathrm{M}}\eta_0/t_{\mathrm{w}} = 2.0$, thus meeting the condition. It is noted, however, that vanishing reflections could also be achieved when C^{E} and L^{M} are sufficiently large. Under these circumstances, it can be deduced from (4.16) and (4.17) that the total electric field can be approximately described by $E_y \simeq E_y^{\mathrm{i}}(x,z,t)\mathrm{sign}(z)$. Finally, it is demonstrated in Figure 4.2(b) that the metasurface with parameters $C^{\mathrm{E}}/\eta_0 t_{\mathrm{w}} = 10$ and $L^{\mathrm{M}}\eta_0/t_{\mathrm{w}} = 0.10$, for instance, approximately mimics a perfectly electrically conducting (PEC) surface.

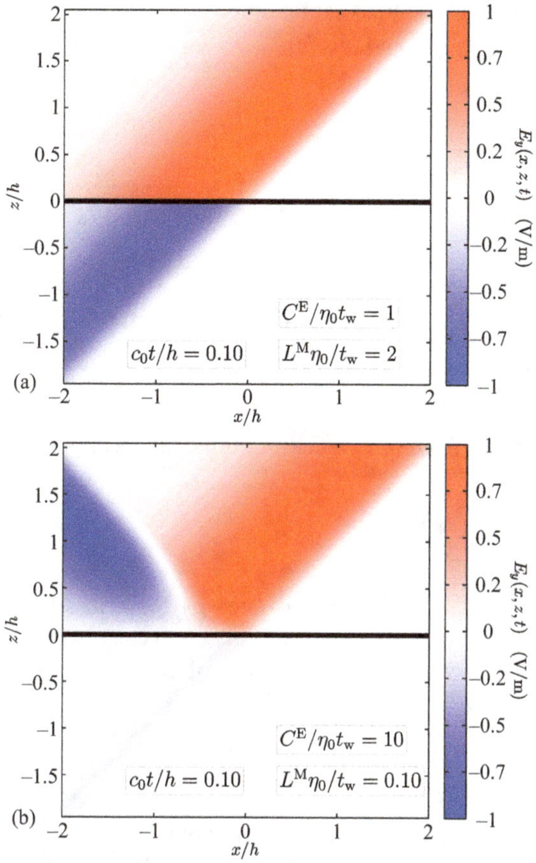

Figure 4.2 *Spatial distributions of the electric-field strength excited by the impulsive plane wave in the presence of the combined sheet with (a) $C^{\mathrm{E}}/\eta_0 t_{\mathrm{w}} = 1.0$ and $L^{\mathrm{M}}\eta_0/t_{\mathrm{w}} = 2.0$; (b) $C^{\mathrm{E}}/\eta_0 t_{\mathrm{w}} = 10$ and $L^{\mathrm{M}}\eta_0/t_{\mathrm{w}} = 0.10$*

4.2 Pulsed EM line-source induced response of a metasurface

The problem configuration analyzed in the previous section can be extended by considering incident EM fields generated by a spatially localized EM source. A problem from this category is the subject matter of this section, where line-source excited pulsed EM fields in the presence of a metasurface are analytically investigated with the aid of the Cagniard–DeHoop (CdH) technique [5].

4.2.1 Electric-line source

The problem under consideration is shown in Figure 4.3. It consists of an infinite, highly contrasting thin sheet and an impulsive electric-line source embedded in the

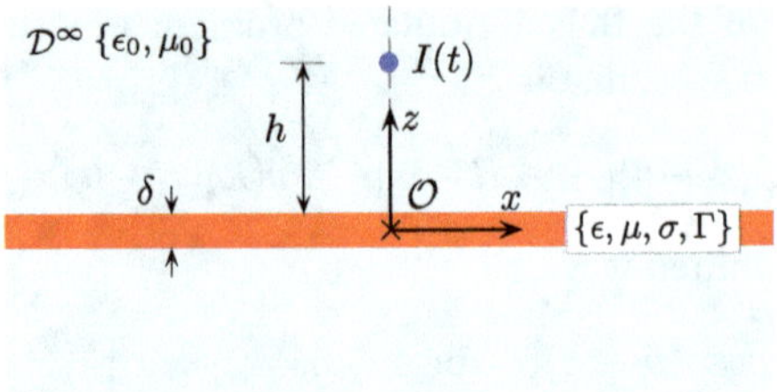

Figure 4.3 Impulsive electric-line source in the presence of a metasurface

homogeneous, isotropic and loss-free medium occupying $\mathcal{D}^\infty$. The exciting line source is in the s-domain defined via the y-oriented electric-current volume density

$$\hat{J}_y(x,z,s) = \hat{I}(s)\delta(x, z-h), \tag{4.32}$$

where $\hat{I}(s)$ is the electric-current source amplitude and $h > 0$ denotes the height of the source above the sheet. In virtue of the shift invariance along the x-direction, the problem is next solved under the wave-slowness representation taken along the x-direction. To show the notation, its definition is given for the non-vanishing electric-field strength

$$\hat{E}_y(x,z,s) = \frac{s}{2\pi \mathrm{i}} \int_{\kappa=-\mathrm{i}\infty}^{\mathrm{i}\infty} \exp(-s\kappa x)\tilde{E}_y(\kappa,z,s)\mathrm{d}\kappa,$$
$$\text{for } \{s \in \mathbb{R}; s > 0\}, \tag{4.33}$$

where we have introduced the imaginary-valued slowness parameter κ. The slowness representation entails $\partial_x \to -s\kappa$. Consequently, the slowness-domain solution can be expressed as

$$\tilde{E}_y(\kappa,z,s) = -s\mu_0\hat{I}(s)\tilde{G}(\kappa,z,s), \tag{4.34a}$$
$$\tilde{H}_x(\kappa,z,s) = -\hat{I}(s)\partial_z\tilde{G}(\kappa,z,s), \tag{4.34b}$$

where the transform-domain Green's function is governed by

$$[\partial_z^2 - s^2p^2(\kappa)]\tilde{G}(\kappa,z,s) = 0, \tag{4.35}$$

in which

$$p(\kappa) = (1/c_0^2 - \kappa^2)^{1/2}, \quad \text{with } \mathrm{Re}(p) \geq 0. \tag{4.36}$$

Equation (4.34a) is further supplemented with the excitation condition

$$\lim_{z\downarrow h}\partial_z\tilde{G}(\kappa,z,s) - \lim_{z\uparrow h}\partial_z\tilde{G}(\kappa,z,s) = -1, \tag{4.37a}$$

and the "boundedness condition" for $|z| \to \infty$. Furthermore, the use of (4.34a) and (4.34b) in the transition conditions (4.2a) and (4.2b) leads to

$$\lim_{z\downarrow 0}\tilde{G} - \lim_{z\uparrow 0}\tilde{G} = \hat{Z}^{\mathrm{M}}(s)\big[\lim_{z\downarrow 0}\partial_z\tilde{G} + \lim_{z\uparrow 0}\partial_z\tilde{G}\big]/2s\mu_0, \tag{4.37b}$$

$$\lim_{z\downarrow 0}\partial_z\tilde{G} - \lim_{z\uparrow 0}\partial_z\tilde{G} = s\mu_0\hat{Y}^{\mathrm{E}}(s)\big[\lim_{z\downarrow 0}\tilde{G} + \lim_{z\uparrow 0}\tilde{G}\big]/2. \tag{4.37c}$$

The general solution of the thus formulated problem is written as a sum of the incident and reflected waves, that is,

$$\tilde{G} = \tilde{A} \exp[-sp(\kappa)|z-h|] + \tilde{A}\tilde{R}_\perp \exp[-sp(\kappa)(z+h)] \quad \text{for } z > 0, \tag{4.38}$$

and as the (total) transmitted wave

$$\tilde{G} = \tilde{A}\tilde{T}_\perp \exp[-sp(\kappa)(h-z)] \quad \text{for } z < 0. \tag{4.39}$$

Pursuing the line of reasoning from the previous subsection, the slowness-domain transmission and reflection coefficients, $\tilde{T}_\perp$ and $\tilde{R}_\perp$, respectively, will be found by employing the transition conditions (4.37b) and (4.37c), while the source-coupling coefficient, $\tilde{A}$, directly follows from the excitation condition (4.37a). In this way, after straightforward algebra, we arrive at (cf. (4.6) and (4.7))

$$\tilde{T}_\perp = \frac{1}{2}\left[\frac{1-\hat{Y}^{\mathrm{E}}(s)/2Y_\perp(\kappa)}{1+\hat{Y}^{\mathrm{E}}(s)/2Y_\perp(\kappa)} + \frac{1-\hat{Z}^{\mathrm{M}}(s)Y_\perp(\kappa)/2}{1+\hat{Z}^{\mathrm{M}}(s)Y_\perp(\kappa)/2}\right], \tag{4.40}$$

$$\tilde{R}_\perp = \frac{1}{2}\left[\frac{1-\hat{Y}^{\mathrm{E}}(s)/2Y_\perp(\kappa)}{1+\hat{Y}^{\mathrm{E}}(s)/2Y_\perp(\kappa)} - \frac{1-\hat{Z}^{\mathrm{M}}(s)Y_\perp(\kappa)/2}{1+\hat{Z}^{\mathrm{M}}(s)Y_\perp(\kappa)/2}\right], \tag{4.41}$$

$$\tilde{A} = 1/2sp(\kappa), \tag{4.42}$$

and recall that $Y_\perp(\kappa) = p(\kappa)/\mu_0$ has the meaning of the slowness-domain wave admittance pertaining to the E-polarized fields. Apparently, (4.6) and (4.7) can be understood as a special case of (4.40) and (4.41), evaluated at $\kappa = \kappa_0$, that is, at the slowness of the incident E-polarized plane wave (4.1). It is seen that both reflection and transmission coefficients have no pole in the complex κ-plane for $\{s \in \mathbb{R}; s > 0\}$, which is a characteristic for the absence of true surface waves [4].

As the total wave motion for the line-source excitation can be viewed as a superposition of plane-wave constituents, the total electric-field strength can be represented via

$$\begin{aligned}\hat{E}_y(x,z,s) = &-\frac{s\mu_0\hat{I}(s)}{2\pi\mathrm{i}}\int_{\kappa=-\mathrm{i}\infty}^{\mathrm{i}\infty} \exp\{-s[\kappa x + p(\kappa)Z^{\mathrm{i}}]\}\frac{\mathrm{d}\kappa}{2p(\kappa)} \\ &-\frac{s\mu_0\hat{I}(s)}{2\pi\mathrm{i}}\int_{\kappa=-\mathrm{i}\infty}^{\mathrm{i}\infty} \exp\{-s[\kappa x + p(\kappa)Z^{\mathrm{r}}]\}\tilde{R}_\perp(\kappa,s)\frac{\mathrm{d}\kappa}{2p(\kappa)} \\ &\text{for all } x \in \mathbb{R}, z > 0 \text{ and } \{s \in \mathbb{R}; s > 0\},\end{aligned} \tag{4.43}$$

where $Z^{\mathrm{i}} = |z-h|$ and $Z^{\mathrm{r}} = z + h$. Furthermore, the transmitted field reads

$$\begin{aligned}\hat{E}_y(x,z,s) = &-\frac{s\mu_0\hat{I}(s)}{2\pi\mathrm{i}}\int_{\kappa=-\mathrm{i}\infty}^{\mathrm{i}\infty} \exp\{-s[\kappa x + p(\kappa)Z^{\mathrm{t}}]\}\tilde{T}_\perp(\kappa,s)\frac{\mathrm{d}\kappa}{2p(\kappa)} \\ &\text{for all } x \in \mathbb{R}, z < 0 \text{ and } \{s \in \mathbb{R}; s > 0\},\end{aligned} \tag{4.44}$$

where $Z^{\mathrm{t}} = h - z$. The wave constituents in (4.43) and (4.44) will be transformed to the TD with the aid of the inversion procedure proposed by de Hoop (see Section 2.1.1). To that end, the integrands are first analytically continued away from

the imaginary axis $\mathrm{Re}(\kappa) = 0$, while keeping $\mathrm{Re}(p) \geq 0$. This implies that branch cuts along $\{\mathrm{Im}(\kappa) = 0, 1/c_0 \leq |\mathrm{Re}(\kappa)| < \infty\}$ are introduced. Subsequently, the integration contour is deformed to the CdH path along which

$$\kappa x + p(\kappa)Z = \tau, \quad \text{for } \{\tau \in \mathbb{R}; \tau > 0\}, \tag{4.45}$$

is satisfied. Here, $Z > 0$ represents the vertical propagation path of either the incident, reflected or transmitted wave, $Z^{\mathrm{i,r,t}}$, respectively. Upon solving (4.45) for κ, we obtain the parametrization of a (hyperbolic) CdH path, say $\mathcal{C} \cup \mathcal{C}^*$, with

$$\mathcal{C} = \left\{\kappa(\tau) = (x/\rho^2)\tau + \mathrm{i}(Z/\rho^2)(\tau^2 - T^2)^{1/2}\right\}$$
$$\text{for } \{T = \rho/c_0 \leq \tau < \infty\}, \tag{4.46}$$

where $\rho = (x^2 + Z^2)^{1/2} > 0$. Employing further Schwartz's reflection principle of complex function theory to combine the integrations in the upper and lower halves of the complex κ-plane, while introducing τ as the variable of integration, we may express the solution via the following integrals:

$$\hat{E}_y(x, z, s) = -\frac{s\mu_0\hat{I}(s)}{2\pi}\int_{\tau=T^{\mathrm{i}}}^{\infty} \exp(-s\tau)\frac{\mathrm{d}\tau}{(\tau^2 - T^{\mathrm{i}2})^{1/2}}$$
$$-\frac{s\mu_0\hat{I}(s)}{2\pi}\int_{\tau=T^{\mathrm{r}}}^{\infty} \exp(-s\tau)\mathrm{Re}\{\tilde{R}_\perp[\kappa(\tau), s]\}\frac{\mathrm{d}\tau}{(\tau^2 - T^{\mathrm{r}2})^{1/2}}$$
$$\text{for all } x \in \mathbb{R}, z > 0 \text{ and } \{s \in \mathbb{R}; s > 0\}, \tag{4.47}$$

and

$$\hat{E}_y(x, z, s) = -\frac{s\mu_0\hat{I}(s)}{2\pi}\int_{\tau=T^{\mathrm{t}}}^{\infty} \exp(-s\tau)\mathrm{Re}\{\tilde{T}_\perp[\kappa(\tau), s]\}\frac{\mathrm{d}\tau}{(\tau^2 - T^{\mathrm{t}2})^{1/2}}$$
$$\text{for all } x \in \mathbb{R}, z < 0 \text{ and } \{s \in \mathbb{R}; s > 0\}. \tag{4.48}$$

The transformation of the integral expressions will be carried out for a loss-free combined layer described by $\hat{Z}^{\mathrm{M}} = sL^{\mathrm{M}}$ and $\hat{Y}^{\mathrm{E}} = sC^{\mathrm{E}}$, again. For this case, we arrive at

$$\hat{E}_y(x, z, s) = -\frac{s\mu_0\hat{I}(s)}{2\pi}\int_{\tau=T^{\mathrm{i}}}^{\infty} \exp(-s\tau)\frac{\mathrm{d}\tau}{(\tau^2 - T^{\mathrm{i}2})^{1/2}}$$
$$-\frac{s\mu_0\hat{I}(s)}{2\pi}\int_{\tau=T^{\mathrm{r}}}^{\infty} \exp(-s\tau)$$
$$\times \mathrm{Re}\left\{\frac{s[\Psi_\perp(\tau) - \Omega_\perp(\tau)]}{s^2 + s[\Psi_\perp(\tau) + \Omega_\perp(\tau)] + \Psi_\perp(\tau)\Omega_\perp(\tau)}\right\}\frac{\mathrm{d}\tau}{(\tau^2 - T^{\mathrm{r}2})^{1/2}}$$
$$\text{for all } x \in \mathbb{R}, z > 0 \text{ and } \{s \in \mathbb{R}; s > 0\}, \tag{4.49}$$

and

$$\hat{E}_y(x,z,s) = \frac{s\mu_0 \hat{I}(s)}{2\pi} \int_{\tau=T^{\mathrm{t}}}^{\infty} \exp(-s\tau)$$
$$\times \operatorname{Re}\left\{ \frac{s^2 - \Psi_\perp(\tau)\Omega_\perp(\tau)}{s^2 + s[\Psi_\perp(\tau)+\Omega_\perp(\tau)] + \Psi_\perp(\tau)\Omega_\perp(\tau)} \right\} \frac{\mathrm{d}\tau}{(\tau^2 - T^{\mathrm{t}2})^{1/2}}$$
$$\text{for all } x \in \mathbb{R},\, z<0 \text{ and } \{s\in\mathbb{R}; s>0\}. \tag{4.50}$$

In (4.49) and (4.50), we used (cf. (4.10) and (4.11))

$$\Psi_\perp(\tau) = 2Y_\perp[\kappa(\tau)]/C^{\mathrm{E}}, \tag{4.51}$$
$$\Omega_\perp(\tau) = 2/L^{\mathrm{M}}Y_\perp[\kappa(\tau)]. \tag{4.52}$$

Finally, employing the Schouten–Van der Pol theorem relying on Lerch's uniqueness theorem [6, p. 1056], the TD counterparts of (4.49) and (4.50) are found as

$$E_y(x,z,t) = -\partial_t I(t) *_t \frac{\mu_0}{2\pi}\frac{\mathrm{H}(t-T^{\mathrm{i}})}{(t^2 - T^{\mathrm{i}2})^{1/2}}$$
$$- \partial_t I(t) *_t \frac{\mu_0}{2\pi}\int_{\tau=T^{\mathrm{r}}}^{t} \operatorname{Re}\{\Psi_\perp(\tau)\exp[-\Psi_\perp(\tau)(t-\tau)]$$
$$- \Omega_\perp(\tau)\exp[-\Omega_\perp(\tau)(t-\tau)]\}\frac{\mathrm{d}\tau}{(\tau^2 - T^{\mathrm{r}2})^{1/2}}$$
$$\text{for all } x\in\mathbb{R},\, z>0 \text{ and } t>0, \tag{4.53}$$

while for the total field below the sheet, we get

$$E_y(x,z,t) = \partial_t I(t) *_t \frac{\mu_0}{2\pi}\frac{\mathrm{H}(t-T^{\mathrm{t}})}{(t^2 - T^{\mathrm{t}2})^{1/2}}$$
$$- \partial_t I(t) *_t \frac{\mu_0}{2\pi}\int_{\tau=T^{\mathrm{t}}}^{t} \operatorname{Re}\{\Psi_\perp(\tau)\exp[-\Psi_\perp(\tau)(t-\tau)]$$
$$+ \Omega_\perp(\tau)\exp[-\Omega_\perp(\tau)(t-\tau)]\}\frac{\mathrm{d}\tau}{(\tau^2 - T^{\mathrm{t}2})^{1/2}}$$
$$\text{for all } x\in\mathbb{R},\, z<0 \text{ and } t>0. \tag{4.54}$$

As the real parts of both functions $\Psi_\perp(\tau)$ and $\Omega_\perp(\tau)$ are positive, the integrands in (4.53) and (4.54) consist of a difference and a sum of two increasing functions of τ having their maximum at $\tau = t$. Accordingly, the reflected-wave constituent in (4.53) would be identically zero if $\Psi_\perp(\tau) = \Omega_\perp(\tau)$, or, in accordance with (4.51) and (4.52), when

$$Y_\perp[\kappa(\tau)] = (C^{\mathrm{E}}/L^{\mathrm{M}})^{1/2} \quad \text{for all } \{T^{\mathrm{r}} \le \tau < t\}. \tag{4.55}$$

Owing to the complex values of $Y_\perp[\kappa(\tau)]$ along $\mathcal{C}$, namely,

$$Y_\perp[\kappa(\tau)] = \eta_0\left[\frac{Z^{\mathrm{r}}}{\rho^{\mathrm{r}2}}c_0\tau - \mathrm{i}\frac{x}{\rho^{\mathrm{r}2}}(c_0^2\tau^2 - \rho^{\mathrm{r}2})^{1/2}\right] \quad \text{for all } \tau \ge T^{\mathrm{r}}, \tag{4.56}$$

the transparency condition (4.55) can hardly be, in general, met in the entire integration interval $\{T^{\mathrm{r}} \leq \tau < t\}$. As a matter of fact, the condition holds true just behind the wave front of the reflected wave as $t \downarrow T^{\mathrm{r}}$ at points satisfying $\eta_0 \cos(\theta) = (C^{\mathrm{E}}/L^{\mathrm{M}})^{1/2}$, where $\cos(\theta) = Z^{\mathrm{r}}/\rho^{\mathrm{r}}$. For sufficiently large values of $\Psi_{\perp}(\tau)$ and $\Omega_{\perp}(\tau)$, however, the integrands "sift out" the values at the upper bound of integration, thus resembling the impulsive Dirac-delta distribution. With this property in mind, it is straightforward to demonstrate that (4.53) and (4.54) yield the correct behavior for the limiting PEC surface described by $L^{\mathrm{M}} \downarrow 0$ and $C^{\mathrm{E}} \to \infty$, for the perfectly magnetically conducting (PMC) surface characterized via $C^{\mathrm{E}} \downarrow 0$ and $L^{\mathrm{M}} \to \infty$, as well as for the non-contrasting case for which $C^{\mathrm{E}} \downarrow 0$ and $L^{\mathrm{M}} \downarrow 0$ (see (4.51) and (4.52)). On the other hand, if both C^{E} and L^{M} are sufficiently large, then the integral terms can be essentially neglected and we end up with the incident and transmitted cylindrical waves on the right-hand sides of (4.53) and (4.54), respectively. Since the resulting transmitted wave below the metasurface is, in fact, equal to the opposite of the incident one, such a sheet virtually behaves as a transparent (non-scattering or zero-reflection) layer, the transition across which merely changes the sign of the incident pulse.

4.2.2 *Magnetic-line source*

In this subsection, the infinite metasurface with combined magnetic and dielectric properties is excited by an impulsive magnetic-line source that can be in the s-domain described via (cf. (4.32))

$$\hat{K}_y(x, z, s) = \hat{K}(s)\delta(x, z - h), \tag{4.57}$$

where $\hat{K}(s)$ is the magnetic-current source amplitude and $h > 0$ denotes the height of the source above the sheet, again. The non-vanishing EM-field components $\{\hat{E}_x, \hat{E}_z, \hat{H}_y\}(x, z, s)$ are in the surrounding homogeneous, isotropic and loss-free medium governed by a y-independent system of Maxwell's equations that are further supplemented with the pertaining cross-layer conditions (4.19a) and (4.19b) and with the condition of causality.

The problem can be tackled by the CdH method as applied in Section 4.2.1. Hence, the EM-field quantities are represented via the wave-slowness representation of the type given by (4.33), so that the horizontal field components can be in the transform domain expressed by (cf. (4.34a) and (4.34b))

$$\tilde{H}_y(\kappa, z, s) = -s\epsilon_0 \hat{K}(s)\tilde{G}(\kappa, z, s), \tag{4.58a}$$

$$\tilde{E}_x(\kappa, z, s) = \hat{K}(s)\partial_z \tilde{G}(\kappa, z, s), \tag{4.58b}$$

where the transform-domain Green's function satisfies (4.35) with the excitation condition (4.37a) and the "boundedness condition" for $|z| \to \infty$, again.

Consequently, by virtue of the transition conditions (4.19a) and (4.19b), the transform-domain Green's function jumps across the layer according to

$$\lim_{z\downarrow 0}\tilde{G} - \lim_{z\uparrow 0}\tilde{G} = \hat{Y}^{\mathrm{E}}(s)\big[\lim_{z\downarrow 0}\partial_z\tilde{G} + \lim_{z\uparrow 0}\partial_z\tilde{G}\big]/2s\epsilon_0, \tag{4.59a}$$

$$\lim_{z\downarrow 0}\partial_z\tilde{G} - \lim_{z\uparrow 0}\partial_z\tilde{G} = s\epsilon_0\hat{Z}^{\mathrm{M}}(s)\big[\lim_{z\downarrow 0}\tilde{G} + \lim_{z\uparrow 0}\tilde{G}\big]/2. \tag{4.59b}$$

The general solution of the transform-domain Green's function is written, again, in the form of a sum of the incident and reflected waves

$$\tilde{G} = \tilde{A}\exp[-sp(\kappa)|z-h|] + \tilde{A}\,\tilde{R}_{\|}\exp[-sp(\kappa)(z+h)] \quad \text{for } z>0, \tag{4.60}$$

and as the (total) transmitted wave

$$\tilde{G} = \tilde{A}\,\tilde{T}_{\|}\exp[-sp(\kappa)(h-z)] \quad \text{for } z<0. \tag{4.61}$$

The use of the general transform-domain solution in the excitation condition (4.37a) and the cross-layer conditions (4.59a) and (4.59b) yields the pertaining coefficients (cf. (4.40) with (4.41) and (4.20) with (4.21))

$$\tilde{T}_{\|} = \frac{1}{2}\left[\frac{1-\hat{Z}^{\mathrm{M}}(s)Y_{\|}(\kappa)/2}{1+\hat{Z}^{\mathrm{M}}(s)Y_{\|}(\kappa)/2} + \frac{1-\hat{Y}^{\mathrm{E}}(s)/2Y_{\|}(\kappa)}{1+\hat{Y}^{\mathrm{E}}(s)/2Y_{\|}(\kappa)}\right], \tag{4.62}$$

$$\tilde{R}_{\|} = \frac{1}{2}\left[\frac{1-\hat{Z}^{\mathrm{M}}(s)Y_{\|}(\kappa)/2}{1+\hat{Z}^{\mathrm{M}}(s)Y_{\|}(\kappa)/2} - \frac{1-\hat{Y}^{\mathrm{E}}(s)/2Y_{\|}(\kappa)}{1+\hat{Y}^{\mathrm{E}}(s)/2Y_{\|}(\kappa)}\right], \tag{4.63}$$

and $\tilde{A} = 1/2sp(\kappa)$, again, with $Y_{\|}(\kappa) = \epsilon_0/p(\kappa)$. Likewise in case of the electric-line excitation, the resulting reflection and transmission coefficients have no pole in the complex κ-plane for $\{s\in\mathbb{R};\, s>0\}$, which is a characteristic for the absence of true (causal) surface waves.

With the transform-domain solution at our disposal, we may further pursue the CdH approach closely described in Section 4.2.1. In this way, the TD solution pertaining to a loss-free metasurface can be written as follows (cf. (4.53)):

$$\begin{aligned} H_y(x,z,t) = {} & -\partial_t K(t) *_t \frac{\epsilon_0}{2\pi}\frac{\mathrm{H}(t-T^{\mathrm{i}})}{(t^2-T^{\mathrm{i}2})^{1/2}} \\ & - \partial_t K(t) *_t \frac{\epsilon_0}{2\pi}\int_{\tau=T^{\mathrm{r}}}^{t} \mathrm{Re}\{\Omega_{\|}(\tau)\exp[-\Omega_{\|}(\tau)(t-\tau)] \\ & - \Psi_{\|}(\tau)\exp[-\Psi_{\|}(\tau)(t-\tau)]\}\frac{\mathrm{d}\tau}{(\tau^2-T^{\mathrm{r}2})^{1/2}} \\ & \text{for all } x\in\mathbb{R},\ z>0 \text{ and } t>0, \end{aligned} \tag{4.64}$$

while for the total field below the metasurface, we find (cf. (4.54))

$$
\begin{aligned}
H_y(x,z,t) = {} & \partial_t K(t) *_t \frac{\epsilon_0}{2\pi} \frac{\mathrm{H}(t-T^{\mathrm{t}})}{(t^2 - T^{\mathrm{t}2})^{1/2}} \\
& - \partial_t K(t) *_t \frac{\epsilon_0}{2\pi} \int_{\tau=T^{\mathrm{t}}}^{t} \mathrm{Re}\{\Omega_{\|}(\tau)\exp[-\Omega_{\|}(\tau)(t-\tau)] \\
& + \Psi_{\|}(\tau)\exp[-\Psi_{\|}(\tau)(t-\tau)]\} \frac{\mathrm{d}\tau}{(\tau^2 - T^{\mathrm{t}2})^{1/2}} \\
& \text{for all } x \in \mathbb{R},\ z < 0 \text{ and } t > 0,
\end{aligned}
\tag{4.65}
$$

where we used (cf. (4.24) with (4.24))

$$\Psi_{\|}(\tau) = 2Y_{\|}[\kappa(\tau)]/C^{\mathrm{E}}, \tag{4.66}$$

$$\Omega_{\|}(\tau) = 2/L^{\mathrm{M}}Y_{\|}[\kappa(\tau)], \tag{4.67}$$

and recall that we take the values of κ along $\mathcal{C}$ specified via (4.46). It is now seen from (4.64) that the reflected field would be zero if $\Omega_{\|}(\tau) = \Psi_{\|}(\tau)$, or, equivalently, if (cf. (4.55))

$$Y_{\|}[\kappa(\tau)] = (C^{\mathrm{E}}/L^{\mathrm{M}})^{1/2} \quad \text{for all } \{T^{\mathrm{r}} \le \tau < t\}, \tag{4.68}$$

where (cf. (4.56))

$$Y_{\|}[\kappa(\tau)] = \eta_0 \left[\frac{Z^{\mathrm{r}}}{\rho^{\mathrm{r}2}} c_0 \tau - \mathrm{i}\frac{x}{\rho^{\mathrm{r}2}} (c_0^2\tau^2 - \rho^{\mathrm{r}2})^{1/2} \right]^{-1}. \tag{4.69}$$

Again, in accordance with the discussion given below (4.55), the transparency condition (4.68) is met just behind the wave front of the reflected wave as $t \downarrow T^{\mathrm{r}}$ at points satisfying $\eta_0/\cos(\theta) = (C^{\mathrm{E}}/L^{\mathrm{M}})^{1/2}$, where $\cos(\theta) = Z^{\mathrm{r}}/\rho^{\mathrm{r}}$. Also, if both C^{E} and L^{M} are sufficiently large, then the metasurface just essentially inverts the sign of the incident cylindrical wave, again.

4.2.3 *Illustrative example*

The feasibility of achieving a transparent sheet will be analyzed numerically by evaluating (4.53) and (4.54). Hence, we calculate the TD response of two metasurfaces to the electric-line source whose signature is described by the unipolar power-exponential pulse (cf. (4.31))

$$I(t) = I_{\mathrm{m}}(t/t_{\mathrm{r}})^n \exp[-n(t/t_{\mathrm{r}} - 1)]\mathrm{H}(t), \tag{4.70}$$

where we take amplitude $I_{\mathrm{m}} = 1.0\,\mathrm{mA}$ and $n = 2$. Its pulse-time width is related to the source height via $c_0 t_{\mathrm{w}} = h$ and the pulse-rise time follows from $t_{\mathrm{r}} = 4\exp(-2)t_{\mathrm{w}} n^{n+1}\exp(-n)/\Gamma(n+1) \simeq 0.5413\,t_{\mathrm{w}}$.

First, we calculate the total-field distribution in the presence of metasurface characterized by $C^{\mathrm{E}}/\eta_0 t_{\mathrm{w}} = 1.0$ and $L^{\mathrm{M}}\eta_0/t_{\mathrm{w}} = 2.0$. With reference to the

discussion given in the previous section, we may then expect relatively small reflections in the early time response close to points where $\cos(\theta) = Z^{\mathrm{r}}/\rho^{\mathrm{r}} = \pi/4$. Figure 4.4(a) shows the resulting field distribution at $c_0 t/h = 2.0$, where the effect of reflected fields is clearly visible.

Second, we analyze the response of a metasurface with relatively large constitutive parameters, $C^{\mathrm{E}}/\eta_0 t_{\mathrm{w}} = 10$ and $L^{\mathrm{M}}\eta_0/t_{\mathrm{w}} = 10$, so that the second (integral) terms in (4.53) and (4.54) are relatively small. The corresponding spatial distribution of the electric field (see Figure 4.4(b)) supports the anticipated reflection and transmission behavior of the metasurface.

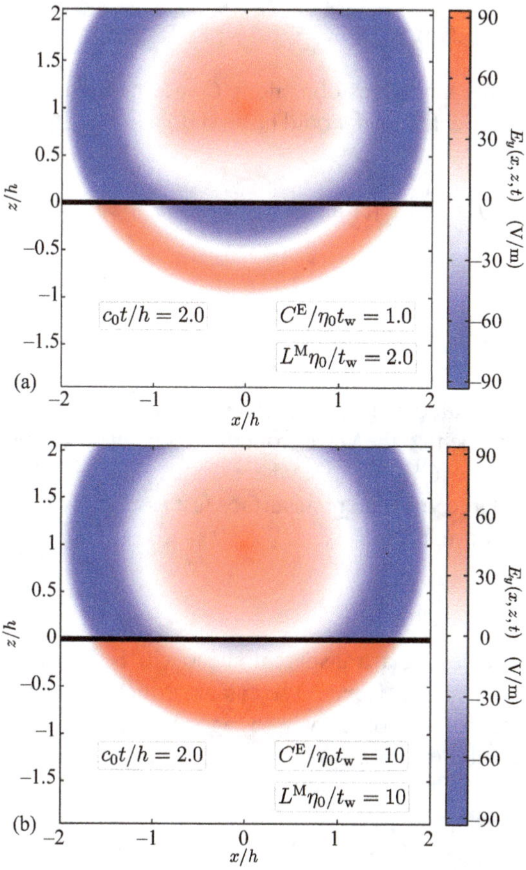

Figure 4.4 Spatial distribution of the electric-field strength excited by the impulsive electric-line source in the presence of the metasurface with (a) $C^{\mathrm{E}}/\eta_0 t_{\mathrm{w}} = 1.0$ and $L^{\mathrm{M}}\eta_0/t_{\mathrm{w}} = 2.0$; (b) $C^{\mathrm{E}}/\eta_0 t_{\mathrm{w}} = 10$ and $L^{\mathrm{M}}\eta_0/t_{\mathrm{w}} = 10$

4.3 Dipole-excited pulsed EM signal transfers via a metasurface

The transition conditions introduced in Section 3.3 are next employed to solve three-dimensional (3-D) problems concerning the pulsed EM-field signal transfers via metasurfaces. Namely, we shall analyze the pulsed, EM-field loop-to-loop and wire-to-wire signal transfers through a highly contrasting layer with both magnetic and dielectric constitutive properties. For applications of the derived closed-form analytical expressions in the field of EMC we refer the reader to [7].

4.3.1 Loop-to-loop pulsed EM signal transfer

The problem configuration to be analyzed in this subsection is depicted in Figure 4.5. It consists of a horizontal transmitting loop, $\mathcal{L}^{\mathrm{T}}$, located on one side of the metasurface at $z = h^{\mathrm{T}} > 0$ and of a horizontal receiving loop, $\mathcal{L}^{\mathrm{R}}$, located on the other side of the metasurface at $z = -h^{\mathrm{R}} < 0$. The EM properties of the metasurface with combined magnetic and dielectric properties is described by parameters $\{\epsilon, \mu, \sigma, \Gamma\}$, again. The surrounding medium is in its EM characteristics homogeneous, isotropic and loss-free. Accordingly, its EM constitutive parameters are (real-valued, scalar and positive) constants $\{\epsilon_0, \mu_0\}$ with the corresponding EM wave speed $c_0 = (\epsilon_0 \mu_0)^{-1/2} > 0$.

Owing to the rotational symmetry of the problem configuration, the non-vanishing tangential (with respect to the contrasting plane) EM-field components are the azimuthal component of the electric-field strength, $\hat{E}_\phi = \hat{E}_\phi(r, z, s)$, and the radial component of the magnetic-field strength, $\hat{H}_r = \hat{H}_r(r, z, s)$. As the sheet is located in an unbounded, homogeneous and isotropic medium, the latter field components can be represented via [6, Section 26.3]

$$\hat{E}_\phi(r, z, s) = \partial_r \hat{\Phi}_z^K(r, z, s), \tag{4.71a}$$

$$\hat{H}_r(r, z, s) = \partial_r \partial_z \hat{\Phi}_z^K(r, z, s)/s\mu_0, \tag{4.71b}$$

where [6, p. 759]

$$\hat{\Phi}_z^K(r, z, s) = s\mu_0 \hat{I}^{\mathrm{T}}(s) \mathcal{A}^{\mathrm{T}} \hat{G}(r, z, s) \tag{4.72}$$

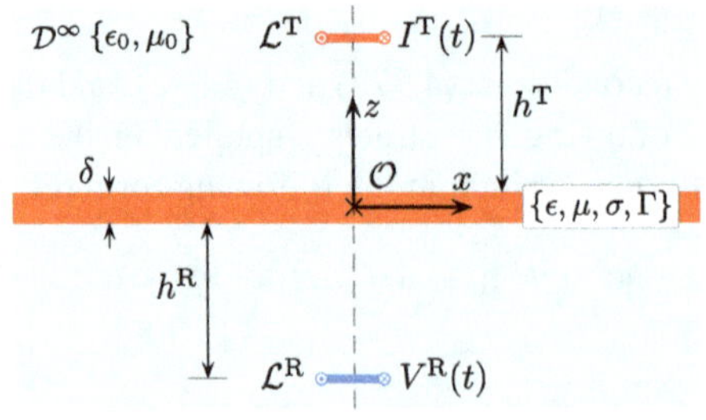

Figure 4.5 A metasurface in between two horizontal transmitting and receiving loop antennas

has the meaning of (the z-component of) the magnetic-current source potential pertaining to a relatively small, horizontal transmitting loop $\mathcal{L}^{\mathrm{T}}$ that is excited by the electric current, $\hat{I}^{\mathrm{T}}(s)$, and whose area is denoted by $\mathcal{A}^{\mathrm{T}}$. Furthermore, $\hat{G}(r, z, s)$ is the corresponding (bounded) Green's function in the s-domain that will be determined with the aid of jump conditions (3.14a) and (3.14b). The goal of this section is to derive a closed-form expression for the TD voltage induced across the ports of the receiving loop $\mathcal{L}^{\mathrm{R}}$, whose s-domain counterpart can be found from

$$\hat{V}^{\mathrm{R}}(s) = -s\mu_0 \mathcal{A}^{\mathrm{R}} \hat{H}_z(0, -h^{\mathrm{R}}, s), \tag{4.73}$$

where $\mathcal{A}^{\mathrm{R}}$ is the area of the (relatively small) receiving loop and the vertical component of the magnetic-field strength can be represented by [6, Section 26.3]

$$\hat{H}_z(r, z, s) = -s\epsilon_0 \hat{\Phi}_z^K(r, z, s) + \partial_z^2 \hat{\Phi}_z^K(r, z, s)/s\mu_0. \tag{4.74}$$

The problem under consideration will be tackled with the aid of the CdH technique, again. To that end, the EM fields are represented via a two-dimensional wave-slowness representation in the plane parallel to the sheet. To show the notation, its definition is given for the non-vanishing electric-field strength (cf. (4.33))

$$\hat{E}_\phi(r, z, s) = \left(\frac{s}{2\pi \mathrm{i}}\right)^2 \int_{\kappa=-\mathrm{i}\infty}^{\mathrm{i}\infty} \mathrm{d}\kappa \int_{\sigma=-\mathrm{i}\infty}^{\mathrm{i}\infty} \exp[-s(\kappa x + \sigma y)] \tilde{E}_\phi(\kappa, \sigma, z, s) \mathrm{d}\sigma, \tag{4.75}$$

with the real-valued and positive Laplace-transform parameter $\{s \in \mathbb{R}; s > 0\}$. Under the representation, the (transform-domain) Green's function is governed by (cf. (4.35))

$$[\partial_z^2 - s^2 p^2(\kappa, \sigma)] \tilde{G}(\kappa, \sigma, z, s) = 0, \tag{4.76}$$

in which (cf. (4.36))

$$p(\kappa, \sigma) = (1/c_0^2 - \kappa^2 - \sigma^2)^{1/2} \quad \text{with } \mathrm{Re}(p) \geq 0. \tag{4.77}$$

Equation (4.76) is further supplemented with the excitation condition (cf. (4.37a))

$$\lim_{z \downarrow h^{\mathrm{T}}} \partial_z \tilde{G}(\kappa, \sigma, z, s) - \lim_{z \uparrow h^{\mathrm{T}}} \partial_z \tilde{G}(\kappa, \sigma, z, s) = -1, \tag{4.78}$$

along with the transition conditions (4.37b) and (4.37c), and the "boundedness condition" as $|z| \to \infty$. Following the strategy applied in the previous section, the transform-domain solution is written in the following form (cf. (4.38) and (4.39)):

$$\tilde{G} = \tilde{A} \exp[-sp(\kappa, \sigma)|z - h^{\mathrm{T}}|] + \tilde{A}\tilde{R}_\perp \exp[-sp(\kappa, \sigma)(z + h^{\mathrm{T}})] \quad \text{for } z > 0, \tag{4.79}$$

and

$$\tilde{G} = \tilde{A}\tilde{T}_\perp \exp[-sp(\kappa, \sigma)(h^{\mathrm{T}} - z)] \quad \text{for } z < 0. \tag{4.80}$$

Subsequently, making use of (4.79) and (4.80) in the excitation and jump conditions, we readily determine coefficients $\tilde{A}$, $\tilde{R}_{\perp}$ and $\tilde{T}_{\perp}$, whose form is similar to the one represented by (4.40)–(4.42). Indeed, the corresponding transmission coefficient, for instance, reads (cf. (4.40))

$$\tilde{T}_{\perp} = \frac{1}{2}\left[\frac{1 - \hat{Y}^{\mathrm{E}}(s)/2\tilde{Y}_{\perp}(\kappa,\sigma)}{1 + \hat{Y}^{\mathrm{E}}(s)/2\tilde{Y}_{\perp}(\kappa,\sigma)} + \frac{1 - \hat{Z}^{\mathrm{M}}(s)\tilde{Y}_{\perp}(\kappa,\sigma)/2}{1 + \hat{Z}^{\mathrm{M}}(s)\tilde{Y}_{\perp}(\kappa,\sigma)/2}\right], \tag{4.81}$$

where we introduced $\tilde{Y}_{\perp}(\kappa,\sigma) = p(\kappa,\sigma)/\mu_0$. Using the transmission coefficient and employing further the transform-domain counterparts of (4.73) and (4.74), the voltage response in the s-domain can be represented via

$$\hat{V}^{\mathrm{R}}(s) = -\frac{s^4\mu_0\hat{I}^{\mathrm{T}}(s)\mathcal{A}^{\mathrm{T}}\mathcal{A}^{\mathrm{R}}}{4\pi^2}\int_{\kappa=-\mathrm{i}\infty}^{\mathrm{i}\infty}\mathrm{d}\kappa \int_{\sigma=-\mathrm{i}\infty}^{\mathrm{i}\infty}\exp[-sp(\kappa,\sigma)(h^{\mathrm{T}}+h^{\mathrm{R}})]\tilde{T}_{\perp}(\kappa,\sigma,s)\frac{\kappa^2+\sigma^2}{2p(\kappa,\sigma)}\mathrm{d}\sigma. \tag{4.82}$$

Noting that the integrand in (4.82) is a function of $\kappa^2 + \sigma^2$, we shall express the variables of integration through the radial slowness parameter

$$\kappa = \gamma\cos(\psi), \tag{4.83a}$$
$$\sigma = \gamma\sin(\psi), \tag{4.83b}$$

with $\{0 \le \psi < 2\pi\}$ and $\{\mathrm{Re}(\gamma) = 0, 0 \le \mathrm{Im}(\gamma) < \infty\}$. Under the substitution, $\kappa^2 + \sigma^2 = \gamma^2$ and $\mathrm{d}\kappa\mathrm{d}\sigma = \gamma\mathrm{d}\gamma\mathrm{d}\psi$, which yields

$$\hat{V}^{\mathrm{R}}(s) = -\frac{s^4\mu_0\hat{I}^{\mathrm{T}}(s)\mathcal{A}^{\mathrm{T}}\mathcal{A}^{\mathrm{R}}}{2\pi}\int_{\gamma=0}^{\mathrm{i}\infty}\exp[-s\bar{p}(\gamma)(h^{\mathrm{T}}+h^{\mathrm{R}})]\bar{T}_{\perp}(\gamma,s)\frac{\gamma^3\mathrm{d}\gamma}{2\bar{p}(\gamma)}, \tag{4.84}$$

where $\bar{p}(\gamma) = (1/c_0^2 - \gamma^2)^{1/2}$ with $\mathrm{Re}[\bar{p}(\gamma)] \ge 0$ and $\bar{T}_{\perp}(\gamma)$ immediately follows upon substituting (4.83a) and (4.83b) in (4.81). In the following step, we substitute

$$\bar{p}(\gamma)(h^{\mathrm{T}}+h^{\mathrm{R}}) = \tau \quad \text{for } \{\tau \in \mathbb{R}; \tau > 0\}, \tag{4.85}$$

from which we find that

$$\gamma(\tau) = \mathrm{i}[\tau^2/(h^{\mathrm{T}}+h^{\mathrm{R}})^2 - 1/c_0^2]^{1/2} \quad \text{for } \{T = (h^{\mathrm{T}}+h^{\mathrm{R}})/c_0 \le \tau < \infty\}, \tag{4.86}$$

where T has the meaning of the pulse travel time. Since from (4.85) we have

$$\gamma\mathrm{d}\gamma/\mathrm{d}\tau = -\bar{p}(\gamma)/(h^{\mathrm{T}}+h^{\mathrm{R}}), \tag{4.87}$$

we may rewrite (4.84) using the new variable of integration as

$$\hat{V}^{\mathrm{R}}(s) = -\frac{s^4\mu_0\hat{I}^{\mathrm{T}}(s)\mathcal{A}^{\mathrm{T}}\mathcal{A}^{\mathrm{R}}}{4\pi(h^{\mathrm{T}}+h^{\mathrm{R}})^3}\int_{\tau=T}^{\infty}\exp(-s\tau)\bar{T}_{\perp}[\gamma(\tau),s](\tau^2-T^2)\mathrm{d}\tau. \tag{4.88}$$

Assuming a loss-free metasurface characterized by $\hat{Z}^{\mathrm{M}} = sL^{\mathrm{M}}$ and $\hat{Y}^{\mathrm{E}} = sC^{\mathrm{E}}$, again, the expression can be cast into the following form (cf. (4.50)):

$$\hat{V}^{\mathrm{R}}(s) = \frac{s^4 \mu_0 \hat{I}^{\mathrm{T}}(s) \mathcal{A}^{\mathrm{T}} \mathcal{A}^{\mathrm{R}}}{4\pi(h^{\mathrm{T}} + h^{\mathrm{R}})^3} \int_{\tau=T}^{\infty} \exp(-s\tau) \\ \times \frac{s^2 - \bar{\Psi}_\perp(\tau)\bar{\Omega}_\perp(\tau)}{s^2 + s[\bar{\Psi}_\perp(\tau) + \bar{\Omega}_\perp(\tau)] + \bar{\Psi}_\perp(\tau)\bar{\Omega}_\perp(\tau)} (\tau^2 - T^2)\mathrm{d}\tau, \tag{4.89}$$

in which (cf. (4.51) and (4.52))

$$\bar{\Psi}_\perp(\tau) = (2\eta_0/C^{\mathrm{E}})[c_0\tau/(h^{\mathrm{T}} + h^{\mathrm{R}})], \tag{4.90}$$

$$\bar{\Omega}_\perp(\tau) = (2/\eta_0 L^{\mathrm{M}})[(h^{\mathrm{T}} + h^{\mathrm{R}})/c_0\tau]. \tag{4.91}$$

With the aid of the Schouten–Van der Pol theorem [6, p. 1056], the inverse of (4.89) readily follows (cf. (4.54))

$$V^{\mathrm{R}}(t) = -V_0^{\mathrm{R}}(t) - \frac{\mu_0 \mathcal{A}^{\mathrm{T}} \mathcal{A}^{\mathrm{R}} \partial_t^4 I^{\mathrm{T}}(t)}{4\pi(h^{\mathrm{T}} + h^{\mathrm{R}})^3} \ast_t \int_{\tau=T}^{t} \{\bar{\Psi}_\perp(\tau) \exp[-\bar{\Psi}_\perp(\tau)(t-\tau)] \\ + \bar{\Omega}_\perp(\tau) \exp[-\bar{\Omega}_\perp(\tau)(t-\tau)]\}(\tau^2 - T^2)\mathrm{d}\tau, \tag{4.92}$$

where

$$V_0^{\mathrm{R}}(t) = -\frac{\mu_0 \mathcal{A}^{\mathrm{T}} \mathcal{A}^{\mathrm{R}} \partial_t^4 I^{\mathrm{T}}(t)}{4\pi(h^{\mathrm{T}} + h^{\mathrm{R}})^3} \ast_t (t^2 - T^2)\mathrm{H}(t-T) \\ = -\frac{1}{2\pi\eta_0} \frac{\mathcal{A}^{\mathrm{T}} \mathcal{A}^{\mathrm{R}}}{(h^{\mathrm{T}} + h^{\mathrm{R}})^4} \left[T\partial_t I^{\mathrm{T}}(t-T) + T^2 \partial_t^2 I^{\mathrm{T}}(t-T)\right] \\ = -\frac{\mu_0 \mathcal{A}^{\mathrm{T}} \mathcal{A}^{\mathrm{R}}}{2\pi} \left[\frac{\partial_t I^{\mathrm{T}}(t-T)}{(h^{\mathrm{T}} + h^{\mathrm{R}})^3} + \frac{\partial_t^2 I^{\mathrm{T}}(t-T)}{c_0(h^{\mathrm{T}} + h^{\mathrm{R}})^2}\right] \tag{4.93}$$

can be associated with the "Thévenin equivalent generator voltage" pertaining to the pulsed EM signal transfer between two mutually parallel loops [8, Section IX]. Clearly, (4.93) is composed of near- and intermediate-field constituents that are proportional to (distance)$^{-3}$ and (distance)$^{-2}$, respectively. As far as their temporal behavior is concerned, the constituents are related to the first and second time derivatives of the excitation electric-current pulse, respectively. Consequently, the near-field interaction can be described via the coupling inductance (see (16.18)).

4.3.2 *Wire-to-wire pulsed EM signal transfer*

Figure 4.6 shows the problem configuration analyzed in this subsection. It consists of a vertical transmitting wire, $\mathcal{W}^{\mathrm{T}}$, located on one side of the metasurface at $z = h^{\mathrm{T}} > 0$ and of a vertical receiving wire, $\mathcal{W}^{\mathrm{R}}$, located on the other side of the metasurface at $z = -h^{\mathrm{R}} < 0$. The remaining parameters are the same as in the problem configuration analyzed in Section 4.3.1.

Now, the non-vanishing EM-field components parallel to the metasurface are the radial component of the electric-field strength, $\hat{E}_r = \hat{E}_r(r, z, s)$, and the azimuthal

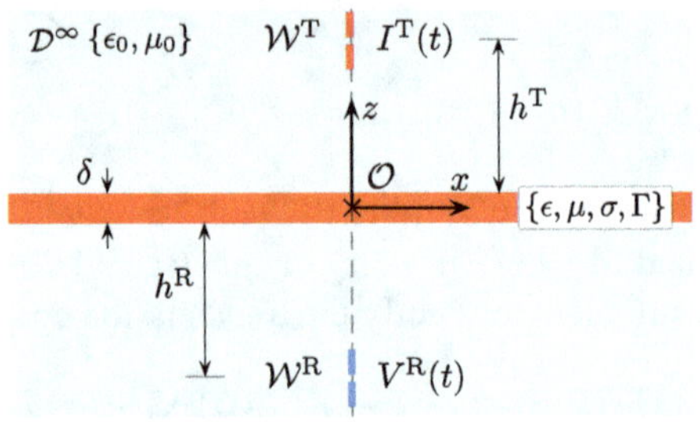

Figure 4.6 A metasurface in between two vertical transmitting and receiving wire antennas

component of the magnetic-field strength, $\hat{H}_\phi = \hat{H}_\phi(r, z, s)$, that are ϕ-independent. In line with [6, Section 26.3], the following s-domain source-type representations for the tangential EM-field components apply (cf. (4.71a) and (4.71b)):

$$\hat{H}_\phi(r, z, s) = -\partial_r \hat{\Phi}^J_z(r, z, s), \tag{4.94a}$$

$$\hat{E}_r(r, z, s) = \partial_r \partial_z \hat{\Phi}^J_z(r, z, s)/s\epsilon_0, \tag{4.94b}$$

where [6, p. 752]

$$\hat{\Phi}^J_z(r, z, s) = \hat{I}^T(s)\ell^{\mathrm{T}}\hat{G}(r, z, s) \tag{4.95}$$

is (the z-component of) the electric-current source potential of a relatively short and thin conducting wire carrying a uniform electric current $\hat{I}^{\mathrm{T}}(s)$ and $\ell^{\mathrm{T}} > 0$ denotes its length. The pertaining Green's function $\hat{G}(r, z, s)$ will be determined with the aid of cross-layer conditions (3.13a) and (3.13b). The main objective of our analysis is the determination of the TD voltage response induced across the port of the receiving wire. Its s-domain image will be expressed via

$$\hat{V}^{\mathrm{R}}(s) = -\ell^{\mathrm{R}}\hat{E}_z(0, -h^{\mathrm{R}}, s), \tag{4.96}$$

where $\ell^{\mathrm{R}} > 0$ denotes the length of the (relatively short) receiving wire and the vertical component of the electric-field strength can be represented by [6, Section 26.3]

$$\hat{E}_z(r, z, s) = -s\mu_0\hat{\Phi}^J_z(r, z, s) + \partial_z^2\hat{\Phi}^J_z(r, z, s)/s\epsilon_0. \tag{4.97}$$

The formulated problem will be next solved using the CdH technique along the lines closely described in Section 4.3.1. Hence, in the first step, we employ the two-dimensional wave-slowness representation of the type specified via (4.75) to determine the corresponding transform-domain Green's function $\tilde{G}(\kappa, \sigma, z, s)$. The Green's function that satisfies (4.76), the excitation condition (4.78) and the corresponding cross-layer conditions (4.59a) and (4.59b) then immediately follows in the following form (cf. (4.79) and (4.80)):

$$\tilde{G} = \tilde{A}\exp[-sp(\kappa,\sigma)|z - h^{\mathrm{T}}|] + \tilde{A}\tilde{R}_{\|}\exp[-sp(\kappa,\sigma)(z + h^{\mathrm{T}})]$$
$$\text{for } z > 0, \tag{4.98}$$

and

$$\tilde{G} = \tilde{A}\tilde{T}_{\|} \exp[-sp(\kappa,\sigma)(h^{\mathrm{T}} - z)] \quad \text{for } z < 0, \tag{4.99}$$

where $\tilde{R}_{\|}$ and $\tilde{T}_{\|}$ have the meaning of the (transform-domain) reflection and transmission coefficients and $\tilde{A} = 1/2sp(\kappa,\sigma)$, again (see (4.77)). To determine the voltage response, it is sufficient to specify the transmission coefficient (cf. (4.62))

$$\tilde{T}_{\|} = \frac{1}{2}\left[\frac{1 - \hat{Z}^{\mathrm{M}}(s)\tilde{Y}_{\|}(\kappa,\sigma)/2}{1 + \hat{Z}^{\mathrm{M}}(s)\tilde{Y}_{\|}(\kappa,\sigma)/2} + \frac{1 - \hat{Y}^{\mathrm{E}}(s)/2\tilde{Y}_{\|}(\kappa,\sigma)}{1 + \hat{Y}^{\mathrm{E}}(s)/2\tilde{Y}_{\|}(\kappa,\sigma)}\right], \tag{4.100}$$

where we used $\tilde{Y}_{\|}(\kappa,\sigma) = \epsilon_0/p(\kappa,\sigma)$. With the transform-domain solution at our disposal, we may collect the results and express the induced voltage in the receiving wire in the following form (cf. (4.82)):

$$\hat{V}^{\mathrm{R}}(s) = -\frac{s^2\hat{I}^{\mathrm{T}}(s)\ell^{\mathrm{T}}\ell^{\mathrm{R}}}{4\pi^2\epsilon_0}\int_{\kappa=-\mathrm{i}\infty}^{\mathrm{i}\infty}\mathrm{d}\kappa \int_{\sigma=-\mathrm{i}\infty}^{\mathrm{i}\infty} \exp[-sp(\kappa,\sigma)(h^{\mathrm{T}} + h^{\mathrm{R}})]\tilde{T}_{\|}(\kappa,\sigma,s)\frac{\kappa^2+\sigma^2}{2p(\kappa,\sigma)}\mathrm{d}\sigma. \tag{4.101}$$

To find the TD original $V^{\mathrm{R}}(t)$, we next carry out the substitutions described via (4.83a), (4.83b) and (4.85) and cast (4.101) into the following form (cf. (4.88)):

$$\hat{V}^{\mathrm{R}}(s) = -\frac{s^2\hat{I}^{\mathrm{T}}(s)\ell^{\mathrm{T}}\ell^{\mathrm{R}}}{4\pi\epsilon_0(h^{\mathrm{T}}+h^{\mathrm{R}})^3}\int_{\tau=T}^{\infty}\exp(-s\tau)\bar{T}_{\|}[\gamma(\tau),s](\tau^2 - T^2)\mathrm{d}\tau, \tag{4.102}$$

where $T = (h^{\mathrm{T}} + h^{\mathrm{R}})/c_0$. For the sake of simplicity, we shall next restrict ourselves to a loss-free metasurface, again. Then, making use of $\hat{Z}^{\mathrm{M}} = sL^{\mathrm{M}}$ and $\hat{Y}^{\mathrm{E}} = sC^{\mathrm{E}}$ in (4.100), the integral expression can be rewritten as (cf. (4.89))

$$\hat{V}^{\mathrm{R}}(s) = \frac{s^2\hat{I}^{\mathrm{T}}(s)\ell^{\mathrm{T}}\ell^{\mathrm{R}}}{4\pi\epsilon_0(h^{\mathrm{T}}+h^{\mathrm{R}})^3}\int_{\tau=T}^{\infty}\exp(-s\tau) \times \frac{s^2 - \bar{\Psi}_{\|}(\tau)\bar{\Omega}_{\|}(\tau)}{s^2 + s[\bar{\Psi}_{\|}(\tau)+\bar{\Omega}_{\|}(\tau)] + \bar{\Psi}_{\|}(\tau)\bar{\Omega}_{\|}(\tau)}(\tau^2 - T^2)\mathrm{d}\tau, \tag{4.103}$$

where (cf. (4.90), (4.91) and (4.66), (4.67))

$$\bar{\Psi}_{\|}(\tau) = (2\eta_0/C^{\mathrm{E}})[(h^{\mathrm{T}}+h^{\mathrm{R}})/c_0\tau], \tag{4.104}$$

$$\bar{\Omega}_{\|}(\tau) = (2/\eta_0 L^{\mathrm{M}})[c_0\tau/(h^{\mathrm{T}}+h^{\mathrm{R}})]. \tag{4.105}$$

The final transformation back to the TD is performed with the aid of the Schouten–Van der Pol theorem [6, p. 1056]. This way yields (cf. (4.92))

$$V^{\mathrm{R}}(t) = -V_0^{\mathrm{R}}(t) - \frac{\ell^{\mathrm{T}}\ell^{\mathrm{R}}\partial_t^2 I^{\mathrm{T}}(t)}{4\pi\epsilon_0(h^{\mathrm{T}}+h^{\mathrm{R}})^3} *_t \int_{\tau=T}^{t}\{\bar{\Omega}_{\|}(\tau)\exp[-\bar{\Omega}_{\|}(\tau)(t-\tau)] + \bar{\Psi}_{\|}(\tau)\exp[-\bar{\Psi}_{\|}(\tau)(t-\tau)]\}(\tau^2 - T^2)\mathrm{d}\tau, \tag{4.106}$$

in which $V_0^{\mathrm{R}}(t)$ denotes the voltage response if the layer were absent. It can described by (cf. (4.93))

$$V_0^{\mathrm{R}}(t) = -\frac{\ell^{\mathrm{T}}\ell^{\mathrm{R}}\partial_t^2 I^{\mathrm{T}}(t)}{4\pi\epsilon_0(h^{\mathrm{T}}+h^{\mathrm{R}})^3} *_t (t^2 - T^2)\mathrm{H}(t-T)$$
$$= -\frac{1}{2\pi\eta_0}\frac{\ell^{\mathrm{T}}\ell^{\mathrm{R}}}{(h^{\mathrm{T}}+h^{\mathrm{R}})^2}\Big[T^{-1}\partial_t^{-1}I^{\mathrm{T}}(t-T) + I^{\mathrm{T}}(t-T)\Big], \tag{4.107}$$

where ∂_t^{-1} denotes the time-integration operator (see (1.3)). Such as the voltage pulse transmitted between the horizontal loops (see (4.93)), the voltage response $V_0^{\mathrm{R}}(t)$ is composed of near- and intermediate-field constituents that are proportional to (distance)$^{-3}$ and (distance)$^{-2}$, respectively. The near- and intermediate-field TD wave constituents, however, are now related to the time-integrated electric-current pulse and its scaled copy, respectively, which represents a fundamental difference with respect to the loop-to-loop signal transfer. Apparently, the dominant term in the (near-field) limit $T \downarrow 0$ has the form $V_0^{\mathrm{R}}(t) \simeq -\partial_t^{-1}I^{\mathrm{T}}(t)/C^{\mathrm{C}}$, where $C^{\mathrm{C}} = 2\pi\epsilon_0(h^{\mathrm{T}}+h^{\mathrm{R}})^3/\ell^{\mathrm{T}}\ell^{\mathrm{R}}$can be interpreted as the coupling capacitance (see (16.18)).

4.3.3 *Illustrative example*

Following the line of reasoning pursued in Section 4.2, it can be demonstrated that the right-hand side of (4.92) approaches zero for both limiting PEC and PMC cases. Also, if both $C^{\mathrm{E}} \downarrow 0$ and $L^{\mathrm{M}} \downarrow 0$, the integral term on the right-hand side of (4.92) tends to $2V_0^{\mathrm{R}}(t)$, thus yielding $V^{\mathrm{R}}(t) = V_0^{\mathrm{R}}(t)$ in total, as expected in the absence of the sheet. Again, an interesting case arises when both functions $\bar{\Psi}_\perp(\tau)$ and $\bar{\Omega}_\perp(\tau)$ (see (4.90) and (4.91)) are sufficiently small such that the integral term tends to zero. Under such circumstances, the metasurface is virtually transparent and the induced voltage is equal to $V^{\mathrm{R}}(t) = -V_0^{\mathrm{R}}(t)$, that is, to the opposite of the response that would be excited in the absence of the sheet. Similar conclusions can be drawn for the results presented in Section 4.3.2.

To demonstrate the property of transparency, we shall evaluate (4.92) for a metasurface described by $C^{\mathrm{E}}/\eta_0 t_{\mathrm{w}} = L^{\mathrm{M}}\eta_0/t_{\mathrm{w}} = 10$, again. The receiving loop is located below the sheet at $z = -h^{\mathrm{R}} = -h^{\mathrm{T}}/2 = -4r^{\mathrm{T,R}}$, where $r^{\mathrm{T}} = r^{\mathrm{R}}$ denote the radii of the transmitting and receiving loops, so that $\mathcal{A}^{\mathrm{T}}\mathcal{A}^{\mathrm{R}}/(h^{\mathrm{T}}+h^{\mathrm{R}})^4 = \pi^2/12^4$. The excitation electric-current pulse, $I^{\mathrm{T}}(t)$, is described by the right-hand side of (4.70), where we take $I_{\mathrm{m}} = 1.0\,\mathrm{mA}$, $n = 6$ and $c_0 t_{\mathrm{w}} = h^{\mathrm{T}} + h^{\mathrm{R}}$ with $t_{\mathrm{r}} = t_{\mathrm{w}} n^{n+1}\exp(-n)/\Gamma(n+1) \simeq 0.9637\,t_{\mathrm{w}}$. Note that the loops are then relatively small with respect to the spatial support of the excitation pulse. The corresponding excitation source signature is shown in Figure 4.7(a). Figure 4.7(b) shows the voltage response in the absence and in the presence of the metasurface. Apparently, the presence of the analyzed metasurface inverts the polarity of the voltage response as predicted in the discussion above.

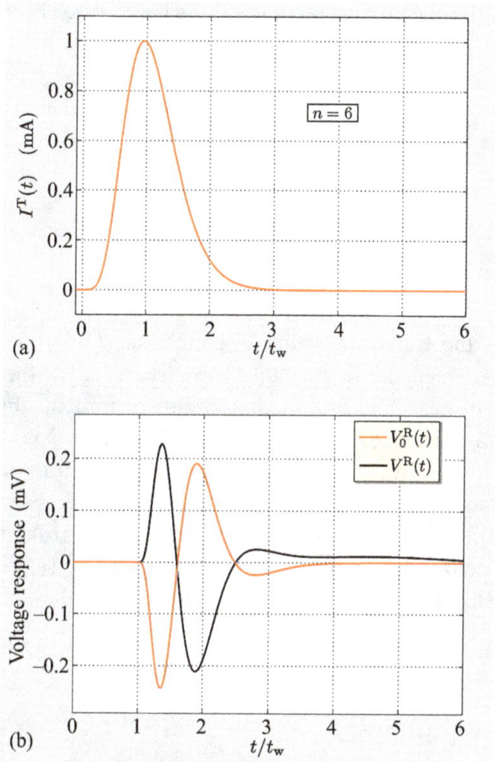

Figure 4.7 (a) Excitation electric-current pulse. (b) Induced voltage responses

4.4 Concluding remarks

The modeling approaches pursued throughout this chapter can be readily extended to analyzing the pulsed EM response of more complex structures. To provide an example, we shall briefly analyze the impact of metasurfaces deposited on a dielectric slab. Finally, we will show that the (approximate) results achieved with the aid of thin-sheet, high-contrast cross-layer conditions can be arrived at via a limit of the exact solution, thus demonstrating their validity.

4.4.1 Transmission via a dielectric slab coated by metasurfaces

In this section, the E-polarized plane-wave response of a dielectric slab that is covered by two metasurfaces with combined magneto-dielectric properties is briefly analyzed (see Figure 4.8). In general, the TD electric-field strength below the covered slab can be expressed through (cf. (4.17))

$$E_y(x, z, t) = \mathbb{T}_\perp(t) *_t e^{\mathrm{i}}[t - \kappa_0 x + p(\kappa_0)z]$$
$$\text{for all } x \in \mathbb{R}, z < -d/2 \text{ and } t > 0, \tag{4.108}$$

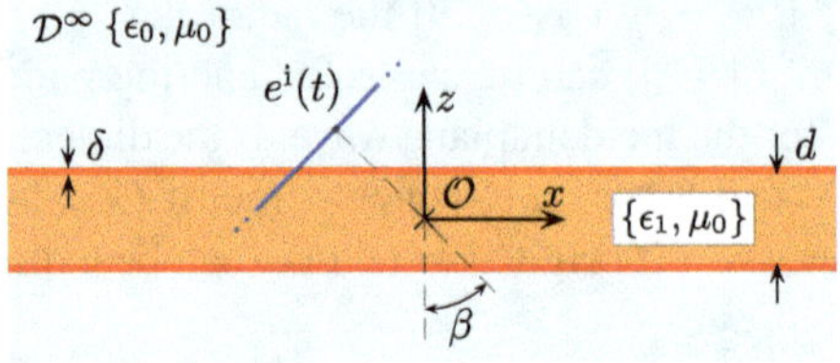

Figure 4.8 A dielectric slab covered by metasurfaces

in which the (generalized) transmission coefficient, $\mathbb{T}_\perp(t)$, follows upon applying the saltus-type conditions (4.2a) and (4.2b) across the interfaces at $z = \pm d/2$. Its s-domain counterpart applying to loss-free metasurfaces can be then written in the following form:

$$\hat{\mathbb{T}}_\perp(s) = -\frac{\hat{T}_\perp}{\hat{\lambda}_\perp}\left[\frac{s - \Psi^{\mathrm{s}}_\perp}{s + \Psi_\perp} + \hat{\Gamma}_0 \frac{s + \Psi^{\mathrm{s}}_\perp}{s + \Psi_\perp}\right] \frac{\exp\{-s[p_1(\kappa_1) - p(\kappa_0)]d\}}{1 - \hat{\Gamma}_0^2 \exp[-2sp_1(\kappa_1)d]}, \tag{4.109}$$

where we used $p_1(\kappa) = (1/c_1^2 - \kappa^2)^{1/2}$, $c_1 = (\mu_0\epsilon_1)^{-1/2} > 0$ and (4.10)–(4.12) along with the corresponding quantities pertaining to the dielectric slab, that is,

$$\Psi^{\mathrm{s}}_\perp = 2Y^{\mathrm{s}}_\perp(\kappa_1)/C^{\mathrm{E}}, \tag{4.110}$$

$$\Omega^{\mathrm{s}}_\perp = 2/L^{\mathrm{M}}Y^{\mathrm{s}}_\perp(\kappa_1), \tag{4.111}$$

where $Y^{\mathrm{s}}_\perp(\kappa_1) = p_1(\kappa_1)/\mu_0 = \eta_1 \cos(\beta^{\mathrm{s}})$, $\eta_1 = 1/c_1\mu_0$ and $\kappa_1 = \sin(\beta^{\mathrm{s}})/c_1 = \kappa_0$ by virtue of Snell's law. Finally, the reflection coefficient pertaining to the metasurface at the interface of a dielectric half-space can be expressed as (cf. (4.13))

$$\hat{\Gamma}_0 = -\frac{1}{2\hat{\lambda}_\perp}\left(\frac{s - \Psi^{\mathrm{s}}_\perp}{s + \Psi_\perp} - \frac{s - \Omega^{\mathrm{s}}_\perp}{s + \Omega_\perp}\right), \tag{4.112}$$

where

$$\hat{\lambda}_\perp = \frac{1}{2}\left(\frac{s + \Psi^{\mathrm{s}}_\perp}{s + \Psi_\perp} + \frac{s + \Omega^{\mathrm{s}}_\perp}{s + \Omega_\perp}\right). \tag{4.113}$$

Apparently, if $\epsilon_1 = \epsilon_0$, then $\hat{\lambda}_\perp = 1$ and $\hat{\Gamma}_0 = \hat{R}_\perp$ as given by (4.13). In such a case, the transmission coefficient comes down to

$$\begin{aligned}\hat{\mathbb{T}}_\perp(s)|_{\epsilon_1=\epsilon_0} &= \frac{\hat{T}^2_\perp}{1 - \hat{R}^2_\perp \exp[-2sp(\kappa_0)d]} \\ &= \hat{T}^2_\perp\Big\{1 + \hat{R}^2_\perp \exp[-2sp(\kappa_0)d] + \hat{R}^4_\perp \exp[-4sp(\kappa_0)d] + \cdots\Big\},\end{aligned} \tag{4.114}$$

thus representing the wave transmission via successive reflections between the metasurfaces. If, in addition, the condition $C^{\mathrm{E}}/L^{\mathrm{M}} = Y^2_\perp(\kappa_0)$ is satisfied (see (4.8)), the internal reflections vanish identically, and we end up with $\hat{\mathbb{T}}_\perp(s) = \hat{T}^2_\perp$. Under these assumptions, the transmission coefficient can be transformed to the TD at once

$$\mathbb{T}_\perp(t)|_{\epsilon_1=\epsilon_0} = \delta(t) - 4\Psi_\perp(1 - \Psi_\perp t)\exp(-\Psi_\perp t)\mathrm{H}(t), \tag{4.115}$$

provided that $(C^{\mathrm{E}}/L^{\mathrm{M}})^{1/2} = \eta_0 \cos(\beta)$. If the factor $\Psi_\perp = \Omega_\perp$ is relatively small, then the second term in (4.115) can be neglected and the parallel-plate structure is essentially transparent for the incident plane wave. If the dielectric slab is present, we may assume $\Psi_\perp = \Omega_\perp^{\mathrm{s}}$ and $\Psi_\perp^{\mathrm{s}} = \Omega_\perp$, which is met if $C^{\mathrm{E}}/L^{\mathrm{M}} = Y_\perp(\kappa_0) Y_\perp^{\mathrm{s}}(\kappa_1)$. If this condition applies, the TD original of (4.112) can be expressed as

$$\Gamma_0 = (\Omega_\perp + \Psi_\perp)\sin[(\Omega_\perp - \Psi_\perp)t/2]\exp[-(\Omega_\perp + \Psi_\perp)t/2]\mathrm{H}(t), \tag{4.116}$$

which clearly shows that the reflection coefficient approaches zero for sufficiently small coefficients $\Psi_\perp = \Omega_\perp^{\mathrm{s}}$ and $\Omega_\perp = \Psi_\perp^{\mathrm{s}}$. In such circumstances, the TD generalized transmission coefficient mimics the Dirac-delta function, again. Consequently, the EM pulse transmitted through the coated slab approximately resembles a delayed copy of the incident one, thus reducing the distortion due to the (multiple) reflections inside the slab.

4.4.2 *Reflection against a slab of finite thickness*

It is finally noted that the reflection and transmission characteristics of thin, highly contrasting sheets can also be arrived at through a limit of expressions pertaining to layers of a finite thickness $\delta > 0$. This approach is briefly demonstrated on the analysis of electric-line source excited, EM fields reflected against a layer with dielectric and conductive properties described by $\{\epsilon, \sigma\}$ (see Section 4.2.1). Thus, instead of employing the approximate cross-layer conditions applying to a thin highly contrasting sheet (cf. (4.2a) and (4.2b)), i.e.

$$\hat{E}_y^+ - \hat{E}_y^- = 0, \tag{4.117a}$$

$$\hat{H}_x^+ - \hat{H}_x^- = \hat{Y}^{\mathrm{E}}(s)\hat{E}_y(x, 0, s), \tag{4.117b}$$

we apply the general continuity-type conditions at the problem's interfaces along $z = \pm\delta/2$, that is,

$$\lim_{z \downarrow \pm\delta/2} \{\hat{E}_y, \hat{H}_x\}(x, z, s) = \lim_{z \uparrow \pm\delta/2} \{\hat{E}_y, \hat{H}_x\}(x, z, s) \quad \text{for all } x \in \mathbb{R}, \tag{4.118}$$

respectively. Under the wave-slowness representation (4.33), the (generalized) reflection coefficient can be then cast into the following form:

$$\tilde{R}_\perp = \tilde{R}\,\frac{1 - \exp(-2s\hat{p}_1\delta)}{1 - \tilde{R}^2\exp(-2s\hat{p}_1\delta)}, \tag{4.119}$$

where

$$\tilde{R} = (p_0 - \hat{p}_1)/(p_0 + \hat{p}_1), \tag{4.120}$$

with $p_0 = p_0(\kappa) = (1/c_0^2 - \kappa^2)^{1/2}$ and $\hat{p}_1 = \hat{p}_1(\kappa, s) = (1/c_1^2 + \sigma\mu_0/s - \kappa^2)^{1/2}$, $c_1 = (\mu_0\epsilon)^{-1/2} > 0$. To derive the approximate reflection coefficient complying with conditions (4.117a) and (4.117b), we first assume that the layer is very thin

such that the propagation factors can be expanded around $s\hat{p}_1\delta \to 0$ as $\exp(-2s\hat{p}_1\delta) = 1 - 2s\hat{p}_1\delta + O(s^2\hat{p}_1^2\delta^2)$. Further, inside a high-contrast layer, we may neglect the field variations along the direction parallel to the interfaces and hence use $\hat{p}_1(\kappa, s) \simeq \hat{p}_1(0, s)$ and $p_0 \ll \hat{p}_1$. Applying these simplifications to (4.119), we find that

$$\tilde{R}_\perp \simeq -1 + \frac{2Y_\perp(\kappa)}{2Y_\perp(\kappa) + \hat{Y}^{\mathrm{E}}(s)}. \tag{4.121}$$

It is easy to show that the thus obtained reflection coefficient can also be derived directly from (4.42) by letting $\hat{Z}^{\mathrm{M}}(s) = 0$.

References

[1] M. Štumpf. Time-domain modeling of thin high-contrast layers with combined dielectric and magnetic properties. *IEEE Antennas and Wireless Propagation Letters*, 19(6):969–971, 2020.

[2] A. T. de Hoop and L. Jiang. Pulsed EM field response of a thin, high-contrast, finely layered structure with dielectric and conductive properties. *IEEE Transactions on Antennas and Propagation*, 57(8):2260–2269, 2009.

[3] J. A. Stratton. *Electromagnetic Theory*. Classic Reissue. Piscataway, NJ: IEEE Press, 2007.

[4] A. T. de Hoop and L. Jiang. Reflection and transmission of line-source excited pulsed EM fields at a thin, high-contrast layer with dielectric and conductive properties. In *2011 IEEE International Symposium on Antennas and Propagation*, pp. 2404–2406, Spokane, WA, USA, July 2011.

[5] A. T. de Hoop. A modification of Cagniard's method for solving seismic pulse problems. *Applied Scientific Research*, B(8):349–356, 1960.

[6] A. T. de Hoop, *Handbook of Radiation and Scattering of Waves*. London, UK: Academic Press, 1995.

[7] M. Štumpf, G. Antonini, I. E. Lager and G. A. E. Vandenbosch. Pulsed electromagnetic field signal transfer across a thin magneto-dielectric sheet. *IEEE Transactions on Electromagnetic Compatibility*, 63(4):1058–1064, 2021.

[8] A. T. de Hoop, I. E. Lager and V. Tomassetti. The pulsed-field multiport antenna system reciprocity relation and its applications—a time-domain approach. *IEEE Transactions on Antennas and Propagation*, 57(3):594–605, 2009.

Chapter 5

Pulsed EM-field surface phenomena on thin sheets

A considerable success achieved in the beginning of the twentieth century in the experimental wireless transmission of electromagnetic (EM) signals over long distances has triggered the intensive research into the EM wave propagation mechanisms. Besides the explanation based on the EM-field reflection against the Kenelly–Heaviside ionized layer [1], a large amount of attention has been paid, under the influence of the discovery of surface wave phenomena traveling along a free surface of a semi-infinite elastic solid [2, 3], to the existence of EM surface waves propagating along the air–earth interface. These efforts have yielded useful mathematical tools for solving a class of EM boundary-value problems [4], but their interpretation has given rise to long-standing controversies (see [5, 6], for example) and terminology confusion [7]. While such disputes are still not fully settled in the realm of a frequency-domain (FD) analysis, the situation is truly transparent in the time domain (TD), where any wave phenomena do manifest themselves. Indeed, under the causality-preserving Cagniard–DeHoop (CdH)-like representation, any singularity in the complex slowness plane can be directly associated with a physical phenomenon occurring in the resulting wave motion [8]. Hence, the presence of a pole singularity in the pertaining slowness complex plane implies the existence of a true (causal) surface wave (e.g. the Rayleigh wave at the traction-free boundary of a solid or the Scholte wave along a fluid–solid interface [9]). Therefore, the CdH-technique is also employed throughout this chapter, where TD EM surface effects excited along the surface of a thin, highly contrasting layer are analyzed analytically.

5.1 Pulsed EM line-source excited surface effects

Throughout this section, we shall analyze the pulsed EM response excited on the surface of a thin, highly contrasting layer by an impulsive line source (see Figure 5.1). The spatially localized source is located at a height $h > 0$ above the layer of thickness $\delta \downarrow 0$ and is defined by the y-component of either the electric- or magnetic-current volume density. Consequently, the line sources can be in the s-domain defined via (4.32) and (4.57), respectively.

The EM properties of the thin layer are, in general, defined by scalar parameters $\{\epsilon, \mu, \sigma, \Gamma\}$, again. It is next assumed that at least one of these material parameters shows a relatively high contrast with respect to the ones of the surrounding medium $\mathcal{D}^{\infty}$. The pulsed EM response is observed just above the thin layer at a

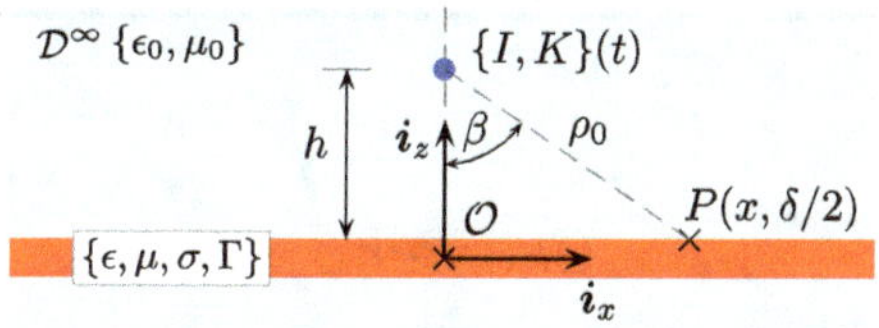

Figure 5.1 Impulsive line source in the presence of a thin highly contrasting sheet

distance $\rho_0 = (x^2 + h^2)^{1/2} > 0$ from the point source. In accordance with Section 4.2, the electric- and magnetic-line sources excite y-independent EM fields with the non-vanishing E- and H-polarized field components, $\{E_y, H_x, H_z\}(x, z, t)$ and $\{H_y, E_x, E_z\}(x, z, t)$, respectively. The two types of excitations will next be studied separately in the ensuing sections.

5.1.1 Electric-line source induced surface effects

The pulsed EM field excited by an impulsive electric-line source in the presence of a metasurface has been fully described in Section 4.2.1 via TD analytical expressions (4.53) and (4.54). In this subsection, we shall discuss some special and limiting cases of (4.53) with regard to the excitation of surface effects. The electric-current line-source excited EM transients excited on the surface of a thin, highly contrasting layer with dielectric and magnetic properties have been studied before in [10].

5.1.1.1 Layer with dielectric and conductive properties

For a thin layer with relatively high contrasts in both electric permittivity ϵ and conductivity σ, we may substitute $\hat{Y}^{\mathrm{E}}(s) = G^{\mathrm{E}} + sC^{\mathrm{E}}$ and $\hat{Z}^{\mathrm{M}}(s) = 0$ in (4.41) and then use (4.43) to express the total electric-field strength at the interface

$$\hat{E}_y(x,0,s) = -\frac{s\mu_0\hat{I}(s)}{2\pi\mathrm{i}}\int_{\kappa=-\mathrm{i}\infty}^{\mathrm{i}\infty}\exp\{-s[\kappa x + p(\kappa)h]\}$$
$$\times\frac{\Psi_\perp(\kappa)}{s + G^{\mathrm{E}}/C^{\mathrm{E}} + \Psi_\perp(\kappa)}\frac{\mathrm{d}\kappa}{2p(\kappa)}$$
$$\text{for all } x\in\mathbb{R} \text{ and } \{s\in\mathbb{R}; s>0\}, \tag{5.1}$$

where $\Psi_\perp(\kappa) = 2Y_\perp(\kappa)/C^{\mathrm{E}}$ (see (4.51)). The transformation of (5.1) to the TD is next carried out with the aid of the CdH technique along the lines followed in Section 4.2.1. Hence, by virtue of Cauchy's theorem and Jordan's lemma, the integration contour in representation (5.1) is deformed into the CdH path that is defined by (cf. (4.45))

$$\kappa x + p(\kappa)h = \tau, \tag{5.2}$$

where τ is the real-valued and positive (time) parameter. Solving (5.2) for the horizontal slowness parameter κ, we arrive at the CdH path $\mathcal{C} \cup \mathcal{C}^*$, where (cf. (4.46))

$$\mathcal{C} = \left\{ \kappa(\tau) = (x/\rho_0^2)\tau + \mathrm{i}(h/\rho_0^2)(\tau^2 - T_0^2)^{1/2} \right\}$$
$$\text{for all } \tau \geq T_0 = \rho_0/c_0. \tag{5.3}$$

During the deformation process, it is noted that the integrand of (5.1) has no pole singularities for $\{s \in \mathbb{R}; s > 0\}$ and $\mathrm{Re}(p) \geq 0$, which implies the absence of true (causal) surface EM waves. As will become clear soon, however, this property does not exclude the excitation of a remarkable surface EM phenomenon. Consequently, upon combining the integrations along $\mathcal{C}$ and its complex conjugate $\mathcal{C}^*$, while introducing τ as the new variable of integration, we arrive at

$$\hat{E}_y(x,0,s) = -\frac{s\mu_0 \hat{I}(s)}{2\pi} \int_{\tau=T_0}^{\infty} \exp(-s\tau)$$
$$\times \frac{\phi_\perp(\tau)[s + G^{\mathrm{E}}/C^{\mathrm{E}} + \phi_\perp(\tau)] + \psi_\perp^2(\tau)}{[s + G^{\mathrm{E}}/C^{\mathrm{E}} + \phi_\perp(\tau)]^2 + \psi_\perp^2(\tau)} \frac{\mathrm{d}\tau}{(\tau^2 - T_0^2)^{1/2}}$$
$$\text{for all } x \in \mathbb{R} \text{ and } \{s \in \mathbb{R}; s > 0\}, \tag{5.4}$$

where, for brevity, we used

$$\Psi_\perp[\kappa(\tau)] = \phi_\perp(\tau) - \mathrm{i}\psi_\perp(\tau), \tag{5.5}$$
$$\phi_\perp(\tau) = (2\eta_0/C^{\mathrm{E}})hc_0\tau/\rho_0^2, \tag{5.6}$$
$$\psi_\perp(\tau) = (2\eta_0/C^{\mathrm{E}})x(c_0^2\tau^2 - \rho_0^2)^{1/2}/\rho_0^2. \tag{5.7}$$

In the final step, we shall employ the Schouten–Van der Pol theorem [11, p. 1056] to transform (5.4) to the TD. This way leads to (cf. (4.53))

$$E_y(x,0,t) = -\partial_t I(t) *_t \frac{\mu_0}{2\pi} \int_{\tau=T_0}^{t} \{\phi_\perp(\tau)\cos[\psi_\perp(\tau)(t-\tau)]$$
$$+ \psi_\perp(\tau)\sin[\psi_\perp(\tau)(t-\tau)]\} \exp\left\{-[\phi_\perp(\tau) + G^{\mathrm{E}}/C^{\mathrm{E}}](t-\tau)\right\} \frac{\mathrm{d}\tau}{(\tau^2 - T_0^2)^{1/2}}$$
$$\text{for all } x \in \mathbb{R} \text{ and } t > T_0. \tag{5.8}$$

As anticipated in [12], interesting surface effects may occur when the line source is located close to the layer. Indeed, under the limit $h \downarrow 0$, we get $\phi_\perp(\tau) \downarrow 0$, so that the first term under the integral sign in (5.8) can be neglected and we end up with

$$\lim_{h \downarrow 0} E_y(x,0,t) = -\partial_t I(t) *_t \frac{\epsilon_0}{C^{\mathrm{E}}} \frac{1}{\eta_0 \pi} \int_{v=1}^{\frac{c_0 t}{|x|}} \sin\left[\frac{2\epsilon_0}{C^{\mathrm{E}}}|x| \left(v^2 - 1\right)^{1/2} \left(\frac{c_0 t}{|x|} - v\right)\right]$$
$$\times \exp\left[-\frac{G^{\mathrm{E}}}{\eta_0} \frac{\epsilon_0}{C^{\mathrm{E}}}|x| \left(\frac{c_0 t}{|x|} - v\right)\right] \mathrm{d}v$$
$$\text{for all } x \in \mathbb{R} \text{ and } t > |x|/c_0. \tag{5.9}$$

A heuristic approximation of the integral with respect to v can be found upon replacing the argument of the sine function with a parabola and then using formula [13, Eq. (7.4.39)] to carry out the thus obtained integral analytically. In this manner, we arrive at the following closed-form TD expression:

$$\begin{aligned}\lim_{h\downarrow 0} E_y(x,0,t) \simeq &- \partial_t I(t) *_t \frac{\epsilon_0}{C^{\mathrm{E}}} \frac{1}{\eta_0 \pi} \left(\frac{2\pi}{\beta(t)}\right)^{1/2} \\ &\left\{ \mathrm{C}\left[\left(\frac{\beta(t)}{2\pi}\right)^{1/2}\left(\frac{c_0 t}{|x|}-1\right)\right] \sin\left[\frac{\beta(t)}{4}\left(\frac{c_0 t}{|x|}-1\right)^2\right] \right. \\ &\left. - \mathrm{S}\left[\left(\frac{\beta(t)}{2\pi}\right)^{1/2}\left(\frac{c_0 t}{|x|}-1\right)\right] \cos\left[\frac{\beta(t)}{4}\left(\frac{c_0 t}{|x|}-1\right)^2\right] \right\} \\ &\times \exp\left(-\frac{G^{\mathrm{E}}}{2\eta_0}\frac{\epsilon_0}{C^{\mathrm{E}}}|x|\frac{c_0 t}{|x|}\right) \mathrm{H}(c_0 t - |x|), \end{aligned} \tag{5.10}$$

where $\mathrm{C}(x)$ and $\mathrm{S}(x)$ denote Fresnel's integrals [13, Eqs. (7.3.1) and (7.3.2)] and

$$\beta(t) = \frac{2\epsilon_0}{C^{\mathrm{E}}}|x| \left[\frac{(c_0 t/|x|+1)^2+1}{(c_0 t/|x|-1)^2+1}\right]^{1/2}. \tag{5.11}$$

To shed some light on the behavior of the oscillatory integral, the argument of the sine function from (5.8) is shown in Figure 5.2 for $c_0 t/\rho_0 = 3.0$. As can be seen, the argument is a concave function of $c_0\tau/\rho_0$ that attains the zero value at the end points of the integration interval. Its maximum is found to be at

$$c_0\tau = c_0\tau_0 = [c_0 t + (c_0^2 t^2 + 8)^{1/2}]/4, \tag{5.12}$$

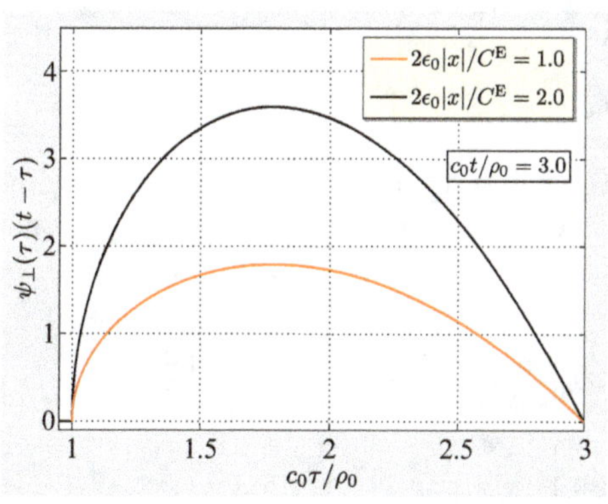

Figure 5.2 The argument of the sine term in (5.8)

where its value is

$$\psi_\perp(\tau_0)(t-\tau_0) = \frac{2\epsilon_0 x}{C^{\mathrm{E}}}\left\{\left[\frac{c_0 t + (c_0^2 t^2 + 8)^{1/2}}{4\rho_0}\right]^2 - 1\right\}^{1/2} \times \left[\frac{3c_0 t - (c_0^2 t^2 + 8)^{1/2}}{4\rho_0}\right]. \tag{5.13}$$

As a matter of fact, (5.12) represents a monotonically increasing function of time. Consequently, the oscillations get faster (along the normalized time axis $t/T_0 = c_0 t/\rho_0 \to c_0 t/|x|$) as both time and $\epsilon_0|x|/C^{\mathrm{E}}$ increase. Simultaneously, the TD response is exponentially attenuated with the decay constant proportional to the factor $(G^{\mathrm{E}}/\eta_0)\epsilon_0|x|/C^{\mathrm{E}}$.

5.1.1.2 Layer with magnetic properties

For a thin sheet with relatively high contrasts in both magnetic permeability μ and parameter Γ, we let $\hat{Y}^{\mathrm{E}}(s) = 0$ and $\hat{Z}^{\mathrm{M}}(s) = R^{\mathrm{M}} + sL^{\mathrm{M}}$ in (4.41) and then use (4.43) to express the electric-field response just above the surface as $z \downarrow \delta/2$, i.e.

$$\hat{E}_y(x, 0^+, s) = 2\hat{E}_y^{\mathrm{i}}(x, 0, s) + \frac{s\mu_0 \hat{I}(s)}{2\pi\mathrm{i}} \int_{\kappa=-\mathrm{i}\infty}^{\mathrm{i}\infty} \exp\{-s[\kappa x + p(\kappa)h]\} \times \frac{\Omega_\perp(\kappa)}{s + R^{\mathrm{M}}/L^{\mathrm{M}} + \Omega_\perp(\kappa)} \frac{\mathrm{d}\kappa}{2p(\kappa)}$$
$$\text{for all } x \in \mathbb{R} \text{ and } \{s \in \mathbb{R}; s > 0\}, \tag{5.14}$$

where $\hat{E}_y^{\mathrm{i}}(x, z, s)$ denotes the incident cylindrical wave corresponding to the first term on the right-hand side of (4.47) and $\Omega_\perp(\kappa) = 2/L^{\mathrm{M}} Y_\perp(\kappa)$ (see (4.52)). The inverse transform of (5.14) to the TD can be accomplished using the CdH approach as applied in Section 5.1.1.1. During the deformation of the integration contour to the CdH path as defined via (5.2), it is noted that the integrand in (5.14) has no poles in the complex κ-plane. This implies, again, the absence of causal surface waves. Hence, upon replacing the variable of integration with the real-valued and positive time parameter τ, we get

$$\hat{E}_y(x, 0^+, s) = 2\hat{E}_y^{\mathrm{i}}(x, 0, s) + \frac{s\mu_0 \hat{I}(s)}{2\pi} \int_{\tau=T_0}^{\infty} \exp(-s\tau) \times \frac{\omega_\perp(\tau)[s + R^{\mathrm{M}}/L^{\mathrm{M}} + \omega_\perp(\tau)] + \chi_\perp^2(\tau)}{[s + R^{\mathrm{M}}/L^{\mathrm{M}} + \omega_\perp(\tau)]^2 + \chi_\perp^2(\tau)} \frac{\mathrm{d}\tau}{(\tau^2 - T_0^2)^{1/2}}$$
$$\text{for all } x \in \mathbb{R} \text{ and } \{s \in \mathbb{R}; s > 0\}, \tag{5.15}$$

where we used (cf. (5.5)–(5.7))

$$\Omega_\perp[\kappa(\tau)] = \omega_\perp(\tau) + \mathrm{i}\chi_\perp(\tau), \tag{5.16}$$

$$\omega_\perp(\tau) = (2/L^{\mathrm{M}}\eta_0)hc_0\tau/(c_0^2\tau^2 - x^2), \tag{5.17}$$

$$\chi_\perp(\tau) = (2/L^{\mathrm{M}}\eta_0)x(c_0^2\tau^2 - \rho_0^2)^{1/2}/(c_0^2\tau^2 - x^2). \tag{5.18}$$

Again, relying on the Schouten–Van der Pol theorem [11, p. 1056], (5.15) can be transformed to the TD and we get (cf. (5.8))

$$E_y(x,0^+,t) = 2E_y^{\mathrm{i}}(x,0,t) + \partial_t I(t) *_t \frac{\mu_0}{2\pi}\int_{\tau=T_0}^{t} \{\omega_\perp(\tau)\cos[\chi_\perp(\tau)(t-\tau)]$$
$$+\chi_\perp(\tau)\sin[\chi_\perp(\tau)(t-\tau)]\}\exp\{-[\omega_\perp(\tau)+R^{\mathrm{M}}/L^{\mathrm{M}}](t-\tau)\}\frac{\mathrm{d}\tau}{(\tau^2-T_0^2)^{1/2}}$$
$$\text{for all } x\in\mathbb{R} \text{ and } t>T_0, \tag{5.19}$$

where $E_y^{\mathrm{i}}(x,0,t)$ can be associated with the first term on the right-hand side of (4.53), that is,

$$E_y^{\mathrm{i}}(x,0,t) = -\partial_t I(t) *_t \frac{\mu_0}{2\pi}\frac{\mathrm{H}(t-T_0)}{(t^2-T_0^2)^{1/2}}. \tag{5.20}$$

Now, under the limit $h\downarrow 0$, we may use $\omega_\perp(\tau)\downarrow 0$ in (5.19) to neglect the first part of the integrand. The argument of the remaining sine function is shown in Figure 5.3 along the normalized axis $\tau/T_0 = c_0\tau/\rho_0$ for the normalized time $c_0 t/\rho_0 = 3.0$ and $h = |x|/4$. As can be seen, the argument is, again, zero at the end points of the range of integration (cf. Figure 5.2). The function attains its maximum shortly beyond the lower end point $c_0\tau/\rho_0 = 1$, namely, at $c_0\tau/\rho_0 = c_0\tau_0/\rho_0 > 1$. The latter can be determined with the aid of Cardano's formula. Further, the function decays rapidly to the zero value at the upper end point $c_0\tau/\rho_0 = c_0 t/\rho_0$. In contrast to (5.12), the position of the maximum is for small values of $h/|x|$ virtually independent of time. Its value grows with time, again, and with the increasing factor $\mu_0|x|/L^{\mathrm{M}}$. In fact, the function becomes unbounded in the limit $h\downarrow 0$. Overall, since the argument of the sine function is in $c_0\tau/\rho_0 > c_0\tau_0/\rho_0$ a rapidly decaying function of $c_0\tau/\rho_0$ with a pronounced maximum located at $c_0\tau/\rho_0 = c_0\tau_0/\rho_0 \downarrow 1$, we conclude that the E-polarized TD field excited just above the surface of a magnetic layer does not show an oscillatory behavior.

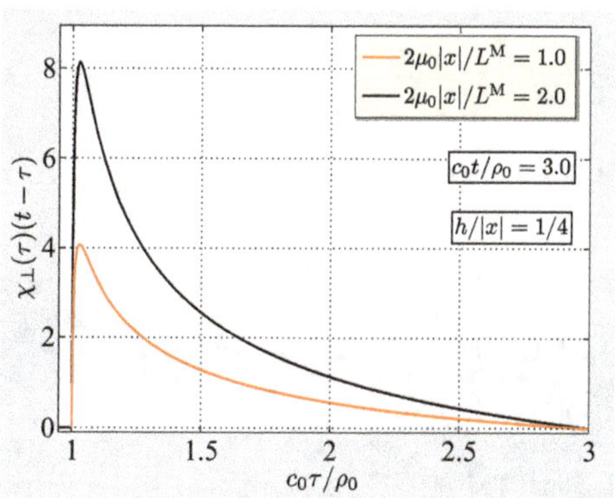

Figure 5.3 The argument of the sine term in (5.19)

5.1.1.3 Plasmonic layer

If the electric-line source is located above a thin layer with the Drude-type plasmonic behavior of its electric conductivity, we substitute $\hat{Y}^{\rm E}(s) = \hat{G}^{\rm E}(s) = \delta\hat{\sigma}(s)$ and $\hat{Z}^{\rm M}(s) = 0$ in (4.41), where the pertaining conduction relaxation function in the s-domain is given by [11, Section 19.5]

$$\hat{\sigma}(s) = \epsilon_0\omega_{\rm pe}^2/(s+\nu_{\rm c}), \tag{5.21}$$

where $\omega_{\rm pe} = (n_{\rm e}{\rm e}^2/\epsilon_0 m_{\rm e})^{1/2}$ is the electron plasma angular frequency, $\nu_{\rm c}$ denotes the collision frequency, $n_{\rm e}$ denotes the number density of conduction electrons, $-{\rm e}$ is the electron charge and $m_{\rm e}$ denotes the electron mass. Upon substituting the thus obtained (transform-domain) reflection coefficient in (4.43), we obtain the following expression for the s-domain electric-field strength just above the layer (cf. (5.14)):

$$\begin{aligned}\hat{E}_y(x,0,s) = \hat{E}_y^{\rm i}(x,0,s) + \frac{s\mu_0\hat{I}(s)}{2\pi{\rm i}}\int_{\kappa=-{\rm i}\infty}^{{\rm i}\infty} \exp\{-s[\kappa x + p(\kappa)h]\} \\ \times \frac{\Lambda_\perp(\kappa)}{s+\nu_{\rm c}+\Lambda_\perp(\kappa)}\frac{{\rm d}\kappa}{2p(\kappa)} \quad \text{for all } x\in\mathbb{R} \text{ and } \{s\in\mathbb{R}; s>0\},\end{aligned} \tag{5.22}$$

where $\hat{E}_y^{\rm i}(x,z,s)$ denotes the incident cylindrical wave, again, and the second (integral) expression represents the reflected wave with $\Lambda_\perp(\kappa) = \delta\epsilon_0\omega_{\rm pe}^2/2Y_\perp(\kappa)$. The transformation of (5.22) back to the TD can be carried out with the aid of the CdH technique, again (see Section 5.1.1.1). In this process, it is noted that the integrand in (5.22) has no poles in the complex κ-plane, which is a characteristic for the absence of true surface waves, again. Hence, after mapping the variable of integration from κ to the (real-valued and positive) time parameter τ, we arrive at (cf. (5.15))

$$\begin{aligned}\hat{E}_y(x,0,s) = \hat{E}_y^{\rm i}(x,0,s) + \frac{s\mu_0\hat{I}(s)}{2\pi}\int_{\tau=T_0}^{\infty} \exp(-s\tau) \\ \times \frac{\lambda_\perp(\tau)[s+\nu_{\rm c}+\lambda_\perp(\tau)]+\xi_\perp^2(\tau)}{[s+\nu_{\rm c}+\lambda_\perp(\tau)]^2+\xi_\perp^2(\tau)}\frac{{\rm d}\tau}{(\tau^2-T_0^2)^{1/2}} \\ \text{for all } x\in\mathbb{R} \text{ and } \{s\in\mathbb{R}; s>0\},\end{aligned} \tag{5.23}$$

where we have taken the values along $\mathcal{C}$ applying to all $\tau \geq T_0 = \rho_0/c_0$ (see (5.3)), that is,

$$\Lambda_\perp[\kappa(\tau)] = \lambda_\perp(\tau) + {\rm i}\xi_\perp(\tau), \tag{5.24}$$

$$\lambda_\perp(\tau) = (\delta\omega_{\rm pe}^2/2c_0)hc_0\tau/(c_0^2\tau^2 - x^2), \tag{5.25}$$

$$\xi_\perp(\tau) = (\delta\omega_{\rm pe}^2/2c_0)x(c_0^2\tau^2-\rho_0^2)^{1/2}/(c_0^2\tau^2-x^2). \tag{5.26}$$

Finally, with the aid of the Schouten–Van der Pol theorem [11, p. 1056], (5.23) can be transformed to the TD. We then obtain

$$
\begin{aligned}
E_y(x,0,t) = E_y^{\mathrm{i}}(x,0,t) + \partial_t I(t) *_t \frac{\mu_0}{2\pi} \int_{\tau=T_0}^{t} \{\lambda_\perp(\tau)\cos[\xi_\perp(\tau)(t-\tau)] \\
+ \xi_\perp(\tau)\sin[\xi_\perp(\tau)(t-\tau)]\} \exp\{-[\lambda_\perp(\tau)+\nu_{\mathrm{c}}](t-\tau)\} \frac{\mathrm{d}\tau}{(\tau^2 - T_0^2)^{1/2}} \\
\text{for all } x \in \mathbb{R} \text{ and } t > T_0,
\end{aligned}
\tag{5.27}
$$

and $E_y^{\mathrm{i}}(x,0,t)$ is given by (5.20). Upon inspection of the resulting TD expression (5.27) with (5.19), it is concluded that the electric-line source excited TD response on the surface of a thin plasmonic layer bears similarities to one just above a magnetic layer. Consequently, based on the analysis presented in Section 5.1.1.2, it may be concluded that the E-polarized TD field excited on the surface of a plasmonic thin layer does not show an oscillatory behavior. This conclusion is in agreement with the results presented before [14, 15].

5.1.2 Magnetic-line source induced surface effects

The pulsed EM field excited by an impulsive magnetic-line source in the presence of a metasurface has been described analytically in Section 4.2.2 (see (4.64) and (4.65)). In this section, we shall follow similar lines of reasoning as in Section 5.1.1 to analyze selected special cases of (4.64) with particular attention to the excitation of surface effects.

5.1.2.1 Layer with dielectric and conductive properties

The H-polarized EM field response induced just above a thin, highly contrasting layer with dielectric and conductive properties bears similarities with the E-polarized EM-field response above a magnetic layer (see Section 5.1.1.2). Indeed, substituting $\hat{Y}^{\mathrm{E}}(s) = G^{\mathrm{E}} + sC^{\mathrm{E}}$ and $\hat{Z}^{\mathrm{M}}(s) = 0$ in (4.63), we get the expression for the pertaining transform-domain reflection coefficient that is subsequently used to express the magnetic-field s-domain response just above the sheet as $z \downarrow \delta/2$, that is (cf. (5.14)),

$$
\begin{aligned}
\hat{H}_y(x,0^+,s) = 2\hat{H}_y^{\mathrm{i}}(x,0,s) + \frac{s\epsilon_0 \hat{K}(s)}{2\pi\mathrm{i}} \int_{\kappa=-\mathrm{i}\infty}^{\mathrm{i}\infty} \exp\{-s[\kappa x + p(\kappa)h]\} \\
\times \frac{\Psi_{||}(\kappa)}{s + G^{\mathrm{E}}/C^{\mathrm{E}} + \Psi_{||}(\kappa)} \frac{\mathrm{d}\kappa}{2p(\kappa)} \\
\text{for all } x \in \mathbb{R} \text{ and } \{s \in \mathbb{R}; s > 0\},
\end{aligned}
\tag{5.28}
$$

where $\hat{H}_y^{\mathrm{i}}(x,z,s)$ denotes the incident cylindrical wave and $\Psi_{\|}(\kappa) = 2Y_{\|}(\kappa)/C^{\mathrm{E}}$ (see (4.66)). Following next the CdH strategy that led to (5.19), the TD counterpart of (5.28) can be written in the following form:

$$H_y(x,0^+,t) = 2H_y^{\mathrm{i}}(x,0,t) + \partial_t K(t) *_t \frac{\epsilon_0}{2\pi}\int_{\tau=T_0}^{t}\{\phi_{\|}(\tau)\cos[\psi_{\|}(\tau)(t-\tau)]$$
$$+\psi_{\|}(\tau)\sin[\psi_{\|}(\tau)(t-\tau)]\}\exp\{-[\phi_{\|}(\tau)+G^{\mathrm{E}}/C^{\mathrm{E}}](t-\tau)\}\frac{\mathrm{d}\tau}{(\tau^2-T_0^2)^{1/2}}$$
$$\text{for all } x\in\mathbb{R} \text{ and } t>T_0, \tag{5.29}$$

with $T_0 = \rho_0/c_0$ and $H_y^{\mathrm{i}}(x,0,t)$ can be associated with the first term on the right-hand side of (4.64), that is,

$$H_y^{\mathrm{i}}(x,0,t) = -\partial_t K(t) *_t \frac{\epsilon_0}{2\pi}\frac{\mathrm{H}(t-T_0)}{(t^2-T_0^2)^{1/2}}, \tag{5.30}$$

and (cf. (5.16)–(5.18))

$$\Psi_{\|}[\kappa(\tau)] = \phi_{\|}(\tau) + \mathrm{i}\psi_{\|}(\tau), \tag{5.31}$$
$$\phi_{\|}(\tau) = (2\eta_0/C^{\mathrm{E}})hc_0\tau/(c_0^2\tau^2 - x^2), \tag{5.32}$$
$$\psi_{\|}(\tau) = (2\eta_0/C^{\mathrm{E}})x(c_0^2\tau^2-\rho_0^2)^{1/2}/(c_0^2\tau^2-x^2). \tag{5.33}$$

Upon noting the similarity of the derived TD expression (5.29) with (5.19), we conclude that the H-polarized TD field excited just above the surface of a thin, highly contrasting layer with dielectric and conductive properties does not exhibit a strongly oscillatory behavior, again. Note that the same conclusion has been drawn for the E-polarized TD response on the surface of a plasmonic sheet (see Section 5.1.1.3).

5.1.2.2 Layer with magnetic properties

In this section, it is demonstrated that the H-polarized EM-field response induced on the surface of a thin, highly contrasting layer with magnetic properties is similar to the E-polarized EM-field response above a lossy dielectric layer (see Section 5.1.1.1). To that end, we substitute $\hat{Y}^{\mathrm{E}}(s)=0$ and $\hat{Z}^{\mathrm{M}}(s) = R^{\mathrm{M}} + sL^{\mathrm{M}}$ in (4.63) and use the result to express the total magnetic-field strength at the interface (cf. (5.1))

$$\hat{H}_y(x,0,s) = -\frac{s\epsilon_0\hat{K}(s)}{2\pi\mathrm{i}}\int_{\kappa=-\mathrm{i}\infty}^{\mathrm{i}\infty}\exp\{-s[\kappa x + p(\kappa)h]\}$$
$$\times\frac{\Omega_{\|}(\kappa)}{s+R^{\mathrm{M}}/L^{\mathrm{M}}+\Omega_{\|}(\kappa)}\frac{\mathrm{d}\kappa}{2p(\kappa)}$$
$$\text{for all } x\in\mathbb{R} \text{ and } \{s\in\mathbb{R}; s>0\}, \tag{5.34}$$

where $\Omega_{\|}(\kappa) = 2/L^{\rm M}Y_{\|}(\kappa)$ (see (4.67)). Further, using the CdH technique that was applied to achieve the inverse of (5.1), the TD original of (5.34) is found in the following form (cf. (5.8)):

$$H_y(x,0,t) = -\partial_t K(t) *_t \frac{\epsilon_0}{2\pi}\int_{\tau=T_0}^{t} \{\omega_{\|}(\tau)\cos[\chi_{\|}(\tau)(t-\tau)]$$

$$+\chi_{\|}(\tau)\sin[\chi_{\|}(\tau)(t-\tau)]\}\exp\{-[\omega_{\|}(\tau)+R^{\rm M}/L^{\rm M}](t-\tau)\}\frac{{\rm d}\tau}{(\tau^2-T_0^2)^{1/2}}$$

$$\text{for all } x\in\mathbb{R} \text{ and } t>T_0, \tag{5.35}$$

in which (cf. (5.5)–(5.7))

$$\Omega_{\|}[\kappa(\tau)] = \omega_{\|}(\tau) - {\rm i}\chi_{\|}(\tau), \tag{5.36}$$

$$\omega_{\|}(\tau) = (2/L^{\rm M}\eta_0)hc_0\tau/\rho_0^2, \tag{5.37}$$

$$\chi_{\|}(\tau) = (2/L^{\rm M}\eta_0)x(c_0^2\tau^2-\rho_0^2)^{1/2}/\rho_0^2. \tag{5.38}$$

Now, if the magnetic-line source is placed just above the surface (cf. (5.9)), then (5.35) boils down to

$$\lim_{h\downarrow 0} H_y(x,0,t) = -\partial_t K(t) *_t \frac{\mu_0}{L^{\rm M}}\frac{\eta_0}{\pi}\int_{v=1}^{\frac{c_0t}{|x|}} \sin\left[\frac{2\mu_0}{L^{\rm M}}|x|\left(v^2-1\right)^{1/2}\left(\frac{c_0t}{|x|}-v\right)\right]$$

$$\times\exp\left[-\eta_0R^{\rm M}\frac{\mu_0}{L^{\rm M}}|x|\left(\frac{c_0t}{|x|}-v\right)\right]{\rm d}v$$

$$\text{for all } x\in\mathbb{R} \text{ and } t>|x|/c_0, \tag{5.39}$$

which, in line with the analysis provided in Section 5.1.1.1, represents an oscillatory TD response. Apparently, the frequency of oscillations (along the normalized time axis $t/T_0 = c_0t/\rho_0 \to c_0t/|x|$) grows as time and the factor $\mu_0|x|/L^{\rm M}$ increase. Furthermore, the amplitude of the TD response is exponentially attenuated with the decay constant proportional to the ratio $\eta_0R^{\rm M}\mu_0|x|/L^{\rm M}$. Again, expression (5.39) can be approximated in terms of Fresnel's integrals (cf. (5.10))

$$\lim_{h\downarrow 0} H_y(x,0,t) \simeq -\partial_t K(t) *_t \frac{\mu_0}{L^{\rm M}}\frac{\eta_0}{\pi}\left(\frac{2\pi}{\beta(t)}\right)^{1/2}$$

$$\left\{{\rm C}\left[\left(\frac{\beta(t)}{2\pi}\right)^{1/2}\left(\frac{c_0t}{|x|}-1\right)\right]\sin\left[\frac{\beta(t)}{4}\left(\frac{c_0t}{|x|}-1\right)^2\right]\right.$$

$$\left.-\,{\rm S}\left[\left(\frac{\beta(t)}{2\pi}\right)^{1/2}\left(\frac{c_0t}{|x|}-1\right)\right]\cos\left[\frac{\beta(t)}{4}\left(\frac{c_0t}{|x|}-1\right)^2\right]\right\}$$

$$\times\exp\left(-\frac{\eta_0R^{\rm M}}{2}\frac{\mu_0}{L^{\rm M}}|x|\frac{c_0t}{|x|}\right){\rm H}(c_0t-|x|), \tag{5.40}$$

where we use (cf. (5.11))

$$\beta(t) = \frac{2\mu_0}{L^{\rm M}}|x|\left[\frac{(c_0t/|x|+1)^2+1}{(c_0t/|x|-1)^2+1}\right]^{1/2}, \tag{5.41}$$

to evaluate the TD formula.

5.1.2.3 Plasmonic layer

In this section, we shall analyze the TD response due to the impulsive magnetic-line source in the presence of a thin layer with the Drude-type plasmonic properties. Such a layer can be described by its equivalent layer's admittance and impedance, $\hat{Y}^{\mathrm{E}}(s) = \hat{G}^{\mathrm{E}}(s) = \delta\hat{\sigma}(s)$ (see (5.21)) and $\hat{Z}^{\mathrm{M}}(s) = 0$, that are now substituted in (4.63) to get the pertaining transform-domain reflection coefficient. Consequently, the magnetic-field strength just above the surface, as $z \downarrow \delta/2$, can be expressed as the superposition of the incident and reflected fields (cf. (5.22))

$$\begin{aligned}\hat{H}_y(x,0^+,s) = \hat{H}^{\mathrm{i}}_y(x,0,s) - \frac{s\epsilon_0\hat{K}(s)}{2\pi\mathrm{i}}\int_{\kappa=-\mathrm{i}\infty}^{\mathrm{i}\infty}\exp\{-s[\kappa x + p(\kappa)h]\}\\ \times\frac{\Lambda_{\|}(\kappa)}{s+\nu_{\mathrm{c}}+\Lambda_{\|}(\kappa)}\frac{\mathrm{d}\kappa}{2p(\kappa)}\\ \text{for all } x\in\mathbb{R} \text{ and } \{s\in\mathbb{R}; s>0\},\end{aligned} \tag{5.42}$$

where $\Lambda_{\|}(\kappa) = \delta\epsilon_0\omega_{\mathrm{pe}}^2/2Y_{\|}(\kappa)$. It is noted that for the real-valued and positive Laplace-transform parameter s and for $\mathrm{Re}(\Lambda_{\|}) \geq 0$ and $\nu_{\mathrm{c}} > 0$, the integrand has no poles in the complex κ-plane. This indicates the absence of true (causal) surface waves, again. Consequently, the s-domain expression (5.42) is transformed back to the TD via the CdH technique as described in Section 5.1.1.1. This procedure yields (cf. (5.35))

$$\begin{aligned}H_y(x,0^+,t) = H^{\mathrm{i}}_y(x,0,t) - \partial_t K(t) *_t \frac{\epsilon_0}{2\pi}\int_{\tau=T_0}^{t}\{\lambda_{\|}(\tau)\cos[\xi_{\|}(\tau)(t-\tau)]\\ +\xi_{\|}(\tau)\sin[\xi_{\|}(\tau)(t-\tau)]\}\exp\{-[\lambda_{\|}(\tau)+\nu_{\mathrm{c}}](t-\tau)\}\frac{\mathrm{d}\tau}{(\tau^2-T_0^2)^{1/2}}\\ \text{for all } x\in\mathbb{R} \text{ and } t>T_0,\end{aligned} \tag{5.43}$$

where the incident wave field is given by (5.30) and (cf. (5.24)–(5.26))

$$\Lambda_{\|}[\kappa(\tau)] = \lambda_{\|}(\tau) - \mathrm{i}\xi_{\|}(\tau), \tag{5.44}$$

$$\lambda_{\|}(\tau) = (\delta\omega_{\mathrm{pe}}^2/2c_0)hc_0\tau/\rho_0^2, \tag{5.45}$$

$$\xi_{\|}(\tau) = (\delta\omega_{\mathrm{pe}}^2/2c_0)x(c_0^2\tau^2-\rho_0^2)^{1/2}/\rho_0^2. \tag{5.46}$$

Upon a closer inspection of the final TD result (5.43) with (5.35) from the previous section, it is concluded that the H-polarized response induced on the surface of a highly contrasting thin layer with magnetic properties bears a resemblance to the *reflected*-field response observed just above a plasmonic layer. Consequently, the reflected TD field against the surface may strongly oscillate (see also [14, 15]).

This can be clearly seen from the response induced by the line source located just above the surface. Hence, under the limit $h \downarrow 0$, we may write (cf. (5.39))

$$\lim_{h\downarrow 0} H_y(x,0^+,t) = \lim_{h\downarrow 0} H_y^{\mathrm{i}}(x,0,t) - \partial_t K(t) *_t \frac{\delta\omega_{\mathrm{pe}}^2}{4c_0^2}\frac{\eta_0}{\pi}\int_{v=1}^{\frac{c_0 t}{|x|}} \sin\left[\frac{\delta\omega_{\mathrm{pe}}^2}{2c_0^2}|x|\left(v^2-1\right)^{1/2}\left(\frac{c_0 t}{|x|}-v\right)\right] \times \exp\left[-\frac{\nu_{\mathrm{c}}}{c_0}|x|\left(\frac{c_0 t}{|x|}-v\right)\right]\mathrm{d}v \quad \text{for all } x\in\mathbb{R} \text{ and } t>|x|/c_0. \tag{5.47}$$

Equation (5.47) represents the sum of the H-polarized incident and reflected fields. The reflected wave constituent, interpreting it as a function of the normalized time $c_0 t/|x|$, is an oscillatory TD response whose oscillations get faster as time and the parameter $\delta|x|\omega_{\mathrm{pe}}^2/c_0^2$ increase. Simultaneously, its amplitude is exponentially attenuated with the decay constant, $\nu_{\mathrm{c}}|x|/c_0$, that is proportional to the collision frequency and the pulse travel time. As the reflected field above a plasmonic layer has the form that is similar to the H-polarized response on the surface of a magnetic sheet (see (5.39)), it can be, again, approximated in terms of Fresnel's integrals (cf. (5.40))

$$\begin{aligned}\lim_{h\downarrow 0} H_y(x,0^+,t) \simeq\ & \lim_{h\downarrow 0} H_y^{\mathrm{i}}(x,0,t) - \partial_t K(t) *_t \frac{\delta\omega_{\mathrm{pe}}^2}{4c_0^2}\frac{\eta_0}{\pi}\left(\frac{2\pi}{\beta(t)}\right)^{1/2} \\ & \left\{\mathrm{C}\left[\left(\frac{\beta(t)}{2\pi}\right)^{1/2}\left(\frac{c_0 t}{|x|}-1\right)\right]\sin\left[\frac{\beta(t)}{4}\left(\frac{c_0 t}{|x|}-1\right)^2\right]\right. \\ & \left. -\,\mathrm{S}\left[\left(\frac{\beta(t)}{2\pi}\right)^{1/2}\left(\frac{c_0 t}{|x|}-1\right)\right]\cos\left[\frac{\beta(t)}{4}\left(\frac{c_0 t}{|x|}-1\right)^2\right]\right\} \\ & \times \exp\left(-\frac{\nu_{\mathrm{c}}}{2c_0}|x|\frac{c_0 t}{|x|}\right)\mathrm{H}(c_0 t-|x|),\end{aligned} \tag{5.48}$$

where we now use (cf. (5.41))

$$\beta(t) = \frac{\delta\omega_{\mathrm{pe}}^2}{2c_0^2}|x|\left[\frac{(c_0 t/|x|+1)^2+1}{(c_0 t/|x|-1)^2+1}\right]^{1/2}, \tag{5.49}$$

to evaluate the approximate TD formula.

5.2 Pulsed EM loop-excited surface effects above a layer with dielectric and conductive properties

In the previous sections, we studied the circumstances under which surface EM transients on the surface of a thin, highly contrasting layer can be excited by a two-dimensional EM line source. In this part, we shall employ these results to demonstrate the existence of the surface space–time EM phenomenon by analyzing three-dimensional problem configurations. To that end, we shall employ the

conclusions drawn in Section 5.1.1.1, where we demonstrated that the E-polarized pulsed EM-field response shows the oscillatory behavior on the surface of a thin sheet with dielectric and conductive properties. Since the reflection and transmission of the EM fields radiated from a small horizontal loop antenna bear a similarity with the ones pertaining to the electric-line source (see Section 4.3.1), we next analyze the pulsed EM response activated by a small transmitting loop antenna, $\mathcal{L}^{\mathrm{T}}$, located at $\{0,0,h^{\mathrm{T}}>0\}$ above a thin, highly contrasting layer described by (real-valued, scalar and positive) electric permittivity and conductivity $\{\epsilon,\sigma\}$ (see Figure 5.4). The thus excited EM-field response will be probed by either a small vertical loop (or, equivalently, a horizontal magnetic dipole), say $\mathcal{L}^{\mathrm{R}}$, or by a horizontal short wire (or, equivalently, a horizontal electric dipole), further denoted $\mathcal{W}^{\mathrm{R}}$. The receiving dipole antennas are located at $\{x,y,h^{\mathrm{R}}>0\}$, where $h^{\mathrm{R}}>0$ denotes their height above the surface. Following the solution methodology applied in Section 4.3.1, the nonzero tangential components of the EM field radiated from the horizontal loop are represented via (4.71a) and (4.71b). The pertaining EM fields induced above the layer's surface will be next determined with the aid of the s-domain cross-layer conditions (see (3.9a) and (3.9b))

$$\hat{E}_\phi^+ - \hat{E}_\phi^- = 0, \tag{5.50a}$$

$$\hat{H}_r^+ - \hat{H}_r^- = \hat{Y}^{\mathrm{E}}(s)\hat{E}_\phi(r,0,s), \tag{5.50b}$$

as $\delta \downarrow 0$ for all $r=(x^2+y^2)^{1/2}>0$, where $\hat{Y}^{\mathrm{E}}(s)$ is the equivalent layer admittance (see (3.15b)).

5.2.1 Receiving horizontal magnetic dipole

In the first case, the pulsed EM field excited by $\mathcal{L}^{\mathrm{T}}$ will be probed by a horizontal magnetic dipole antenna represented by a vertically oriented small loop (see Figure 5.4). Consequently, the voltage induced in the receiving loop can be found from (cf. (4.73))

$$\hat{V}^{\mathrm{R}}(s) = -s\mu_0 \mathcal{A}^{\mathrm{R}}\hat{H}_r(r,h^{\mathrm{R}},s), \tag{5.51}$$

where $\mathcal{A}^{\mathrm{R}}$ is the area of the receiving loop and $\hat{H}_r(r,h^{\mathrm{R}},s)$ denotes the radial component of the magnetic-field strength at the location of the receiver $\{r>0,z=h^{\mathrm{R}}>0\}$.

Thanks to the spatial shift invariance of the problem configuration, the problem is solved under the wave-slowness representation (4.75), again. The pertaining transform-domain coefficients, $\tilde{A}$ and $\tilde{R}_\perp$, can be found from the excitation condition (4.78) and the slowness-domain counterparts of jump conditions (5.50a) and (5.50b), respectively. In this way, one arrives at (cf. (5.1))

$$\tilde{R}_\perp(\kappa,\sigma) = -1 + \frac{\tilde{\Psi}_\perp(\kappa,\sigma)}{s + G^{\mathrm{E}}/C^{\mathrm{E}} + \tilde{\Psi}_\perp(\kappa,\sigma)}, \tag{5.52}$$

where $\tilde{\Psi}_\perp(\kappa,\sigma) = 2\tilde{Y}_\perp(\kappa,\sigma)/C^{\mathrm{E}} = 2p(\kappa,\sigma)/\mu_0 C^{\mathrm{E}}$ and $\tilde{A} = 1/2sp(\kappa,\sigma)$, again, and recall that the vertical slowness parameter $p(\kappa,\sigma)$ is defined by (4.77). Upon

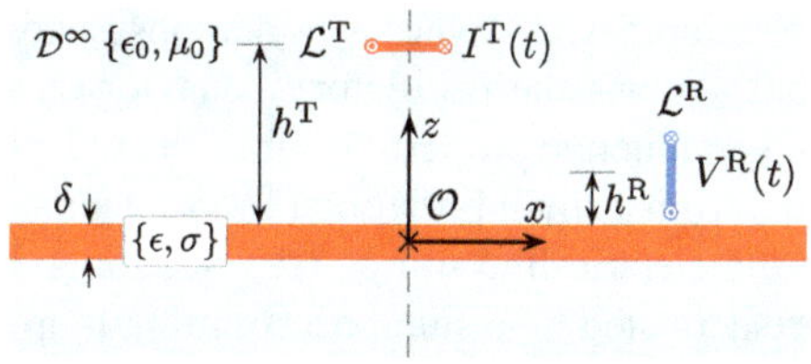

Figure 5.4 Transmitting loop and receiving horizontal magnetic dipole in the presence of a thin highly contrasting sheet

collecting the results, the induced voltage in the receiving loop can be expressed as a sum of two terms

$$\hat{V}^{\rm R}(s) = \hat{V}_0^{\rm R}(s) + \Delta\hat{V}^{\rm R}(s), \tag{5.53}$$

where $\hat{V}_0^{\rm R}(s)$ represents the voltage response that would be excited in the absence of the layer and $\Delta\hat{V}^{\rm R}(s)$ denotes the correction voltage (with respect to the void case) that accounts for the non-vanishing layer's permittivity and conductivity. The two voltage contributions will next be transformed to the TD under the assumption that the receiving loop is located just above the surface, that is, under the limit $h^{\rm R} \downarrow 0$.

5.2.1.1 The voltage in the absence of sheet

The voltage induced by $\mathcal{L}^{\rm T}$ in the receiving loop, $\mathcal{L}^{\rm R}$, in the absence of the layer can be represented via

$$\hat{V}_0^{\rm R}(s) = \frac{s^3\mu_0\hat{I}^{\rm T}(s)\mathcal{A}^{\rm T}\mathcal{A}^{\rm R}}{8\pi^2}\int_{\kappa=-\mathrm{i}\infty}^{\mathrm{i}\infty}\mathrm{d}\kappa$$
$$\int_{\sigma=-\mathrm{i}\infty}^{\mathrm{i}\infty}\exp\{-s[\kappa x + \sigma y + p(\kappa,\sigma)h^{\rm T}]\}\tilde{\partial}_r(\kappa,\sigma)\mathrm{d}\sigma, \tag{5.54}$$

as $h^{\rm R} \downarrow 0$, where $\tilde{\partial}_r(\kappa,\sigma)$ denotes the differentiation with respect to r in the transform domain. It can be verified using the closed-form formulas given in [11, Section 26.10] that (5.54), in fact, represents the simple closed-form expression

$$\hat{V}_0^{\rm R}(s) = \frac{s\mu_0\hat{I}^{\rm T}(s)\mathcal{A}^{\rm T}\mathcal{A}^{\rm R}}{4\pi R_0^3}\frac{rh^{\rm T}}{R_0^2}\left(3 + 3sT + s^2T^2\right)\exp(-sT), \tag{5.55}$$

as $h^{\rm R} \downarrow 0$, where $R_0 = (r^2 + h^{\rm T2})^{1/2} > 0$ denotes the distance between the transmitting and receiving loops and $T = R_0/c_0$ is the corresponding pulse travel time. In the limit $h^{\rm T} \downarrow 0$, the mutually orthogonal loops are located at the same vertical level in vacuum, which leads to the zero voltage response, as expected. The s-domain expression (5.55) can be easily transformed to the TD, which yields at once (cf. (4.93))

$$V_0^{\rm R}(t) = \frac{\mu_0\mathcal{A}^{\rm T}\mathcal{A}^{\rm R}}{4\pi R_0^3}\frac{rh^{\rm T}}{R_0^2}\left[3\partial_t I^{\rm T}(t-T) + 3T\partial_t^2 I^{\rm T}(t-T) + T^2\partial_t^3 I^{\rm T}(t-T)\right], \tag{5.56}$$

as $h^{\mathrm{R}} \downarrow 0$ for $t > 0$. As can be seen from (5.56), the induced voltage response is composed of near-, intermediate- and far-field constituents that are proportional to (distance)$^{-3}$, (distance)$^{-2}$ and (distance)$^{-1}$, respectively. As far as their temporal behavior is concerned, the constituents are related to the first, second and third time derivatives of the excitation electric-current pulse, respectively. As $T \downarrow 0$, we may write $V_0^{\mathrm{R}}(t) = L^{\mathrm{C}}\partial_t I^{\mathrm{T}}(t) + O(T)$, where $L^{\mathrm{C}} = (3\mu_0/4\pi)\mathcal{A}^{\mathrm{T}}\mathcal{A}^{\mathrm{R}} r h^{\mathrm{T}}/R_0^5$ has the meaning of (near-field) coupling inductance.

5.2.1.2 The correction voltage

The difference of the actual voltage response, $\hat{V}^{\mathrm{R}}(s)$, and the one that would be excited in the absence of the layer, $\hat{V}_0^{\mathrm{R}}(s)$, can be represented via (cf. (4.82))

$$\Delta\hat{V}^{\mathrm{R}}(s) = \frac{s^3\mu_0\hat{I}^{\mathrm{T}}(s)\mathcal{A}^{\mathrm{T}}\mathcal{A}^{\mathrm{R}}}{8\pi^2}\int_{\kappa=-\mathrm{i}\infty}^{\mathrm{i}\infty}\mathrm{d}\kappa \int_{\sigma=-\mathrm{i}\infty}^{\mathrm{i}\infty}\exp\{-s[\kappa x + \sigma y + p(\kappa,\sigma)h^{\mathrm{T}}]\}\tilde{\partial}_r(\kappa,\sigma) \times\left[1 - \frac{\tilde{\Psi}_{\perp}(\kappa,\sigma)}{s + G^{\mathrm{E}}/C^{\mathrm{E}} + \tilde{\Psi}_{\perp}(\kappa,\sigma)}\right]\mathrm{d}\sigma, \tag{5.57}$$

as $h^{\mathrm{R}} \downarrow 0$, and recall that $\tilde{\partial}_r(\kappa,\sigma)$ denotes the differentiation with respect to r in the transform domain. The transformation of (5.57) to the TD will be carried out using the standard CdH technique (see Section 2.1.1). We shall assume that the transmitting loop is located just above the layer's surface, that is, we take the limit $h^{\mathrm{T}} \downarrow 0$. With reference to the results of Section 5.2.1.1, this assumption implies that the correction voltage is, in fact, equal to the total voltage response. Consequently, the correction voltage under the limits $h^{\mathrm{T,R}} \downarrow 0$ is the effect that can be attributed merely to the presence of the layer.

In the first step, the variables of integration $\{\kappa,\sigma\}$ are replaced with $\{\gamma, q\}$ according to (cf. (2.19a) and (2.19b))

$$\kappa = \gamma\cos(\phi) - \mathrm{i}q\sin(\phi), \tag{2.78a revisited}$$
$$\sigma = \gamma\sin(\phi) + \mathrm{i}q\cos(\phi), \tag{2.78b revisited}$$

with $x = r\cos(\phi)$ and $y = r\sin(\phi)$ for $r > 0$ and $\{0 \le \phi < 2\pi\}$. Under this substitution, $\kappa x + \sigma y = \gamma r$, $\kappa^2 + \sigma^2 = \gamma^2 - q^2$, $\mathrm{d}\kappa\mathrm{d}\sigma = \mathrm{id}\gamma\mathrm{d}q$, $\tilde{\partial}_r(\kappa,\sigma) = -s\gamma$, which yields

$$\Delta\hat{V}^{\mathrm{R}}(s) = \frac{s^4\mu_0\hat{I}^{\mathrm{T}}(s)\mathcal{A}^{\mathrm{T}}\mathcal{A}^{\mathrm{R}}}{8\pi^2\mathrm{i}}\int_{q=-\infty}^{\infty}\mathrm{d}q \int_{\gamma=-\mathrm{i}\infty}^{\mathrm{i}\infty}\exp(-s\gamma r)\frac{s + G^{\mathrm{E}}/C^{\mathrm{E}}}{s + G^{\mathrm{E}}/C^{\mathrm{E}} + \bar{\Psi}_{\perp}(\gamma,q)}\gamma\mathrm{d}\gamma, \tag{5.58}$$

as $h^{\mathrm{R,T}} \downarrow 0$, where $\bar{\Psi}_\perp(\gamma, q) = 2\bar{p}(\gamma, q)/\mu_0 C^{\mathrm{E}}$ with $\bar{p}(\gamma, q) = [\Omega_0^2(q) - \gamma^2]^{1/2}$, where $\Omega_0(q) = (1/c_0^2 + q^2)^{1/2} > 0$ and $\mathrm{Re}[\bar{p}(\gamma, q)] \geq 0$. In the next step, the integration contour in the complex γ-plane is deformed into the CdH path $\mathcal{C} \cup \mathcal{C}^*$

$$\mathcal{C} = \left\{\gamma(\tau) = \tau/r + \mathrm{i}0\right\} \quad \text{for } \{r\,\Omega_0(q) \leq \tau < \infty\}, \tag{5.59}$$

thus representing the loop encircling the branch cut along $\{\Omega_0(q) \leq \mathrm{Re}(\gamma) < \infty, \mathrm{Im}(\gamma) = 0\}$. The deformation is permissible since the contributions from circular arcs at infinity in $\mathrm{Im}(\gamma) \gtrless 0$ cancel each other; that is, Jordan's lemma does not apply in this case. Consequently, combining the contributions of integration from $\mathcal{C}$ and $\mathcal{C}^*$, while introducing τ as the new variable of integration we arrive at

$$\Delta\hat{V}^{\mathrm{R}}(s) = \frac{s^4 \mu_0 \hat{I}^{\mathrm{T}}(s) \mathcal{A}^{\mathrm{T}} \mathcal{A}^{\mathrm{R}}}{4\pi^2 r^2} \int_{q=-\infty}^{\infty} \mathrm{d}q \int_{\tau = r\Omega_0(q)}^{\infty} \exp(-s\tau) \frac{(s + G^{\mathrm{E}}/C^{\mathrm{E}})\psi_\perp(\tau, q)}{(s + G^{\mathrm{E}}/C^{\mathrm{E}})^2 + \psi_\perp^2(\tau, q)} \tau \mathrm{d}\tau, \tag{5.60}$$

where (cf. (5.5) and (5.7))

$$\bar{\Psi}_\perp[\gamma(\tau), q] = -\mathrm{i}\psi_\perp(\tau, q) = -\mathrm{i}(2\eta_0/C^{\mathrm{E}}) \left[c_0^2\tau^2/r^2 - c_0^2\Omega_0^2(q)\right]^{1/2}. \tag{5.61}$$

Upon interchanging the order of the integrations, we next find that

$$\Delta\hat{V}^{\mathrm{R}}(s) = \frac{s^4 \mu_0 \hat{I}^{\mathrm{T}}(s) \mathcal{A}^{\mathrm{T}} \mathcal{A}^{\mathrm{R}}}{4\pi^2 r^2} \int_{\tau=T}^{\infty} \exp(-s\tau)\tau \mathrm{d}\tau \int_{q=-Q(\tau)}^{Q(\tau)} \frac{(s + G^{\mathrm{E}}/C^{\mathrm{E}})\psi_\perp(\tau, q)}{(s + G^{\mathrm{E}}/C^{\mathrm{E}})^2 + \psi_\perp^2(\tau, q)} \mathrm{d}q, \tag{5.62}$$

where $Q(\tau) = (\tau^2/r^2 - 1/c_0^2)^{1/2}$ and $T = r/c_0$ denotes the corresponding pulse travel time. Thanks to $\psi_\perp(\tau, q) = (2/\mu_0 C^{\mathrm{E}})[Q^2(\tau) - q^2]^{1/2}$, the inner integral with respect to q can be carried out analytically, which after a few straightforward steps leads to

$$\Delta\hat{V}^{\mathrm{R}}(s) = \frac{s^4 \mu_0 \hat{I}^{\mathrm{T}}(s) \mathcal{A}^{\mathrm{T}} \mathcal{A}^{\mathrm{R}}}{8\pi r^2 c_0^2} \frac{C^{\mathrm{E}}}{\epsilon_0}(s + G^{\mathrm{E}}/C^{\mathrm{E}}) \int_{\tau=T}^{\infty} \exp(-s\tau) \left\{1 - \frac{s + G^{\mathrm{E}}/C^{\mathrm{E}}}{[(s + G^{\mathrm{E}}/C^{\mathrm{E}})^2 + (2/\mu_0 C^{\mathrm{E}})^2 Q^2(\tau)]^{1/2}}\right\} \tau \mathrm{d}\tau \quad \text{for } \{s \in \mathbb{R}; s > 0\}. \tag{5.63}$$

Expression (5.63) has the desired form that is amenable to the unique inverse transform with the aid of the Schouten–Van der Pol theorem [11, p. 1056]. Hence, making use of [13, (29.3.56)], we finally end up with (cf. (5.9))

$$\Delta V^{\mathrm{R}}(t) = \frac{A^{\mathrm{T}} A^{\mathrm{R}}}{2\pi r^4} \frac{T^4 \partial_t^4 I^{\mathrm{T}}(t)}{C^{\mathrm{E}}} *_t \int_{v=1}^{t/T} \left\{ \mathrm{J}_0 \left[(2\epsilon_0/C^{\mathrm{E}}) r (v^2-1)^{1/2} (t/T - v) \right] \right.$$
$$\left. - \frac{\mathrm{J}_1 \left[(2\epsilon_0/C^{\mathrm{E}}) r (v^2-1)^{1/2} (t/T - v) \right]}{(2\epsilon_0/C^{\mathrm{E}}) r (v^2-1)^{1/2} (t/T - v)} \right\}$$
$$\times \exp\left[-\frac{G^{\mathrm{E}}}{\eta_0} \frac{\epsilon_0}{C^{\mathrm{E}}} r \left(\frac{t}{T} - v \right) \right] (v^2 - 1) v \mathrm{d}v \quad \text{for } t > T = r/c_0, \quad (5.64)$$

as $h^{\mathrm{R,T}} \downarrow 0$, where $\mathrm{J}_{0,1}(x)$ are the Bessel functions of the first kind [13, Section 9]. Owing to the fact that $V_0^{\mathrm{R}}(t) = 0$ as $h^{\mathrm{R}} = h^{\mathrm{T}}$, the final TD expression (5.64) represents the *total* induced voltage in the presence of the screen provided that both transmitting and receiving antennas are located just above the surface. Owing to the occurrence of the Bessel functions, (5.64) may under certain circumstances represent a rapidly oscillating voltage signal whose oscillations (along the normalized time axis t/T) get faster as both time and $\epsilon_0 r/C^{\mathrm{E}}$ increase. The nonzero electric conductivity manifests itself in the exponential decay term, again. Finally, to gain further insights into the phenomenon, it is noted that the voltage signal induced just above a loss-free dielectric layer can be expressed in a simpler form (cf. (5.9)), namely,

$$\Delta V^{\mathrm{R}}(t) = \frac{A^{\mathrm{T}} A^{\mathrm{R}}}{4\pi \epsilon_0 r^5} T^5 \partial_t^5 I^{\mathrm{T}}(t)$$
$$*_t \int_{v=1}^{t/T} \mathrm{J}_1 \left[\frac{2\epsilon_0}{C^{\mathrm{E}}} r (v^2-1)^{1/2} \left(\frac{t}{T} - v \right) \right] (v^2-1)^{1/2} v \mathrm{d}v \quad \text{for } t > T, \quad (5.65)$$

as $h^{\mathrm{R,T}} \downarrow 0$ and $G^{\mathrm{E}} \downarrow 0$. The oscillatory behavior of the voltage response will be demonstrated through illustrative numerical examples in Section 5.3.

5.2.2 *Receiving horizontal electric dipole*

In this subsection, we assume that the pulsed EM field radiated from the transmitting loop is probed by a short-wire antenna whose axis is oriented along the azimuthal direction $\boldsymbol{i}_\phi = -\boldsymbol{i}_x \sin(\phi) + \boldsymbol{i}_y \cos(\phi)$, where $\cos(\phi) = x/r$ and $\sin(\phi) = y/r$ for $r > 0$ and $\{0 \le \phi < 2\pi\}$ (see Figure 5.5). Consequently, the induced voltage in the thus oriented electric dipole is related to the azimuthal component of the electric-field strength via (cf. (4.96) and (5.51))

$$\hat{V}^{\mathrm{R}}(s) = -\ell^{\mathrm{R}} \hat{E}_\phi(r, h^{\mathrm{R}}, s), \quad (5.66)$$

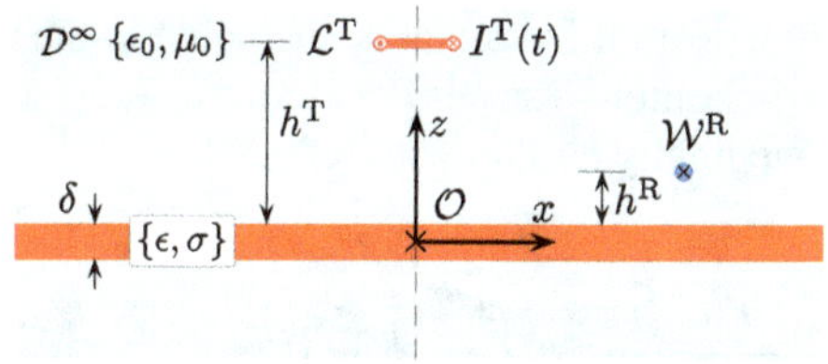

Figure 5.5 Transmitting loop and receiving horizontal electric dipole in the presence of a thin highly contrasting sheet

where $\ell^R > 0$ denotes the length of the (relatively short) receiving wire (see Figure 5.5). Solving the problem under the wave-slowness representation (4.75), again, the voltage response can be expressed as (cf. (5.57))

$$\hat{V}^R(s) = \frac{s^3 \mu_0 \hat{I}^T(s) \mathcal{A}^T \ell^R}{4\pi^2} \int_{\kappa=-i\infty}^{i\infty} d\kappa$$
$$\int_{\sigma=-i\infty}^{i\infty} \exp\{-s[\kappa x + \sigma y + p(\kappa, \sigma) h^T]\} \tilde{\partial}_r(\kappa, \sigma)$$
$$\frac{\tilde{\Psi}_\perp(\kappa, \sigma)}{s + G^E/C^E + \tilde{\Psi}_\perp(\kappa, \sigma)} \frac{d\sigma}{2p(\kappa, \sigma)}, \tag{5.67}$$

as $h^R \downarrow 0$. Equation (5.67) is further transformed to the TD using the CdH technique as specified in Section 5.2.1.2. Again, it is assumed that the transmitting loop is located just above the layer's surface; that is, we take the limit $h^T \downarrow 0$. Consequently, upon transforming the variables of integration according to (2.78a) and (2.78b), we obtain

$$\hat{V}^R(s) = \frac{s^3 \mu_0 \hat{I}^T(s) \mathcal{A}^T \ell^R}{4\pi^2 i} \int_{q=-\infty}^{\infty} dq$$
$$\int_{\gamma=-i\infty}^{i\infty} \exp(-s\gamma r) \frac{\bar{\Psi}_\perp(\gamma, q)}{s + G^E/C^E + \bar{\Psi}_\perp(\gamma, q)} \frac{\gamma d\gamma}{2\bar{p}(\gamma, q)}, \tag{5.68}$$

as $h^{R,T} \downarrow 0$, where $\bar{\Psi}_\perp(\gamma, q)$ and $\bar{p}(\gamma, q)$ are specified just below (5.58). Owing to the property $\bar{p}(\gamma, q) \sim \mp i|\gamma|$ along supplementing arcs in $\mathrm{Im}(\gamma) \gtrless 0$ as $|\gamma| \to \infty$, respectively, their contributions to the integration cancel each other. Then, relying on Cauchy's theorem, the integration path running along $\mathrm{Re}(\gamma) = 0$ can be replaced with the CdH path, $\mathcal{C} \cup \mathcal{C}^*$, that represents the loop encircling the branch cut (see (5.59)). Combining further the integrations just above and just below the branch cut, while introducing the time parameter as the new variable of integration, we arrive at (cf. (5.60))

$$\hat{V}^R(s) = \frac{s^3 \hat{I}^T(s) \mathcal{A}^T \ell^R}{2\pi^2 r^2 C^E} \int_{q=-\infty}^{\infty} dq$$
$$\int_{\tau=r\Omega_0(q)}^{\infty} \exp(-s\tau) \frac{\psi_\perp(\tau, q)}{(s + G^E/C^E)^2 + \psi_\perp^2(\tau, q)} \tau d\tau, \tag{5.69}$$

where $\psi_\perp(\tau, q)$ follows from (5.61). Changing the order of the integrations, we obtain (cf. (5.62))

$$\hat{V}^{\mathrm{R}}(s) = \frac{s^3 \mu_0 \hat{I}^{\mathrm{T}}(s) \mathcal{A}^{\mathrm{T}} \ell^{\mathrm{R}}}{4\pi^2 r^2} \int_{\tau=T}^{\infty} \exp(-s\tau) \tau \mathrm{d}\tau$$
$$\int_{q=-Q(\tau)}^{Q(\tau)} \frac{\psi_\perp^2(\tau, q)}{(s + G^{\mathrm{E}}/C^{\mathrm{E}})^2 + \psi_\perp^2(\tau, q)} \frac{\mathrm{d}q}{[Q^2(\tau) - q^2]^{1/2}}, \tag{5.70}$$

and recall that $Q(\tau) = (\tau^2/r^2 - 1/c_0^2)^{1/2}$. Upon carrying out the inner integration with respect to q analytically, we end up with (cf. (5.63))

$$\hat{V}^{\mathrm{R}}(s) = \frac{s^3 \mu_0 \hat{I}^{\mathrm{T}}(s) \mathcal{A}^{\mathrm{T}} \ell^{\mathrm{R}}}{4\pi r^2}$$
$$\int_{\tau=T}^{\infty} \exp(-s\tau) \left\{ 1 - \frac{s + G^{\mathrm{E}}/C^{\mathrm{E}}}{[(s + G^{\mathrm{E}}/C^{\mathrm{E}})^2 + (2/\mu_0 C^{\mathrm{E}})^2 Q^2(\tau)]^{1/2}} \right\} \tau \mathrm{d}\tau.$$
$$\text{for } \{s \in \mathbb{R}; s > 0\}. \tag{5.71}$$

The resulting s-domain expression (5.71) can be transformed to the TD using the Schouten–Van der Pol theorem [11, p. 1056] and [13, (29.3.56)]. In this way, we finally arrive at (cf. (5.65))

$$V^{\mathrm{R}}(t) = \frac{\mathcal{A}^{\mathrm{T}} \ell^{\mathrm{R}}}{2\pi r^3} \frac{1}{C^{\mathrm{E}}} T^3 \partial_t^3 I^{\mathrm{T}}(t) *_t \int_{v=1}^{t/T} \mathrm{J}_1 \left[\frac{2\epsilon_0}{C^{\mathrm{E}}} r (v^2 - 1)^{1/2} \left(\frac{t}{T} - v \right) \right]$$
$$\times \exp\left[-\frac{G^{\mathrm{E}}}{\eta_0} \frac{\epsilon_0}{C^{\mathrm{E}}} r \left(\frac{t}{T} - v \right) \right] (v^2 - 1)^{1/2} v \mathrm{d}v \quad \text{for all } t > T, \tag{5.72}$$

as $h^{\mathrm{R,T}} \downarrow 0$. Taking the limits $G^{\mathrm{E}} \downarrow 0$ and $C^{\mathrm{E}} \downarrow 0$, the voltage response in the absence layer follows as (cf. (5.56))

$$V_0^{\mathrm{R}}(t) = \frac{\mu_0 \mathcal{A}^{\mathrm{T}} \ell^{\mathrm{R}}}{4\pi r^2} \left[\partial_t I^{\mathrm{T}}(t - T) + T \partial_t^2 I^{\mathrm{T}}(t - T) \right] \quad \text{for all } t > T, \tag{5.73}$$

as $h^{\mathrm{R}} \downarrow 0$. Apparently, the induced voltage pulse expressed by (5.73) consists of two terms that are proportional to (distance)$^{-2}$ and (distance)$^{-1}$. The closed-form TD formula can also be easily deduced using [11, Eqs. (26.10-27) and (26.10-28)] expressing the intermediate- and far-field constituents of the electric-field strength radiated from a small loop antenna. Note that these constituents are proportional to the (delayed) first and second time derivatives of the excitation electric-current pulse, respectively. Under the limit $T \downarrow 0$, the far-field contribution can be neglected, so that $V_0^{\mathrm{R}}(t) = L^{\mathrm{C}} \partial_t I^{\mathrm{T}}(t) + O(T)$, where $L^{\mathrm{C}} = (\mu_0/4\pi) \mathcal{A}^{\mathrm{T}} \ell^{\mathrm{R}} / r^2$ has the meaning of (intermediate-field) coupling inductance.

5.3 Illustrative examples

In the first example, we shall calculate the electric-field strength on the surface a thin sheet with dielectric and conductive properties via (5.9). For the sake of simplicity, the shape of the exciting electric-current pulse is described by the rectangular function

$$I(t) = I_{\mathrm{m}}[\mathrm{H}(t) - \mathrm{H}(t - t_{\mathrm{w}})], \tag{5.74}$$

where we take the unit amplitude, $I_{\mathrm{m}} = 1.0\,\mathrm{mA}$, and the pulse-time width, t_{w}, is chosen such that $c_0 t_{\mathrm{w}} = |x|$. The equivalent layer's capacitance is defined via $C^{\mathrm{E}}/\eta_0 t_{\mathrm{w}} = 1.0$, which implies that $\epsilon_0|x|/C^{\mathrm{E}} = 1.0$. The dissipation in the thin layer is incorporated via its relative layer's conductance $G^{\mathrm{E}}/\eta_0 = 0.20$. The resulting (scaled) E-field response as evaluated using (5.9) in the bounded time window $\{0 \le t/T_0 = c_0 t/|x| \le 20\}$ is shown in Figure 5.6 (exact). It can be seen that the oscillatory response is well approximated by (5.10) (approximate). Finally, to illustrate the effect of the layer, we have also included the corresponding response that would be excited in the absence of the sheet (void), that is (cf. (5.9)),

$$\lim_{h \downarrow 0} E_y(x, 0, t) = -\frac{I_{\mathrm{m}}}{|x|\eta_0}\left\{\frac{\mathrm{H}(t/T_0 - 1)}{(t^2/T_0^2 - 1)^{1/2}} - \frac{\mathrm{H}[(t - t_{\mathrm{w}})/T_0 - 1]}{[(t - t_{\mathrm{w}})^2/T_0^2 - 1]^{1/2}}\right\}, \tag{5.75}$$

as $\epsilon \downarrow \epsilon_0$ and $\sigma \downarrow 0$. Apparently, the TD response (5.75) exhibits the typical inverse-square root behavior as t/T_0 and $(t - t_{\mathrm{w}})/T_0$ approach the value one (see Figure 5.6 and (5.20)).

As the second example, we make use of (5.64) to compute the voltage induced in a vertical receiving loop, $\mathcal{L}^{\mathrm{R}}$, located just above a thin, highly contrasting layer with both dielectric and conductive properties. The (horizontal) transmitting loop is excited by the electric-current pulse, $I^{\mathrm{T}}(t)$, whose shape is described by the right-hand side of (4.70). Here, we take $I_{\mathrm{m}} = 1.0\,\mathrm{mA}$ and $n = 6$, again (see Figure 4.7(a)).

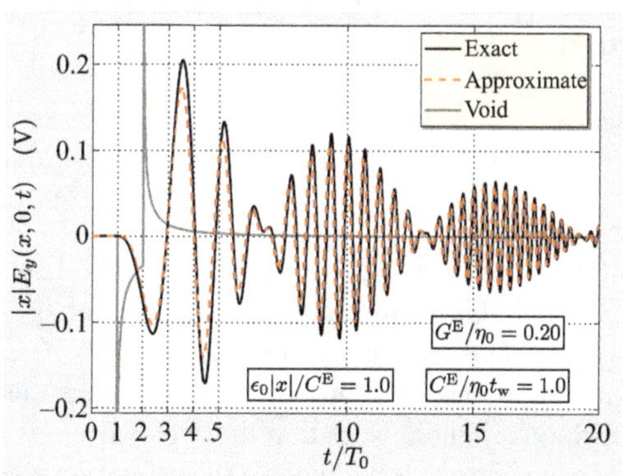

Figure 5.6 The (scaled) total electric-field pulse shapes excited on the surface of the thin layer with dielectric and conductive properties. The response in the absence of layer is denoted by "Void".

Its pulse-time width is related to the horizontal distance between the loops, r, via $c_0 t_\mathrm{w} = r$. The equivalent layer's capacitance is taken to be $C^\mathrm{E}/\eta_0 t_\mathrm{w} = 1.0$, again, which implies that $\epsilon_0 r/C^\mathrm{E} = 1.0$. The electric conductivity of the layer is now chosen such that the equivalent relative layer's conductance is given by $G^\mathrm{E}/\eta_0 = 1.0$. The radius of both transmitting and receiving loops is taken to be a tenth of the distance, $r^\mathrm{T} = r^\mathrm{R} = r/10$, so that the loops are relatively small with respect to the spatial support of the excitation pulse and $\mathcal{A}^\mathrm{T}\mathcal{A}^\mathrm{R}/r^4 = \pi^2/10^4$. The induced voltage pulse, $\Delta V^\mathrm{R}(t) = V^\mathrm{R}(t)$, as calculated in the time window $\{0 \le t/T = c_0 t/r \le 20\}$, is shown in Figure 5.7.

Finally, we employ (5.72) to calculate the voltage response induced across the port of a short horizontal wire, $\mathcal{W}^\mathrm{R}$, that is placed just above the layer, again. The remaining excitation and configurational parameters are adopted from the previous example. Figure 5.8 shows the calculated voltage pulse. To demonstrate the impact of the layer on the response, we have also included the corresponding voltage response

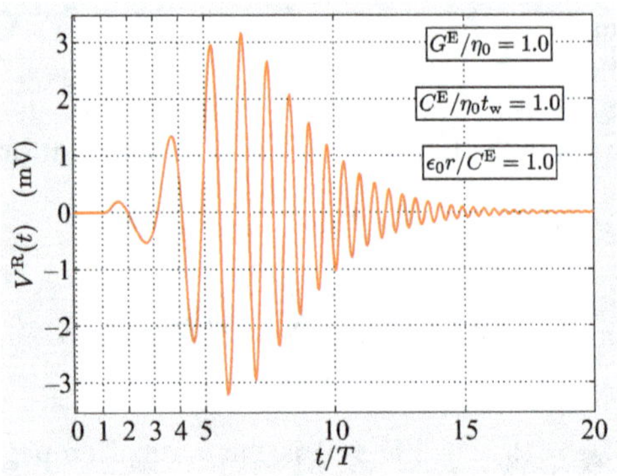

Figure 5.7 The voltage response induced in the small loop $\mathcal{L}^\mathrm{R}$ placed close to the surface of the thin layer with dielectric and conductive properties

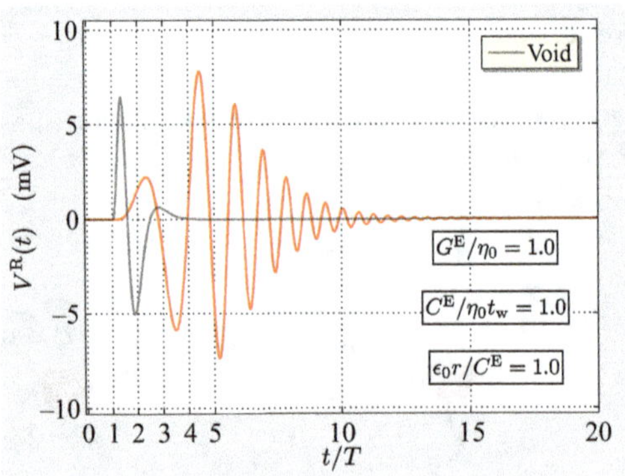

Figure 5.8 The voltage response induced in the short wire $\mathcal{W}^\mathrm{R}$ placed close to the surface of the thin layer with dielectric and conductive properties. The response in the absence of layer is denoted by "Void".

that would be excited in the absence of the layer (void) (see (5.73)). As can be clearly seen, the layer makes the response oscillatory and significantly longer.

5.3.1 Concluding remarks

In order to gain physical insights into the TD EM surface phenomenon described in the previous sections, we shall study the line source excited EM fields in the presence of a dielectric layer of a finite thickness $\delta > 0$. In particular, we shall describe the electric-field strength at the surface of the layer induced by an impulsive electric-line source that is located just above the layer (see Figure 5.9). To that end, the cross-layer conditions (3.4a) and (3.4b) that led to the results presented in Section 5.1.1.1 are replaced with the continuity-type boundary conditions applying at the interfaces along $z = \pm\delta/2$, that is,

$$\lim_{z\downarrow\pm\delta/2}\{E_y, H_x\}(x,z,t) = \lim_{z\uparrow\pm\delta/2}\{E_y, H_x\}(x,z,t)$$
$$\text{for all } x \in \mathbb{R} \text{ and } t > 0. \tag{5.76}$$

Upon applying the wave-slowness representation (4.33), the transform-domain solution can be obtained in the following (geometric-series) form (cf. (2.69b)):

$$\lim_{z\downarrow\delta/2}\tilde{E}_y(\kappa,z,s) = -\left[\mu_0\hat{I}(s)/2p_0(\kappa)\right]\left[1+\tilde{R}(\kappa)\right]$$
$$+\left[\mu_0\hat{I}(s)/2p_0(\kappa)\right]\tilde{R}(\kappa)\left[1-\tilde{R}^2(\kappa)\right]\sum_{N=0}^{\infty}\tilde{R}^{2N}(\kappa)\exp[-sp_1(\kappa)Z_N], \tag{5.77}$$

as $h\downarrow\delta/2$, where $Z_N = 2(N+1)\delta > 0$ is the vertical propagation path and (cf. (4.112))

$$\tilde{R}(\kappa) = [p_0(\kappa)-p_1(\kappa)]/[p_0(\kappa)+p_1(\kappa)] \tag{5.78}$$

has the meaning of reflection coefficient. In its expression, $p_{0,1} = (1/c_{0,1}^2 - \kappa^2)^{1/2}$, with $\mathrm{Re}(p_{0,1}) \geq 0$, are the vertical slowness parameters pertaining to the surrounding medium and the layer, where the EM wave speeds are $c_0 = (\mu_0\epsilon_0)^{-1/2} > c_1$, and $c_1 = (\mu_0\epsilon)^{-1/2} > 0$, respectively.

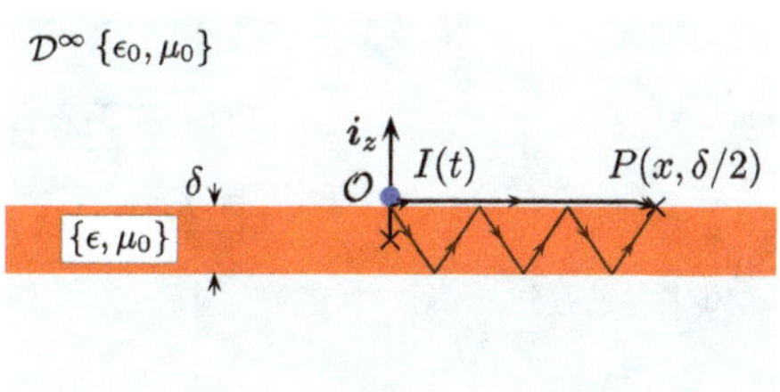

Figure 5.9 Electric-line source induced response on the surface of a dielectric layer

The terms in (5.77) can be interpreted as transform-domain generalized-ray constituents [16] (see also Section 2.2.1). Accordingly, the first term on its right-hand side represents the sum of the direct wave and the wave reflected against the top interface. The second term can be then understood as a superposition of generalized-ray constituents propagating through reflections inside the layer (see Figure 5.9). With reference to (4.33), the generic form of such a generalized-ray constituent can be hence represented via

$$\hat{F}_N(x,s) = \frac{s\mu_0 \hat{I}(s)}{2\mathrm{i}\pi} \int_{\kappa=-\mathrm{i}\infty}^{\mathrm{i}\infty} \exp\{-s[\kappa x + p_1(\kappa) Z_N]\} \tilde{\Pi}_N(\kappa) \frac{\mathrm{d}\kappa}{2p_0(\kappa)}, \tag{5.79}$$

where $\tilde{\Pi}_N(\kappa)$ is the s-independent factor that describes the wave transfer (via reflections and transmissions) across the interfaces. The factor hence contains both $p_0(\kappa)$ and $p_1(\kappa)$ and behaves as $\tilde{\Pi}_N(\kappa) = o(\kappa)$ as $|\kappa| \to \infty$, so that the integrand in (5.79) goes to zero at infinity and Jordan's lemma applies. Owing to the (possibly multiple) reflections against the surrounding medium in which the ray constituent does not propagate, the TD counterpart of (5.79), in general, consists of head- and body-wave contributions, say $F_{N;\mathrm{H}}(x,t)$ and $F_{N;\mathrm{B}}(x,t)$, whose arrival times are $T_\mathrm{H} = |x|/c_0 + Z_N(1/c_1^2 - 1/c_0^2)^{1/2}$ and $T_\mathrm{B} = (x^2 + Z_N^2)^{1/2}/c_1$, respectively. The head-wave contribution occurs only in the region where $|x|/(x^2 + Z_N^2)^{1/2} > c_1/c_0$ (cf. (2.92)).

Now, if the layer under consideration is assumed to be thin (i.e. $\delta \downarrow 0$) and highly dielectric (i.e. $c_1 \ll c_0$), it follows that the (oscillatory) surface response is dominantly composed of head-wave constituents propagated via reflections inside the layer. Mathematically, the head-wave contributions arise from the integration along the so-called head-wave CdH path encircling the segments $\{c_0^{-1} < |\mathrm{Re}(\kappa)| < |x|/(x^2 + Z_N^2)^{1/2} c_1^{-1}, \mathrm{Im}(\kappa) = 0\}$ along the branch cut in the complex κ-plane (see $\mathcal{G}_\mathrm{H} \cup \mathcal{G}_\mathrm{H}^*$ in Figure H.1). In accordance with [17, Appendix A], the TD head-wave contribution can be then expressed as

$$F_{N;\mathrm{H}}(x,t) = \frac{\mu_0}{2\pi} \partial_t I(t) \ast_t \mathrm{Im}\left\{ \tilde{\Pi}_N[\kappa(t)] \frac{p_1[\kappa(t)]}{p_0[\kappa(t)]} \right\} \frac{\mathrm{H}(t - T_\mathrm{H}) - \mathrm{H}(t - T_\mathrm{B})}{(T_\mathrm{B}^2 - t^2)^{1/2}}, \tag{5.80}$$

in which (cf. (H.8))

$$\mathcal{G}_\mathrm{H} = \left\{ \kappa(t) = -\frac{x}{x^2 + Z_N^2} t - \frac{Z_N}{x^2 + Z_N^2} (T_\mathrm{B}^2 - t^2)^{1/2} + \mathrm{i}0 \right\}$$
$$\text{for all } \{T_\mathrm{H} \le t \le T_\mathrm{B}\}. \tag{5.81}$$

Thus, employing (5.80) to transform the electric-field response (5.77) just above a high-dielectric thin layer to the TD, it is found that the oscillatory TD surface effect can also be explained through the superposition of TD head-wave constituents propagated via reflections in the layer.

References

[1] A. E. Kennelly. On the elevation of the electrically-conducting strata of the earth's atmosphere. *Electrical World and Engineer*, 39(11):473, 1902.

[2] Lord Rayleigh. On waves propagated along the plane surface of an elastic solid. *Proceedings of the London Mathematical Society*, 1(1):4–11, 1885.

[3] H. Lamb. On the propagation of tremors over the surface of an elastic solid. *Philosophical Transactions of the Royal Society of London. Series A*, 203(359-371):1–42, 1904.

[4] A. Sommerfeld. Über die Ausbreitung der Wellen in der drahtlosen Telegraphie. *Annalen Der Physik*, 28(4):665–736, 1909.

[5] C. J. Bouwkamp. On Sommerfeld's surface wave. *Physical Review*, 80(2): 294, 1950.

[6] J. R. Wait. The ancient and modern history of EM ground-wave propagation. *IEEE Antennas and Propagation Magazine*, 40(5):7–24, 1998.

[7] S. Schelkunoff. Anatomy of "surface waves". *IRE Transactions on Antennas and Propagation*, 7(5):133–139, 1959.

[8] A. T. de Hoop. Line-source excited pulsed acoustic wave reflection against the mass-loaded boundary of a fluid. In *AIP Conference Proceedings*, volume 1106, pp. 118–129, Växjö, Sweden, June 2008.

[9] A. T. de Hoop and J. H. M. T. Van der Hijden. Generation of acoustic waves by an impulsive point source in a fluid/solid configuration with a plane boundary. *The Journal of the Acoustical Society of America*, 75(6):1709–1715, 1984.

[10] M. Štumpf and I. E. Lager. On the excitation of anomalous EM transients along the surface of a thin highly-contrasting sheet with dielectric and conductive properties. *IEEE Antennas and Wireless Propagation Letters*, 20(1):58–62, 2021.

[11] A. T. de Hoop. *Handbook of Radiation and Scattering of Waves*. London, UK: Academic Press, 1995.

[12] A. T. de Hoop and L. Jiang. Reflection and transmission of line-source excited pulsed EM fields at a thin, high-contrast layer with dielectric and conductive properties. In *2011 IEEE International Symposium on Antennas and Propagation*, pp. 2404–2406, Spokane, WA, USA, July 2011.

[13] M. Abramowitz and I. A. Stegun. *Handbook of Mathematical Functions*. New York, NY: Dover Publications, 1972.

[14] M. Štumpf and G. A. E. Vandenbosch. Line-source excited impulsive EM field response of thin plasmonic metal films. *Photonics and Nanostructures—Fundamentals and Applications*, 11(3):253–260, 2013.

[15] M. Štumpf and G. A. E. Vandenbosch. Impulsive electromagnetic response of thin plasmonic metal sheets. *Radio Science*, 49(8):689–697, 2014.

[16] M. Štumpf, A. T. de Hoop and G. A. E. Vandenbosch. Generalized ray theory for time-domain electromagnetic fields in horizontally layered media. *IEEE Transactions on Antennas and Propagation*, 61(5):2676–2687, 2013.

[17] A. T. de Hoop, M. Štumpf and I. E. Lager. Pulsed electromagnetic field radiation from a wide slot antenna with a dielectric layer. *IEEE Transactions on Antennas and Propagation*, 59(8):2789–2798, 2011.

Chapter 6

Pulsed EM-field diffraction by semi-infinite sheets

Electromagnetic (EM) diffraction by a perfectly electrically conducting (PEC) semi-infinite screen is a canonical problem in EM wave theory whose first exact solution is attributed to Sommerfeld [1]. The need for physical insights into EM diffraction problems that would enable the evaluation of antenna performance and EM propagation in a complex environment called for a general solution methodology capable of handling more involved problem configurations. Significant progress in these efforts has been accomplished via integral-equation formulations [2, 3] that are amenable to the Wiener–Hopf (WH) method [4]. Through this methodology, the original Sommerfeld's half-plane problem has been generalized by incorporating finite conductivity of the diffracting screen [5, 6] or by placing the semi-infinite PEC screen on the planar interface of two media [7]. A detailed summary of major achievements in this field of research is given in Ref. [8].

Despite a myriad of applications relying entirely on pulsed EM fields, the majority of available analytical solutions have been obtained under the assumption of the sinusoidal time dependence only. In these closed-form frequency-domain (FD) solutions, the (real-valued) frequency parameter generally occurs in intricate functional dependencies, which does not allow achieving their time-domain (TD) counterparts analytically without the use of an inverse (fast) Fourier transform. To provide closed-form analytical solutions directly in the TD, we shall next combine the Cagniard–DeHoop (CdH) technique (see Chapter 1) with the WH method. For truly seminal works pioneering the presented methodology, we refer the reader to [9, 10].

6.1 Pulsed EM diffraction by a semi-infinite PEC sheet

We shall analyze the pulsed EM diffraction by a semi-infinite, PEC planar sheet occupying $\{0 < x < \infty, -\infty < y < \infty, -\delta/2 < z < \delta/2\}$ (see Figure 6.1). The screen is located in the homogeneous, isotropic and loss-free embedding $\mathcal{D}^\infty$, the EM properties of which are described by (scalar, real-valued and positive) electric permittivity and magnetic permeability $\{\epsilon_0, \mu_0\}$. The corresponding EM wave speed is given as $c_0 = (\epsilon_0\mu_0)^{-1/2} > 0$. The screen as well as its excitation is assumed to

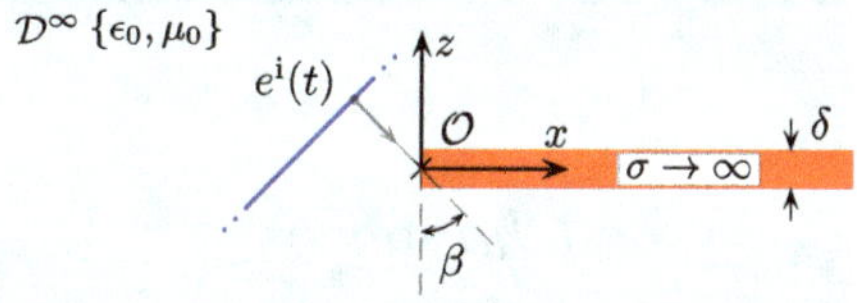

Figure 6.1 Pulsed EM plane wave impinging upon a PEC semi-infinite sheet

be y-independent. Considering the uniform, E-polarized plane-wave excitation, the incident field is in the s-domain described by

$$\hat{E}^{\mathrm{i}}_y(x, z, s) = \hat{e}^{\mathrm{i}}(s) \exp\{-s[\kappa_0 x - p(\kappa_0) z]\}, \tag{4.1 revisited}$$

with $\kappa_0 = \sin(\beta)/c_0$ and $p(\kappa_0) = \cos(\beta)/c_0$ for $\{0 \leq \beta \leq \pi/2\}$. Referring to (3.1), the scattered electric-field strength, $\hat{E}^{\mathrm{s}}_y$, is defined as the difference between the total electric-field strength and the incident one, $\hat{E}^{\mathrm{i}}_y$. The presence of the PEC screen is accounted for via the explicit-type boundary condition on the surface of the sheet

$$\hat{E}^{\mathrm{s}}_y(x, \pm\delta/2, s) = -\hat{E}^{\mathrm{i}}_y(x, 0, s) \quad \text{for all } x > 0, \tag{6.1a}$$

as $\delta \downarrow 0$, while

$$\lim_{z \downarrow \delta/2} \hat{H}^{\mathrm{s}}_x(x, z, s) - \lim_{z \uparrow -\delta/2} \hat{H}^{\mathrm{s}}_x(x, z, s) = 0 \quad \text{for all } x < 0, \tag{6.1b}$$

as $\delta \downarrow 0$, again. It is next postulated that the scattered electric-field strength in the s-domain can be represented via (cf. (4.33))

$$\hat{E}^{\mathrm{s}}_y(x, z, s) = \frac{\hat{e}^{\mathrm{i}}(s)}{2\pi \mathrm{i}} \int_{\kappa \in \mathcal{L}} \exp\{-s[\kappa x + p(\kappa)|z|]\} \tilde{A}(\kappa) \frac{\mathrm{d}\kappa}{2p(\kappa)}$$
$$\text{for all } x \in \mathbb{R}, z \in \mathbb{R} \text{ and } \{s \in \mathbb{R}; s > 0\}, \tag{6.2}$$

thereby representing a wave field traveling away from the screen lying along $z = 0$. In the representation, $\tilde{A}(\kappa)$ is the coefficient to be determined and $\mathcal{L}$ denotes the path of integration dividing the complex κ-plane into two half-planes located to the left and to the right of $\mathcal{L}$. These domains are denoted by $\mathcal{D}^-$ and $\mathcal{D}^+$, respectively. The explicit-type and continuity-type boundary conditions (6.1a) and (6.1b) yield a pair of dual integral equations formulating the WH problem, that is,

$$\frac{1}{2\pi \mathrm{i}} \int_{\kappa \in \mathcal{L}} \exp(-s\kappa x) \tilde{A}(\kappa) \mathrm{d}\kappa = 0 \quad \text{for all } x < 0, \tag{6.3a}$$

and

$$\frac{1}{2\pi \mathrm{i}} \int_{\kappa \in \mathcal{L}} \exp(-s\kappa x) \tilde{A}(\kappa) \frac{\mathrm{d}\kappa}{2p(\kappa)} = -\exp(-s\kappa_0 x) \quad \text{for all } x > 0. \tag{6.3b}$$

Equation (6.3a) implies that $\tilde{A}(\kappa)$ is a regular function in $\mathcal{D}^-$ and of $o(1)$ as $|\kappa| \to \infty$ in $\mathcal{D}^-$. Furthermore, (6.3b) is satisfied if

$$\frac{\tilde{A}(\kappa)}{2p(\kappa)} = \frac{1}{\kappa - \kappa_0} + \tilde{H}^+(\kappa) \quad \text{for } \kappa \in \mathcal{L} \cup \mathcal{D}^+, \tag{6.4}$$

where $\tilde{H}^+(\kappa)$ is some function that is regular in $\mathcal{D}^+$ and has the property $\tilde{H}^+(\kappa) = o(1)$ as $|\kappa| \to \infty$ in $\mathcal{D}^+$, provided that $\mathcal{L}$ is chosen such that $\kappa_0 \in \mathcal{D}^+$ (see Figure 6.2). Note that in the approach developed by de Hoop [10], the WH relation (6.4) is finally required to hold along $\kappa \in \mathcal{L}$ only, and not in a strip of non-vanishing width as is standardly assumed in the relevant literature on the subject. To solve (6.4) for $\tilde{A}$, the vertical slowness parameter $p(\kappa)$ is further factorized into a factor that is regular in $\mathcal{D}^-$, say $p^-(\kappa)$, and a factor regular in $\mathcal{D}^+$, say $p^+(\kappa)$. In this case, the factorization can be carried out by inspection, that is,

$$p(\kappa) = p^-(\kappa) p^+(\kappa) = (1/c_0 - \kappa)^{1/2} (1/c_0 + \kappa)^{1/2}, \tag{6.5}$$

respectively, which allows to cast relation (6.4) into the following form:

$$\frac{\tilde{A}(\kappa)}{2p^-(\kappa)} - \frac{p^+(\kappa_0)}{\kappa - \kappa_0} = p^+(\kappa)\tilde{H}^+(\kappa) + \frac{p^+(\kappa) - p^+(\kappa_0)}{\kappa - \kappa_0} \quad \text{for } \kappa \in \mathcal{L}. \tag{6.6}$$

Since κ_0 is located to the right of $\mathcal{L}$, the left-hand side of (6.6) is regular in $\mathcal{D}^-$, while its right-hand side is regular in $\mathcal{D}^+$. Consequently, by virtue of Liouville's theorem [11, p. 85], the left- and the right-hand sides do represent one and the same entire function that is a constant. Since the left-hand side of (6.6) is of $o(\kappa^{-1/2})$ as $|\kappa| \to \infty$ in $\mathcal{D}^-$, while its right-hand side is of $o(\kappa^{1/2})$ as $|\kappa| \to \infty$ in $\mathcal{D}^+$, the extended form of Liouville's theorem implies that the entire function is zero and we get

$$\tilde{A}(\kappa) = \frac{2p^+(\kappa_0)p^-(\kappa)}{\kappa - \kappa_0}. \tag{6.7}$$

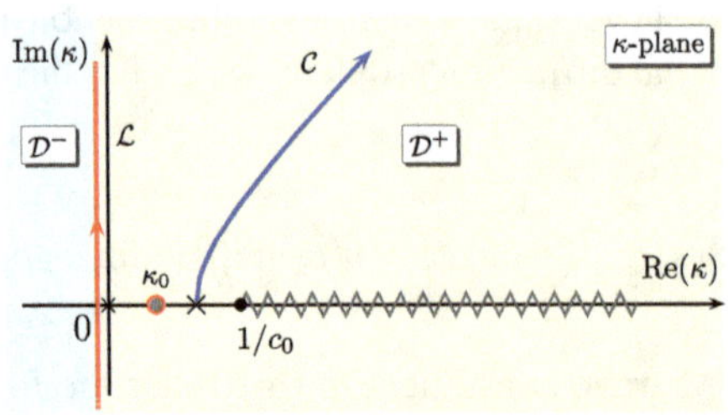

Figure 6.2 Complex κ-plane with the CdH path C

With the coefficient at our disposal, the slowness-domain representation of the scattered field (6.2) is fully determined. Its transformation back to the TD can be carried out with the aid of the standard CdH technique (see Section 2.1.1), which is the subject of the ensuing section.

6.1.1 TD solution

To transform the slowness-integral representation (6.2) with (6.7) to the TD, its integration contour $\mathcal{L}$ is deformed to the CdH path defined by

$$\kappa x + p(\kappa)|z| = \tau \quad \text{for } \{\tau \in \mathbb{R}; \tau > 0\}. \tag{6.8}$$

The deformation is permissible since the contributions from circular arcs joining $\mathcal{L}$ with the CdH path are vanishingly small. Solving then (6.8) for κ, we obtain a hyperbolic CdH path, say $\mathcal{C} \cup \mathcal{C}^*$, with

$$\mathcal{C} = \left\{\kappa(\tau) = (x/\rho^2)\tau + \mathrm{i}(|z|/\rho^2)\left(\tau^2 - \rho^2/c_0^2\right)^{1/2}\right\}$$
$$\text{for all } \tau \geq \rho/c_0, \tag{6.9}$$

with $\rho = (x^2 + z^2)^{1/2} > 0$ being, in fact, the distance from the edge (see Figure 6.1). Collecting the contributions from $\mathcal{C} \cup \mathcal{C}^*$ and from the simple pole singularity at $\kappa = \kappa_0$, while introducing parameter τ as the variable of integration, we may rewrite (6.2) as

$$\hat{E}_y^{\mathrm{s}}(x, z, s) = \frac{\hat{e}^{\mathrm{i}}(s)}{2\pi} \int_{\tau=\rho/c_0}^{\infty} \exp(-s\tau)\mathrm{Re}\{\tilde{A}[\kappa(\tau)]\} \frac{\mathrm{d}\tau}{(\tau^2 - \rho^2/c_0^2)^{1/2}}$$
$$+ \hat{E}_y^{\mathrm{s;g}}(x, z, s) \quad \text{for } \{s \in \mathbb{R}; s > 0\}, \tag{6.10}$$

where $\hat{E}_y^{\mathrm{s;g}}$ represents the pole contribution. The latter component must be taken into account whenever the pole is located in between $\mathcal{L}$ and the CdH path. Since the CdH path intersects the real κ-axis at $\tau = \rho/c_0$, where $\kappa(\tau) = (x/\rho)c_0^{-1}$, the pole contribution appears whenever $c_0\kappa_0 \leq x/\rho$. This contribution, in fact, represents the geometrical solution of the diffraction problem, which has the form of a plane wave (cf. (4.1))

$$\hat{E}_y^{\mathrm{s;g}}(x, z, s) = -\hat{e}^{\mathrm{i}}(s) \exp\{-s[\kappa_0 x + p(\kappa_0)|z|]\} \quad \text{for } x/\rho \geq \sin(\beta). \tag{6.11}$$

Since the diffracted EM wave constituent in (6.10) has the form of Laplace transformation, its inversion back to TD can be readily carried out with the help of Lerch's uniqueness theorem of the one-sided Laplace transformation [12, Appendix].

Pursuing this line of reasoning, we arrive at the following expression for the *total* electric-field strength

$$E_y(x,z,t) = \frac{e^{\mathrm{i}}(t)}{2\pi} *_t \mathrm{Re}\left\{\frac{2p^+(\kappa_0)p^-[\kappa(t)]}{\kappa(t)-\kappa_0}\right\}\frac{\mathrm{H}(t-\rho/c_0)}{(t^2-\rho^2/c_0^2)^{1/2}}$$
$$- e^{\mathrm{i}}[t-\kappa_0 x - p(\kappa_0)|z|]\mathrm{H}[x-\rho\sin(\beta)] + e^{\mathrm{i}}[t-\kappa_0 x + p(\kappa_0)z]$$
$$\text{for all } x\in\mathbb{R}, z\in\mathbb{R} \text{ and } t>0, \tag{6.12}$$

where we used the TD counterparts of the incident plane wave (4.1) and of the geometrical solution (6.11). In conclusion, the first term on the right-hand side of (6.12) represents a diffracted cylindrical wave emanating from the edge of the sheet, the second term is a plane wave representing the geometrical solution of the diffraction problem, and, finally, the third term is the incident plane wave. Apparently, the incident plane wave and the geometrical solution cancel each other in the shadow region where $\{x/\rho > \sin(\beta), z<0\}$ (see Figure 6.3). On the other hand, in the symmetrical region above the sheet, their superposition leads to $e^{\mathrm{i}}[t-\kappa_0 x + p(\kappa_0)z] - e^{\mathrm{i}}[t - \kappa_0 x - p(\kappa_0)z]$.

The solution of the diffraction problem has been previously found by Sommerfeld with the aid of multivalued wave functions [1]. A link between the CdH solution and Sommerfeld's one can be seen upon providing an explicit expression for the diffracted wave amplitude. This is exactly the main objective of Appendix A, where such an expression is derived.

6.1.2 *Illustrative example*

To provide an illustrative example, we shall analyze the electric-field distribution in the presence of a PEC semi-infinite sheet that is irradiated by the E-polarized plane wave (see (4.1)), whose angle of incidence is $\beta=\pi/4$. Its pulse shape is described by the rectangular function (cf. (4.31) and (5.74))

$$e^{\mathrm{i}}(t) = e_{\mathrm{m}}[\mathrm{H}(t) - \mathrm{H}(t-t_{\mathrm{w}})], \tag{6.13}$$

where we take the unit amplitude $e_{\mathrm{m}} = 1.0\,\mathrm{V/m}$, again, and the pulse-time width, t_{w}, is related to $h>0$ through $c_0 t_{\mathrm{w}} = h/2$. The total electric-field spatial distributions as evaluated via (6.12) at two observation times $c_0 t/h = 0.50$ and $c_0 t/h = 1.50$ are shown in Figure 6.4(a) and (b), respectively.

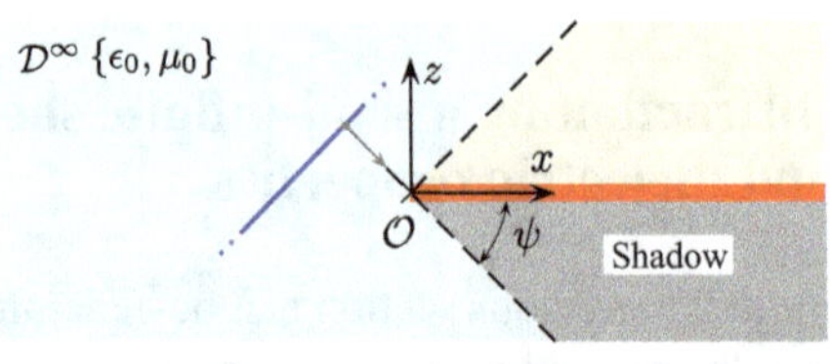

Figure 6.3 Geometrical solution of the diffraction problem

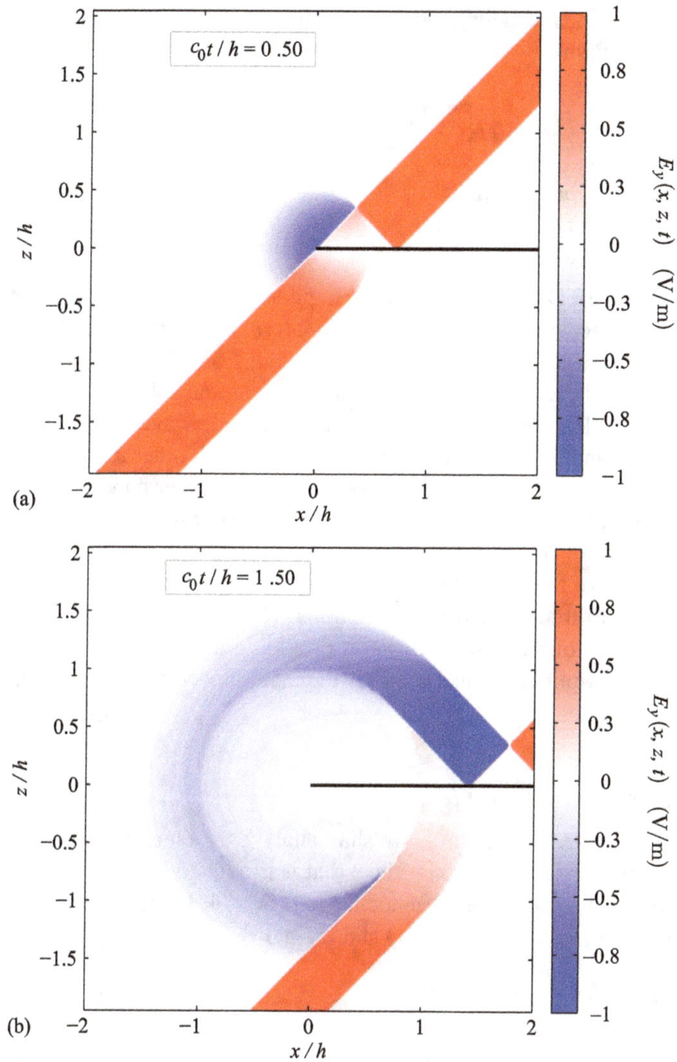

Figure 6.4 Spatial distributions of the electric-field strength excited by the impulsive plane wave in the presence of the semi-infinite PEC sheet as observed at (a) $c_0t/h = 0.50$ and (b) $c_0t/h = 1.50$

6.2 Pulsed EM diffraction by a semi-infinite sheet with conductive and dielectric properties

The analysis carried out in the previous section can be generalized by considering a semi-infinite thin sheet with conductive and dielectric properties. Again, for the sake of simplicity, we shall limit ourselves to analyzing the two-dimensional problem

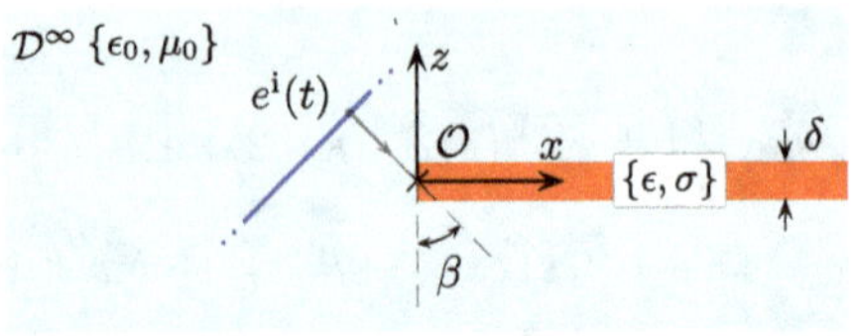

Figure 6.5 Pulsed EM plane wave impinging upon a penetrable semi-infinite sheet

configuration (see Figure 6.5) in which the EM wave fields are activated by the E-polarized plane wave represented by (4.1).

In accordance with (3.7a) and (3.7b), the presence of the semi-infinite, EM-penetrable thin sheet is accounted for via

$$\lim_{z\downarrow\delta/2} \hat{E}^s_y(x,z,s) - \lim_{z\uparrow-\delta/2} \hat{E}^s_y(x,z,s) = 0, \tag{6.14a}$$

$$\lim_{z\downarrow\delta/2} \hat{H}^s_x(x,z,s) - \lim_{z\uparrow-\delta/2} \hat{H}^s_x(x,z,s) = \hat{Y}^{\mathrm{E}}(s)\hat{E}_y(x,0,s)\mathrm{H}(x) \quad \text{for all } x \in \mathbb{R}, \tag{6.14b}$$

as $\delta \downarrow 0$, where we used (3.15b) to denote the equivalent layer's admittance of the electric type. Pursuing the line of reasoning from Section 6.2, the scattered electric-field strength is represented in the following form:

$$\hat{E}^s_y(x,z,s) = \frac{\hat{e}^i(s)}{2\pi i}\int_{\kappa\in\mathcal{L}} \exp\{-s[\kappa x + p(\kappa)z]\}\tilde{A}(\kappa,s)\frac{d\kappa}{2p(\kappa)} \quad \text{for all } x \in \mathbb{R}, z \geq 0 \text{ and } \{s \in \mathbb{R}; s > 0\}. \tag{6.15}$$

Likewise, the field below the screen is represented via

$$\hat{E}^s_y(x,z,s) = \frac{\hat{e}^i(s)}{2\pi i}\int_{\kappa\in\mathcal{L}} \exp\{-s[\kappa x - p(\kappa)z]\}\tilde{B}(\kappa,s)\frac{d\kappa}{2p(\kappa)} \quad \text{for all } x \in \mathbb{R}, z \leq 0 \text{ and } \{s \in \mathbb{R}; s > 0\}. \tag{6.16}$$

In the representation, $\mathcal{L}$ denotes the path of integration dividing the complex κ-plane into two half-planes located to the left and to the right of the path. The integration path is assumed to be parallel to $\mathrm{Re}(\kappa) = 0$ as $|\kappa| \to \infty$, but its location is to be determined later.

Next, making use of (6.15) and (6.16) in the continuity-type condition (6.14a), we readily infer that

$$\tilde{A}(\kappa,s) = \tilde{B}(\kappa,s) \quad \text{for } \kappa \in \mathcal{L}. \tag{6.17}$$

Furthermore, the saltus-type condition (6.14b) yields a pair of dual integral equations

$$\frac{1}{2\pi i}\int_{\kappa\in\mathcal{L}} \exp(-s\kappa x)\tilde{A}(\kappa,s)d\kappa = 0 \quad \text{for } x < 0, \tag{6.18a}$$

and

$$\frac{1}{\mu_0}\frac{1}{2\pi\mathrm{i}}\int_{\kappa\in\mathcal{L}}\exp(-s\kappa x)\{1+[\hat{Y}^{\mathrm{E}}(s)Z_{\perp}(\kappa)/2]\}\tilde{A}(\kappa,s)\mathrm{d}\kappa$$
$$= -\hat{Y}^{\mathrm{E}}(s)\exp(-s\kappa_0 x) \quad \text{for } x<0, \tag{6.18b}$$

where we used $Z_{\perp}(\kappa)=\mu_0/p(\kappa)$. Equation (6.18a) then implies that $\tilde{A}(\kappa,s)$ is a regular function in $\mathcal{D}^-$ and of $o(1)$ as $|\kappa|\to\infty$ in $\mathcal{D}^-$. Furthermore, (6.18b) holds true if

$$\mu_0^{-1}[\hat{Z}^{\mathrm{E}}(s)+Z_{\perp}(\kappa)/2]\tilde{A}(\kappa,s)=[\tilde{Q}^+(\kappa,s)/\tilde{Q}^+(\kappa_0,s)]/(\kappa-\kappa_0)$$
$$\text{for } \kappa\in\mathcal{L}\cup\mathcal{D}^+, \tag{6.19}$$

where $\tilde{Q}^+(\kappa,s)$ is a regular function for $\kappa\in\mathcal{D}^+$ and of $o(\kappa)$ as $|\kappa|\to\infty$ in $\mathcal{D}^+$. The problem can be handled along the lines introduced by Senior [5] (see Section 6.2.2). Alternatively, one may assume $\hat{Z}^{\mathrm{E}}(s)>0$ and start from the following WH equation (cf. (6.19))

$$\mu_0^{-1}\hat{Z}^{\mathrm{E}}(s)\tilde{A}(\kappa,s)/\tilde{K}(\kappa,s)=[\tilde{Q}^+(\kappa,s)/\tilde{Q}^+(\kappa_0,s)]/(\kappa-\kappa_0), \tag{6.20}$$

which is required to hold for $\kappa\in\mathcal{L}$ only, and where we defined

$$\tilde{K}(\kappa,s)=\hat{Z}^{\mathrm{E}}(s)/[\hat{Z}^{\mathrm{E}}(s)+Z_{\perp}(\kappa)/2]. \tag{6.21}$$

Now, upon introducing normalized variables

$$\Pi=c_0\kappa, \tag{6.22}$$

$$\hat{\Lambda}=\zeta_0/2\hat{Z}^{\mathrm{E}}(s), \tag{6.23}$$

with $\zeta_0=1/\eta_0=c_0\mu_0$, (6.21) can be cast into the following form:

$$\bar{K}(\Pi,s)=(1-\Pi^2)^{1/2}/[\hat{\Lambda}+(1-\Pi^2)^{1/2}]. \tag{6.24}$$

Since $\bar{K}(\Pi,s)=1+O(|\Pi|^{-1})$ as $|\Pi|\to\infty$, the formulas of Plemelj can be applied to $\log\bar{K}(\Pi,s)$ to find

$$\bar{K}(\Pi,s)=\bar{K}^+(\Pi,s)\bar{K}^-(\Pi,s) \quad \text{for } \Pi\in\bar{\mathcal{L}}, \tag{6.25}$$

where $\bar{\mathcal{L}}$ lies in the normalized complex Π-plane and corresponds to $\mathcal{L}$ via (6.22). Accordingly, we write

$$\log\bar{K}^{\mp}(\Pi,s)=\tfrac{1}{2}\log\bar{K}(\Pi,s)\pm\frac{1}{2\mathrm{i}\pi}\mathrm{PV}\int_{w\in\bar{\mathcal{L}}}\log\bar{K}(w,s)\frac{\mathrm{d}w}{w-\Pi}$$
$$\text{for } \Pi\in\bar{\mathcal{L}}, \tag{6.26}$$

where PV denotes the Cauchy principal value of the integral. The corresponding integral, that is,

$$\log\bar{K}^{\mp}(\Pi,s)=\pm\frac{1}{2\mathrm{i}\pi}\int_{w\in\bar{\mathcal{L}}}\log\left[\frac{(1-w^2)^{1/2}}{\hat{\Lambda}+(1-w^2)^{1/2}}\right]\frac{\mathrm{d}w}{w-\Pi}$$
$$\text{for } \Pi\in\bar{\mathcal{D}}^{\mp}, \tag{6.27}$$

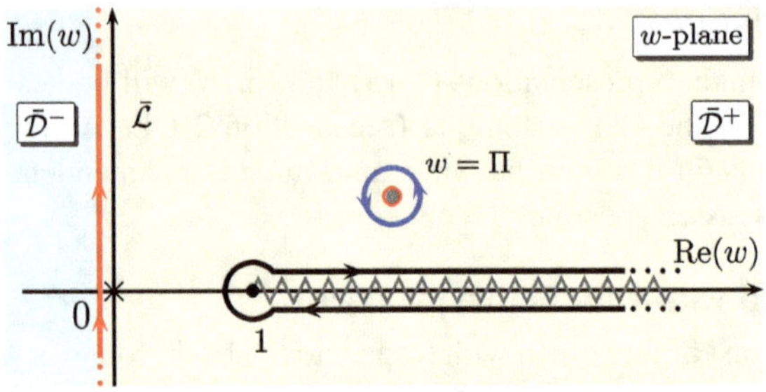

Figure 6.6 Complex w-plane for the evaluation of $\bar{K}^+(\Pi, s)$ with $\Pi \in \bar{\mathcal{D}}^+$

can be evaluated upon supplementing $\bar{\mathcal{L}}$ by a semicircular arc that extends to infinity in $\bar{\mathcal{D}}^{\mp}$ and using Cauchy's theorem (see Figure 6.6). In this way, we after a few straightforward steps arrive at

$$\log \bar{K}^+(\Pi, s) = -\frac{1}{\pi}\int_{w=1}^{\infty} \tan^{-1}\left[\frac{\hat{\Lambda}}{(w^2-1)^{1/2}}\right]\frac{\mathrm{d}w}{w+\Pi} \quad \text{for } \Pi \in \bar{\mathcal{D}}^+, \tag{6.28a}$$

and, likewise, noting that $\bar{K}^-(\Pi, s) = \bar{K}^+(-\Pi, s)$, we have

$$\log \bar{K}^-(\Pi, s) = -\frac{1}{\pi}\int_{w=1}^{\infty} \tan^{-1}\left[\frac{\hat{\Lambda}}{(w^2-1)^{1/2}}\right]\frac{\mathrm{d}w}{w-\Pi} \quad \text{for } \Pi \in \bar{\mathcal{D}}^-. \tag{6.28b}$$

The (improper) integrals over the unbounded domain of integration can be further developed to ease their numerical handling (see (6.58)).

With the factorization at our disposal, we make use of (6.25) in the WH equation (6.20) and get

$$\begin{aligned}&\mu_0^{-1}\hat{Z}^{\mathrm{E}}(s)\tilde{A}(\kappa, s)/\tilde{K}^-(\kappa, s) - \tilde{K}^+(\kappa_0, s)/(\kappa-\kappa_0)\\ &= \left[\tilde{K}^+(\kappa, s)\tilde{Q}^+(\kappa, s)/\tilde{Q}^+(\kappa_0, s) - \tilde{K}^+(\kappa_0, s)\right]/(\kappa-\kappa_0) \quad \text{for } \kappa \in \mathcal{L}. \end{aligned} \tag{6.29}$$

If the pole singularity at $\kappa = \kappa_0$ is located to the right of $\mathcal{L}$ (cf. Figure 6.2), then the left-hand side of the latter equation is regular in $\mathcal{D}^-$, while its right-hand side is regular in $\mathcal{D}^+$. Hence, in virtue of Liouville's theorem [11, p. 85], the left- and the right-hand sides do represent one and the same entire function that is a constant. Since $\tilde{K}^{\pm}(\kappa, s) = O(1)$ as $|\kappa| \to \infty$ in $\mathcal{D}^{\pm}$, respectively, the entire function is zero and we get

$$\tilde{A}(\kappa, s) = \mu_0\hat{Y}^{\mathrm{E}}(s)\frac{\tilde{K}^+(\kappa_0, s)\tilde{K}^-(\kappa, s)}{\kappa-\kappa_0} \quad \text{for } \kappa \in \mathcal{L}. \tag{6.30}$$

Equations (6.30) with (6.17) fully determine the complex slowness representations (6.15) and (6.16). Their transformation back to TD is discussed in the following text.

6.2.1 TD solution

The slowness domain representations (6.15) and (6.16) will be next transformed to TD with the aid of the CdH technique (see Section 2.1.1). As the transformation procedure depends on whether the wave amplitude is s-dependent or not, the two cases shall be discussed separately.

6.2.1.1 Screen with conductive properties

If the (high) contrast of the screen with respect to the background is due to its electric conductivity only, then the equivalent layer's admittance (see (3.15b)) is equal to its (real-valued and positive) conductance G^{E}. In this case, the wave amplitude $\tilde{A}$ is no longer s-dependent and we get (cf. (6.15) and (6.16))

$$\hat{E}_y^{\mathrm{s}}(x,z,s) = \frac{\hat{e}^{\mathrm{i}}(s)\,G^{\mathrm{E}}}{2\pi\mathrm{i}\;2\eta_0}\int_{\kappa\in\mathcal{L}}\frac{\tilde{K}^{+}(\kappa_0)\tilde{K}^{-}(\kappa)}{c_0 p(\kappa)}\exp\{-s[\kappa x + p(\kappa)|z|]\}\frac{\mathrm{d}\kappa}{\kappa-\kappa_0}$$
$$\text{for all } x\in\mathbb{R},\, z\in\mathbb{R} \text{ and } \{s\in\mathbb{R}; s>0\}, \tag{6.31}$$

in which $\tilde{K}^{\pm}$ follow from (6.28a) and (6.28b), i.e.

$$\tilde{K}^{+}(\kappa) = \exp\left\{-\frac{1}{\pi}\int_{w=1}^{\infty}\tan^{-1}\left[\frac{G^{\mathrm{E}}/2\eta_0}{(w^2-1)^{1/2}}\right]\frac{\mathrm{d}w}{w+c_0\kappa}\right\}, \tag{6.32a}$$

$$\tilde{K}^{-}(\kappa) = \exp\left\{-\frac{1}{\pi}\int_{w=1}^{\infty}\tan^{-1}\left[\frac{G^{\mathrm{E}}/2\eta_0}{(w^2-1)^{1/2}}\right]\frac{\mathrm{d}w}{w-c_0\kappa}\right\}. \tag{6.32b}$$

It is noted, again, that the (improper) integrals can be replaced with the alternative expressions according to (6.58). Yet alternatively, it will be demonstrated in Section 6.2.3 that (6.32a) and (6.32b) can be approximated by (6.59a) and (6.59b), respectively. Following the CdH methodology applied in Section 6.1.1, the corresponding scattered field can be expressed as (cf. (6.10))

$$\hat{E}_y^{\mathrm{s}}(x,z,s) = \frac{\hat{e}^{\mathrm{i}}(s)}{2\pi}\frac{G^{\mathrm{E}}}{\eta_0}\int_{\tau=\rho/c_0}^{\infty}\exp(-s\tau)$$
$$\mathrm{Re}\left\{\frac{\tilde{K}^{+}(\kappa_0)\tilde{K}^{-}[\kappa(\tau)]}{c_0\kappa(\tau)-c_0\kappa_0}\right\}\frac{\mathrm{d}\tau}{(\tau^2-\rho^2/c_0^2)^{1/2}} + \hat{E}_y^{\mathrm{s;g}}(x,z,s) \quad \text{for } \{s\in\mathbb{R}; s>0\}, \tag{6.33}$$

where we take the values of $\kappa(\tau)$ along $\mathcal{C}$ as given by (6.9) and $\hat{E}_y^{\mathrm{s;g}}$ has the meaning of the geometrical solution. The latter component, in fact, arises from the pole singularity at $\kappa=\kappa_0$ (cf. (6.11)) and can be thus written as

$$\hat{E}_y^{\mathrm{s;g}}(x,z,s) = -\frac{G^{\mathrm{E}}}{G^{\mathrm{E}}+2Y_{\perp}(\kappa_0)}\hat{e}^{\mathrm{i}}(s)\exp\{-s[\kappa_0 x + p(\kappa_0)|z|]\}$$
$$\text{for } x/\rho \geq \sin(\beta), \tag{6.34}$$

and recall that $Y_{\perp}(\kappa_0) = p(\kappa_0)/\mu_0 = \eta_0\cos(\beta)$. Since the integral in (6.33) has the form of Laplace transform (see (1.4)), its inversion back to TD can be accomplished

by inspection via Lerch's uniqueness theorem [12, Appendix]. Transforming also the plane-wave constituents (see (4.1) and (6.34)), we finally end up with the TD *total* electric-field strength (cf. (6.12))

$$E_y(x,z,t) = \frac{e^{\mathrm{i}}(t)}{2\pi}\frac{G^{\mathrm{E}}}{\eta_0} *_t \mathrm{Re}\left\{\frac{\tilde{K}^{+}(\kappa_0)\tilde{K}^{-}[\kappa(t)]}{c_0\kappa(t)-c_0\kappa_0}\right\}\frac{\mathrm{H}(t-\rho/c_0)}{(t^2-\rho^2/c_0^2)^{1/2}}$$
$$+ E_y^{\mathrm{s;g}}(x,z,t) + e^{\mathrm{i}}[t-\kappa_0 x + p(\kappa_0)z] \quad \text{for all } x\in\mathbb{R}, z\in\mathbb{R} \text{ and } t>0, \quad (6.35)$$

where

$$E_y^{\mathrm{s;g}}(x,z,t) = -\frac{G^{\mathrm{E}}}{G^{\mathrm{E}}+2Y_{\perp}(\kappa_0)}e^{\mathrm{i}}[t-\kappa_0 x - p(\kappa_0)|z|]\mathrm{H}[x-\rho\sin(\beta)]. \quad (6.36)$$

Again, the first term on the right-hand side of (6.35) can be interpreted as a diffracted cylindrical wave emanating from the edge of the screen, the second term specified by (6.36) is a plane wave representing the geometrical solution of the diffraction problem, and, finally, the third term is the incident plane wave. Apparently, the two plane-wave constituents would form the geometrical shadow (see Figure 6.3) in the limit $G^{\mathrm{E}} \to \infty$, which agrees with the result pertaining to the ideal PEC screen (see (6.12)).

6.2.1.2 Screen with conductive and dielectric properties

If the screen shows a high contrast in both electric conductivity and permittivity, the wave amplitude in the slowness representations (6.15) and (6.16) is s-dependent (see (6.30) with (6.28a) and (6.28b) with (3.15b)), which renders their analytical transformation to the TD inaccessible. A way out of the difficulty is an ingenious approximate factorization due to Koiter (see [13] and [14]) that replaces (6.24) along $\bar{\mathcal{L}}$ with an approximate function that can be factorized by inspection. The approximate function, say $\bar{K}^{\dagger}(\Pi,s)$, is chosen such that its value at $\Pi=0$ and its asymptotic behavior as $\Pi\to\pm\mathrm{i}\infty$ match the corresponding behavior of $\bar{K}(\Pi,s)$. Following this strategy, we take

$$\bar{K}^{\dagger}(\Pi,s) = (1-\Pi^2)^{1/2}/\left[(\hat{\Lambda}+1)^2-\Pi^2\right]^{1/2} \quad \text{for } \Pi\in\bar{\mathcal{L}}. \quad (6.37)$$

It is observed that the difference $\bar{K}-\bar{K}^{\dagger}$ is of $O(\Pi^2)$ as $\Pi\to 0$ and of $O(\Pi^{-1})$ as $\Pi\to\pm\mathrm{i}\infty$. Moreover, as $\hat{\Lambda}\to\infty$, which applies to PEC screens, both functions tend to $(1-\Pi^2)^{1/2}/\hat{\Lambda}$. Upon factorizing the latter and substituting the result in (6.30), we obtain the wave amplitude pertaining to the PEC screen (see (6.7)), as expected. Figure 6.7 shows the relative error of the approximation along a section of $\bar{\mathcal{L}}$ for two values of $\hat{\Lambda}$. Hence, applying the approximation and factorizing the result, we obtain (cf. (6.32a)–(6.32b))

$$\tilde{K}^{+}(\kappa,s) \overset{\kappa}{\simeq} \left(\frac{2\eta_0}{C^{\mathrm{E}}}\right)^{1/2}\frac{(1+c_0\kappa)^{1/2}}{[s+G^{\mathrm{E}}/C^{\mathrm{E}}+(2\eta_0/C^{\mathrm{E}})(1+c_0\kappa)]^{1/2}}, \quad (6.38\text{a})$$

$$\tilde{K}^{-}(\kappa,s) \overset{\kappa}{\simeq} \left(\frac{2\eta_0}{C^{\mathrm{E}}}\right)^{1/2}\frac{(1-c_0\kappa)^{1/2}}{[s+G^{\mathrm{E}}/C^{\mathrm{E}}+(2\eta_0/C^{\mathrm{E}})(1-c_0\kappa)]^{1/2}}, \quad (6.38\text{b})$$

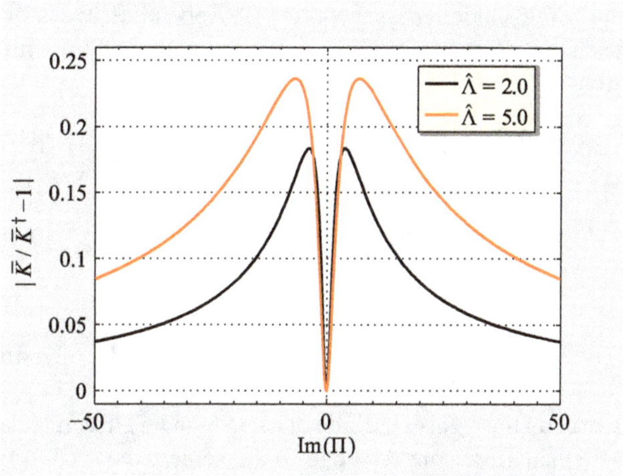

Figure 6.7 Relative error of the approximate function $\bar{K}^{\dagger}$ along $\bar{\mathcal{L}}$

where $\overset{\kappa}{\simeq}$ denotes the Koiter approximation. The latter expressions are subsequently substituted in (6.30) to approximate the diffracted-wave amplitude, which is in turn used to specify the wave-slowness representations (6.15) and (6.16). Following further the CdH procedure described in Section 6.1.1, we get (cf. (6.10))

$$\hat{E}^{\mathrm{s}}_y(x,z,s) = \frac{\hat{e}^{\mathrm{i}}(s)}{2\pi}\int_{\tau=\rho/c_0}^{\infty} \exp(-s\tau)\mathrm{Re}\{\tilde{A}[\kappa(\tau),s]\}\frac{\mathrm{d}\tau}{(\tau^2-\rho^2/c_0^2)^{1/2}} + \hat{E}^{\mathrm{s;g}}_y(x,z,s) \quad \text{for } \{s\in\mathbb{R}; s>0\}, \tag{6.39}$$

where $\hat{E}^{\mathrm{s;g}}_y$ is, again, associated with the geometrical solution of the diffraction problem. Accordingly, upon evaluating the pole contribution, we find (cf. (3.15b) and (6.34))

$$\hat{E}^{\mathrm{s;g}}_y(x,z,s) = -\frac{sC^{\mathrm{E}}+G^{\mathrm{E}}}{sC^{\mathrm{E}}+G^{\mathrm{E}}+2Y_{\perp}(\kappa_0)}\hat{e}^{\mathrm{i}}(s)\exp\{-s[\kappa_0 x + p(\kappa_0)|z|]\} \quad \text{for } x/\rho \geq \sin(\beta). \tag{6.40}$$

Application of some elementary rules of the one-sided time Laplace transformation and using Lerch's theorem on the uniqueness of the transformation at real, positive, transform parameter [12, Appendix], we finally end up with

$$E_y(x,z,t) = \frac{e^{\mathrm{i}}(t)}{2\pi} \ast_t \int_{\tau=\rho/c_0}^{\infty} \mathrm{Re}\{A[\kappa(\tau),t-\tau]\}\frac{\mathrm{d}\tau}{(\tau^2-\rho^2/c_0^2)^{1/2}} + E^{\mathrm{s;g}}_y(x,z,t) + e^{\mathrm{i}}[t-\kappa_0 x + p(\kappa_0)z] \quad \text{for all } x\in\mathbb{R}, z\in\mathbb{R} \text{ and } t>0, \tag{6.41}$$

where

$$A(\kappa,t) \overset{\kappa}{\simeq} \frac{2p^+(\kappa_0)p^-(\kappa)}{\kappa-\kappa_0}\,\delta(t)$$
$$-\frac{2p^+(\kappa_0)p^-(\kappa)}{\kappa-\kappa_0}\frac{\eta_0}{C^{\mathrm{E}}}\Big\{\lambda(\kappa,\kappa_0)\mathrm{I}_0[(\eta_0/C^{\mathrm{E}})\omega(\kappa,\kappa_0)t]$$
$$-\,\omega(\kappa,\kappa_0)\mathrm{I}_1[(\eta_0/C^{\mathrm{E}})\omega(\kappa,\kappa_0)t]\Big\}$$
$$\times\exp[-(\eta_0/C^{\mathrm{E}})\lambda(\kappa,\kappa_0)t]\exp[-(G^{\mathrm{E}}/C^{\mathrm{E}})t]\mathrm{H}(t). \tag{6.42}$$

In (6.42), $\mathrm{I}_0(x)$ and $\mathrm{I}_1(x)$ denote the modified Bessel functions of the first kind and

$$\omega(\kappa,\kappa_0) = c_0[p^-(\kappa)]^2 - c_0[p^+(\kappa_0)]^2, \tag{6.43}$$

$$\lambda(\kappa,\kappa_0) = c_0[p^-(\kappa)]^2 + c_0[p^+(\kappa_0)]^2, \tag{6.44}$$

with $p^{\pm}(\kappa) = (1/c_0 \pm \kappa)^{1/2}$ (see (6.5)). For numerical implementation purposes, it is useful to introduce in (6.42) the modified *scaled* Bessel functions via $\mathbb{I}_{0,1}(x) = \mathrm{I}_{0,1}(x)\exp(-x)$ (see [15, Figure 9.8]) and use the fact that $\lambda(\kappa,\kappa_0) - \omega(\kappa,\kappa_0) = 2c_0[p^+(\kappa_0)]^2$ (see (6.95)). Finally, the TD counterpart of the geometrical solution (6.40) follows as

$$E_y^{\mathrm{s;g}}(x,z,t) = -e^{\mathrm{i}}[t-\kappa_0 x - p(\kappa_0)|z|]$$
$$*_t\left\{\delta(t) - \frac{2Y_\perp(\kappa_0)}{C^{\mathrm{E}}}\exp\left[-\frac{G^{\mathrm{E}}+2Y_\perp(\kappa_0)}{C^{\mathrm{E}}}t\right]\mathrm{H}(t)\right\}\mathrm{H}[x-\rho\sin(\beta)]. \tag{6.45}$$

Employing the "sifting" property of the Dirac-delta distribution, it is seen that the early-time part of the total electric-field response (6.41) is identical with the one corresponding to the PEC screen (see (6.12)). This is to be expected, since any relaxational phenomena are generally negligible just behind the wave front.

6.2.2 *Alternative factorization*

To demonstrate the relation of our results to the factorization procedure introduced in Ref. [5] (see also [4, Exercise 2.10]), the WH equation (6.20) is rewritten as

$$\mu_0^{-1}\tilde{A}(\kappa,s)/\tilde{H}(\kappa,s) = \left[\tilde{Q}^+(\kappa,s)/\tilde{Q}^+(\kappa_0,s)\right]/(\kappa-\kappa_0), \tag{6.46}$$

where (cf. (6.21))

$$\tilde{H}(\kappa,s) = \left[\hat{Z}^{\mathrm{E}}(s) + Z_\perp(\kappa)/2\right]^{-1} \tag{6.47}$$

is to be factorized. Employing further (6.22) and (6.23), we may cast the expression for $\tilde{H}(\kappa, s)$ into the following form:

$$\bar{H}(\Pi, s) = \left\{\hat{Z}^{\mathrm{E}}(s)\left[1 + \hat{\Lambda}/(1-\Pi^2)^{1/2}\right]\right\}^{-1}. \tag{6.48}$$

For an ideal PEC screen, the latter reduces to $\bar{H}(\Pi, s) = (2/\zeta_0)(1-\Pi^2)^{1/2}$, which can be easily factorized by inspection (see (6.5)). To handle the general case, however, we next define

$$\bar{F}(\Pi, s) = \frac{\mathrm{d}}{\mathrm{d}\Pi}\log\bar{H}(\Pi, s) = \frac{\hat{\Lambda}\Pi}{(\Pi^2-1)[\hat{\Lambda} + (1-\Pi^2)^{1/2}]} \tag{6.49}$$

that meets the condition for the additive splitting. Accordingly, we write

$$\bar{F}(\Pi, s) = \bar{F}^{+}(\Pi, s) + \bar{F}^{-}(\Pi, s) \quad \text{for } \Pi \in \bar{\mathcal{L}}, \tag{6.50}$$

where $\bar{F}^{\pm}(\Pi, s)$ are found through the formulas of Plemelj (see Figure 6.6)

$$\tilde{F}^{\mp}(\Pi, s) = \pm\frac{\hat{\Lambda}}{2\mathrm{i}\pi}\int_{w\in\bar{\mathcal{L}}}\frac{w}{w^2-1}\frac{1}{\hat{\Lambda}+(1-w^2)^{1/2}}\frac{\mathrm{d}w}{w-\Pi} \quad \text{for } \Pi \in \mathcal{D}^{\mp}. \tag{6.51}$$

Again, the integrand in (6.51) shows the pole singularities at $w = \{\pm 1, \Pi\}$ as well as branch cuts extending along $\{1 \le |\mathrm{Re}(w)| < \infty, \mathrm{Im}(w) = 0\}$. To demonstrate the procedure, we shall next evaluate the integration for $\tilde{F}^{+}(\Pi, s)$. Collecting the contributions as indicated in Figure 6.6, we get

$$\begin{aligned}\tilde{F}^{+}(\Pi, s) = {} & \frac{1}{2}\frac{1}{1-\Pi} - \frac{\Pi}{1-\Pi^2}\frac{\hat{\Lambda}}{\hat{\Lambda}+(1-\Pi^2)^{1/2}} \\ & - \frac{\hat{\Lambda}}{\pi}\int_{w=1}^{\infty}\frac{w}{(w^2-1)^{1/2}}\frac{1}{\hat{\Lambda}^2-1+w^2}\frac{\mathrm{d}w}{w-\Pi}.\end{aligned} \tag{6.52}$$

The integral in (6.52) corresponds to the integration around the loop enclosing the branch cut and can be evaluated in closed form. Substituting the result in (6.52) and integrating the contributions with respect to Π, we after a few steps of algebra end up with

$$\begin{aligned}\log\bar{H}^{+}(\Pi, s) = {} & -\frac{1}{2}\log\left\{\frac{\hat{Z}^{\mathrm{E}}(s)[\hat{\Lambda}+(1-\Pi^2)^{1/2}]}{1+\Pi}\right\} \\ & + \frac{1}{2\pi\mathrm{i}}\log\left[\frac{(\hat{\Lambda}^2-1)^{1/2}-\mathrm{i}\Pi}{(\hat{\Lambda}^2-1)^{1/2}+\mathrm{i}\Pi}\right]\cosh^{-1}(\hat{\Lambda}) \\ & - \frac{\hat{\Lambda}}{\pi}\int_{u=0}^{\Pi}\frac{u\sin^{-1}(u)}{\hat{\Lambda}^2-1+u^2}\frac{\mathrm{d}u}{(1-u^2)^{1/2}}.\end{aligned} \tag{6.53}$$

Through substitutions $u = \sin(\nu)$ and $\hat{\Lambda} = \cos(\hat{\chi})$, the integral in (6.53) can be rewritten to an alternative form and we then find

$$\log \bar{H}^{+}(\Pi, s) = -\frac{1}{2}\log\left\{\frac{\hat{Z}^{\mathrm{E}}(s)\big[\hat{\Lambda} + (1-\Pi^2)^{1/2}\big]}{1+\Pi}\right\}$$
$$+\frac{1}{2\pi\mathrm{i}}\log\left[\frac{(\hat{\Lambda}^2-1)^{1/2}-\mathrm{i}\Pi}{(\hat{\Lambda}^2-1)^{1/2}+\mathrm{i}\Pi}\right]\cosh^{-1}(\hat{\Lambda})$$
$$-\frac{1}{2\pi}\int_{\nu=0}^{\sin^{-1}(\Pi)}\left[\frac{\nu}{\sin(\nu+\hat{\chi})}+\frac{\nu}{\sin(\nu-\hat{\chi})}\right]\mathrm{d}\nu. \tag{6.54}$$

Expressing $\bar{H}^{+}(\Pi, s)$, we finally end up with

$$\bar{H}^{+}(\Pi, s) = \left[\frac{1}{\hat{Z}^{\mathrm{E}}(s)}\frac{1+\Pi}{\hat{\Lambda}+(1-\Pi^2)^{1/2}}\right]^{1/2}\left[\frac{(\hat{\Lambda}^2-1)^{1/2}-\mathrm{i}\Pi}{(\hat{\Lambda}^2-1)^{1/2}+\mathrm{i}\Pi}\right]^{\cosh^{-1}(\hat{\Lambda})/2\mathrm{i}\pi}$$
$$\exp\left\{-\frac{1}{2\pi}\int_{\nu=0}^{\sin^{-1}(\Pi)}\left[\frac{\nu}{\sin(\nu+\hat{\chi})}+\frac{\nu}{\sin(\nu-\hat{\chi})}\right]\mathrm{d}\nu\right\}. \tag{6.55a}$$

In a similar fashion, we may get

$$\bar{H}^{-}(\Pi, s) = \left[\frac{1}{\hat{Z}^{\mathrm{E}}(s)}\frac{1-\Pi}{\hat{\Lambda}+(1-\Pi^2)^{1/2}}\right]^{1/2}\left[\frac{(\hat{\Lambda}^2-1)^{1/2}-\mathrm{i}\Pi}{(\hat{\Lambda}^2-1)^{1/2}+\mathrm{i}\Pi}\right]^{-\cosh^{-1}(\hat{\Lambda})/2\mathrm{i}\pi}$$
$$\exp\left\{\frac{1}{2\pi}\int_{\nu=0}^{\sin^{-1}(\Pi)}\left[\frac{\nu}{\sin(\nu+\hat{\chi})}+\frac{\nu}{\sin(\nu-\hat{\chi})}\right]\mathrm{d}\nu\right\}. \tag{6.55b}$$

Clearly, $\bar{H}^{-}(\Pi, s) = \bar{H}^{+}(-\Pi, s)$ and the identity $\bar{H}(\Pi, s) = \bar{H}^{+}(\Pi, s)\bar{H}^{-}(\Pi, s)$ holds true (see (6.48)). The integrals can be further expressed in terms of Malyuzhinets' functions (see [6] and [8, p. 56]), but since these special functions are not available in standard computing environments such as MATLAB®, we shall retain the present (integral) form that can be readily evaluated using standard integration routines.

Finally, with the factorization at our disposal, the WH equation (6.46) is written in the following form (cf. (6.29)):

$$\mu_0^{-1}\tilde{A}(\kappa, s)/\tilde{H}^{-}(\kappa, s) - \tilde{H}^{+}(\kappa_0, s)/(\kappa-\kappa_0)$$
$$= \big[\tilde{H}^{+}(\kappa, s)\tilde{Q}^{+}(\kappa, s)/\tilde{Q}^{+}(\kappa_0, s) - \tilde{H}^{+}(\kappa_0, s)\big]/(\kappa-\kappa_0) \quad \text{for } \kappa \in \mathcal{L}. \tag{6.56}$$

Employing Liouville's theorem [11, p. 85], again, we end up with (cf. (6.30))

$$\tilde{A}(\kappa, s) = \mu_0 \tilde{H}^+(\kappa_0, s)\tilde{H}^-(\kappa, s)/(\kappa - \kappa_0) \quad \text{for } \kappa \in \mathcal{L}, \tag{6.57}$$

which can be substituted in the complex slowness representations (6.15) and (6.16). Finally, upon comparison of (6.21) with (6.47) and (6.28a) with (6.54), we get an alternative expression for the integral over the unbounded domain

$$\begin{aligned}
&\frac{1}{\pi}\int_{w=1}^{\infty} \tan^{-1}\left[\frac{\hat{\Lambda}}{(w^2-1)^{1/2}}\right]\frac{\mathrm{d}w}{w+\Pi} \\
&= \frac{1}{2}\log\left\{\frac{\hat{\Lambda} + (1-\Pi^2)^{1/2}}{1+\Pi}\right\} - \frac{1}{2\pi \mathrm{i}}\log\left[\frac{(\hat{\Lambda}^2-1)^{1/2} - \mathrm{i}\Pi}{(\hat{\Lambda}^2-1)^{1/2} + \mathrm{i}\Pi}\right]\cosh^{-1}(\hat{\Lambda}) \\
&+ \frac{1}{2\pi}\int_{\nu=0}^{\sin^{-1}(\Pi)}\left[\frac{\nu}{\sin(\nu+\hat{\chi})} + \frac{\nu}{\sin(\nu-\hat{\chi})}\right]\mathrm{d}\nu.
\end{aligned} \tag{6.58}$$

A similar relation applies to the factor that is regular in $\bar{\mathcal{D}}^-$.

6.2.3 Illustrative examples

In this section, we shall analyze the pulsed EM scattering of EM-penetrable semi-infinite screens. As to their excitation, we adopt the impulsive EM plane wave considered in Section 6.1.2.

In the first example, we analyze the TD response of a conductive screen characterized by its layer conductance ratio $G^{\mathrm{E}}/\eta_0 = 20$ (see (3.5a)). The total electric-field strength is calculated using (6.35), where we use (6.32a) and (6.32b) to evaluate the wave amplitude of the diffracted cylindrical wave. In addition, the latter integral expressions are simplified with the aid of Koiter's approximate factorization [13], whose application to screens with conductive and dielectric characteristics was discussed in Section 6.2.1.2. Pursuing this line of reasoning, (6.32a) and (6.32b) can be approximately replaced by

$$\tilde{K}^+(\kappa) \overset{\kappa}{\simeq} \frac{(1+c_0\kappa)^{1/2}}{(G^{\mathrm{E}}/2\eta_0 + 1 + c_0\kappa)^{1/2}}, \tag{6.59a}$$

$$\tilde{K}^-(\kappa) \overset{\kappa}{\simeq} \frac{(1-c_0\kappa)^{1/2}}{(G^{\mathrm{E}}/2\eta_0 + 1 - c_0\kappa)^{1/2}}. \tag{6.59b}$$

It is observed that for a highly conducting surface, (6.59a) and (6.59b) have the following asymptotic form:

$$\tilde{K}^\pm(\kappa) \overset{\kappa}{\simeq} \frac{(1\pm c_0\kappa)^{1/2}}{(G^{\mathrm{E}}/2\eta_0)^{1/2}}\left[1 + O(1/G^{\mathrm{E}})\right] \quad \text{as } G^{\mathrm{E}} \to \infty, \tag{6.60}$$

which through (6.30) yields the diffracted-wave amplitude (6.7) pertaining to the PEC sheet. To further validate the approximate expressions (6.59a) and (6.59b),

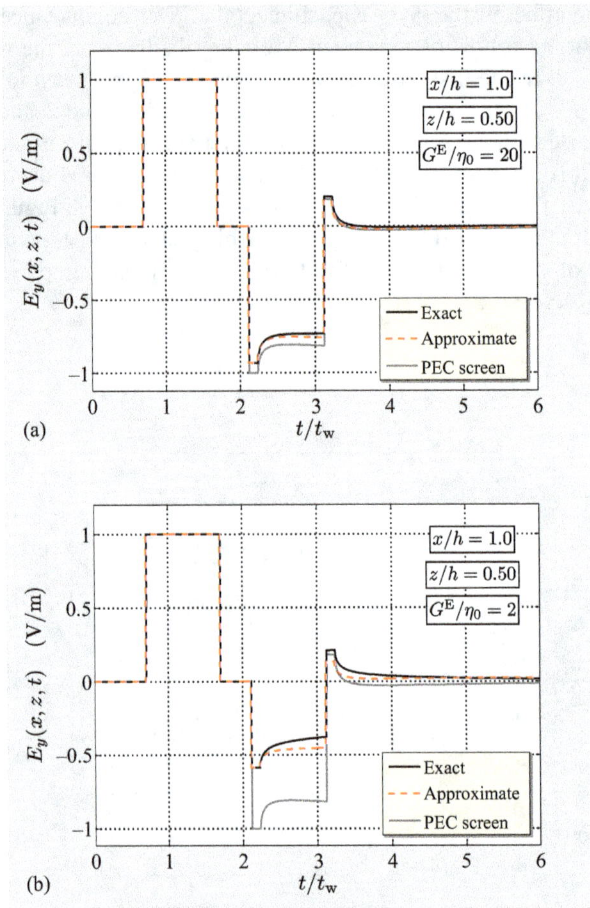

Figure 6.8 The total electric-field pulse shape calculated for (a) $G^E/\eta_0 = 20$ and (b) $G^E/\eta_0 = 2$ using (6.35) exactly (via (6.32a) and (6.32b)), approximately (via (6.59a) and (6.59b)) and using (6.12) pertaining to the PEC screen

we have evaluated the total electric-field strength at the field point $x/h = 1.0$ and $z/h = 0.5$. Figure 6.8 demonstrates that the significant simplification introduced through (6.59a) and (6.59b) is relevant for the chosen model parameters. As can be further seen from Figure 6.8(a), the PEC model analyzed in Section 6.1 may be applicable to relatively highly conducting screens as well. This is not the case for the second example, where we analyze the screen with a lower-layer conductance ratio, $G^E/\eta_0 = 2$. The corresponding Figure 6.8(b) clearly shows that while the discrepancies between the exact solution and its approximation may still be tolerable, the PEC model is no longer appropriate in this case.

In the next example, we analyze the TD electric field in the presence of a thin semi-infinite screen with both conductive and dielectric properties. In order to

demonstrate the effect of the layer capacitance, the layer conductance is kept constant throughout the following examples. We take, for instance, the relatively low value $G^{\mathrm{E}}/\eta_0 = 2$. In the first case, the layer capacitance is chosen to be relatively high such that $C^{\mathrm{E}}/\eta_0 t_{\mathrm{w}} = 10$. Despite the relatively low conductance, the corresponding response then does not significantly differ from the one of the PEC screen (see Figure 6.9(a)). In the second case, the layer capacitance is decreased so that $C^{\mathrm{E}}/\eta_0 t_{\mathrm{w}} = 0.10$. Comparing the resulting pulse shape shown in Figure 6.9(b) with Figure 6.8(b), it is concluded that the non-vanishing capacitance manifests itself in the occurrence of spikes at the wave front arrival corresponding to $t/t_{\mathrm{w}} = \sqrt{5}$ and the ensuing zero-crossing.

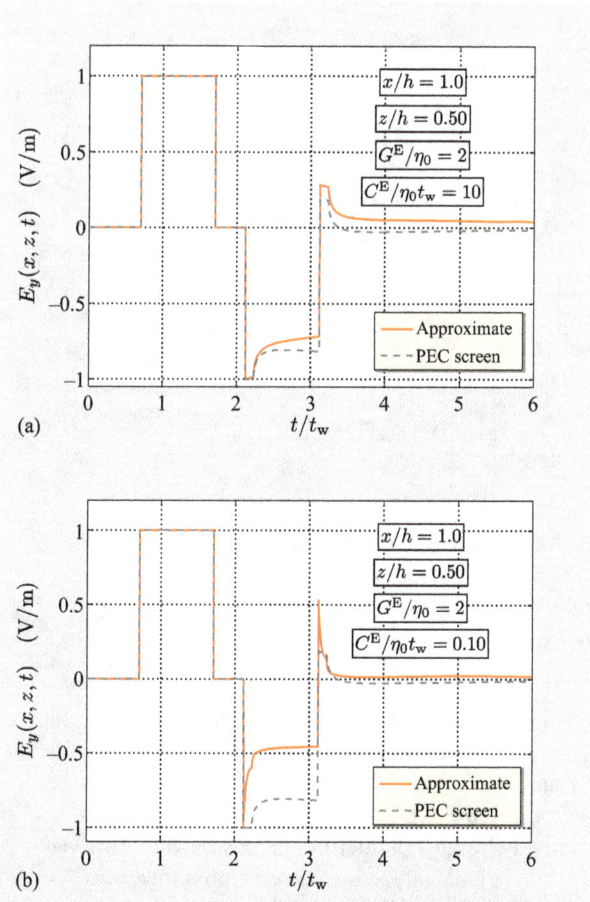

Figure 6.9 The total electric-field pulse shape calculated approximately for $G^{\mathrm{E}}/\eta_0 = 2$ and (a) $C^{\mathrm{E}}/\eta_0 t_{\mathrm{w}} = 10$ and (b) $C^{\mathrm{E}}/\eta_0 t_{\mathrm{w}} = 0.10$ using (6.41) and using (6.12) pertaining to the PEC screen

6.3 Pulsed EM diffraction by the junction of two coplanar semi-infinite sheets with conductive and dielectric properties

The main objective of this section is to describe the pulsed EM diffraction by two semi-infinite coplanar impedance sheets as shown in Figure 6.10.

The thin screens that occupy domains $\mathcal{S}_A = \{-\infty < x < 0, -\delta/2 < z < \delta/2\}$ and $\mathcal{S}_B = \{0 < x < \infty, -\delta/2 < z < \delta/2\}$ are located in the homogeneous, isotropic and loss-free embedding $\mathcal{D}^\infty$. The contrasting layers $\mathcal{S}_{A,B}$ are described by their electric permittivity $\epsilon_{A,B}$ and electric conductivity $\sigma_{A,B}$, respectively. The screens, as well as their plane-wave excitation defined via (4.1), are y-independent.

To model the interaction of the thin layer with the exciting EM plane wave, we shall employ the following transition conditions in terms of the total EM fields (cf. (6.14a) and (6.14b)):

$$\lim_{z\downarrow\delta/2} \hat{E}_y(x,z,s) - \lim_{z\uparrow-\delta/2} \hat{E}_y(x,z,s) = 0, \tag{6.61a}$$

$$\begin{aligned} &\lim_{z\downarrow\delta/2} \hat{H}_x(x,z,s) - \lim_{z\uparrow-\delta/2} \hat{H}_x(x,z,s) \\ &= \hat{Y}_A^{\mathrm{E}}(s)\hat{E}_y(x,0,s)[1-\mathrm{H}(x)] + \hat{Y}_B^{\mathrm{E}}(s)\hat{E}_y(x,0,s)\mathrm{H}(x) \\ &\quad \text{for all } x \in \mathbb{R}, \end{aligned} \tag{6.61b}$$

as $\delta \downarrow 0$, where $\hat{Y}_{A,B}^{\mathrm{E}}(s)$ are the equivalent layers' admittances pertaining to $\mathcal{S}_{A,B}$, respectively (see 3.15b).

We next in part follow the solution strategy proposed in [8, Section 3.7.1]. Accordingly, the original problem configuration is replaced with the equivalent one (see Figure 6.11), in which the layer is composed of an infinite layer of admittance

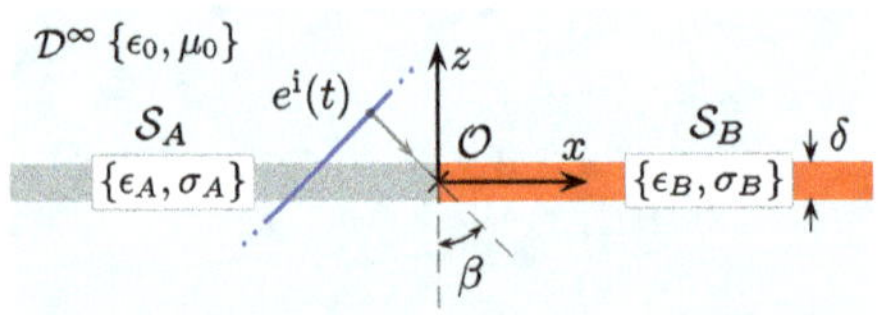

Figure 6.10 Pulsed EM plane wave impinging upon an impedance sheet junction

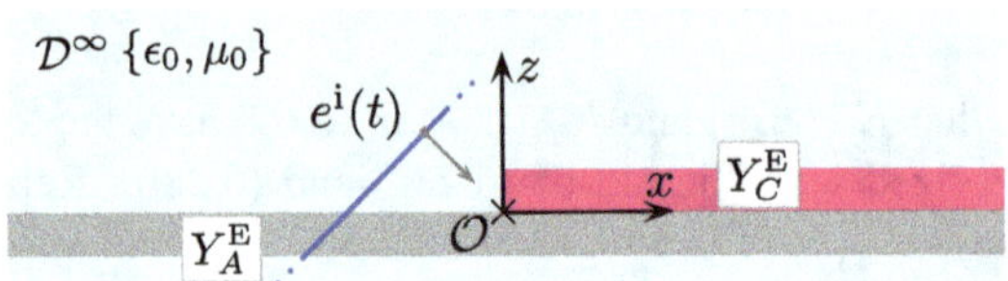

Figure 6.11 The equivalent problem configuration

$\hat{Y}_A^{\mathrm{E}}(s)$ and a semi-infinite layer in $x > 0$ of the equivalent layer admittance, $\hat{Y}_C^{\mathrm{E}}(s)$, chosen such that $\hat{Y}_A^{\mathrm{E}}(s) + \hat{Y}_C^{\mathrm{E}}(s) = \hat{Y}_B^{\mathrm{E}}(s)$. In view of the decomposition depicted in Figure 6.11, we define the EM field scattered by the equivalent semi-infinite sheet as the difference between the total EM field and the excitation one (denoted by superscript ^{e}) as (cf. (3.1))

$$\{\hat{E}_y^{\mathrm{s}}, \hat{H}_x^{\mathrm{s}}, \hat{H}_z^{\mathrm{s}}\}(x, z, s) = \{\hat{E}_y, \hat{H}_x, \hat{H}_z\}(x, z, s) - \{\hat{E}_y^{\mathrm{e}}, \hat{H}_x^{\mathrm{e}}, \hat{H}_z^{\mathrm{e}}\}(x, z, s), \tag{6.62}$$

where the excitation field is, as a matter of fact, the total EM field in absence of the semi-infinite layer. Hence, the excitation EM field satisfies

$$\lim_{z\downarrow\delta/2} \hat{H}_x^{\mathrm{e}}(x, z, s) - \lim_{z\uparrow-\delta/2} \hat{H}_x^{\mathrm{e}}(x, z, s) = \hat{Y}_A^{\mathrm{E}}(s)\hat{E}_y^{\mathrm{e}}(x, 0, s) \quad \text{for all } x \in \mathbb{R}, \tag{6.63}$$

and the electric-field strength in the layer, as occurs on the right-hand side of (6.63), follows from

$$\hat{E}_y^{\mathrm{e}}(x, 0, s) = \frac{2Y_\perp(\kappa_0)}{2Y_\perp(\kappa_0) + \hat{Y}_A^{\mathrm{E}}(s)} \hat{e}^{\mathrm{i}}(s) \exp(-s\kappa_0 x). \tag{6.64}$$

Postulating that the scattered field can be represented via complex slowness integrals (6.15) and (6.16), again, the continuity of $\hat{E}_y^{\mathrm{s}}(x, z, s)$ across $z = 0$ implies the equality (6.17), again. Consequently, the use of the slowness representation in condition (6.61b) yields the following pair of dual integral equations (cf. (6.18a) and (6.18b)):

$$\frac{1}{2\pi\mathrm{i}} \int_{\kappa\in\mathcal{L}} \exp(-s\kappa x) \frac{\tilde{A}(\kappa, s)}{\tilde{K}_A(\kappa, s)} \mathrm{d}\kappa = 0 \quad \text{for } x < 0, \tag{6.65}$$

and

$$\frac{1}{\mu_0} \frac{1}{2\pi\mathrm{i}} \int_{\kappa\in\mathcal{L}} \exp(-s\kappa x) \frac{\tilde{A}(\kappa, s)}{\tilde{K}_B(\kappa, s)} \mathrm{d}\kappa = -\tilde{K}_A(\kappa_0, s)\hat{Y}_C^{\mathrm{E}}(s) \exp(-s\kappa_0 x) \quad \text{for } x > 0, \tag{6.66}$$

where we used

$$\tilde{K}_{A,B}(\kappa, s) = \hat{Z}_{A,B}^{\mathrm{E}}(s) / \left[\hat{Z}_{A,B}^{\mathrm{E}}(s) + Z_\perp(\kappa)/2\right], \tag{6.67}$$

respectively. Note that $\tilde{K}_A(\kappa, s)$ and $\tilde{K}_B(\kappa, s)$ in (6.67) have the form of (6.21) and can be hence factorized with the aid of (6.28a) and (6.28b). Relation (6.65) then implies that $\tilde{A}(\kappa, s)/\tilde{K}_A(\kappa, s)$ is a regular function in $\mathcal{D}^-$ and of $o(1)$ as $|\kappa| \to \infty$ in $\mathcal{D}^-$. Next, (6.66) holds true if (cf. (6.20))

$$\frac{\hat{Z}_C^{\mathrm{E}}(s)}{\mu_0} \frac{\tilde{A}(\kappa, s)}{\tilde{K}_B(\kappa, s)} = \frac{\tilde{K}_A^-(\kappa_0, s)\tilde{K}_A^+(\kappa, s)}{\kappa - \kappa_0} \quad \text{for } \kappa \in \mathcal{L} \cup \mathcal{D}^+, \tag{6.68}$$

since $\tilde{K}_A^+(\kappa, s) = o(\kappa)$ as $|\kappa| \to \infty$ in $\mathcal{D}^+$. As the factorization of both $\tilde{K}_{A,B}(\kappa, s)$ is available, the WH equation (6.68) can be cast into the following form:

$$\frac{\hat{Z}_C^{\mathrm{E}}(s)}{\mu_0} \frac{\tilde{A}(\kappa, s)}{\tilde{K}_A(\kappa, s)} \frac{\tilde{K}_A^-(\kappa, s)}{\tilde{K}_A^-(\kappa_0, s)} \frac{1}{\tilde{K}_B^-(\kappa, s)} - \frac{\tilde{K}_B^+(\kappa_0, s)}{\kappa - \kappa_0} = \frac{\tilde{K}_B^+(\kappa, s) - \tilde{K}_B^+(\kappa_0, s)}{\kappa - \kappa_0}$$

$$\text{for } \kappa \in \mathcal{L}. \tag{6.69}$$

For $\kappa_0 = \sin(\beta)/c_0$ lying in $\mathcal{D}^+$, the left-hand side of (6.69) is regular in $\mathcal{D}^-$, while its right-hand side is regular in $\mathcal{D}^+$. Hence, by virtue of Liouville's theorem [11, p. 85], the left- and right-hand sides of the equation do represent the one and the same entire function that must be a constant. It can be shown that this constant is zero, so that we find

$$\tilde{A}(\kappa, s) = \mu_0 \hat{Y}_C^{\mathrm{E}}(s) \frac{\tilde{K}_A^+(\kappa, s)\tilde{K}_A^-(\kappa_0, s)\tilde{K}_B^+(\kappa_0, s)\tilde{K}_B^-(\kappa, s)}{\kappa - \kappa_0} \quad \text{for } \kappa \in \mathcal{L}. \tag{6.70}$$

Making use of the amplitude (6.70) in (6.15) and (6.16) with (6.17), the slowness representation of the scattered field is fully determined. Recall, however, that the thus specified scattered field describes the impact of the additional semi-infinite layer in the equivalent problem configuration (see Figure 6.11). Accordingly, the total EM field in the problem configuration can be found via (6.62) by adding the excitation field, that is, the total field in the absence of the equivalent semi-infinite layer. Finally observe that if the medium in $\mathcal{S}_A$ does not contrast with respect to the background medium, then $\hat{Y}_A^{\mathrm{E}}(s) = 0$ and (6.70) boils down to

$$\tilde{A}(\kappa, s) = \mu_0 \hat{Y}_B^{\mathrm{E}}(s) \frac{\tilde{K}_B^+(\kappa_0, s)\tilde{K}_B^-(\kappa, s)}{\kappa - \kappa_0} \quad \text{for } \kappa \in \mathcal{L}, \tag{6.71}$$

which is fully compatible with the results applying to a semi-infinite sheet with conductive and dielectric properties (see (6.30)).

6.3.1 TD solution

We next provide TD closed-form expressions describing the plane-wave excited EM fields in the presence of junction of two resistive coplanar screens. To that end, we shall largely follow the CdH procedure introduced in Section 6.1.1. Hence, the TD scattered electric-field strength due to the equivalent semi-infinite sheet of admittance $\hat{Y}_C^{\mathrm{E}}(s) = G_B^{\mathrm{E}} - G_A^{\mathrm{E}}$ can be expressed as a superposition of the edge-diffracted cylindrical wave and the pertaining geometrical solution (cf. (6.35))

$$E_y^{\mathrm{s}}(x, z, t) = \frac{e^{\mathrm{i}}(t)}{2\pi} \frac{G_B^{\mathrm{E}} - G_A^{\mathrm{E}}}{\eta_0} \overset{*}{_t} \mathrm{Re}\left\{ \frac{\tilde{K}_A^+[\kappa(t)]\tilde{K}_A^-(\kappa_0)\tilde{K}_B^+(\kappa_0)\tilde{K}_B^-[\kappa(t)]}{c_0\kappa(t) - c_0\kappa_0} \right\}$$
$$\times \frac{\mathrm{H}(t - \rho/c_0)}{(t^2 - \rho^2/c_0^2)^{1/2}} + E_y^{\mathrm{s;g}}(x, z, t). \tag{6.72}$$

In deriving the TD formula, we employed the fact that the diffracted-wave amplitude as given by (6.70) is for purely resistive sheets s-independent. Furthermore, recall that the values of κ along the CdH path are specified by (6.9) and the diffracted-wave amplitude can be computed via (cf. (6.32a) and (6.32b))

$$\tilde{K}^{\pm}_{A,B}(\kappa) = \exp\left\{ -\frac{1}{\pi}\int_{w=1}^{\infty} \tan^{-1}\left[\frac{G^{\mathrm{E}}_{A,B}/2\eta_0}{(w^2-1)^{1/2}}\right]\frac{\mathrm{d}w}{w \pm c_0\kappa}\right\}, \tag{6.73}$$

respectively. The geometrical solution can be obtained upon evaluating the pole contribution at $\kappa = \kappa_0$ and transforming the result to the TD. In this manner, we end up with (cf. (6.36))

$$E^{\mathrm{s;g}}_y(x,z,t) = -\frac{G^{\mathrm{E}}_B - G^{\mathrm{E}}_A}{2Y_\perp(\kappa_0)} T_A T_B\, e^{\mathrm{i}}[t-\kappa_0 x - p(\kappa_0)|z|]\mathrm{H}[x-\rho\sin(\beta)], \tag{6.74}$$

where we used the transmission coefficients

$$T_{A,B} = \frac{2Y_\perp(\kappa_0)}{2Y_\perp(\kappa_0) + G^{\mathrm{E}}_{A,B}}, \tag{6.75}$$

pertaining to infinite sheets of conductivity $\sigma_{A,B}$, respectively. Apparently, if $G^{\mathrm{E}}_A = G^{\mathrm{E}}_B$, the scattered EM field is zero and the total EM field in the configuration is equal to the excitation one. Accordingly, the latter field can be described by

$$E^{\mathrm{e}}_y(x,z,t) = e^{\mathrm{i}}[t-\kappa_0 x + p(\kappa_0)z] - (1-T_A)e^{\mathrm{i}}[t-\kappa_0 x - p(\kappa_0)z] \quad \text{for } z \geq 0, \tag{6.76}$$

and

$$E^{\mathrm{e}}_y(x,z,t) = T_A e^{\mathrm{i}}[t-\kappa_0 x + p(\kappa_0)z] \quad \text{for } z \leq 0. \tag{6.77}$$

Finally, the total field is found as $E_y = E^{\mathrm{e}}_y + E^{\mathrm{s}}_y$ using (6.72), (6.74) with (6.76) or (6.77). This result is used in the following section.

6.3.2 Illustrative examples

To provide an illustrative example, we shall plot the spatial electric-field distribution in the presence of a junction of two resistive coplanar screens. For this purpose, we employ the TD formulas introduced in the previous section. Likewise in Section 6.1.2, the layer is irradiated by an impulsive EM plane wave with the unit rectangular pulse shape (see (6.13)) and $\beta = \pi/4$. In the example that follows we take $G^{\mathrm{E}}_A/\eta_0 = 0.2$ and $G^{\mathrm{E}}_B/\eta_0 = 20$, so that relatively poorly conducting $\mathcal{S}_A$

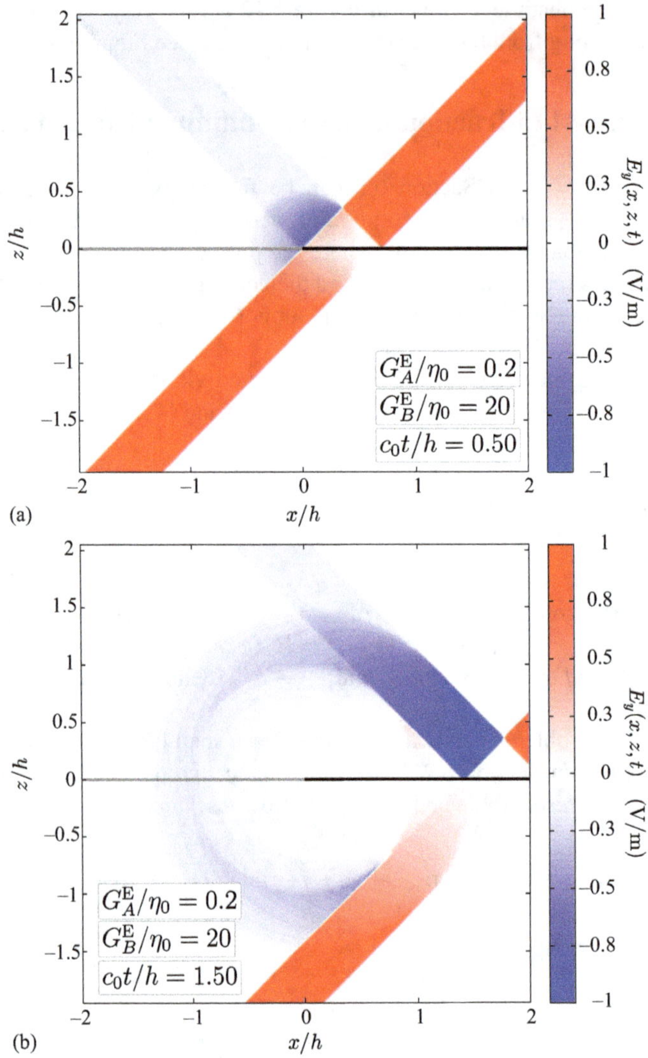

Figure 6.12 Spatial distributions of the electric-field strength excited by the impulsive plane wave in the presence of a junction of two resistive sheets as observed at (a) $c_0t/h = 0.50$ and (b) $c_0t/h = 1.50$

adjoins with a good conductor $\mathcal{S}_B$. Figure 6.12(a) and (b) shows the resulting total electric-field spatial distributions as observed at $c_0t/h = 0.50$ and $c_0t/h = 1.50$.

In contrast to the corresponding result pertaining to the edge diffraction by a semi-infinite PEC screen (see Figure 6.4), in Figure 6.12 we may observe the EM field reflected from $\mathcal{S}_A$. It is also noted that since the edge-diffracted cylindrical wave at a junction is proportional to the contrast of the adjoining sheets (see (6.72)), the

cylindrical wave constituent shown in Figure 6.12 is relatively weaker with respect to the one emanating from the edge of the PEC screen (see Figure 6.4).

6.4 Pulsed EM diffraction by a semi-infinite metasurface

The analysis presented in Section 6.2 can be further extended by incorporating (highly contrasting) magnetic constitutive properties of the semi-infinite thin layer. This is the main goal of this section, where the pulsed EM plane-wave response of a metasurface with combined magnetic and dielectric properties (see Figure 6.13) is derived. Likewise in Section 6.2, the layer is supposed to be activated via the uniform EM plane wave represented via (4.1).

The scattering problem can be then formulated via the cross-layer conditions (3.10a) and (3.10b) in the following form (cf. (4.2a) and (4.2b)):

$$\hat{E}_y^+ - \hat{E}_y^- = \hat{Z}^{\mathrm{M}}(s)\langle H_x\rangle_-^+ \mathrm{H}(x), \tag{6.78a}$$

$$\hat{H}_x^+ - \hat{H}_x^- = \hat{Y}^{\mathrm{E}}(s)\langle E_y\rangle_-^+ \mathrm{H}(x) \quad \text{for all } x \in \mathbb{R}, \tag{6.78b}$$

applying to the tangential components of the (y-independent) electric- and magnetic-field strengths, $\hat{E}_y = \hat{E}_y(x, z, s)$ and $\hat{H}_x = \hat{H}_x(x, z, s)$, respectively, in the limits as $z \downarrow \delta/2$ (denoted by superscript $^+$) and $z \uparrow -\delta/2$ (denoted by superscript $^-$) with $\delta \downarrow 0$.

To solve the problem analytically in the TD, we shall largely follow the solution strategy described in Section 6.2. Therefore, the scattered electric-field strength is supposed to be representable via the complex slowness integrals

$$\hat{E}_y^{\mathrm{s}}(x, z, s) = \frac{\hat{e}^{\mathrm{i}}(s)}{2\pi \mathrm{i}} \int_{\kappa \in \mathcal{L}} \exp\{-s[\kappa x + p(\kappa) z]\} \tilde{A}(\kappa, s) \frac{\mathrm{d}\kappa}{2p(\kappa)}$$
$$\text{for all } x \in \mathbb{R}, z > 0 \text{ and } \{s \in \mathbb{R}; s > 0\}, \tag{6.15 revisited}$$

and

$$\hat{E}_y^{\mathrm{s}}(x, z, s) = \frac{\hat{e}^{\mathrm{i}}(s)}{2\pi \mathrm{i}} \int_{\kappa \in \mathcal{L}} \exp\{-s[\kappa x - p(\kappa) z]\} \tilde{B}(\kappa, s) \frac{\mathrm{d}\kappa}{2p(\kappa)}$$
$$\text{for all } x \in \mathbb{R}, z < 0 \text{ and } \{s \in \mathbb{R}; s > 0\}. \tag{6.16 revisited}$$

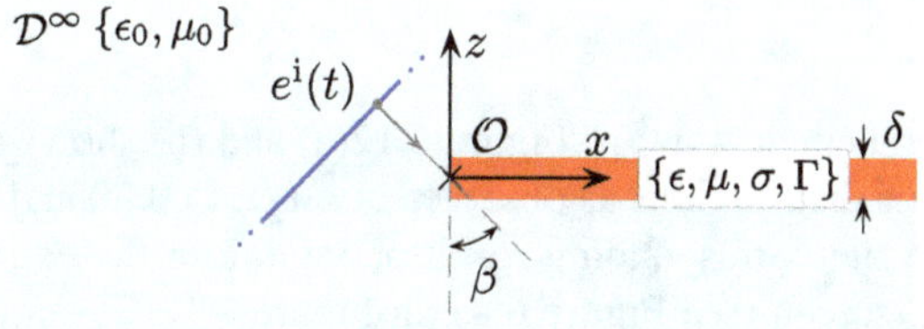

Figure 6.13 Pulsed EM plane wave impinging upon a semi-infinite metasurface

Upon invoking the continuity of the incident field across the layer, combination of the saltus-type conditions (6.78a) and (6.78b) with the complex slowness representations yields four integral relations. First, making use of (6.78a), we obtain (cf. (6.18a))

$$\frac{1}{2\pi \mathrm{i}}\int_{\kappa\in\mathcal{L}} \exp(-s\kappa x)\tilde{U}(\kappa,s)\mathrm{d}\kappa = 0 \quad \text{for } x<0, \tag{6.79a}$$

and (cf. (6.18b))

$$\frac{1}{\mu_0}\frac{1}{2\pi \mathrm{i}}\int_{\kappa\in\mathcal{L}} \exp(-s\kappa x)c_0 p(\kappa)\frac{\tilde{U}(\kappa,s)}{\tilde{K}_M(\kappa,s)}\mathrm{d}\kappa = -2Y_\perp(\kappa_0)\exp(-s\kappa_0 x) \quad \text{for } x>0, \tag{6.79b}$$

where we used

$$\tilde{U}(\kappa,s) = \left[\tilde{B}(\kappa,s) - \tilde{A}(\kappa,s)\right]/2c_0 p(\kappa), \tag{6.80}$$

$$\tilde{K}_M(\kappa,s) = \hat{Z}^{\mathrm{M}}(s)/\left[\hat{Z}^{\mathrm{M}}(s) + 2Z_\perp(\kappa)\right]. \tag{6.81}$$

Second, we employ the second cross-layer condition (6.78b) and get

$$\frac{1}{2\pi \mathrm{i}}\int_{\kappa\in\mathcal{L}} \exp(-s\kappa x)\tilde{V}(\kappa,s)\mathrm{d}\kappa = 0 \quad \text{for } x<0, \tag{6.82a}$$

and

$$\frac{1}{\mu_0}\frac{1}{2\pi \mathrm{i}}\int_{\kappa\in\mathcal{L}} \exp(-s\kappa x)\frac{\tilde{V}(\kappa,s)}{\tilde{K}_E(\kappa,s)}\mathrm{d}\kappa = -\hat{Y}^{\mathrm{E}}(s)\exp(-s\kappa_0 x) \quad \text{for } x>0, \tag{6.82b}$$

where we used

$$\tilde{V}(\kappa,s) = \left[\tilde{B}(\kappa,s) + \tilde{A}(\kappa,s)\right]/2, \tag{6.83}$$

$$\tilde{K}_E(\kappa,s) = \hat{Z}^{\mathrm{E}}(s)/\left[\hat{Z}^{\mathrm{E}}(s) + Z_\perp(\kappa)/2\right]. \tag{6.84}$$

Equations (6.79a), (6.79b) and (6.82a), (6.82b) constitute a WH problem that is further solved.

First, it is observed that (6.79a) is satisfied if $\tilde{U}(\kappa,s)$ is a regular function in $\mathcal{D}^-$ and of $o(1)$ as $|\kappa|\to\infty$ in $\mathcal{D}^-$. Furthermore, (6.79b) holds true if

$$\frac{c_0 p(\kappa)}{\mu_0}\frac{\tilde{U}(\kappa,s)}{\tilde{K}_M(\kappa,s)} = 2Y_\perp(\kappa_0)\frac{\tilde{Q}^+(\kappa,s)}{\tilde{Q}^+(\kappa_0,s)}\frac{1}{\kappa-\kappa_0} \quad \text{for } \kappa\in\mathcal{L}\cup\mathcal{D}^+, \tag{6.85}$$

where $\tilde{Q}^+(\kappa,s)$ is some function that is regular in $\mathcal{D}^+$ and of $o(\kappa)$ as $|\kappa|\to\infty$ in $\mathcal{D}^+$, provided that the integration contour $\mathcal{L}$ is chosen such that $\kappa_0\in\mathcal{D}^+$. Similar conclusions are next drawn from (6.82a) and (6.82b). Hence, (6.82a) implies that $\tilde{V}(\kappa,s)$ is a regular function in $\mathcal{D}^-$, while (6.82b) is satisfied if

$$\frac{1}{\mu_0}\frac{\tilde{V}(\kappa,s)}{\tilde{K}_E(\kappa,s)} = \hat{Y}^{\mathrm{E}}(s)\frac{\tilde{H}^+(\kappa,s)}{\tilde{H}^+(\kappa_0,s)}\frac{1}{\kappa-\kappa_0} \quad \text{for } \kappa\in\mathcal{L}\cup\mathcal{D}^+, \tag{6.86}$$

where $\tilde{H}^+(\kappa, s)$ is some function that is regular in $\mathcal{D}^+$ and of $o(\kappa)$ as $|\kappa| \to \infty$ in $\mathcal{D}^+$. Equations (6.85) and (6.86) represent WH relations that can be solved through the factorization technique (e.g. [4]). To that end, functions $\tilde{K}_E(\kappa, s)$ and $\tilde{K}_M(\kappa, s)$ need to be factorized into a function that is regular in $\mathcal{D}^-$ and a function that is regular in $\mathcal{D}^+$. As both $\tilde{K}_E(\kappa, s)$ and $\tilde{K}_M(\kappa, s)$ can be readily cast into the form of (6.24), we may employ (6.28a) and (6.28b) to get

$$\tilde{K}^{\pm}_{E,M}(\kappa, s) = \exp\left\{ -\frac{1}{\pi} \int_{w=1}^{\infty} \tan^{-1}\left[\frac{\hat{\Lambda}^{\mathrm{E,M}}(s)}{(w^2-1)^{1/2}}\right] \frac{\mathrm{d}w}{w \pm c_0\kappa} \right\} \quad \text{for } \kappa \in \mathcal{D}^{\pm}, \tag{6.87}$$

respectively, where $\hat{\Lambda}^{\mathrm{E}}(s) = \hat{Y}^{\mathrm{E}}(s)/2\eta_0$ and $\hat{\Lambda}^{\mathrm{M}}(s) = 2\zeta_0/\hat{Z}^{\mathrm{M}}(s)$. It is noted that the approximate factorization due to Koiter is applicable in this case too (see Section 6.2.1.2). Consequently, through (6.5) and (6.87), the first WH relation is cast into the following form:

$$\frac{p^-(\kappa)}{\mu_0} \frac{\tilde{U}(\kappa, s)}{\tilde{K}^-_M(\kappa, s)} - \frac{2Y_\perp(\kappa_0)}{c_0\kappa - c_0\kappa_0} \frac{\tilde{K}^+_M(\kappa_0, s)}{p^+(\kappa_0)} = \frac{2Y_\perp(\kappa_0)}{c_0\kappa - c_0\kappa_0} \frac{\tilde{Q}^+(\kappa, s)\tilde{K}^+_M(\kappa, s)/p^+(\kappa) - \tilde{Q}^+(\kappa_0, s)\tilde{K}^+_M(\kappa_0, s)/p^+(\kappa_0)}{\tilde{Q}^+(\kappa_0, s)} \quad \text{for } \kappa \in \mathcal{L}. \tag{6.88}$$

Owing to the fact that $\kappa_0 \in \mathcal{D}^+$, we conclude that the left-hand side of relation (6.88) is regular in $\mathcal{D}^-$, while its right-hand side is regular in $\mathcal{D}^+$. This implies that the left- and the right-hand sides do represent one and the same entire function that is a constant (see [11, p. 85]). Since the left-hand side of (6.88) is of $o(\kappa^{1/2})$ as $|\kappa| \to \infty$ in $\mathcal{D}^-$, while its right-hand side is of $o(\kappa^{-1/2})$ as $|\kappa| \to \infty$ in $\mathcal{D}^+$, the extended form of Liouville's theorem implies that the entire function is zero, which yields

$$\tilde{U}(\kappa, s) = \frac{2p^-(\kappa_0)}{p^-(\kappa)} \frac{\tilde{K}^+_M(\kappa_0, s)\tilde{K}^-_M(\kappa, s)}{c_0\kappa - c_0\kappa_0} \quad \text{for } \kappa \in \mathcal{L}. \tag{6.89}$$

In a similar manner, we shall handle the second WH relation (6.86). Hence, the relation is first rewritten as

$$\frac{1}{\mu_0} \frac{\tilde{V}(\kappa, s)}{\tilde{K}^-_E(\kappa, s)} - \frac{\hat{Y}^{\mathrm{E}}(s)\tilde{K}^+_E(\kappa_0, s)}{\kappa - \kappa_0} = \frac{\hat{Y}^{\mathrm{E}}(s)}{\kappa - \kappa_0} \frac{\tilde{H}^+(\kappa, s)\tilde{K}^+_E(\kappa, s) - \tilde{H}^+(\kappa_0, s)\tilde{K}^+_E(\kappa_0, s)}{\tilde{H}^+(\kappa_0, s)} \quad \text{for } \kappa \in \mathcal{L}. \tag{6.90}$$

Since the left-hand side of (6.90) is regular in $\mathcal{D}^-$, while its right-hand side is regular in $\mathcal{D}^+$, they must represent one and the same function that is a constant. Since this constant is zero, we obtain

$$\tilde{V}(\kappa, s) = \mu_0 \hat{Y}^{\mathrm{E}}(s) \frac{\tilde{K}^+_E(\kappa_0, s)\tilde{K}^-_E(\kappa, s)}{\kappa - \kappa_0} \quad \text{for } \kappa \in \mathcal{L}. \tag{6.91}$$

Subsequently, upon combining (6.80) and (6.83) with (6.89) and (6.91), we arrive at the desired expressions for the diffracted-wave amplitudes

$$\tilde{A}(\kappa, s) = \mu_0 \hat{Y}^{\mathrm{E}}(s) \frac{\tilde{K}_E^{+}(\kappa_0, s)\tilde{K}_E^{-}(\kappa, s)}{\kappa - \kappa_0} - 2p^{-}(\kappa_0)p^{+}(\kappa) \frac{\tilde{K}_M^{+}(\kappa_0, s)\tilde{K}_M^{-}(\kappa, s)}{\kappa - \kappa_0} \tag{6.92}$$

$$\tilde{B}(\kappa, s) = \mu_0 \hat{Y}^{\mathrm{E}}(s) \frac{\tilde{K}_E^{+}(\kappa_0, s)\tilde{K}_E^{-}(\kappa, s)}{\kappa - \kappa_0} + 2p^{-}(\kappa_0)p^{+}(\kappa) \frac{\tilde{K}_M^{+}(\kappa_0, s)\tilde{K}_M^{-}(\kappa, s)}{\kappa - \kappa_0}$$

$$\text{for } \kappa \in \mathcal{L}. \tag{6.93}$$

It is immediately seen that the first term in both (6.92) and (6.93) is identical with the one applying to a sheet supporting electric currents only (see (6.30)). Consequently, the second term represents the effect of (relatively high) magnetic constitutive parameters of the metasurface. Apparently, this term changes its sign as a consequence of the jump of E_y when crossing the metasurface (see (6.78a)). With the wave amplitudes at our disposal, we may proceed further with the inverse of the s-domain scattered fields (6.15) and (6.16) to the TD.

6.4.1 TD solution

Substitution of (6.92) and (6.93) in (6.15) and (6.16) leads to the complex wave-slowness integrals that are amenable to transformation by the CdH technique. Hence, following the strategy described in Sections 6.1.1 and 6.2.1.2, the EM field scattered by the semi-infinite combined sheet can be expressed via (cf. (6.41))

$$E_y^{\mathrm{s}}(x, z, t) = \frac{e^{\mathrm{i}}(t)}{2\pi} *_t \int_{\tau=\rho/c_0}^{\infty} \mathrm{Re}\{A[\kappa(\tau), t - \tau]\} \frac{\mathrm{d}\tau}{(\tau^2 - \rho^2/c_0^2)^{1/2}}$$
$$+ E_y^{\mathrm{s;g}}(x, z, t) \quad \text{for all } x \in \mathbb{R}, z > 0 \text{ and } t > 0, \tag{6.94}$$

with

$$\begin{aligned} A(\kappa, t) \overset{\kappa}{\simeq}\ & \frac{2p^{+}(\kappa_0)p^{-}(\kappa)}{\kappa - \kappa_0}\delta(t) - \frac{2p^{-}(\kappa_0)p^{+}(\kappa)}{\kappa - \kappa_0}\delta(t) \\ & - \frac{2p^{+}(\kappa_0)p^{-}(\kappa)}{\kappa - \kappa_0}\frac{\eta_0}{C^{\mathrm{E}}}\Bigg\{\lambda(\kappa, \kappa_0)\mathbb{I}_0\left[\frac{\eta_0}{C^{\mathrm{E}}}\omega(\kappa, \kappa_0)t\right] \\ & - \omega(\kappa, \kappa_0)\mathbb{I}_1\left[\frac{\eta_0}{C^{\mathrm{E}}}\omega(\kappa, \kappa_0)t\right]\Bigg\}\exp\left\{-\frac{2\eta_0}{C^{\mathrm{E}}}c_0[p^{+}(\kappa_0)]^2 t\right\}\mathrm{H}(t) \\ & + \frac{2p^{-}(\kappa_0)p^{+}(\kappa)}{\kappa - \kappa_0}\frac{\zeta_0}{L^{\mathrm{M}}}\frac{1}{c_0^2[p^{+}(\kappa_0)p^{-}(\kappa)]^2}\Bigg\{\lambda(\kappa, \kappa_0)\mathbb{I}_0\left[\frac{\zeta_0}{L^{\mathrm{M}}}\frac{\omega(-\kappa_0, -\kappa)t}{c_0^2[p^{+}(\kappa_0)p^{-}(\kappa)]^2}\right] \\ & - \omega(-\kappa_0, -\kappa)\mathbb{I}_1\left[\frac{\zeta_0}{L^{\mathrm{M}}}\frac{\omega(-\kappa_0, -\kappa)t}{c_0^2[p^{+}(\kappa_0)p^{-}(\kappa)]^2}\right]\Bigg\}\exp\left\{-\frac{2\zeta_0}{L^{\mathrm{M}}}\frac{t}{c_0[p^{+}(\kappa_0)]^2}\right\}\mathrm{H}(t), \end{aligned} \tag{6.95}$$

where $\overset{\kappa}{\simeq}$ denotes Koiter's approximation, functions $\omega(\kappa,\kappa_0)$ and $\lambda(\kappa,\kappa_0)$ are defined via (6.43) and (6.44), respectively, and $\mathbb{I}_{0,1}(x) = \mathrm{I}_{0,1}(x)\exp(-x)$ (see [15, Figure 9.8]). With reference to (6.16) and (6.93), a similar expression applies to the TD scattered field below the layer. Hence, we write

$$E_y^{\mathrm{s}}(x,z,t) = \frac{e^{\mathrm{i}}(t)}{2\pi} *_t \int_{\tau=\rho/c_0}^{\infty} \mathrm{Re}\{B[\kappa(\tau), t-\tau]\} \frac{\mathrm{d}\tau}{(\tau^2-\rho^2/c_0^2)^{1/2}}$$

$$+ E_y^{\mathrm{s;g}}(x,z,t) \quad \text{for all } x \in \mathbb{R}, z < 0 \text{ and } t > 0, \tag{6.96}$$

where

$$B(\kappa,t) \overset{\kappa}{\simeq} \frac{2p^+(\kappa_0)p^-(\kappa)}{\kappa-\kappa_0}\delta(t) + \frac{2p^-(\kappa_0)p^+(\kappa)}{\kappa-\kappa_0}\delta(t)$$
$$- \frac{2p^+(\kappa_0)p^-(\kappa)}{\kappa-\kappa_0}\frac{\eta_0}{C^{\mathrm{E}}}\left\{\lambda(\kappa,\kappa_0)\mathbb{I}_0\left[\frac{\eta_0}{C^{\mathrm{E}}}\omega(\kappa,\kappa_0)t\right]\right.$$
$$\left. - \omega(\kappa,\kappa_0)\mathbb{I}_1\left[\frac{\eta_0}{C^{\mathrm{E}}}\omega(\kappa,\kappa_0)t\right]\right\}\exp\left\{-\frac{2\eta_0}{C^{\mathrm{E}}}c_0[p^+(\kappa_0)]^2 t\right\}\mathrm{H}(t)$$
$$- \frac{2p^-(\kappa_0)p^+(\kappa)}{\kappa-\kappa_0}\frac{\zeta_0}{L^{\mathrm{M}}}\frac{1}{c_0^2[p^+(\kappa_0)p^-(\kappa)]^2}\left\{\lambda(\kappa,\kappa_0)\mathbb{I}_0\left[\frac{\zeta_0}{L^{\mathrm{M}}}\frac{\omega(-\kappa_0,-\kappa)t}{c_0^2[p^+(\kappa_0)p^-(\kappa)]^2}\right]\right.$$
$$\left. - \omega(-\kappa_0,-\kappa)\mathbb{I}_1\left[\frac{\zeta_0}{L^{\mathrm{M}}}\frac{\omega(-\kappa_0,-\kappa)t}{c_0^2[p^+(\kappa_0)p^-(\kappa)]^2}\right]\right\}\exp\left\{-\frac{2\zeta_0}{L^{\mathrm{M}}}\frac{t}{c_0[p^+(\kappa_0)]^2}\right\}\mathrm{H}(t). \tag{6.97}$$

To fully determine the scattered field, it remains to specify the corresponding geometrical solution $E_y^{\mathrm{s;g}}(x,z,t)$. To this end, we evaluate the contribution of the pole singularity at $\kappa = \kappa_0$ and find (cf. (6.45) and (8.47))

$$E_y^{\mathrm{s;g}}(x,z,t) = -e^{\mathrm{i}}[t-\kappa_0 x - p(\kappa_0)|z|]$$
$$*_t \left\{\delta(t) - \frac{2Y_\perp(\kappa_0)}{C^{\mathrm{E}}}\exp\left[-\frac{2Y_\perp(\kappa_0)}{C^{\mathrm{E}}}t\right]\mathrm{H}(t)\right\}\mathrm{H}[x-\rho\sin(\beta)]$$
$$+ e^{\mathrm{i}}[t-\kappa_0 x - p(\kappa_0)|z|]\mathrm{sign}(z)$$
$$*_t \left\{\delta(t) - \frac{2Z_\perp(\kappa_0)}{L^{\mathrm{M}}}\exp\left[-\frac{2Z_\perp(\kappa_0)}{L^{\mathrm{M}}}t\right]\mathrm{H}(t)\right\}\mathrm{H}[x-\rho\sin(\beta)]$$
$$\text{for all } x \in \mathbb{R}, z \in \mathbb{R} \text{ and } t > 0, \tag{6.98}$$

and recall that $Y_{\perp}(\kappa_0) = 1/Z_{\perp}(\kappa_0) = \eta_0 \cos(\beta)$. It is seen that if the (transparency) condition $(C^{\mathrm{E}}/L^{\mathrm{M}})^{1/2} = Y_{\perp}(\kappa_0)$ is met (cf. (4.8)), the geometrical solution vanishes identically throughout $z > 0$. The closed-form analytical TD expressions introduced in this section will next be used to calculate the space–time distribution of the electric-field strength in the presence of a loss-free scalar metasurface.

6.4.2 Illustrative examples

For the sake of illustration, we first take the limits $C^{\mathrm{E}} \to \infty$ and $L^{\mathrm{M}} \to \infty$ and plot the electric-field distributions pertaining to PEC and perfectly magnetically conducting (PMC) semi-infinite screens, respectively, and their combination. To that end, we assume the impulsive uniform EM plane wave with the rectangular pulse shape (see (6.13)) and the angle of incidence $\beta = \pi/4$, again. Figure 6.14 shows the resulting electric-field spatial distributions as observed at

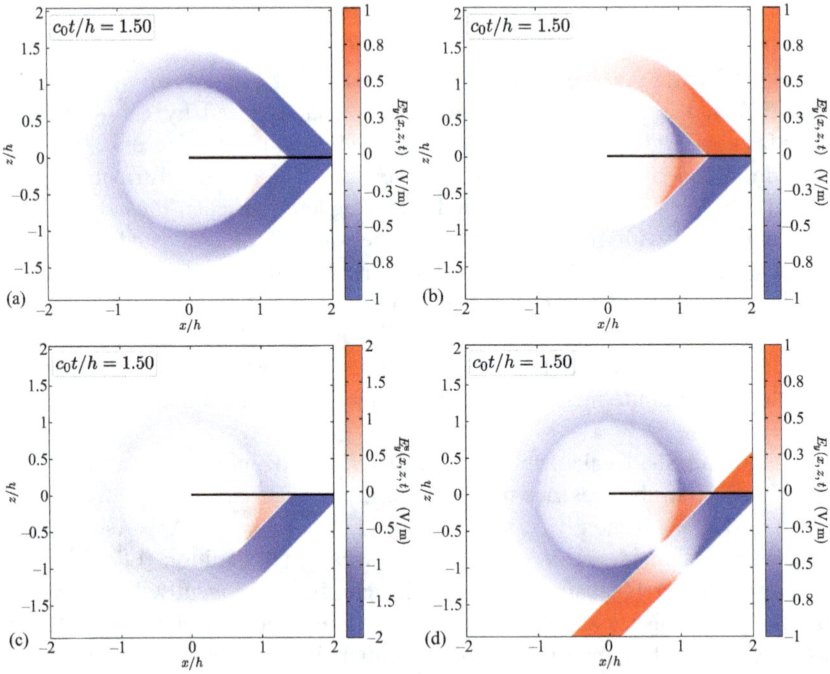

Figure 6.14 Spatial distributions of the electric-field strength at $c_0t/h = 1.50$ as excited by the impulsive plane wave: (a) the scattered field in the presence of a PEC screen ($C^{\mathrm{E}} \to \infty$); (b) the scattered field in the presence of a PMC screen ($L^{\mathrm{M}} \to \infty$); (c) the scattered field in the presence of a metasurface ($C^{\mathrm{E}} \to \infty$ and $L^{\mathrm{M}} \to \infty$); (d) the total field in the presence of a metasurface ($C^{\mathrm{E}} \to \infty$ and $L^{\mathrm{M}} \to \infty$)

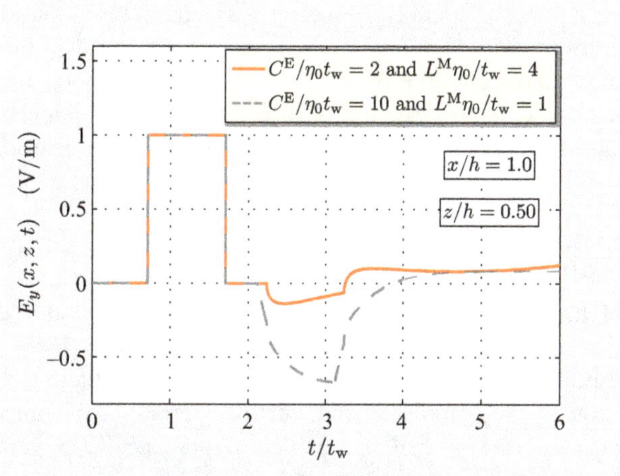

Figure 6.15 The total electric-field pulse shape calculated using (6.94) with (6.98)

$c_0 t/h = 1.50$. In Figure 6.14(a), we may observe the *scattered* field (cf. Figure 6.4(b)) in the presence of a PEC screen characterized by $C^{\mathrm{E}} \to \infty$. In a similar fashion, Figure 6.14(b) shows the *scattered* field in the presence of a PMC screen described by $L^{\mathrm{M}} \to \infty$. As expected, the latter field distribution jumps across the PMC screen due to the induced magnetic currents, thus showing the "odd" symmetry with respect to $z = 0$. The sum of the scattered fields from Figure 6.14(a) and (b) is consequently shown in Figure 6.14(c) and, finally, the *total* electric-field distribution in the presence of the metasurface (with $C^{\mathrm{E}} \to \infty$ and $L^{\mathrm{M}} \to \infty$) is given in Figure 6.14(d). Much like the pulsed field distribution in the presence of an infinite metasurface (see Figure 4.2), the transition across the semi-infinite metasurface changes the sign of the impinging pulsed EM field.

Second, we shall calculate the total electric-field pulse shape at $x/h = 1.0$ and $z/h = 0.50$ in the presence of two metasurfaces described by $C^{\mathrm{E}}/\eta_0 t_{\mathrm{w}} = 2.0$ with $L^{\mathrm{M}}\eta_0/t_{\mathrm{w}} = 4.0$ and $C^{\mathrm{E}}/\eta_0 t_{\mathrm{w}} = 10$ with $L^{\mathrm{M}}\eta_0/t_{\mathrm{w}} = 1.0$. Apparently, assuming the angle of incidence $\beta = \pi/4$, the transparency condition $(C^{\mathrm{E}}/L^{\mathrm{M}})^{1/2} = \eta_0 \cos(\beta)$ (see (4.8)) is satisfied for the first layer only. Consequently, the geometrical solution (6.98) vanishes identically above the "transparent" sheet and the relatively weak reflected fields come from the edge-diffracted cylindrical wave only (see Figure 6.15). It is interesting to observe that the relatively strong reflected field above the second layer, dominantly formed by the geometrical solution (6.98), appears at the field point earlier than the cylindrical wave front emanating from the edge of the "transparent" sheet. To further illustrate the phenomena, we plot the sum of the geometrical solution (6.98) and the incident field in the vicinity of the metasurfaces. Figure 6.16(a) then shows the reflection-free field distribution in the presence of the "transparent" sheet, while Figure 6.16(b) illustrates a relatively strong (geometrical) reflected field above the second layer at $c_0 t/h = 1.0$.

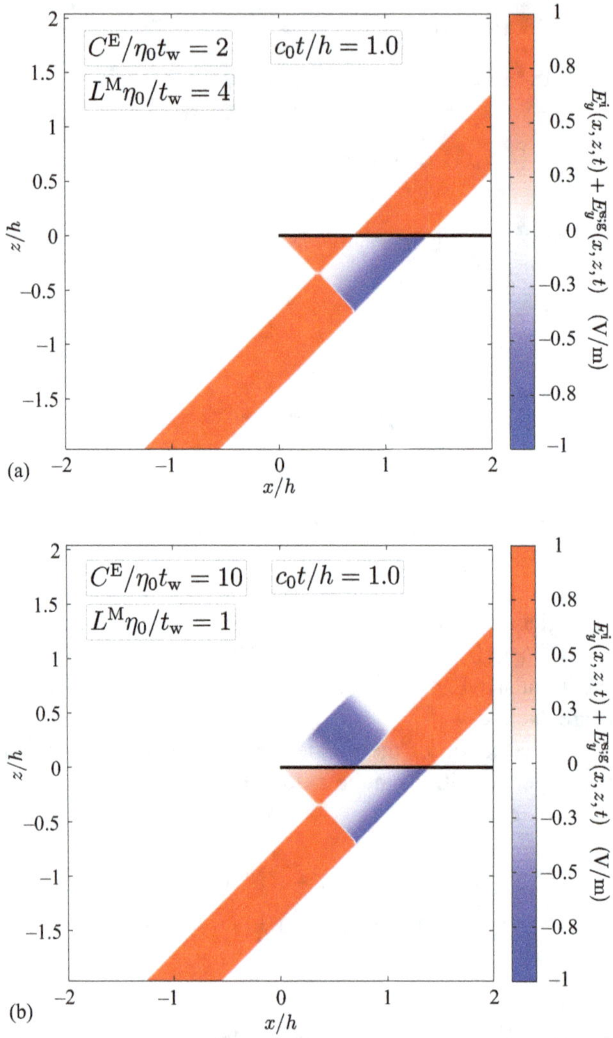

Figure 6.16 *Spatial distributions of* $E_y^{\mathrm{i}} + E_y^{\mathrm{s;g}}$ *at* $c_0 t/h = 1.0$ *in the presence of two metasurfaces with (a)* $C^{\mathrm{E}}/\eta_0 t_{\mathrm{w}} = 2.0$ *and* $L^{\mathrm{M}}\eta_0/t_{\mathrm{w}} = 4.0$*; (b)* $C^{\mathrm{E}}/\eta_0 t_{\mathrm{w}} = 10$ *and* $L^{\mathrm{M}}\eta_0/t_{\mathrm{w}} = 1.0$

6.5 Pulsed EM diffraction by the junction of two coplanar semi-infinite metasurfaces

In this section, we shall study the pulsed EM diffraction by the junction of two coplanar semi-infinite metasurfaces located in the homogeneous, isotropic and loss medium in $\mathcal{D}^\infty$ (see Figure 6.17). The metasurfaces occupy domains $\mathcal{S}_A = \{-\infty < x < 0, -\delta/2 < z < \delta/2\}$ and $\mathcal{S}_B = \{0 < x < \infty, -\delta/2 < z < \delta/2\}$. Their EM

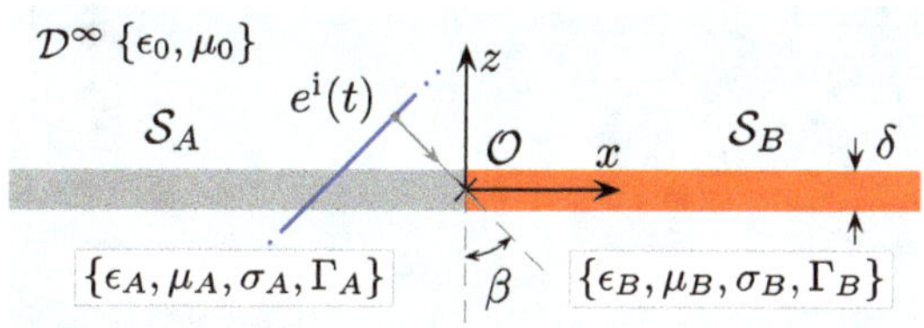

Figure 6.17 Pulsed EM plane wave impinging upon the junction of two metasurfaces

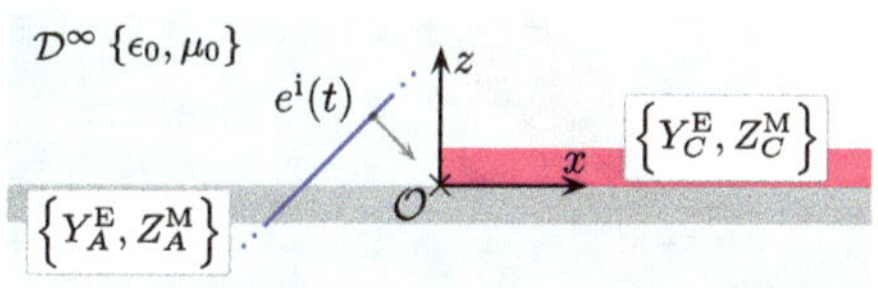

Figure 6.18 The equivalent problem configuration

constitutive properties are described by relatively high, real-valued and scalar parameters $\{\epsilon_{A,B}, \mu_{A,B}, \sigma_{A,B}, \Gamma_{A,B}\}$, respectively. The structure under consideration is irradiated by the impulsive EM plane wave defined by (4.1), so that only E-polarized transient fields are excited.

To model the interaction of the incident EM field with the junction, we shall employ the cross-layer conditions in the s-domain that can be written as (cf. (6.78a), (6.78b) with (6.61a), (6.61b))

$$\hat{E}_y^+ - \hat{E}_y^- = \hat{Z}_A^{\mathrm{M}}(s)\langle H_x\rangle_-^+[1 - \mathrm{H}(x)] + \hat{Z}_B^{\mathrm{M}}(s)\langle H_x\rangle_-^+\mathrm{H}(x), \tag{6.99a}$$

$$\hat{H}_x^+ - \hat{H}_x^- = \hat{Y}_A^{\mathrm{E}}(s)\langle E_y\rangle_-^+[1 - \mathrm{H}(x)] + \hat{Y}_B^{\mathrm{E}}(s)\langle E_y\rangle_-^+\mathrm{H}(x) \quad \text{for all } x \in \mathbb{R}, \tag{6.99b}$$

where $\hat{Y}_{A,B}^{\mathrm{E}}(s)$ and $\hat{Z}_{A,B}^{\mathrm{M}}(s)$ are the equivalent layer admittances and impedances pertaining to $\mathcal{S}_{A,B}$, respectively (see (3.15a) and (3.15b)).

Likewise in Section 6.3, we shall not attempt to solve the present problem directly, but instead, we shall tackle the equivalent problem configuration depicted in Figure 6.18. Hence, the original junction of two metasurfaces is represented by a semi-infinite layer of parameters $\{\hat{Y}_C^{\mathrm{E}}, \hat{Z}_C^{\mathrm{M}}\}$ that is in $x > 0$ superimposed on an infinite sheet described by $\{\hat{Y}_A^{\mathrm{E}}, \hat{Z}_A^{\mathrm{M}}\}$. To represent the original problem configuration, the parameters of the equivalent semi-infinite layer are chosen such that $\{\hat{Y}_C^{\mathrm{E}}, \hat{Z}_C^{\mathrm{M}}\} = \{\hat{Y}_B^{\mathrm{E}} - \hat{Y}_A^{\mathrm{E}}, \hat{Z}_B^{\mathrm{M}} - \hat{Z}_A^{\mathrm{M}}\}$. Consequently, by virtue of the decomposition, the scattered EM field is defined as

$$\{\hat{E}_y^{\mathrm{s}}, \hat{H}_x^{\mathrm{s}}, \hat{H}_z^{\mathrm{s}}\}(x, z, s) = \{\hat{E}_y, \hat{H}_x, \hat{H}_z\}(x, z, s) - \{\hat{E}_y^{\mathrm{e}}, \hat{H}_x^{\mathrm{e}}, \hat{H}_z^{\mathrm{e}}\}(x, z, s), \tag{6.62 revisited}$$

where superscript ^{e} denotes the excitation fields, that is, the total fields in the absence of the equivalent semi-infinite sheet. Therefore, the tangential components of the excitation fields satisfy the cross-layer conditions (cf. (4.2a) and (4.2b))

$$\hat{E}_y^{\mathrm{e};+} - \hat{E}_y^{\mathrm{e};-} = \hat{Z}_A^{\mathrm{M}}(s)\langle H_x^{\mathrm{e}}\rangle_-^+, \tag{6.100a}$$

$$\hat{H}_x^{\mathrm{e};+} - \hat{H}_x^{\mathrm{e};-} = \hat{Y}_A^{\mathrm{E}}(s)\langle E_y^{\mathrm{e}}\rangle_-^+ \quad \text{for all } x \in \mathbb{R}. \tag{6.100b}$$

Employing the latter conditions, it is found that (cf. (6.64))

$$\langle E_y^{\mathrm{e}}\rangle_-^+ = \frac{2Y_\perp(\kappa_0)}{2Y_\perp(\kappa_0) + \hat{Y}_A^{\mathrm{E}}(s)}\hat{e}^{\mathrm{i}}(s)\exp(-s\kappa_0 x), \tag{6.101}$$

$$\langle H_x^{\mathrm{e}}\rangle_-^+ = \frac{2Z_\perp(\kappa_0)}{2Z_\perp(\kappa_0) + \hat{Z}_A^{\mathrm{M}}(s)}Y_\perp(\kappa_0)\hat{e}^{\mathrm{i}}(s)\exp(-s\kappa_0 x). \tag{6.102}$$

Assuming now that the scattered field can be represented via (6.15) and (6.16), again, the use of (6.62) with (6.101) and (6.102) in the cross-layer conditions (6.99a) and (6.99b) yields the following integral equations:

$$\frac{1}{2\pi\mathrm{i}}\int_{\kappa\in\mathcal{L}}\exp(-s\kappa x)\frac{\tilde{U}(\kappa,s)}{\tilde{K}_{M;A}(\kappa,s)}\mathrm{d}\kappa = 0 \quad \text{for } x < 0, \tag{6.103a}$$

$$\frac{\hat{Z}_B^{\mathrm{M}}(s)}{4\mu_0}\frac{1}{2\pi\mathrm{i}}\int_{\kappa\in\mathcal{L}}\exp(-s\kappa x)\frac{\tilde{U}(\kappa,s)}{\tilde{K}_{M;B}(\kappa,s)}\mathrm{d}\kappa = -\frac{2\hat{Z}_C^{\mathrm{M}}(s)}{2Z_\perp(\kappa_0) + \hat{Z}_A^{\mathrm{M}}(s)}\exp(-s\kappa_0 x) \quad \text{for } x > 0, \tag{6.103b}$$

$$\frac{1}{2\pi\mathrm{i}}\int_{\kappa\in\mathcal{L}}\exp(-s\kappa x)\frac{\tilde{V}(\kappa,s)}{\tilde{K}_{E;A}(\kappa,s)}\mathrm{d}\kappa = 0 \quad \text{for } x < 0, \tag{6.103c}$$

and, finally,

$$\frac{\hat{Z}_B^{\mathrm{E}}(s)}{2\mu_0}\frac{1}{2\pi\mathrm{i}}\int_{\kappa\in\mathcal{L}}\exp(-s\kappa x)\frac{\tilde{V}(\kappa,s)}{\tilde{K}_{E;B}(\kappa,s)}\mathrm{d}\kappa = -\frac{\hat{Z}_A^{\mathrm{E}}(s)}{\hat{Z}_C^{\mathrm{E}}(s)}\frac{2\hat{Z}_B^{\mathrm{E}}(s)}{Z_\perp(\kappa_0) + 2\hat{Z}_A^{\mathrm{E}}(s)}\exp(-s\kappa_0 x) \quad \text{for } x > 0. \tag{6.103d}$$

In (6.103a)–(6.103d), we used

$$\tilde{U}(\kappa,s) = \tilde{B}(\kappa,s) - \tilde{A}(\kappa,s), \tag{6.104}$$

$$\tilde{V}(\kappa,s) = \tilde{B}(\kappa,s) + \tilde{A}(\kappa,s), \tag{6.105}$$

and (cf. (6.67), (6.81) and (6.84))

$$\tilde{K}_{E;A,B}(\kappa, s) = \hat{Z}^{\mathrm{E}}_{A,B}(s)/\big[\hat{Z}^{\mathrm{E}}_{A,B}(s) + Z_\perp(\kappa)/2\big], \tag{6.106}$$

$$\tilde{K}_{M;A,B}(\kappa, s) = \hat{Z}^{\mathrm{M}}_{A,B}(s)/\big[\hat{Z}^{\mathrm{M}}_{A,B}(s) + 2Z_\perp(\kappa)\big], \tag{6.107}$$

respectively. Equations (6.103a)–(6.103d) do represent a WH problem that is solved in the following text.

First, it is concluded that (6.103a) is satisfied if

$$\frac{\tilde{U}(\kappa, s)}{\tilde{K}_{M;A}(\kappa, s)} = \tilde{F}^-(\kappa, s) \quad \text{for } \kappa \in \mathcal{L} \cup \mathcal{D}^-, \tag{6.108}$$

where $\tilde{F}^-(\kappa, s)$ is a regular function in $\mathcal{D}^-$ and of $o(1)$ as $|\kappa| \to \infty$ in $\mathcal{D}^-$. Second, (6.103b) holds true if

$$\frac{\tilde{U}(\kappa, s)}{\tilde{K}_{M;B}(\kappa, s)} = \frac{4\mu_0}{\hat{Z}^{\mathrm{M}}_B(s)} \frac{2\hat{Z}^{\mathrm{M}}_C(s)}{2Z_\perp(\kappa_0) + \hat{Z}^{\mathrm{M}}_A(s)} \frac{1}{\kappa - \kappa_0} + \tilde{G}^+(\kappa, s) \quad \text{for } \kappa \in \mathcal{L} \cup \mathcal{D}^+, \tag{6.109}$$

in which $\tilde{G}^+(\kappa, s)$ is a regular function in $\mathcal{D}^+$ and of $o(1)$ as $|\kappa| \to \infty$ in $\mathcal{D}^+$. Third, (6.103c) is satisfied if

$$\frac{\tilde{V}(\kappa, s)}{\tilde{K}_{E;A}(\kappa, s)} = \tilde{Q}^-(\kappa, s) \quad \text{for } \kappa \in \mathcal{L} \cup \mathcal{D}^-, \tag{6.110}$$

where $\tilde{Q}^-(\kappa, s)$ is regular in $\mathcal{D}^-$ has the property $\tilde{Q}^-(\kappa, s) = o(1)$ as $|\kappa| \to \infty$ in $\mathcal{D}^-$. Finally, from (6.103d), it is found that

$$\frac{\tilde{V}(\kappa, s)}{\tilde{K}_{E;B}(\kappa, s)} = \frac{2\mu_0}{\hat{Z}^{\mathrm{E}}_B(s)} \frac{\hat{Z}^{\mathrm{E}}_A(s)}{\hat{Z}^{\mathrm{E}}_C(s)} \frac{2\hat{Z}^{\mathrm{E}}_B(s)}{Z_\perp(\kappa_0) + 2\hat{Z}^{\mathrm{E}}_A(s)} \frac{1}{\kappa - \kappa_0} + \tilde{H}^+(\kappa, s) \quad \text{for } \kappa \in \mathcal{L} \cup \mathcal{D}^+, \tag{6.111}$$

where $\tilde{H}^+(\kappa, s)$ is some function regular in $\mathcal{D}^+$ with the property $\tilde{H}^+(\kappa, s) = o(1)$ as $|\kappa| \to \infty$ in $\mathcal{D}^+$. The WH relations (6.108)–(6.111) are next tackled with the aid of the factorization technique (e.g. [4]). Starting with (6.108) and (6.109), their combination leads to

$$\frac{\tilde{K}_{M;A}(\kappa, s)}{\tilde{K}_{M;B}(\kappa, s)} \tilde{F}^-(\kappa, s) = \frac{2\hat{P}_M(s)}{\kappa - \kappa_0} + \tilde{G}^+(\kappa, s) \quad \text{for } \kappa \in \mathcal{L}, \tag{6.112}$$

where the s-dependent factor on the right-hand side can be expressed as

$$\hat{P}_M(s) = 4\mu_0 \big[\hat{Y}^{\mathrm{M}}_A(s) - \hat{Y}^{\mathrm{M}}_B(s)\big] \tilde{K}_{M;A}(\kappa_0, s). \tag{6.113}$$

Since the expressions for $\tilde{K}_{M;A,B}(\kappa, s)$ have the form that is amenable to the factorization procedure (see (6.107) and Section 6.2), relation (6.112) can be rewritten as

$$\frac{\tilde{K}^-_{M;A}(\kappa,s)}{\tilde{K}^-_{M;B}(\kappa,s)}\tilde{F}^-(\kappa,s) - \frac{2\hat{P}_M(s)}{\kappa-\kappa_0}\frac{\tilde{K}^+_{M;B}(\kappa_0,s)}{\tilde{K}^+_{M;A}(\kappa_0,s)}$$
$$= 2\hat{P}_M(s)\frac{\tilde{K}^+_{M;B}(\kappa,s)/\tilde{K}^-_{M;A}(\kappa,s) - \tilde{K}^+_{M;B}(\kappa_0,s)/\tilde{K}^-_{M;A}(\kappa_0,s)}{\kappa-\kappa_0}$$
$$+ \frac{\tilde{K}^+_{M;B}(\kappa,s)}{\tilde{K}^+_{M;A}(\kappa,s)}\tilde{G}^+(\kappa,s) \quad \text{for } \kappa \in \mathcal{L}. \tag{6.114}$$

Since the integration path $\mathcal{L}$ was chosen such that $\kappa_0 \in \mathcal{D}^+$, it is seen that the left-hand side of (6.114) is regular in $\mathcal{D}^-$, while its right-hand side is regular in $\mathcal{D}^+$. Consequently, the left- and the right-hand sides of the latter relation do represent one and the same entire function that is regular throughout the complex κ-plane [11, p. 85]. It can be easily shown that the entire function is zero, which through (6.108) and (6.113) yields

$$\tilde{U}(\kappa,s) = 8\mu_0\left[\hat{Y}^{\mathrm{M}}_A(s) - \hat{Y}^{\mathrm{M}}_B(s)\right]$$
$$\times \frac{\tilde{K}^+_{M;A}(\kappa,s)\tilde{K}^-_{M;A}(\kappa_0,s)\tilde{K}^+_{M;B}(\kappa_0,s)\tilde{K}^-_{M;B}(\kappa,s)}{\kappa-\kappa_0} \quad \text{for } \kappa \in \mathcal{L}. \tag{6.115}$$

In a similar manner, we shall next express $\tilde{V}(\kappa, s)$. To that end, we combine (6.110) with (6.111) and get

$$\frac{\tilde{K}_{E;A}(\kappa,s)}{\tilde{K}_{E;B}(\kappa,s)}\tilde{Q}^-(\kappa,s) = \frac{\hat{P}_E(s)}{\kappa-\kappa_0} + \tilde{H}^+(\kappa,s) \quad \text{for } \kappa \in \mathcal{L}, \tag{6.116}$$

where

$$\hat{P}_E(s) = 2\mu_0\left[\hat{Y}^{\mathrm{E}}_B(s) - \hat{Y}^{\mathrm{E}}_A(s)\right]\tilde{K}_{E;A}(\kappa_0,s). \tag{6.117}$$

Again, since the factorization of $\tilde{K}_{E;A,B}(\kappa, s)$ as defined by (6.106) is available (see Section 6.2), relation (6.116) can be cast into the following form (cf. (6.114)):

$$\frac{\tilde{K}^-_{E;A}(\kappa,s)}{\tilde{K}^-_{E;B}(\kappa,s)}\tilde{Q}^-(\kappa,s) - \frac{\hat{P}_E(s)}{\kappa-\kappa_0}\frac{\tilde{K}^+_{E;B}(\kappa_0,s)}{\tilde{K}^+_{E;A}(\kappa_0,s)}$$
$$= \hat{P}_E(s)\frac{\tilde{K}^+_{E;B}(\kappa,s)/\tilde{K}^-_{E;A}(\kappa,s) - \tilde{K}^+_{E;B}(\kappa_0,s)/\tilde{K}^-_{E;A}(\kappa_0,s)}{\kappa-\kappa_0}$$
$$+ \frac{\tilde{K}^+_{E;B}(\kappa,s)}{\tilde{K}^+_{E;A}(\kappa,s)}\tilde{H}^+(\kappa,s) \quad \text{for } \kappa \in \mathcal{L}. \tag{6.118}$$

Upon observing that the left-hand side of (6.118) is regular in $\mathcal{D}^-$, while its right-hand side is regular in $\mathcal{D}^+$, Liouville's theorem can be employed to express $\tilde{Q}^-(\kappa, s)$. The latter is then substituted in (6.110), which gives

$$\tilde{V}(\kappa, s) = 2\mu_0\left[\hat{Y}_B^{\mathrm{E}}(s) - \hat{Y}_A^{\mathrm{E}}(s)\right] \times \frac{\tilde{K}_{E;A}^+(\kappa, s)\tilde{K}_{E;A}^-(\kappa_0, s)\tilde{K}_{E;B}^+(\kappa_0, s)\tilde{K}_{E;B}^-(\kappa, s)}{\kappa - \kappa_0} \quad \text{for } \kappa \in \mathcal{L}. \tag{6.119}$$

Finally, upon combining (6.104) and (6.105) with (6.115) and (6.119), we arrive at the following expressions for the scattered-wave amplitudes:

$$\begin{aligned}\tilde{A}(\kappa, s) = {}& \mu_0\left[\hat{Y}_B^{\mathrm{E}}(s) - \hat{Y}_A^{\mathrm{E}}(s)\right] \\ & \times \frac{\tilde{K}_{E;A}^+(\kappa, s)\tilde{K}_{E;A}^-(\kappa_0, s)\tilde{K}_{E;B}^+(\kappa_0, s)\tilde{K}_{E;B}^-(\kappa, s)}{\kappa - \kappa_0} \\ & - 4\mu_0\left[\hat{Y}_A^{\mathrm{M}}(s) - \hat{Y}_B^{\mathrm{M}}(s)\right] \\ & \times \frac{\tilde{K}_{M;A}^+(\kappa, s)\tilde{K}_{M;A}^-(\kappa_0, s)\tilde{K}_{M;B}^+(\kappa_0, s)\tilde{K}_{M;B}^-(\kappa, s)}{\kappa - \kappa_0} \\ & \text{for } \kappa \in \mathcal{L},\end{aligned} \tag{6.120}$$

and in a similar fashion,

$$\begin{aligned}\tilde{B}(\kappa, s) = {}& \mu_0\left[\hat{Y}_B^{\mathrm{E}}(s) - \hat{Y}_A^{\mathrm{E}}(s)\right] \\ & \times \frac{\tilde{K}_{E;A}^+(\kappa, s)\tilde{K}_{E;A}^-(\kappa_0, s)\tilde{K}_{E;B}^+(\kappa_0, s)\tilde{K}_{E;B}^-(\kappa, s)}{\kappa - \kappa_0} \\ & + 4\mu_0\left[\hat{Y}_A^{\mathrm{M}}(s) - \hat{Y}_B^{\mathrm{M}}(s)\right] \\ & \times \frac{\tilde{K}_{M;A}^+(\kappa, s)\tilde{K}_{M;A}^-(\kappa_0, s)\tilde{K}_{M;B}^+(\kappa_0, s)\tilde{K}_{M;B}^-(\kappa, s)}{\kappa - \kappa_0} \\ & \text{for } \kappa \in \mathcal{L}.\end{aligned} \tag{6.121}$$

As the transform-domain solution is fully specified by (6.120) and (6.121), we may proceed further with its inverse to the TD. For the sake of simplicity, we shall next limit ourselves to the junction of two loss-free metasurfaces.

6.5.1 TD solution

We shall consider the junction of two loss-free metasurfaces, $\mathcal{S}_A$ and $\mathcal{S}_B$, described by their equivalent layer admittances, $\hat{Y}_{A,B}^{\mathrm{E}}(s) = sC_{A,B}^{\mathrm{E}}$, and their equivalent layer impedances $\hat{Z}_{A,B}^{\mathrm{M}}(s) = sL_{A,B}^{\mathrm{M}}$, respectively. To find TD expressions for the scattered EM fields in closed form, we shall make use of Koiter's approximation (see Section 6.2.1.2), again, which makes it possible to write (cf. (6.94))

$$E_y^{\mathrm{s}}(x, z, t) = \frac{\partial_t e^{\mathrm{i}}(t)}{2\pi} *_t \int_{\tau=\rho/c_0}^{\infty} \mathrm{Re}\{\bar{A}[\kappa(\tau), t - \tau]\} \frac{\mathrm{d}\tau}{(\tau^2 - \rho^2/c_0^2)^{1/2}} + E_y^{\mathrm{s;g}}(x, z, t) \quad \text{for all } x \in \mathbb{R}, z > 0 \text{ and } t > 0, \tag{6.122}$$

where (the integral of) the wave amplitudes can be written in the following form:

$$\bar{A}(\kappa,t) = \partial_t^{-1}A(\kappa,t) \overset{\kappa}{\simeq} \frac{2p(\kappa_0)p(\kappa)}{\kappa-\kappa_0}\frac{2}{\mu_0}\frac{C_B^{\mathrm{E}}-C_A^{\mathrm{E}}}{C_A^{\mathrm{E}}C_B^{\mathrm{E}}}$$

$$\mathbb{I}_0\left[\frac{\eta_0}{C_B^{\mathrm{E}}}\omega(\kappa,\kappa_0)t\right]\exp\left\{-\frac{2\eta_0}{C_B^{\mathrm{E}}}c_0[p^+(\kappa_0)]^2t\right\}\mathrm{H}(t)$$

$$*_t\ \mathbb{I}_0\left[\frac{\eta_0}{C_A^{\mathrm{E}}}\omega(-\kappa,-\kappa_0)t\right]\exp\left\{-\frac{2\eta_0}{C_A^{\mathrm{E}}}c_0[p^+(-\kappa_0)]^2t\right\}\mathrm{H}(t)$$

$$-\frac{4\mu_0}{\kappa-\kappa_0}\frac{L_B^{\mathrm{M}}-L_A^{\mathrm{M}}}{L_A^{\mathrm{M}}L_B^{\mathrm{M}}}\mathbb{I}_0\left[\frac{\zeta_0}{L_B^{\mathrm{M}}}\frac{\omega(-\kappa_0,-\kappa)t}{c_0^2[p^+(\kappa_0)p^-(\kappa)]^2}\right]\exp\left\{-\frac{2\zeta_0}{L_B^{\mathrm{M}}}\frac{t}{c_0[p^+(\kappa_0)]^2}\right\}\mathrm{H}(t)$$

$$*_t\ \mathbb{I}_0\left[\frac{\zeta_0}{L_A^{\mathrm{M}}}\frac{\omega(\kappa_0,\kappa)t}{c_0^2[p^+(-\kappa_0)p^-(-\kappa)]^2}\right]\exp\left\{-\frac{2\zeta_0}{L_A^{\mathrm{M}}}\frac{t}{c_0[p^+(-\kappa_0)]^2}\right\}\mathrm{H}(t). \quad (6.123)$$

In a similar fashion, we write (cf. (6.96))

$$E_y^{\mathrm{s}}(x,z,t) = \frac{\partial_t e^{\mathrm{i}}(t)}{2\pi} *_t \int_{\tau=\rho/c_0}^{\infty}\mathrm{Re}\{\bar{B}[\kappa(\tau),t-\tau]\}\frac{\mathrm{d}\tau}{(\tau^2-\rho^2/c_0^2)^{1/2}}$$

$$+E_y^{\mathrm{s;g}}(x,z,t) \quad \text{for all } x\in\mathbb{R}, z<0 \text{ and } t>0, \quad (6.124)$$

where

$$\bar{B}(\kappa,t) = \partial_t^{-1}B(\kappa,t) \overset{\kappa}{\simeq} \frac{2p(\kappa_0)p(\kappa)}{\kappa-\kappa_0}\frac{2}{\mu_0}\frac{C_B^{\mathrm{E}}-C_A^{\mathrm{E}}}{C_A^{\mathrm{E}}C_B^{\mathrm{E}}}$$

$$\mathbb{I}_0\left[\frac{\eta_0}{C_B^{\mathrm{E}}}\omega(\kappa,\kappa_0)t\right]\exp\left\{-\frac{2\eta_0}{C_B^{\mathrm{E}}}c_0[p^+(\kappa_0)]^2t\right\}\mathrm{H}(t)$$

$$*_t\ \mathbb{I}_0\left[\frac{\eta_0}{C_A^{\mathrm{E}}}\omega(-\kappa,-\kappa_0)t\right]\exp\left\{-\frac{2\eta_0}{C_A^{\mathrm{E}}}c_0[p^+(-\kappa_0)]^2t\right\}\mathrm{H}(t)$$

$$+\frac{4\mu_0}{\kappa-\kappa_0}\frac{L_B^{\mathrm{M}}-L_A^{\mathrm{M}}}{L_A^{\mathrm{M}}L_B^{\mathrm{M}}}\mathbb{I}_0\left[\frac{\zeta_0}{L_B^{\mathrm{M}}}\frac{\omega(-\kappa_0,-\kappa)t}{c_0^2[p^+(\kappa_0)p^-(\kappa)]^2}\right]\exp\left\{-\frac{2\zeta_0}{L_B^{\mathrm{M}}}\frac{t}{c_0[p^+(\kappa_0)]^2}\right\}\mathrm{H}(t)$$

$$*_t\ \mathbb{I}_0\left[\frac{\zeta_0}{L_A^{\mathrm{M}}}\frac{\omega(\kappa_0,\kappa)t}{c_0^2[p^+(-\kappa_0)p^-(-\kappa)]^2}\right]\exp\left\{-\frac{2\zeta_0}{L_A^{\mathrm{M}}}\frac{t}{c_0[p^+(-\kappa_0)]^2}\right\}\mathrm{H}(t) \quad (6.125)$$

is the pertaining wave amplitude. Finally, the geometrical solution of the diffraction problem follows upon evaluating the contribution of the pole singularity at $\kappa=\kappa_0$ and transforming the result to the TD. In this way, we arrive at (cf. (6.98))

$$E_y^{\mathrm{s;g}}(x,z,t) = -e^{\mathrm{i}}[t-\kappa_0x-p(\kappa_0)|z|]\mathrm{H}[x-\rho\sin(\beta)]$$

$$*_t\left\{\frac{2Y_\perp(\kappa_0)}{C_A^{\mathrm{E}}}\exp\left[-\frac{2Y_\perp(\kappa_0)}{C_A^{\mathrm{E}}}t\right]-\frac{2Y_\perp(\kappa_0)}{C_B^{\mathrm{E}}}\exp\left[-\frac{2Y_\perp(\kappa_0)}{C_B^{\mathrm{E}}}t\right]\right\}\mathrm{H}(t)$$

$$+ e^{\mathrm{i}}[t - \kappa_0 x - p(\kappa_0)|z|]\mathrm{H}[x - \rho\sin(\beta)]\mathrm{sign}(z)$$

$$*_t \left\{ \frac{2Z_\perp(\kappa_0)}{L_A^{\mathrm{M}}} \exp\left[- \frac{2Z_\perp(\kappa_0)}{L_A^{\mathrm{M}}} t \right] - \frac{2Z_\perp(\kappa_0)}{L_B^{\mathrm{M}}} \exp\left[- \frac{2Z_\perp(\kappa_0)}{L_B^{\mathrm{M}}} t \right] \right\} \mathrm{H}(t)$$

$$\text{for all } x \in \mathbb{R}, z \in \mathbb{R} \text{ and } t > 0. \tag{6.126}$$

Apparently, if $(C_A^{\mathrm{E}}/L_A^{\mathrm{M}})^{1/2} = (C_B^{\mathrm{E}}/L_B^{\mathrm{M}})^{1/2} = Y_\perp(\kappa_0)$, the geometrical solution vanishes identically throughout $z > 0$. Some special cases of the TD geometrical solution (6.126) are illustrated graphically in the next section.

6.5.2 *Illustrative examples*

We further demonstrate potentialities of two coplanar metasurfaces to shape pulsed EM fields. To that end, for the sake of simplicity, we shall evaluate the sum of the geometrical solution (6.126) and the pertaining excitation field. Recall that the latter field corresponds to the total field in the presence of an infinite metasurface of the parameters pertaining to $\mathcal{S}_A$ and can be hence found with the help of (4.16) and (4.17). All analyzed junctions are irradiated by the impulsive EM plane wave defined by its rectangular pulse shape (see (6.13)) and $\beta = \pi/4$, again.

In the first example, it is demonstrated that the junction of two metasurfaces may mimic the scattering from (special) semi-infinite layers analyzed before. Indeed, Figure 6.19(a) and (b) shows the field distribution in the presence of a highly conducting semi-infinite sheet $\mathcal{S}_A$ and $\mathcal{S}_B$, respectively (cf. Figure 6.4). Figure 6.19(c) and (d) then clearly shows that the choice $C_A^{\mathrm{E}}/\eta_0 t_{\mathrm{w}} = 0.01$, $L_A^{\mathrm{M}}\eta_0/t_{\mathrm{w}} = 0.02$ in combination with $C_B^{\mathrm{E}}/\eta_0 t_{\mathrm{w}} = 2.0$, $L_B^{\mathrm{M}}\eta_0/t_{\mathrm{w}} = 4.0$ and $C_B^{\mathrm{E}}/\eta_0 t_{\mathrm{w}} = 10$, $L_B^{\mathrm{M}}\eta_0/t_{\mathrm{w}} = 1.0$ resembles the field distribution in the presence of the semi-infinite metasurface as plotted in Figure 6.16(a) and (b), respectively.

Next, we shall plot the field distribution in the presence of two coplanar metasurfaces with non-vanishing equivalent Kirchhoff circuit parameters. Figure 6.20(a) shows the reflection-free field in the vicinity of the junction of two "transparent" semi-infinite metasurfaces with identical EM properties. As a matter of fact, the geometrical solution (6.126) vanishes identically in this case and the resulting field distribution thus corresponds to the excitation field. The ensuing Figure 6.20(b) shows the field distribution in the presence of two "transparent" metasurfaces, again, but with different EM properties. Consequently, the transmitted field distribution below the layer is affected by the jump in the EM properties. If screen $\mathcal{S}_B$ is not "transparent" anymore, the reflected field above this screen occurs. This is illustrated in Figure 6.20(c), where the tangential electric-field distribution is significantly reduced in the close neighborhood of the relatively highly electrically conducting screen $\mathcal{S}_B$. If, in addition, the parameters of $\mathcal{S}_A$ do not meet the transparency condition (see (4.8)), we then get the relatively complex field distribution shown in Figure 6.20(d), where the reflected field from the relatively highly magnetically conducting screen $\mathcal{S}_A$ can be observed.

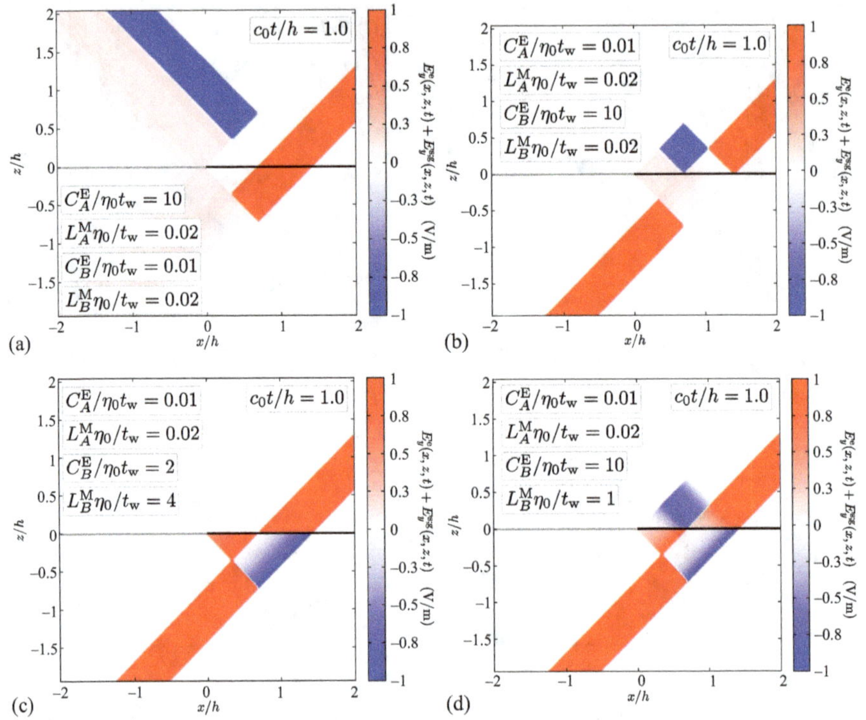

Figure 6.19 *Spatial distributions of $E_y^e + E_y^{s;g}$ at $c_0t/h = 1.0$ in the presence of the junction of two metasurfaces with (a) $C_A^E/\eta_0 t_w = 10$, $L_A^M \eta_0/t_w = 0.02$ and $C_B^E/\eta_0 t_w = 0.01$, $L_B^M \eta_0/t_w = 0.02$; (b) $C_A^E/\eta_0 t_w = 0.01$, $L_A^M \eta_0/t_w = 0.02$ and $C_B^E/\eta_0 t_w = 10$, $L_B^M \eta_0/t_w = 0.02$; (c) $C_A^E/\eta_0 t_w = 0.01$, $L_A^M \eta_0/t_w = 0.02$ and $C_B^E/\eta_0 t_w = 2.0$, $L_B^M \eta_0/t_w = 4.0$; (d) $C_A^E/\eta_0 t_w = 0.01$, $L_A^M \eta_0/t_w = 0.02$ and $C_B^E/\eta_0 t_w = 10$, $L_B^M \eta_0/t_w = 1.0$*

6.6 Kirchhoff diffraction by semi-infinite sheets

The TD EM wave fields diffracted on semi-infinite thin sheets can also be described approximately. This is briefly demonstrated in this section through the Kirchhoff approximation and the classic CdH technique.

6.6.1 PEC sheets

The problem of edge EM diffraction by a semi-infinite screen has been in Section 6.1 formulated using the explicit-type boundary condition (6.1a) and the continuity-type condition (6.1b). In order to solve the problem with the aid of the Kirchhoff approximation, the presence of the sheet is accounted for through a saltus-type boundary condition applying to the tangential component of the magnetic-field strength. Namely, it is assumed that the magnetic-field component exhibits a jump

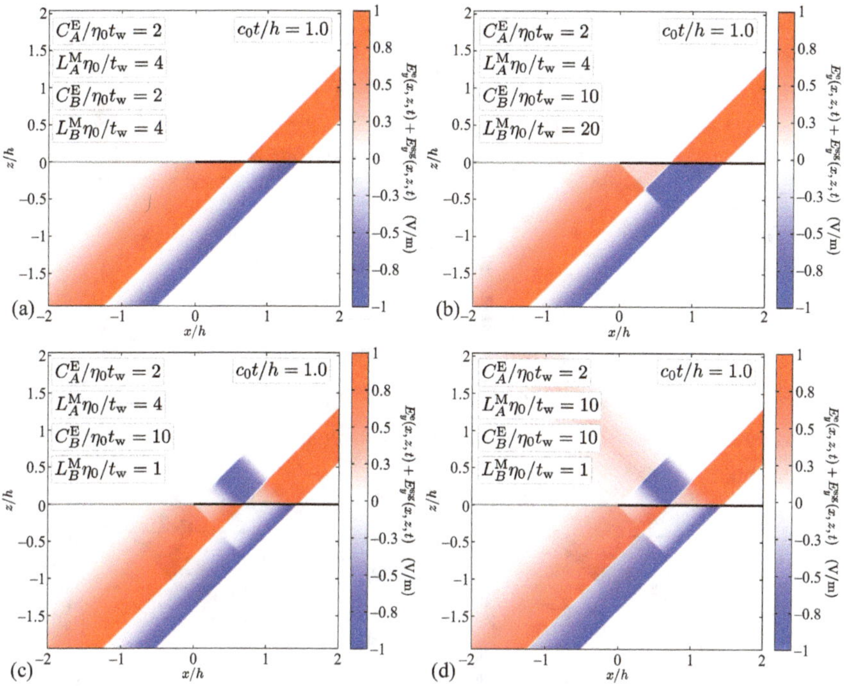

Figure 6.20 *Spatial distributions of $E_y^e + E_y^{s;g}$ at $c_0 t/h = 1.0$ in the presence of the junction of two metasurfaces with (a) $C_A^E/\eta_0 t_w = 2.0$, $L_A^M \eta_0/t_w = 4.0$ and $C_B^E/\eta_0 t_w = 2.0$, $L_B^M \eta_0/t_w = 4.0$; (b) $C_A^E/\eta_0 t_w = 2.0$, $L_A^M \eta_0/t_w = 4.0$ and $C_B^E/\eta_0 t_w = 10$, $L_B^M \eta_0/t_w = 20$; (c) $C_A^E/\eta_0 t_w = 2.0$, $L_A^M \eta_0/t_w = 4.0$ and $C_B^E/\eta_0 t_w = 10$, $L_B^M \eta_0/t_w = 1.0$; (d) $C_A^E/\eta_0 t_w = 2.0$, $L_A^M \eta_0/t_w = 10$ and $C_B^E/\eta_0 t_w = 10$, $L_B^M \eta_0/t_w = 1.0$*

across the layer, the magnitude of which corresponds to an equivalent sheet with the same EM properties but of infinite width. Pursuing this line of reasoning, conditions (6.1a) and (6.1b) are replaced with

$$\lim_{z\downarrow\delta/2} \hat{H}_x^s(x,z,s) - \lim_{z\uparrow-\delta/2} \hat{H}_x^s(x,z,s) = 2\hat{H}_x^i(x,0,s)\mathrm{H}(x) \quad \text{for all } x \in \mathbb{R}, \tag{6.127}$$

as $\delta \downarrow 0$. The transition condition (6.127) states that the x-component of the scattered magnetic-field strength is continuous across $\{-\infty < x < 0, z = 0\}$ (see (6.1b)), and jumps by amount $2Y_\perp(\kappa_0)\hat{e}^i(s)\exp(-s\kappa_0 x)$ when crossing the screen in $x > 0$. With the aid of the residue theorem [11, Section 3.11], it is straightforward to rewrite (6.127) as

$$\begin{aligned}&\lim_{z\downarrow\delta/2} \hat{H}_x^s(x,z,s) - \lim_{z\uparrow-\delta/2} \hat{H}_x^s(x,z,s)\\ &= -2Y_\perp(\kappa_0)\frac{\hat{e}^i(s)}{2\pi \mathrm{i}}\int_{\kappa\in\mathcal{L}} \exp(-s\kappa x)\frac{\mathrm{d}\kappa}{\kappa-\kappa_0} \quad \text{for all } x \in \mathbb{R},\end{aligned} \tag{6.128}$$

where the integration contour, $\mathcal{L}$, is depicted in Figure 6.2. Assuming next that the scattered magnetic field can be represented via the slowness representation of the type given by (6.2), its use on the left-hand side of (6.128) yields the (Kirchhoff-approximate) expression for the diffracted-wave amplitude (cf. (6.7))

$$\tilde{A}(\kappa) \overset{k}{\simeq} \frac{2p(\kappa_0)}{\kappa - \kappa_0}, \tag{6.129}$$

where $\overset{k}{\simeq}$ denotes the Kirchhoff approximation. Apparently, since κ_0 is located to the right of $\mathcal{L}$, $\tilde{A}(\kappa)$ is a regular function for $\kappa \in \mathcal{L} \cup \mathcal{D}^-$, as required by the continuity of $\hat{H}_x^{\mathrm{s}}$ across $\{-\infty < x < 0, z = 0\}$. To arrive at the TD expression, the diffracted-wave amplitude (6.129) is substituted in (6.2) and the resulting representation is transformed with the help of the CdH procedure (see Section 6.1.1). By doing so, we end up with the approximate expression for the *total* electric-field strength (cf. (6.12))

$$\begin{aligned} E_y(x,z,t) &\overset{k}{\simeq} \frac{e^{\mathrm{i}}(t)}{2\pi} *_t \mathrm{Re}\left\{\frac{2p(\kappa_0)}{\kappa(t) - \kappa_0}\right\} \frac{\mathrm{H}(t - \rho/c_0)}{(t^2 - \rho^2/c_0^2)^{1/2}} \\ &- e^{\mathrm{i}}[t - \kappa_0 x - p(\kappa_0)|z|]\mathrm{H}[x - \rho\sin(\beta)] + e^{\mathrm{i}}[t - \kappa_0 x + p(\kappa_0) z] \\ &\text{for all } x \in \mathbb{R}, z \in \mathbb{R} \text{ and } t > 0, \end{aligned} \tag{6.130}$$

and recall that $p(\kappa_0) = \cos(\beta)/c_0$. Using the parameters introduced in Section A, the approximate diffracted-wave amplitude can also be expressed explicitly (cf. (A.12)):

$$\mathrm{Re}\{\tilde{A}[\kappa(\tau)]\} \overset{k}{\simeq} \frac{2[(c_0\tau/\rho)\cos(\theta) - \cos(\psi)]\sin(\psi)}{[c_0\tau/\rho - \cos(\psi + \theta)][c_0\tau/\rho - \cos(\psi - \theta)]}. \tag{6.131}$$

This result can be substituted in (6.130) to evaluate the Kirchhoff edge diffraction in the TD.

6.6.2 *EM penetrable sheets*

To account for the presence of a thin sheet showing a high contrast in its electric conductivity and permittivity (with respect to the unbounded, isotropic and homogeneous background), we may apply the (modified) Kirchhoff approximation. Namely, we shall assume that the tangential magnetic-field component suffers a jump across the layer much as the screen was infinite in its width, again. In this manner, we arrive at (cf. (6.127))

$$\begin{aligned} &\lim_{z \downarrow \delta/2} \hat{H}_x^{\mathrm{s}}(x,z,s) - \lim_{z \uparrow -\delta/2} \hat{H}_x^{\mathrm{s}}(x,z,s) = 2\hat{\Gamma}^{\mathrm{E}}(s)\hat{H}_x^{\mathrm{i}}(x,0,s)\mathrm{H}(x) \\ &\text{for all } x \in \mathbb{R}, \end{aligned} \tag{6.132}$$

as $\delta \downarrow 0$, where

$$\hat{\Gamma}^{\mathrm{E}}(s) = \hat{Y}^{\mathrm{E}}(s)/\left[\hat{Y}^{\mathrm{E}}(s) + 2Y_\perp(\kappa_0)\right]. \tag{6.133}$$

Apparently, if $\hat{Y}^{\mathrm{E}}(s) \to \infty$, which applies to a PEC sheet analyzed in the previous section, then $\hat{\Gamma}^{\mathrm{E}}(s) \to 1$ and (6.132) reduces to (6.127). Consequently, upon inspection of (6.132) with (6.127), it is concluded that the EM field *scattered* by a penetrable sheet can be easily deduced from the corresponding scattered field pertaining to the PEC screen, that is,

$$E_y^{\mathrm{s}}(x,z,t) \overset{k}{\simeq} \Gamma^{\mathrm{E}}(t) *_t E_y^{\mathrm{s;PEC}}(x,z,t), \tag{6.134}$$

where $E_y^{\mathrm{s;PEC}}$ immediately follows from (6.130) upon dropping the last term representing the incident plane wave, and $\Gamma^{\mathrm{E}}(t)$ is the TD counterpart of the coefficient represented by (6.133). For a conductive sheet, the coefficient is s-independent, so that

$$\Gamma^{\mathrm{E}}(t) = \{G^{\mathrm{E}}/[G^{\mathrm{E}} + 2Y_\perp(\kappa_0)]\}\delta(t). \tag{6.135}$$

By virtue of the "sifting" property of the Dirac-delta distribution, the use of (6.135) in (6.134) results in a mere multiplication of $E_y^{\mathrm{s;PEC}}$ by the coefficient $\{G^{\mathrm{E}}/[G^{\mathrm{E}} + 2Y_\perp(\kappa_0)]$. The resulting geometrical solution of the diffraction problem conforms with the exact one achieved with the WH method (see (6.36)). Finally, the sheet with both conductive and dielectric properties can be handled via the coefficient

$$\Gamma^{\mathrm{E}}(t) = \delta(t) - \frac{2Y_\perp(\kappa_0)}{C^{\mathrm{E}}}\exp\left[-\frac{G^{\mathrm{E}} + 2Y_\perp(\kappa_0)}{C^{\mathrm{E}}}t\right]\mathrm{H}(t). \tag{6.136}$$

Owing to the presence of Dirac-delta distribution, the substitution of (6.136) in (6.134) yields the sum of an instantaneously reacting part, $E_y^{\mathrm{s;PEC}}$, and a relaxational contribution describing the effect of the finite layer conductance, G^{E}, and capacitance, C^{E}. The thus obtained geometrical solution agrees, again, with the one found previously (see (6.45)).

6.6.3 Illustrative examples

In this section, we shall provide illustrative examples demonstrating the validity of the results based on Kirchhoff's approximation. To that end, we shall adopt the plane-wave excitation used in Sections 6.1.2 and 6.2.3, again.

In the first example, we assess the validity of the approximation pertaining to the PEC semi-infinite screen. Accordingly, the corresponding TD approximation (6.130) is used to calculate the electric-field pulse shape at $x/h = 1.0$ and $z/h = 0.5$ and the result is compared with the corresponding exact solution given by (6.12). Figure 6.21 clearly demonstrates that the Kirchhoff approximation (6.130) is appropriate just in the early-time part of the response. Afterwards, significant discrepancies with respect to the exact WH-based solution can be observed.

As the second example, we analyze the field in the presence of the purely conductive screen with $G^{\mathrm{E}}/\eta_0 = 2$. It was demonstrated in Figure 6.8(b) that the PEC model is not appropriate in this case. Figure 6.22 shows that for the chosen parameters the Kirchhoff approximation (denoted by (k)), described by (6.134) and (6.135),

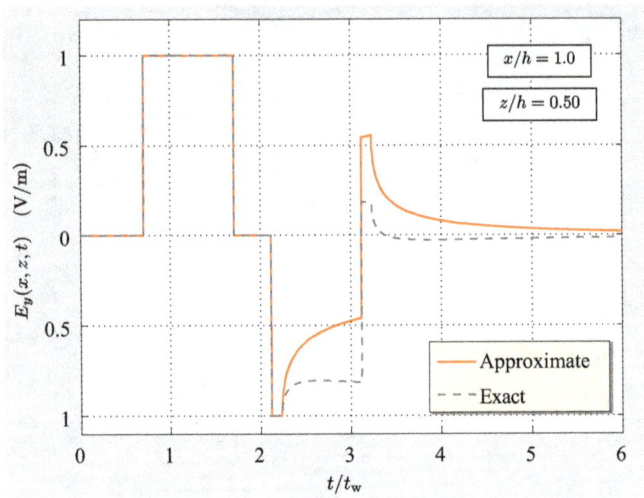

Figure 6.21 *The total electric-field pulse shape calculated using the Kirchhoff approximation (k) (6.130) and using (6.12) pertaining to the PEC screen*

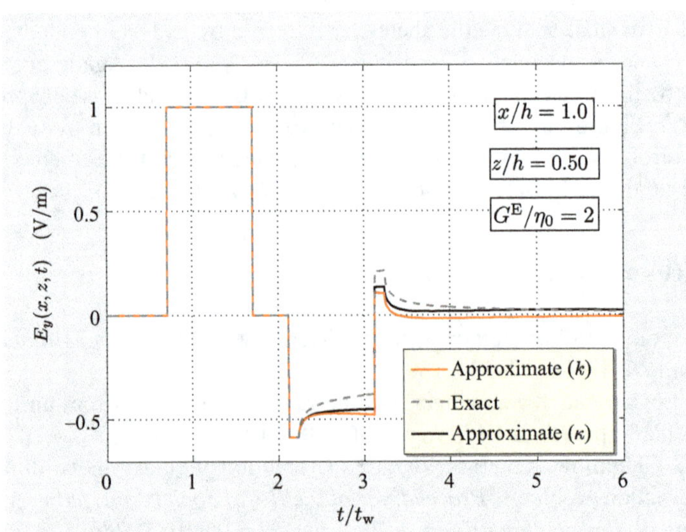

Figure 6.22 *The total electric-field pulse shape calculated using the Kirchhoff approximation (k) (6.134) with (6.135), using the exact solution (6.35) with (6.32a) and (6.32b), and its Koiter approximation (κ) with (6.59a) and (6.59b)*

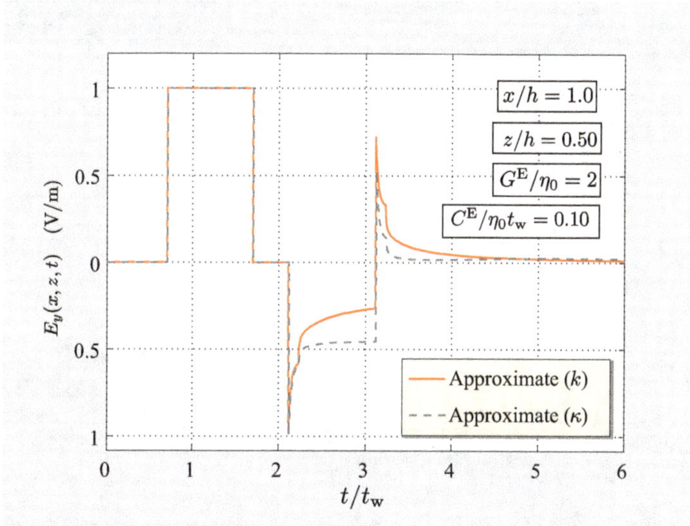

Figure 6.23 The total electric-field pulse shape calculated using the Kirchhoff approximation (k) (6.134) with (6.136), and using the Koiter approximation (κ) (6.41) with (6.42)

is potentially applicable, but the Koiter approximation (denoted by (κ)), described by (6.35) with (6.59a) and (6.59b), is more accurate.

Finally, we shall analyze the sheet characterized by its layer conductance ratio $G^E/\eta_0 = 2$ and by its relative capacitance $C^E/\eta_0 t_w = 0.10$. Again, according to Figure 6.9(b), the replacement of this layer with its PEC equivalent is not appropriate. As shown in Figure 6.23, the Kirchhoff approximation (k), given by (6.134) with (6.136), correlates well with the Koiter approximation (κ) specified via (6.41) with (6.42) in the early-time part of the response only (cf. Figure 6.21).

References

[1] A. Sommerfeld. Mathematische Theorie der Diffraction. *Mathematische Annalen*, 47(2):317–374, 1896.

[2] H. Levine and J. Schwinger. On the radiation of sound from an unflanged circular pipe. *Physical Review*, 73(4):383, 1948.

[3] P. C. Clemmow. A method for the exact solution of a class of two-dimensional diffraction problems. *Proceedings of the Royal Society of London A: Mathematical, Physical and Engineering Sciences*, 205(1081):286–308, 1951.

[4] B. Noble. *Methods Based on the Wiener-Hopf Technique for the Solution of Partial Differential Equations*. London, UK: Pergamon Press, 1958.

[5] T. B. A. Senior. Diffraction by a semi-infinite metallic sheet. *Proceedings of the Royal Society of London A: Mathematical, Physical and Engineering Sciences*, 213(1115):436–458, 1952.

[6] T. B. A. Senior. Half plane edge diffraction. *Radio Science*, 10(6):645–650, 1975.

[7] P. C. Clemmow. Radio propagation over a flat earth across a boundary separating two different media. *Philosophical Transactions of the Royal Society of London A: Mathematical, Physical and Engineering Sciences*, 246(905):1–55, 1953.

[8] T. B. A. Senior and J. L. Volakis. *Approximate Boundary Conditions in Electromagnetics*. London, UK: IEE, 1995.

[9] A. T. de Hoop. *Representation Theorems for the Displacement in an Elastic Solid and Their Application to Elastodynamic Diffraction Theory*. PhD thesis, Delft University of Technology, the Netherlands, 1958.

[10] A. T. de Hoop and M. V. de Hoop. Transient elastic P and SV wave edge diffraction by a semi-infinite perfect-slip fracture in an isotropic dispersive solid: a canonical problem. *Journal of Geophysical Research*, 108(B4), 2003.

[11] E. C. Titchmarsh. *The Theory of Functions. 2nd edition*. London, UK: Oxford University Press, 1939.

[12] M. Štumpf. *Electromagnetic Reciprocity in Antenna Theory*. Hoboken, NJ: John Wiley & Sons, Inc., 2018.

[13] W. T. Koiter. Approximate solution of Wiener–Hopf type integral equations with applications. *Proc. Koninkl. Nederl. Akademie van Wetenschappen, Series B*, 57:558–579, 1954.

[14] R. Du Cloux. Pulsed electromagnetic radiation from a line source in the presence of a semi-infinite screen in the plane interface of two different media. *Wave Motion*, 6(5):459–476, 1984.

[15] M. Abramowitz and I. A. Stegun. *Handbook of Mathematical Functions*. New York, NY: Dover Publications, 1972.

Chapter 7
Pulsed EM-field scattering by narrow metastrips

Electromagnetic (EM) scattering by an infinitely long, narrow perfectly electrically conducting (PEC) strip has been approximately analyzed in the real frequency domain (FD) by Schelkunoff using the concept of external (per-unit-length) impedance [1, Section 8.5]. This efficient approach has been later generalized in Ref. [2] with the aid of the time-domain (TD) EM reciprocity theorem (see [3]), thus formulating the scattering problem in the form of an integral equation of the time-convolution type. This line of reasoning is also pursued in this chapter, where an efficient computational methodology capable of achieving the transient electric- and magnetic-current distributions induced along a narrow metastrip is introduced. The combined effect of (magneto-)dielectric and conductive properties of EM-penetrable metastrips is incorporated using the concept of thin-sheet, high-contrast transition conditions as introduced in Chapter 3.

7.1 Pulsed EM scattering by a narrow PEC strip

First, we shall analyze the pulsed EM scattering by a narrow PEC strip occupying a bounded domain $\{-w/2 < x < w/2, -\infty < y < \infty, -\delta/2 < z < \delta/2\}$ (see Figure 7.1). The strip is located in the homogeneous, isotropic and loss-free embedding $\mathcal{D}^\infty$ that is described by its electric permittivity and magnetic permeability$\{\epsilon_0, \mu_0\}$. The narrow strip is assumed to be irradiated by the uniform, E-polarized plane wave

$$\hat{E}^{\mathrm{i}}_y(x, z, s) = \hat{e}^{\mathrm{i}}(s) \exp\{-s[\kappa_0 x - p(\kappa_0) z]\}, \qquad \text{(4.1 revisited)}$$

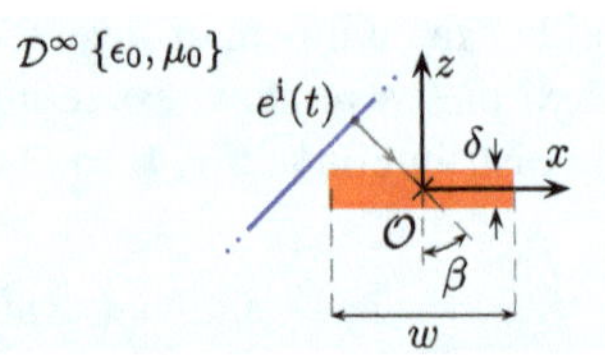

Figure 7.1 Pulsed EM plane wave impinging upon a narrow plane strip

where $\kappa_0 = \sin(\beta)/c_0$ and $p(\kappa_0) = \cos(\beta)/c_0$ for $\{0 \le \beta \le \pi/2\}$. The width of the strip, $w > 0$, as well as its thickness, $\delta > 0$, is assumed to be (relatively) small with respect to the spatial support of the excitation pulse.

The scattering problem is next formulated with the aid of the EM reciprocity theorem of the time-convolution type [4, Section 1.4.1] that is applied to the (unbounded) domain exterior to the strip and to EM-field states associated with the scattered EM field (denoted by superscript ^{s}) and with the testing field (denoted by superscript ^{T}). The testing field is chosen to be a causal EM wave field whose non-vanishing field components, $\{E_y^{\text{T}}, H_x^{\text{T}}, H_z^{\text{T}}\}$, are governed by the y-dependent set of EM-field equations applying to the unbounded embedding. Since the interrelated EM-field states are each other's adjoint [4, Section 1.4.1], the reciprocity theorem yields

$$\int_{x=-w/2}^{w/2} \hat{E}_y^{\text{T}}(x,0,s)\partial\hat{J}_y^{\text{s}}(x,s)\mathrm{d}x = \int_{x=-w/2}^{w/2} \hat{E}_y^{\text{s}}(x,0,s)\partial\hat{J}_y^{\text{T}}(x,s)\mathrm{d}x, \tag{7.1}$$

where $\partial\hat{J}_y(x,s)$ denotes the axial electric-current surface density (in A/m). Owing to the relatively small width of the strip, we shall further postulate that the *spatial* distribution of the surface density can be approximated in a simple manner. To that end, we shall consider either the uniformly distributed electric-current surface density (denoted by superscript $^{\sqcap}$), that is,

$$\partial\hat{J}_y^{\sqcap}(x,s) = \frac{\hat{I}(s)}{w}\left[\mathrm{H}(x+w/2) - \mathrm{H}(x-w/2)\right], \tag{7.2}$$

or the inverse square-root distribution (denoted by superscript $^{\cup}$)

$$\partial\hat{J}_y^{\cup}(x,s) = \frac{\hat{I}(s)}{w}\frac{2}{\pi}\frac{\mathrm{H}(x+w/2) - \mathrm{H}(x-w/2)}{[1-(2x/w)^2]^{1/2}}, \tag{7.3}$$

thus complying with the pertaining edge condition.

The testing EM field is determined by solving the field equations supplemented with the continuity-type condition

$$\lim_{z\downarrow\delta/2} \hat{E}_y^{\text{T}}(x,z,s) - \lim_{z\uparrow-\delta/2} \hat{E}_y^{\text{T}}(x,z,s) = 0 \quad \text{for all } x \in \mathbb{R}, \tag{7.4a}$$

and the excitation condition

$$\lim_{z\downarrow\delta/2} \hat{H}_x^{\text{T}}(x,z,s) - \lim_{z\uparrow-\delta/2} \hat{H}_x^{\text{T}}(x,z,s) = \partial\hat{J}_y^{\text{T}}(x,s) \quad \text{for all } x \in \mathbb{R}, \tag{7.4b}$$

as $\delta \downarrow 0$, where $\partial\hat{J}_y^{\text{T}}(x,s)$ shows either the uniform or inverse square-root behavior (see (7.2) and (7.3)). Employing next the wave-slowness representation of the type specified by (4.33), the testing-field distribution at $z = 0$ can be represented by

$$\hat{E}_y^{\text{T};\sqcap}(x,0,s) = -\frac{s\mu_0\hat{I}^{\text{T};\sqcap}(s)}{2\pi\mathrm{i}}\int_{\kappa=-\mathrm{i}\infty}^{\mathrm{i}\infty} \exp(-s\kappa x)\,\mathrm{i}_0(s\kappa w/2)\frac{\mathrm{d}\kappa}{2p(\kappa)}$$
$$\text{for } \{s \in \mathbb{R}; s > 0\}, \tag{7.5}$$

assuming the uniform distribution, where $\mathrm{i}_0(x)$ denotes the modified spherical Bessel function of the first kind. Alternatively, postulating the inverse square-root testing current distribution, the testing field can be represented via

$$\hat{E}_y^{\mathrm{T};\cup}(x,0,s) = -\frac{s\mu_0 \hat{I}^{\mathrm{T};\cup}(s)}{2\pi\mathrm{i}} \int_{\kappa=-\mathrm{i}\infty}^{\mathrm{i}\infty} \exp(-s\kappa x)\, \mathrm{I}_0(s\kappa w/2) \frac{\mathrm{d}\kappa}{2p(\kappa)}$$
$$\text{for } \{s \in \mathbb{R}; s > 0\}, \tag{7.6}$$

where $\mathrm{I}_0(x)$ denotes the modified Bessel function of the first kind. The thus represented testing-field distributions can be further substituted in the left-hand side of the reciprocity relation (7.1), which yields an interaction quantity in which we change the order of the integrations and, subsequently, evaluate the inner integral with respect to x. In the ensuing step, we employ the explicit-type boundary condition, $\hat{E}_y^{\mathrm{s}}(x,0,s) = -\hat{E}_y^{\mathrm{i}}(x,0,s) = -\hat{e}^{\mathrm{i}}(s)\exp(-s\kappa_0 x)$, with the corresponding testing-current distribution in the right-hand side of (7.1) to get a quantity that is expressible in terms of the modified Bessel function. Finally, invoking the condition that the resulting relation has to hold for arbitrary values of the testing-current amplitude $\hat{I}^{\mathrm{T}}(s)$, we end up with

$$\hat{Z}_{\mathrm{ext}}^{\sqcap}(s)\, \hat{I}^{\mathrm{s};\sqcap}(s) = \hat{e}^{\mathrm{i}}(s)\, \mathrm{i}_0(s\kappa_0 w/2), \tag{7.7a}$$
$$\hat{Z}_{\mathrm{ext}}^{\cup}(s)\, \hat{I}^{\mathrm{s};\cup}(s) = \hat{e}^{\mathrm{i}}(s)\, \mathrm{I}_0(s\kappa_0 w/2), \tag{7.7b}$$

where $\hat{I}^{\mathrm{s}}(s)$ is the s-domain induced electric current and $\hat{Z}_{\mathrm{ext}}(s)$ has the meaning of *external* impedance (per unit length). The latter parameter can be expressed via the following complex slowness integrals:

$$\hat{Z}_{\mathrm{ext}}^{\sqcap}(s) = \frac{s\mu_0}{2\pi\mathrm{i}} \int_{\kappa=-\mathrm{i}\infty}^{\mathrm{i}\infty} \mathrm{i}_0^2(s\kappa w/2) \frac{\mathrm{d}\kappa}{2p(\kappa)}, \tag{7.8a}$$
$$\hat{Z}_{\mathrm{ext}}^{\cup}(s) = \frac{s\mu_0}{2\pi\mathrm{i}} \int_{\kappa=-\mathrm{i}\infty}^{\mathrm{i}\infty} \mathrm{I}_0^2(s\kappa w/2) \frac{\mathrm{d}\kappa}{2p(\kappa)}, \tag{7.8b}$$

pertaining to the uniform and inverse square-root testing-current distributions, as given by (7.2) and (7.3), respectively. It is noted that owing to the small-argument expansions, $\mathrm{i}_0(x) = 1 + O(x^2)$ and $\mathrm{I}_0(x) = 1 + O(x^2)$, as $x \downarrow 0$, both (7.7a) and (7.7b) can be cast into the following form:

$$\hat{Z}_{\mathrm{ext}}(s)\, \hat{I}^{\mathrm{s}}(s) = \hat{e}^{\mathrm{i}}(s), \tag{7.9}$$

as $w \downarrow 0$ or/and for the normal plane-wave incidence.

7.1.1 Alternative expressions for the external impedance

Expressions (7.8a) and (7.8b) for the external impedance can be cast into their alternative forms using integral representations of the modified Bessel functions.

For instance, we may substitute [5, Eq. (9.6.16)] in (7.8a) and change the order of integration to get

$$\hat{Z}^{\cup}_{\rm ext}(s) = \frac{s\mu_0}{2\pi^3 \mathrm{i}} \int_{u=0}^{\pi} \mathrm{d}u \int_{v=0}^{\pi} \mathrm{d}v \int_{\kappa=-\mathrm{i}\infty}^{\mathrm{i}\infty}$$

$$\exp\{-s\kappa w[\cos(u) + \cos(v)]/2\} \frac{\mathrm{d}\kappa}{2p(\kappa)}. \tag{7.10}$$

Associating now $\kappa w[\cos(u) + \cos(v)]/2$ with the real-valued and positive time parameter τ, the integration contour in the complex κ-plane can be deformed to the loop encircling the branch cut $\{1/c_0 < |\mathrm{Re}(\kappa)| < \infty, \mathrm{Im}(\kappa) = 0\}$. In the next step, the integration with respect to κ is transformed to the integration with respect τ, while combining the integrations just above and just below the branch cut. In this manner, we arrive at

$$\hat{Z}^{\cup}_{\rm ext}(s) = \frac{s\mu_0}{2\pi^3} \int_{u=0}^{\pi} \mathrm{d}u \int_{v=0}^{\pi} \mathrm{d}v \int_{\tau=|D|/c_0}^{\infty} \exp(-s\tau) \frac{\mathrm{d}\tau}{(\tau^2 - D^2/c_0^2)^{1/2}}, \tag{7.11}$$

where $D = D(u, v) = (w/2)[\cos(u) + \cos(v)]$. The inner integral with respect to τ can be associated with the zero-order modified Bessel function of the second kind (see [5, Eq. (9.6.23)]), so that we finally end up with [2, Eq. (9)]

$$\hat{Z}^{\cup}_{\rm ext}(s) = \frac{s\mu_0}{2\pi^3} \int_{u=0}^{\pi} \mathrm{d}u \int_{v=0}^{\pi} \mathrm{K}_0[sw|\cos(u) - \cos(v)|/2c_0] \mathrm{d}v, \tag{7.12a}$$

where we used substitution $v = \pi - t$ with $\cos(v) = -\cos(t)$ and $t = v$, again. Likewise, we may find an alternative expression applying to the uniform electric-current distribution [2, Eq. (8)]

$$\hat{Z}^{\sqcap}_{\rm ext}(s) = \frac{s\mu_0}{2\pi} \int_{u=0}^{1} \mathrm{d}u \int_{v=0}^{1} \mathrm{K}_0(sw|u - v|/c_0) \mathrm{d}v. \tag{7.12b}$$

The integrals in (7.12a) and (7.12a) can be evaluated with the aid of the small-argument expansion of the modified Bessel function [5, Eq. (9.6.54)] and the following integral formulas:

$$\int_{u=0}^{1} \mathrm{d}u \int_{v=0}^{1} \log(w|u - v|) \mathrm{d}v = \log(w) - 3/2, \tag{7.13}$$

$$\frac{1}{\pi^2} \int_{u=0}^{\pi} \mathrm{d}u \int_{v=0}^{\pi} \log[w|\cos(u) - \cos(v)|] \mathrm{d}v = \log(w) - \log(2). \tag{7.14}$$

To establish (7.14), we used the following identity [6, p. 69]:

$$\log[|\cos(u) - \cos(v)|] = -\log(2) - \sum_{m=1}^{\infty} (2/m) \cos(mu) \cos(mv). \tag{7.15}$$

Hence, making use of integral formulas (7.13) and (7.14), we arrive at compact expressions, viz

$$\hat{Z}^{\sqcap}_{\mathrm{ext}}(s) \simeq -(s\mu_0/2\pi)[\log(sw/c_0) - 3/2 - \log(2) + \gamma], \tag{7.16a}$$

$$\hat{Z}^{\sqcup}_{\mathrm{ext}}(s) \simeq -(s\mu_0/2\pi)[\log(sw/c_0) - 3\ln(2) + \gamma], \tag{7.16b}$$

as $sw/c_0 \ll 1$ with $\gamma = 0.57721\ldots$ representing Euler's constant. Now, taking the limit, $\{s = \mathrm{i}\omega + \delta; \omega \in \mathbb{R}, \delta \downarrow 0\}$, we get the corresponding expressions in the real-FD [2, Eqs. (13) and (14)]:

$$\hat{Z}^{\sqcap}_{\mathrm{ext}}(\mathrm{i}\omega) \simeq \omega\mu_0/4 + (\mathrm{i}\omega\mu_0/2\pi)\left[\log(\lambda/w) + 3/2 - \gamma - \log(\pi)\right], \tag{7.17a}$$

$$\hat{Z}^{\sqcup}_{\mathrm{ext}}(\mathrm{i}\omega) \simeq \omega\mu_0/4 + (\mathrm{i}\omega\mu_0/2\pi)\left[\log(\lambda/w) + 2\log(2) - \gamma - \log(\pi)\right], \tag{7.17b}$$

provided that $w/\lambda \ll 1$ with $\lambda = 2\pi c_0/\omega$. Apparently, (7.17a) corresponds to Schelkunoff's formula [1, Eq. (8.5-12)], in which the change to the inverse square-root electric-current distribution amounts to replacing $3/2 - \gamma - \log(\pi) \simeq -0.222$ with $2\log(2) - \gamma - \log(\pi) \simeq -0.336$.

7.1.2 TD solution

The TD counterpart of (7.9) will next be solved numerically through an updating, marching-on-in-time scheme. To that end, the time axis, $\{t \in \mathbb{R}; t > 0\}$, is discretized uniformly with the constant time step, $\{t_k = k\Delta t; \Delta t > 0, k = 1, 2, \ldots, M\}$. Furthermore, the (unknown) induced electric current $I^{\mathrm{s}}(t)$ is expanded in a piecewise-linear basis as

$$I^{\mathrm{s}}(t) \simeq \sum_{k=1}^{M} i_k \Lambda_k(t), \tag{7.18}$$

where i_k denotes the coefficients to be found and $\Lambda_k(t)$ is the triangular function defined via

$$\Lambda_k(t) = \begin{cases} 1 + (t - t_k)/\Delta t & \text{for } t \in [t_{k-1}, t_k], \\ 1 - (t - t_k)/\Delta t & \text{for } t \in [t_k, t_{k+1}]. \end{cases} \tag{7.19}$$

The unknown current coefficients can be obtained through a step-by-step updating procedure that results from the transformation of (7.9) to the TD, in which we use the piecewise linear current expansion (7.18) and (the TD counterpart of) the representation (7.8a) for the external impedance (see Section 7.4). In this way, we end up with (cf. (7.44))

$$i_m = \Phi_1^{-1}\left[c_0\Delta t\eta_0 e^{\mathrm{i}}(t_m) - \sum_{k=1}^{m-1} i_k\left(\Phi_{m-k+1} - 2\Phi_{m-k} + \Phi_{m-k-1}\right)\right]$$
$$\text{for all } m = \{1, \ldots, M\}, \tag{7.20}$$

where we used $\Phi_k = \Phi(t_k)$, $\Phi_1^{-1} = 1/\Phi(\Delta t)$ and the TD function $\Phi(t)$ can be achieved through a generic integral as specified in Appendix B. Hence, with reference to (7.8a) and (B.7), we find

$$\Phi(t) = \left[\Upsilon^B(w,t) - 2\Upsilon^B(0,t) + \Upsilon^B(-w,t)\right]/w^2. \tag{7.21}$$

An alternative marching-on-in-time solution concerning the pulsed EM scattering by a narrow PEC strip has been presented in Ref. [2].

7.2 Pulsed EM scattering by a narrow strip with conductive and dielectric properties

In order to incorporate bounded conductive and dielectric constitutive parameters of the narrow strip under consideration (see Figure 7.1), we start with the reciprocity relation (7.1), where the scattered electric-field strength on its right-hand side is expressed as the difference between the total and incident fields. Consequently, the total electric-field distribution along the strip is expressed through the pertaining transition condition (cf. (3.15b) and (6.14b)), i.e.

$$\partial\hat{J}_y^{\mathrm{s}}(x,s) = \hat{Y}^{\mathrm{E}}(s)\hat{E}_y(x,0,s)[\mathrm{H}(x+w/2) - \mathrm{H}(x-w/2)] \quad \text{for all } x \in \mathbb{R}, \tag{7.22}$$

and we end up with

$$\int_{x=-w/2}^{w/2} \hat{E}_y^{\mathrm{T}}(x,0,s)\partial\hat{J}_y^{\mathrm{s}}(x,s)\mathrm{d}x = -\int_{x=-w/2}^{w/2} \hat{E}_y^{\mathrm{i}}(x,0,s)\partial\hat{J}_y^{\mathrm{T}}(x,s)\mathrm{d}x + \hat{Z}^{\mathrm{E}}(s)\int_{x=-w/2}^{w/2} \partial\hat{J}_y^{\mathrm{s}}(x,s)\partial\hat{J}_y^{\mathrm{T}}(x,s)\mathrm{d}x. \tag{7.23}$$

Here, recall that $\hat{Z}^{\mathrm{E}}(s) = 1/\hat{Y}^{\mathrm{E}}(s)$. Assuming further the uniform electric-current distribution (see (7.2)), we may follow the procedure that led to (7.7a), and obtain

$$\left[\hat{Z}_{\mathrm{ext}}^{\sqcap}(s) + \hat{Z}^{\mathrm{E}}(s)/w\right]\hat{I}^{\mathrm{s};\sqcap}(s) = \hat{e}^{\mathrm{i}}(s)\,\mathrm{i}_0(s\kappa_0 w/2). \tag{7.24}$$

Owing to the relatively small width of the strip, the right-hand side of (7.24) can be replaced by $\hat{e}^{\mathrm{i}}(s)$, again. The thus approximated equation will be next solved numerically in the TD.

7.2.1 TD solution

We next follow the solution strategy introduced in Section 7.1.2 to calculate the electric-current pulse induced in narrow EM-penetrable strips. As the resulting solution depends on the actual constitutive characteristics of the strip, we shall discuss several important cases in separate sections.

7.2.1.1 Strip with conductive properties

If the strip shows a high EM contrast with respect to the surrounding medium in its electric conductivity only, the s-domain layer admittance is equal to $\hat{Y}^{\mathrm{E}}(s) = 1/\hat{Z}^{\mathrm{E}}(s) = G^{\mathrm{E}}$ (see (3.5a)), which is an s-independent parameter. Subsequently, upon expanding the desired electric-current surface density according to (7.18), the unknown current coefficients follow from the pertaining TD counterpart of (7.24) using the marching-on-in-time solution (7.20) in which $\Phi(t)$ as given by (7.21) is replaced with

$$\begin{aligned}\Phi(t) = {} & [\Upsilon^B(w,t) - 2\Upsilon^B(0,t) + \Upsilon^B(-w,t)]/w^2 \\ & + (c_0 t/w)(\eta_0/G^{\mathrm{E}})\mathrm{H}(t).\end{aligned} \tag{7.25}$$

Obviously, (7.21) can be viewed as the limit of (7.25) as $G^{\mathrm{E}} \to \infty$.

7.2.1.2 Strip with conductive and dielectric properties

If the strip is made of a medium showing relatively high both electric conductivity and permittivity, the equivalent layer admittance is given by $\hat{Y}^{\mathrm{E}}(s) = 1/\hat{Z}^{\mathrm{E}}(s) = G^{\mathrm{E}} + sC^{\mathrm{E}}$ (see (3.5a) and (3.5b)). The updating scheme for achieving the unknown current coefficients can be, again, cast into the form of (7.20), where we replace $\Phi(t)$ as given by (7.21) with

$$\begin{aligned}\Phi(t) = {} & [\Upsilon^B(w,t) - 2\Upsilon^B(0,t) + \Upsilon^B(-w,t)]/w^2 \\ & + (c_0/w\alpha^{\mathrm{E}})(\eta_0/G^{\mathrm{E}})\left[\alpha^{\mathrm{E}} t - 1 + \exp(-\alpha^{\mathrm{E}} t)\right]\mathrm{H}(t),\end{aligned} \tag{7.26}$$

where $\alpha^{\mathrm{E}} = G^{\mathrm{E}}/C^{\mathrm{E}} > 0$ (see (3.19)). For a vanishing equivalent layer capacitance, the parameter α^{E} tends to infinity and (7.26) reduces to (7.25), as expected. Furthermore, in the late-time limit as $t \to \infty$, we may replace $\alpha^{\mathrm{E}} t - 1 + \exp(-\alpha^{\mathrm{E}} t)$ in (7.26) with $\alpha^{\mathrm{E}} t$, which yields $\Phi(t)$ pertaining to a conductive strip, again (see (7.25)). The parameter α^{E} approaches infinity also if $G^{\mathrm{E}} \to \infty$, which leads to (7.21) applying to the PEC strip. On the other hand, for a bounded value $G^{\mathrm{E}} = O(1)$ and $C^{\mathrm{E}} \to \infty$ parameter α^{E} approaches the value zero and (7.26) reduces to (7.21), again, thus modeling the response of a PEC strip.

7.2.1.3 Strip with plasmonic properties

If the relatively high electric conductivity of the strip is described via the Drude-type conduction relaxation function (5.21), the layer admittance is no longer s-independent, but is given by $\hat{Y}^{\mathrm{E}}(s) = 1/\hat{Z}^{\mathrm{E}}(s) = \hat{G}^{\mathrm{E}}(s) = \delta\epsilon_0\omega_{\mathrm{pe}}^2/(s+\nu_{\mathrm{c}})$, where $\delta > 0$ is the thickness of the strip, and recall that ω_{pe} and ν_{c} denote the electron plasma angular frequency and the collision frequency, respectively. The desired electric-current coefficients are, again, attainable via the updating scheme (7.20), in which we use

$$\begin{aligned}\Phi(t) = {} & [\Upsilon^B(w,t) - 2\Upsilon^B(0,t) + \Upsilon^B(-w,t)]/w^2 \\ & + (c_0/w\nu_{\mathrm{c}})[\eta_0/\hat{G}^{\mathrm{E}}(0)](\nu_{\mathrm{c}} t + 1)\mathrm{H}(t).\end{aligned} \tag{7.27}$$

In the late-time limit, we may replace $\nu_c t + 1$ in (7.27) with $\nu_c t$, which yields $\Phi(t)$ in the form of (7.25), where the "static conductance" $\hat{G}^{\mathrm{E}}(0) = \delta\epsilon_0\omega_{\mathrm{pe}}^2/\nu_{\mathrm{c}}$ plays the role of G^{E} representing a resistive strip.

7.3 Pulsed EM scattering by a narrow metastrip

In this section, we shall analyze the pulsed EM response of a narrow metastrip. To that end, the starting EM reciprocity relation (7.1) is generalized to account for the jump discontinuity of both (the tangential components of) electric- and magnetic-field strengths across the metastrip. Accordingly, upon introducing the equivalent magnetic-current surface density denoted by $\partial\hat{K}_x$, we write

$$\int_{x=-w/2}^{w/2} \langle\hat{E}_y^{\mathrm{T}}\rangle_-^+(x,s)\partial\hat{J}_y^{\mathrm{s}}(x,s)\mathrm{d}x - \int_{x=-w/2}^{w/2} \langle\hat{H}_x^{\mathrm{T}}\rangle_-^+(x,s)\partial\hat{K}_x^{\mathrm{s}}(x,s)\mathrm{d}x$$
$$= \int_{x=-w/2}^{w/2} \langle\hat{E}_y^{\mathrm{s}}\rangle_-^+(x,s)\partial\hat{J}_y^{\mathrm{T}}(x,s)\mathrm{d}x - \int_{x=-w/2}^{w/2} \langle\hat{H}_x^{\mathrm{s}}\rangle_-^+(x,s)\partial\hat{K}_x^{\mathrm{T}}(x,s)\mathrm{d}x, \tag{7.28}$$

and recall that operator $\langle\rangle_-^+$ has been defined by (3.11). The testing fields are determined using the excitation conditions (cf. (7.2), (7.4a) and (7.4b))

$$\partial\hat{K}_x^{\mathrm{T}}(x,s) = \frac{2\hat{K}^{\mathrm{T}}(s)}{w}\left[\mathrm{H}(x+w/4) - \mathrm{H}(x-w/4)\right], \tag{7.29a}$$

$$\partial\hat{J}_y^{\mathrm{T}}(x,s) = \frac{\hat{J}^{\mathrm{T}}(s)}{w}\left[\mathrm{H}(x+w/2) - \mathrm{H}(x-w/2)\right],$$
$$\text{for all } x \in \mathbb{R}. \tag{7.29b}$$

It is seen in (7.29a) that the support of the testing magnetic-current surface density extends along $\{-w/4 < x < w/4\}$ only, such that the current expansion complies with the edge condition $\partial\hat{K}_x^{\mathrm{T}}(\pm w/2, s) = 0$. The corresponding testing fields can be then represented via (cf. (7.5))

$$\langle\hat{E}_y^{\mathrm{T}}\rangle_-^+(x,s) = -\frac{s\hat{I}^{\mathrm{T}}(s)}{2\pi\mathrm{i}}\int_{\kappa=-\mathrm{i}\infty}^{\mathrm{i}\infty} \exp(-s\kappa x)\mathrm{i}_0(s\kappa w/2)\frac{\mu_0}{2p(\kappa)}\mathrm{d}\kappa, \tag{7.30}$$

$$\langle\hat{H}_x^{\mathrm{T}}\rangle_-^+(x,s) = -\frac{s\hat{K}^{\mathrm{T}}(s)}{2\pi\mathrm{i}}\int_{\kappa=-\mathrm{i}\infty}^{\mathrm{i}\infty} \exp(-s\kappa x)\mathrm{i}_0(s\kappa w/4)\frac{p(\kappa)}{2\mu_0}\mathrm{d}\kappa,$$
$$\text{for } \{s \in \mathbb{R}; s > 0\}. \tag{7.31}$$

We shall next proceed in a similar fashion with the *induced* current densities. Hence, we postulate the uniform spatial distribution for the induced electric-current surface

density and the triangular distribution for the magnetic-current surface density, that is,

$$\partial\hat{J}^{\mathrm{s}}_y(x,s) = \frac{\hat{I}^{\mathrm{s}}(s)}{w}\big[\mathrm{H}(x+w/2) - \mathrm{H}(x-w/2)\big], \tag{7.32a}$$

$$\partial\hat{K}^{\mathrm{s}}_x(x,s) = \frac{2\hat{K}^{\mathrm{s}}(s)}{w}(1-2|x|/w)\big[\mathrm{H}(x+w/2) - \mathrm{H}(x-w/2)\big], \tag{7.32b}$$

thus meeting the end conditions $\partial\hat{K}^{\mathrm{s}}_x(\pm w/2,s) = 0$, again. The property that $\langle\hat{E}^{\mathrm{T}}_y\rangle^+_-$ and $\langle\hat{H}^{\mathrm{T}}_x\rangle^+_-$ are generated by $\partial\hat{J}^{\mathrm{T}}_y$ and $\partial\hat{K}^{\mathrm{T}}_x$ alone, respectively, makes it possible to decouple the reciprocity relation (7.28) to two mutually independent equations (cf. (7.24))

$$\big[\hat{Z}_{\mathrm{ext}}(s) + \hat{Z}^{\mathrm{E}}(s)/w\big]\,\hat{I}^{\mathrm{s}}(s) = \hat{e}^{\mathrm{i}}(s)\,\mathrm{i}_0(s\kappa_0 w/2), \tag{7.33a}$$

$$\big[\hat{Y}_{\mathrm{ext}}(s) + 3\hat{Y}^{\mathrm{M}}(s)/2w\big]\,\hat{K}^{\mathrm{s}}(s) = Y_{\perp}(\kappa_0)\hat{e}^{\mathrm{i}}(s)\,\mathrm{i}_0(s\kappa_0 w/4), \tag{7.33b}$$

where $Y_{\perp}(\kappa_0) = \eta_0\cos(\beta)$ and the pertaining integral representations for the external impedance and admittance follow as (see (7.8a))

$$\hat{Z}_{\mathrm{ext}}(s) = \frac{s}{2\pi\mathrm{i}}\int_{\kappa=-\mathrm{i}\infty}^{\mathrm{i}\infty} \mathrm{i}_0^2(s\kappa w/2)\frac{\mu_0}{2p(\kappa)}\mathrm{d}\kappa, \tag{7.34}$$

$$\hat{Y}_{\mathrm{ext}}(s) = \frac{s}{2\pi\mathrm{i}}\int_{\kappa=-\mathrm{i}\infty}^{\mathrm{i}\infty} \mathrm{i}_0^3(s\kappa w/4)\frac{p(\kappa)}{2\mu_0}\mathrm{d}\kappa. \tag{7.35}$$

Since the width of the metastrip is assumed to be relatively small, the modified Bessel functions on the right-hand sides of (7.33a) and (7.33b) can be replaced by number one, whatever the angle of incidence. The thus obtained s-domain relations can be solved either directly in the FD or in the TD via the marching-on-in-time technique (see Section 7.1.2, for instance).

With the electric and magnetic currents induced in a metastrip at our disposal, we may next express the scattered electric-field strength at the field point, say (x_0, y_0), by summing their contributions via

$$\begin{aligned}\hat{E}^{\mathrm{s}}_y(x_0,z_0,s) = &-\frac{s\mu_0}{2\pi}\int_{x=-w/2}^{w/2}\mathrm{K}_0[s\rho_0(x)/c_0]\partial\hat{J}^{\mathrm{s}}_y(x,s)\mathrm{d}x\\ &+\frac{s}{2\pi c_0}\int_{x=-w/2}^{w/2}\frac{z_0}{\rho_0(x)}\mathrm{K}_1[s\rho_0(x)/c_0]\partial\hat{K}^{\mathrm{s}}_x(x,s)\mathrm{d}x,\end{aligned} \tag{7.36}$$

where $\rho_0(x) = [(x-x_0)^2 + z_0^2]^{1/2} > 0$. Owing to the small width, representation (7.36) can be roughly approximated by

$$\hat{E}^{\mathrm{s}}_y(x_0,z_0,s) \simeq -\frac{s\mu_0\hat{I}^{\mathrm{s}}(s)}{2\pi}\mathrm{K}_0(sr_0/c_0) + \frac{s\mu_0\eta_0\hat{K}^{\mathrm{s}}(s)}{2\pi}\frac{z_0}{r_0}\mathrm{K}_1(sr_0/c_0), \tag{7.37}$$

where $r_0 = \rho_0(0) = (x_0^2 + z_0^2)^{1/2} > 0$. As a matter of fact, (7.37) represents the superposition of electric-field strengths radiated from electric and magnetic line sources located at the origin whose source amplitudes are $\hat{I}^{\mathrm{s}}(s)$ and $\hat{K}^{\mathrm{s}}(s)$, respectively.

7.3.1 TD solution

As the solution of (7.33a) has been presented in Sections 7.1.2 and 7.2.1, we shall next solve the TD counterpart of (7.33b) for the TD magnetic current $K^{\mathrm{s}}(t)$. For that purpose, the desired magnetic-current pulse is first expanded in a series of triangular functions as (cf. (7.18))

$$K^{\mathrm{s}}(t) \simeq \sum_{k=1}^{M} v_k \Lambda_k(t), \tag{7.38}$$

where v_k are unknown coefficients to be found and $\Lambda_k(t)$ was defined via (7.19). Substituting then the temporal expansion (7.38) in the TD counterpart of (7.33b), the coefficients follow via an updating procedure (cf. (7.20))

$$v_m = \Theta_1^{-1} \left[c_0 \Delta t e^{\mathrm{i}}(t_m) \cos(\beta) - \sum_{k=1}^{m-1} v_k \left(\Theta_{m-k+1} - 2\Theta_{m-k} + \Theta_{m-k-1} \right) \right]$$
$$\text{for all } m = \{1, \ldots, M\}, \tag{7.39}$$

where we used the shorthand notation, $\Theta_k = \Theta(t_k)$, and the TD function $\Theta(t)$ can be found from (cf. (7.26))

$$\begin{aligned} \Theta(t) = {} & 8[\Xi(3w/4, t) - 3\Xi(w/4, t) + 3\Xi(-w/4, t) - \Xi(-3w/4, t)]/w^3 \\ & + (3c_0/2w\alpha^{\mathrm{M}})(\zeta_0/R^{\mathrm{M}}) \left[\alpha^{\mathrm{M}} t - 1 + \exp(-\alpha^{\mathrm{M}} t) \right] \mathrm{H}(t). \end{aligned} \tag{7.40}$$

Here, $\alpha^{\mathrm{M}} = R^{\mathrm{M}}/L^{\mathrm{M}} > 0$ (see (3.21)) and TD function $\Xi(x, t)$ can be found in Appendix E. For a vanishing equivalent layer inductance, the parameter α^{M} tends to infinity so that the second term in (7.40) approaches $(3c_0 t/2w)(\zeta_0/R^{\mathrm{M}})\mathrm{H}(t)$, which is also the dominant term in the late-time limit (cf. (7.25)). If either $R^{\mathrm{M}} \to \infty$ or $L^{\mathrm{M}} \to \infty$, the second term in (7.40) is zero, thus modeling the response of a perfectly magnetically conducting (PMC) strip. On the other hand, if $R^{\mathrm{M}} \downarrow 0$ the second term approaches $(3c_0 t/4w)(\zeta_0 t/L^{\mathrm{M}})\ \mathrm{H}(t)$, thereby modeling the response of a loss-free metastrip.

7.4 Concluding remarks

The TD induced current response of a narrow strip has been calculated via the marching-on-in-time technique applied to the TD original of (cf. (7.9) and (7.18))

$$\sum_{k=1}^{M} i_k \hat{Z}(s)[\exp(-st_{k-1}) - 2\exp(-st_k) + \exp(-st_{k+1})] = \hat{V}(s), \tag{7.41}$$

where $\hat{Z}(s)$ denotes the impedance function in the s-domain and $\hat{V}(s)$ represents the (known) excitation. Next, we shall derive two recurrence relations enabling the calculation of the current coefficients i_k for all $k = \{1, \ldots, M\}$ along the discretized time axis.

First, employing [5, Eq. (29.2.15)], (7.41) can be directly transformed to the TD and evaluated at $t = t_m = m\Delta t$. In this way, we find

$$\sum_{k=1}^{m}(Z_{m-k+1} - 2Z_{m-k} + Z_{m-k-1})\, i_k = V_m, \tag{7.42}$$

where, on account of $Z(t) = 0$ for $t < 0$, we truncated the upper limit of the summation. Next, we used the shorthand notation $Z(m\Delta t) = Z_m$ and $V(m\Delta t) = V_m$. It is seen that the impedance terms can be associated with the second-order central difference [5, Eq. (25.1.2)]

$$\delta^2 Z_n = Z_{n+1} - 2Z_n + Z_{n-1}. \tag{7.43}$$

Consequently, upon separating the actual current term at $t = t_m = m\Delta t$, we end up with an explicit solution

$$i_m = Z_1^{-1}\left(V_m - \sum_{k=1}^{m-1}\delta^2 Z_{m-k}\, i_k\right) \quad \text{for all } m = \{1, \ldots, M\}, \tag{7.44}$$

where Z_1^{-1} denotes the inverse of $Z_1 = Z(\Delta t)$. In this manner, the actual current term, i_m, is expressed in terms of the corresponding excitation term and the past current samples. Yet alternatively, notice that we may rewrite the starting equation to an equivalent form

$$\sum_{k=1}^{M} i_k \hat{Z}(s) \exp(-st_{k-1}) = \frac{\hat{V}(s)}{[1 - \exp(-s\Delta t)]^2}. \tag{7.45}$$

Then, using the (convergent) geometric-series expansion

$$\frac{1}{[1 - \exp(-s\Delta t)]^2} = \sum_{n=0}^{\infty}(n+1)\exp(-sn\Delta t) \tag{7.46}$$

the pertaining TD solution can be expressed as

$$i_m = Z_1^{-1}\left[\sum_{n=0}^{m-1}(n+1)V_{m-n} - \sum_{k=1}^{m-1} i_k Z_{m-k+1}\right] \quad \text{for all } m = \{1, \ldots, M\}, \tag{7.47}$$

where we employed the property of causality, again, and the zero initial condition, $V(0) = 0$. These marching-on-in-time solutions are used throughout the book.

References

[1] S. A. Schelkunoff. *Electromagnetic Waves*. New York, NY: D. Van Nostrand Company, Inc., 9th printing edition, 1957.

[2] M. Štumpf. Coupling of impulsive EM plane-wave fields to narrow conductive strips: an analysis based on the concept of external impedance. *IEEE Transactions on Electromagnetic Compatibility*, 60(2):548–551, 2018.
[3] M. Štumpf. Evaluating the ground impedance—a new methodology based on EM reciprocity. *IEEE Transactions on Electromagnetic Compatibility*, 61(3):928–934, 2019.
[4] M. Štumpf. *Electromagnetic Reciprocity in Antenna Theory*. Hoboken, NJ: John Wiley & Sons, Inc., 2018.
[5] M. Abramowitz and I. A. Stegun. *Handbook of Mathematical Functions*. New York, NY: Dover Publications, 1972.
[6] C. J. Bouwkamp. Diffraction theory: a critique of some recent developments. Research report no. EM–50. Technical report, Washington Square Coll. of Arts and Science, New York Univ., 1953.

Chapter 8
Pulsed EM-field scattering by bounded metasurfaces in a homogeneous embedding

Closed-form solutions of pulsed electromagnetic (EM) scattering problems in terms of analytic functions are attainable for a limited number of configurations only. Several examples from this category including infinite, semi-infinite and narrow thin screens are presented in Chapters 4, 6 and 7, respectively. Whenever the problem under consideration cannot be plausibly approximated by one of the limiting cases, one has to resort to a numerical technique. An efficient computational method capable of handling pulsed EM scattering by bounded planar screens is termed as the Cagniard–DeHoop method of moments (CdH-MoM) (see [1, Chapters 2 and 14]). Its application to analyzing the pulsed EM scattering by two-dimensional (2-D) planar structures located on a homogeneous background is hence the main subject matter of this chapter.

8.1 Pulsed EM scattering by a bounded PEC screen

In this section, we shall analyze the pulsed EM scattering by a bounded perfectly electrically conducting (PEC) screen that occupies domain $\{a < x < b, -\infty < y < \infty, -\delta/2 < z < \delta/2\}$, where $\delta > 0$ denotes its (relatively small) thickness and $w = b - a > 0$ is its (bounded) width (see Figure 8.1). The analyzed screen is located in the homogeneous, isotropic and loss-free embedding $\mathcal{D}^\infty$ that is described by parameters $\{\epsilon_0, \mu_0\}$ and is excited by the uniform, E-polarized plane wave, i.e.

$$\hat{E}_y^{\rm i}(x, z, s) = \hat{e}^{\rm i}(s)\exp\{-s[\kappa_0 x - p(\kappa_0)z]\}, \qquad \text{(4.1 revisited)}$$

where $\kappa_0 = \sin(\beta)/c_0$ and $p(\kappa_0) = \cos(\beta)/c_0$ for $\{0 \le \beta \le \pi/2\}$. In contrast to the problems analyzed in Chapter 7, the screen's width w does not have to be small with

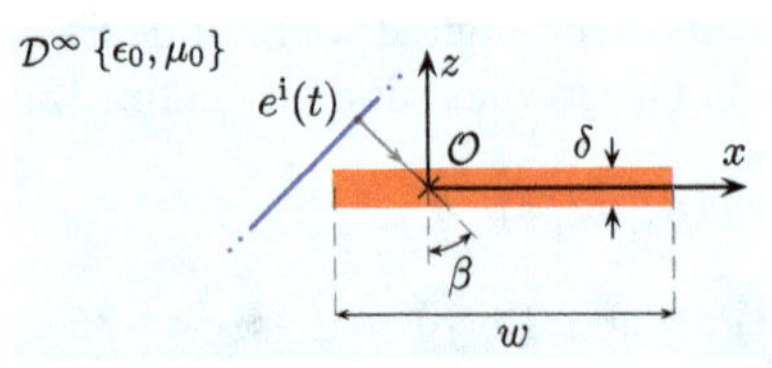

Figure 8.1 Pulsed EM plane wave impinging upon a bounded plane screen

respect to the spatial support of the excitation pulse. Accordingly, the spatial distribution of the induced electric-current surface density is no longer generally trivial (e.g. (7.2) and (7.3)), yet it is supposed to be representable in a piecewise-linear fashion. In this respect, it is noted that by virtue of the pertaining edge condition, the induced electric-current surface density cannot grow more rapidly than $\rho^{-1/2}$ as $\rho \downarrow 0$, where ρ represents the distance from the edge [2, Chapter 4].

8.1.1 *Problem formulation*

Pursuing the line of reasoning of Section 7.1, the scattering problem will be formulated through the following reciprocity relation (cf. (7.1)):

$$\int_{x=a}^{b} \hat{E}_y^{\mathrm{T}}(x,0,s)\partial\hat{J}_y^{\mathrm{s}}(x,s)\mathrm{d}x = \int_{x=a}^{b} \hat{E}_y^{\mathrm{s}}(x,0,s)\partial\hat{J}_y^{\mathrm{T}}(x,s)\mathrm{d}x, \tag{8.1}$$

thus interrelating the scattered EM-field state (denoted by superscript ^{s}) with the testing one (denoted by superscript ^{T}). For the problem configuration at hand, the relevant testing EM-field quantities $\{\hat{E}_y^{\mathrm{T}}, \hat{H}_x^{\mathrm{T}}, \hat{H}_z^{\mathrm{T}}\}(x,z,s)$ are governed by an y-independent EM-field equations applying to $\mathcal{D}^{\infty}$. To determine the testing electric-field distribution $\hat{E}_y^{\mathrm{T}}$ on the screen, the EM-field equations are further supplemented with the transition condition

$$\begin{aligned}&\lim_{z\downarrow\delta/2} \hat{H}_x^{\mathrm{T}}(x,z,s) - \lim_{z\uparrow-\delta/2} \hat{H}_x^{\mathrm{T}}(x,z,s)\\ &= \partial\hat{J}_y^{\mathrm{T}}(x,s)\big[\mathrm{H}(x-a) - \mathrm{H}(x-b)\big] \quad \text{for all } x \in \mathbb{R},\end{aligned} \tag{8.2}$$

as $\delta \downarrow 0$, thereby coupling the testing field with its source.

The thus formulated problem will be next solved with the aid of the CdH-MoM whose point of departure is the reciprocity relation in the slowness domain (see e.g. [1, Section 2.2]). To that end, we employ the wave-slowness representation (4.33) to represent the electric-field strengths in the s-domain reciprocity relation (8.1). Subsequently, upon changing the order of integration in the resulting expression, the reciprocity relation is cast into the following form:

$$\begin{aligned}&\frac{s}{2\pi\mathrm{i}}\int_{\kappa=-\mathrm{i}\infty}^{\mathrm{i}\infty} \tilde{E}_y^{\mathrm{T}}(\kappa,0,s)\partial\tilde{J}_y^{\mathrm{s}}(-\kappa,s)\mathrm{d}\kappa = \frac{s}{2\pi\mathrm{i}}\int_{\kappa=-\mathrm{i}\infty}^{\mathrm{i}\infty} \tilde{E}_y^{\mathrm{s}}(\kappa,0,s)\partial\tilde{J}_y^{\mathrm{T}}(-\kappa,s)\mathrm{d}\kappa\\ &\text{for } \{s\in\mathbb{R}; s>0\}.\end{aligned} \tag{8.3}$$

The pertaining expression for the transform-domain testing electric-field strength, $\tilde{E}_y^{\mathrm{T}}$, immediately follows upon solving the source-free EM-field equations with the transition condition (8.2) in the slowness domain. In this way, we arrive at

$$\tilde{E}_y^{\mathrm{T}}(\kappa,0,s) = -[\mu_0/2p(\kappa)]\partial\tilde{J}_y^{\mathrm{T}}(\kappa,s), \tag{8.4}$$

with $p(\kappa) = (1/c_0^2 - \kappa^2)^{1/2}$ and $\mathrm{Re}(p) \geq 0$ (see (4.36)). The slowness-domain reciprocity relation (8.3) with the testing source-field (8.4) will next be solved numerically with the aid of the CdH technique (see Chapter 1).

8.1.2 Problem solution

To solve the EM scattering problem numerically, the space–time solution space has to be properly discretized. For that purpose, the spatial domain occupied by the screen is divided into $N - 1$ segments of a constant length $\Delta = w/(N - 1)$, where N denotes the number of nodes along the uniform grid (see Figure 8.2 for $N = 7$).

The position of the nodes is specified by

$$x_n = a + (n - 1)\Delta \quad \text{for } n = \{1, \ldots, N\}, \tag{8.5}$$

from which the end points follow as $x_1 = a$ and $x_N = a + w = b$. In a similar fashion, the time axis $\{t \in \mathbb{R}; t > 0\}$ is discretized uniformly with the constant time step $\{t_k = k\Delta t; \Delta t > 0, k = 1, 2, \ldots, M\}$. With the discretized solution space at our disposal, the unknown induced electric-current surface density is expanded in a series of triangular functions

$$\partial J_y^{\mathrm{s}}(x, t) \simeq \sum_{n=1}^{N} \sum_{k=1}^{M} j_k^{[n]} \Lambda^{[n]}(x) \Lambda_k(t), \tag{8.6}$$

where $j_k^{[n]}$ are unknown coefficients (in A/m) to be found and the spatial basis function (see Figure 8.2) can be defined via

$$\Lambda^{[n]}(x) = \begin{cases} 1 + (x - x_n)/\Delta & \text{for } x \in [x_{n-1}, x_n], \\ 1 - (x - x_n)/\Delta & \text{for } x \in [x_n, x_{n+1}], \end{cases}$$
$$\text{for } n = \{2, \ldots, N - 1\}, \tag{8.7a}$$

with

$$\Lambda^{[1]}(x) = 1 - (x - x_1)/\Delta \quad \text{for } x \in [x_1, x_1 + \Delta], \tag{8.7b}$$
$$\Lambda^{[N]}(x) = 1 + (x - x_N)/\Delta \quad \text{for } x \in [x_N - \Delta, x_N], \tag{8.7c}$$

thus accounting for the nonzero induced current along the edge (see Section 8.1). Furthermore, the temporal triangular function, $\Lambda_k(t)$, is described, in a similar way,

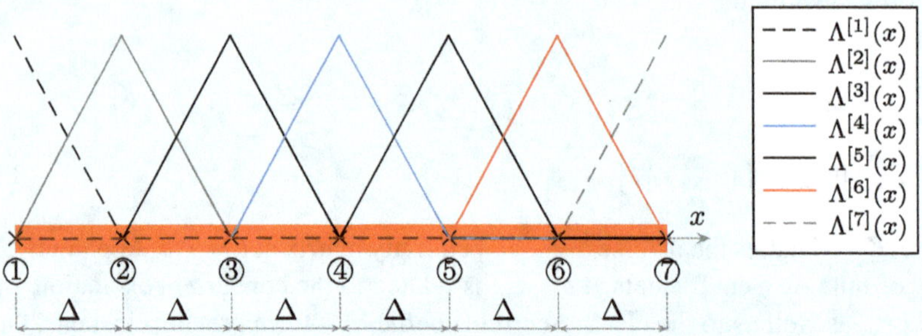

Figure 8.2 Uniform discretization of the spatial solution domain

by (7.19). It remains to specify the testing current density. The space–time testing electric-current distribution is chosen to be a piecewise-constant function of x and the Dirac-delta impulsive function in t, that is,

$$\partial J_y^{\mathrm{T}}(x,t) = \Pi^{[S]}(x)\delta(t) \quad \text{for all } S = \{1,\ldots,N\}, \tag{8.8}$$

where

$$\Pi^{[n]}(x) = \begin{cases} 1 & \text{for } x \in [x_n - \Delta/2, x_n + \Delta/2], \\ 0 & \text{elsewhere}\,, \end{cases}$$
$$\text{for all } n = \{2,\ldots,N-1\}, \tag{8.9a}$$

with

$$\Pi^{[1]}(x) = \begin{cases} 1 & \text{for } x \in [x_1, x_1 + \Delta/2], \\ 0 & \text{elsewhere}, \end{cases} \tag{8.9b}$$

$$\Pi^{[N]}(x) = \begin{cases} 1 & \text{for } x \in [x_N - \Delta/2, x_N], \\ 0 & \text{elsewhere}, \end{cases} \tag{8.9c}$$

over the (halves of the) outer segments. Substituting the transform-domain counterparts of (8.6) and (8.8) in the reciprocity relation (8.3) supplemented with (8.4), we arrive at a system of equations in the s-domain, constituents of which are amenable to handling via the CdH technique. Pursuing this approach, we end up with (cf. (7.42))

$$\sum_{k=1}^{m} \left(\underline{\boldsymbol{\mathcal{Z}}}_{m-k+1} - 2\underline{\boldsymbol{\mathcal{Z}}}_{m-k} + \underline{\boldsymbol{\mathcal{Z}}}_{m-k-1}\right) \cdot \boldsymbol{J}_k = \boldsymbol{V}_m, \tag{8.10}$$

where $\underline{\boldsymbol{\mathcal{Z}}}_k = \underline{\boldsymbol{\mathcal{Z}}}(t_k)$ is a 2-D $[N \times N]$ array at $t = t_k = k\Delta t$, elements of which are closely specified in Appendix B. Furthermore, $\boldsymbol{J}_k$ is a one-dimensional (1-D) $[N \times 1]$ array of the unknown coefficients at $t = t_k$, again, and $\boldsymbol{V}_m$ on the right-hand side is a 1-D $[N \times 1]$ excitation array at $t = t_m$. Subsequently, the convolution-type system of equations represented by (8.10) can be solved for $\boldsymbol{J}_m$ using the marching-on-in-time technique, which yields (cf. (7.44))

$$\boldsymbol{J}_m = \underline{\boldsymbol{\mathcal{Z}}}_1^{-1} \cdot \left[\boldsymbol{V}_m - \sum_{k=1}^{m-1} \left(\underline{\boldsymbol{\mathcal{Z}}}_{m-k+1} - 2\underline{\boldsymbol{\mathcal{Z}}}_{m-k} + \underline{\boldsymbol{\mathcal{Z}}}_{m-k-1}\right) \cdot \boldsymbol{J}_k\right]$$
$$\text{for all } m = \{1,\ldots,M\}, \tag{8.11}$$

where $\underline{\boldsymbol{\mathcal{Z}}}_1^{-1}$ denotes the inverse of the impeditivity matrix at $t = \Delta t$. Apparently, the array of unknown coefficients at $t = t_m$ is related to the concurrent excitation array at $t = t_m$ as well as to the electric-current coefficients from previous instants. Once the iterative procedure is executed, the electric-current coefficients are known and the pertaining (space–time) distribution of the induced current follows from (8.6).

The elements of the excitation array, $V^{[S]}$, can be determined upon substituting the (transform-domain) explicit-type boundary condition on the PEC screen (see (6.1a)), i.e.

$$\tilde{E}_y^{\mathrm{s}}(\kappa, 0, s) = -\tilde{E}_y^{\mathrm{i}}(\kappa, 0, s), \tag{8.12}$$

in the right-hand side of the starting reciprocity relation (8.3). Considering the EM plane-wave excitation represented by (4.1), the evaluation of the resulting slowness-domain integral leads to

$$V^{[S]}(t) = \frac{1}{\kappa_0}\left\{\int_{\tau=0}^{t-\kappa_0(x_S-\Delta/2)} e^{\mathrm{i}}(\tau)\mathrm{d}\tau - \int_{\tau=0}^{t-\kappa_0(x_S+\Delta/2)} e^{\mathrm{i}}(\tau)\mathrm{d}\tau\right\}$$
$$\text{for all } S = \{2, \ldots, N-1\}, \tag{8.13a}$$

with

$$V^{[1]}(t) = \frac{1}{\kappa_0}\left\{\int_{\tau=0}^{t-\kappa_0 x_1} e^{\mathrm{i}}(\tau)\mathrm{d}\tau - \int_{\tau=0}^{t-\kappa_0(x_1+\Delta/2)} e^{\mathrm{i}}(\tau)\mathrm{d}\tau\right\}, \tag{8.13b}$$

$$V^{[N]}(t) = \frac{1}{\kappa_0}\left\{\int_{\tau=0}^{t-\kappa_0(x_N-\Delta/2)} e^{\mathrm{i}}(\tau)\mathrm{d}\tau - \int_{\tau=0}^{t-\kappa_0 x_N} e^{\mathrm{i}}(\tau)\mathrm{d}\tau\right\}, \tag{8.13c}$$

applying to the edge nodes. For the normal incidence, $\beta = 0$ (see Figure 8.1) and hence $\kappa_0 = 0$. The corresponding limit of (8.13a), (8.13b) and (8.13c) has the simple form, i.e.

$$V^{[S]}(t) = \begin{cases} e^{\mathrm{i}}(t)\Delta & \text{for } S = \{2, \ldots, N-1\}, \\ e^{\mathrm{i}}(t)\Delta/2 & \text{for } S = \{1, N\}. \end{cases} \tag{8.14}$$

A brief description of the derivation that led to (8.13a)–(8.14) can be found in Appendix C.

8.1.3 Illustrative examples

We shall calculate the electric-current distribution induced on a bounded PEC screen located along $\{a = -w/2 < x < w/2 = b, z = 0\}$. The excitation EM plane wave (see (4.1)) impinging normally upon the screen has the pulse shape of a bipolar triangle that can be described by

$$e^{\mathrm{i}}(t) = \frac{2e_{\mathrm{m}}}{t_{\mathrm{w}}}\left[t\,\mathrm{H}(t) - 2\left(t - \frac{t_{\mathrm{w}}}{2}\right)\mathrm{H}\left(t - \frac{t_{\mathrm{w}}}{2}\right)\right.$$
$$\left. + 2\left(t - \frac{3t_{\mathrm{w}}}{2}\right)\mathrm{H}\left(t - \frac{3t_{\mathrm{w}}}{2}\right) - (t - 2t_{\mathrm{w}})\,\mathrm{H}(t - 2t_{\mathrm{w}})\right], \tag{8.15}$$

where we take the unit amplitude $e_{\mathrm{m}} = 1.0\,\mathrm{V/m}$ (see Figure 8.3).

In the first example, the screen is taken to be relatively narrow with respect to the spatial support of the plane-wave signature such that the approximate solution given

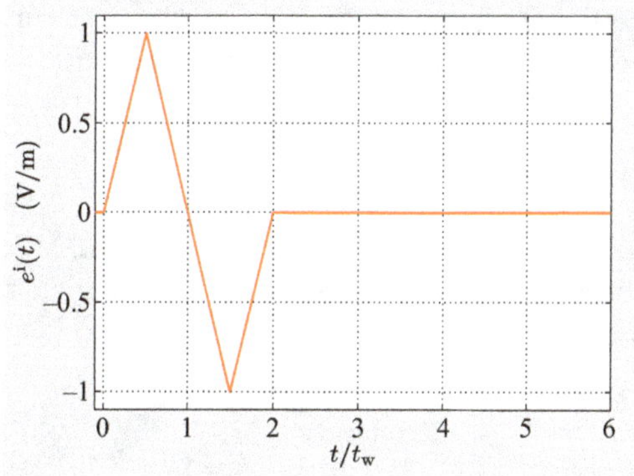

Figure 8.3 Incident plane-wave pulse shape

in Section 7.1.2 applies. Accordingly, in the corresponding CdH-MoM solution, we take $N = 2$, thus approximating the (spatial distribution of the) induced current by two half-triangles only (see (8.7b) and (8.7c)). The calculated pulse shapes of the electric-current surface density at the center of the PEC screen for $c_0 t_{\mathrm{w}}/w = 10$ are shown in Figure 8.4(a). Despite the different models, the resulting pulse shapes that bear a resemblance to the time integral of the exciting pulse shape do overlap each other.

In the second example, we choose $c_0 t_{\mathrm{w}}/w = 1.0$, so that the solution introduced in Section 7.1 is no longer applicable. In this case, the CdH-MoM solution will be compared with the Kirchhoff approximation that makes it possible to express the induced electric current using the tangential component of the incident magnetic-field strength (cf. (6.127)), i.e.

$$\partial J_y^{\mathrm{s}}(x,t) \overset{k}{\simeq} 2Y_{\perp}(\kappa_0) e^{\mathrm{i}}(t - \kappa_0 x), \tag{8.16}$$

and recall that $Y_{\perp}(\kappa_0) = \eta_0 \cos(\beta)$. Figure 8.4(b) shows that the CdH-MoM-based solution with $N = 11$ nodes correlates well with the Kirchhoff approximation, in particular, in its early-time part.

8.2 Pulsed EM scattering by a bounded screen with conductive and dielectric properties

To incorporate finite electric conductivity and permittivity in the model of a bounded screen, we largely follow the strategy applied in Section 7.2 to an EM-penetrable narrow strip. Hence, in the starting reciprocity relation (8.3), we employ the (transform-domain) transition condition (cf. (3.15b) and (7.22))

$$\tilde{E}_y^{\mathrm{s}}(\kappa, 0, s) = -\tilde{E}_y^{\mathrm{i}}(\kappa, 0, s) + \hat{Z}^{\mathrm{E}}(s)\partial \tilde{J}_y^{\mathrm{s}}(\kappa, s), \tag{8.17}$$

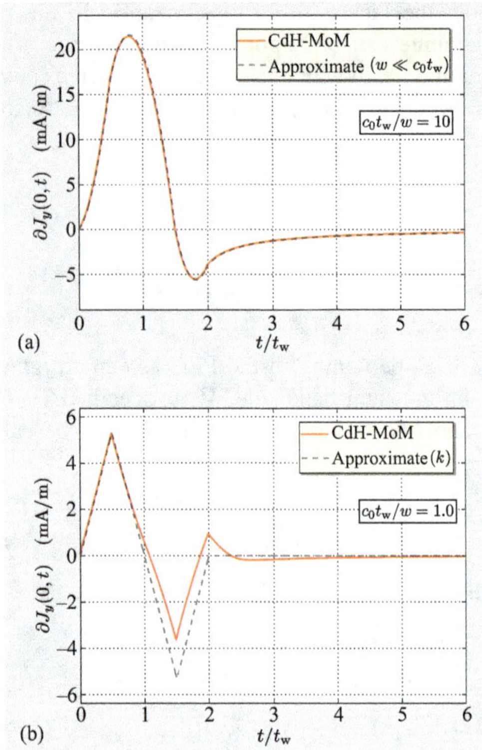

Figure 8.4 Plane-wave induced electric-current surface density at the center of the PEC screen. (a) The CdH-MoM solution (8.11) and the narrow-strip approximation (7.20) with (7.21) for $c_0 t_w / w = 10$. (b) The CdH-MoM solution (8.11) and the Kirchhoff approximation (8.16) for $c_0 t_w / w = 1.0$.

and get the following generalized reciprocity relation:

$$\begin{aligned} &\frac{s\mu_0}{2\pi i}\int_{\kappa=-i\infty}^{i\infty}\partial\tilde{J}_y^{\mathrm{T}}(\kappa,s)\partial\tilde{J}_y^{\mathrm{s}}(-\kappa,s)\frac{\mathrm{d}\kappa}{2p(\kappa)} \\ &\qquad+\frac{s\hat{Z}^{\mathrm{E}}(s)}{2\pi i}\int_{\kappa=-i\infty}^{i\infty}\partial\tilde{J}_y^{\mathrm{s}}(\kappa,s)\partial\tilde{J}_y^{\mathrm{T}}(-\kappa,s)\mathrm{d}\kappa \\ &\qquad\qquad=\frac{s}{2\pi i}\int_{\kappa=-i\infty}^{i\infty}\tilde{E}_y^{\mathrm{i}}(\kappa,0,s)\partial\tilde{J}_y^{\mathrm{T}}(-\kappa,s)\mathrm{d}\kappa \\ &\qquad\qquad\text{for } \{s\in\mathbb{R}; s>0\}. \end{aligned} \tag{8.18}$$

The second term on the left-hand side of (8.18) accounts for the screen's electric conductivity and permittivity, whose incorporation is the main subject of this section. Obviously, this term vanishes identically as $\hat{Z}^{\mathrm{E}}(s) \downarrow 0$, which is characteristic of a PEC screen analyzed in Section 8.1.

Upon introducing the (space–time) discretization in the reciprocity relation (8.18), the CdH technique can be applied, again, to arrive at a system of equations of the convolution type, in which the effect of screen's constitutive parameters manifests itself in a modification of the impeditivity array. Accordingly, we obtain (cf. (8.10))

$$\sum_{k=1}^{m} \left(\bar{\underline{\mathcal{Z}}}_{m-k+1} - 2\bar{\underline{\mathcal{Z}}}_{m-k} + \bar{\underline{\mathcal{Z}}}_{m-k-1}\right) \cdot \boldsymbol{J}_k = \boldsymbol{V}_m, \tag{8.19}$$

with

$$\bar{\underline{\mathcal{Z}}} = \underline{\mathcal{Z}} + \underline{\mathcal{R}}, \tag{8.20}$$

where the elements of $\underline{\mathcal{Z}}$ pertaining to the PEC screen are given in Appendix B, the excitation array on the right-hand side, $\boldsymbol{V}$, is specified via (8.13a)–(8.14) (see also Appendix C), and finally, the elements of $\underline{\mathcal{R}}$ can be calculated from (see Appendix D)

$$\begin{aligned} \mathcal{R}^{[S,n]}(t) = {} & \Gamma^A(x_n - x_S + 3\Delta/2, t) - 3\Gamma^A(x_n - x_S + \Delta/2, t) \\ & + 3\Gamma^A(x_n - x_S - \Delta/2, t) - \Gamma^A(x_n - x_S - 3\Delta/2, t) \\ & \text{for all } S = \{2, \ldots, N-1\} \text{ and } n = \{2, \ldots, N-1\}. \end{aligned} \tag{8.21a}$$

Furthermore, considering the edge nodes, we write

$$\mathcal{R}^{[S,1]}(t) = \Gamma^A(\Delta/2, t)\delta_{S,2}, \tag{8.21b}$$

$$\begin{aligned} \mathcal{R}^{[S,N]}(t) = {} & \left[2\Gamma^A(\Delta/2, t) - \Gamma^A(3\Delta/2, t)\right. \\ & \left. + \Gamma^B(3\Delta/2, t) - \Gamma^B(\Delta/2, t)\right]\delta_{S,N-1}, \\ & \text{for all } S = \{2, \ldots, N-1\}, \end{aligned} \tag{8.21c}$$

with $\delta_{n,k} = 1$ for $n = k$ and $\delta_{n,k} = 0$ otherwise, and

$$\mathcal{R}^{[1,1]}(t) = \Gamma^A(\Delta, t) - \Gamma^A(\Delta/2, t), \tag{8.21d}$$

$$\mathcal{R}^{[N,1]}(t) = 0. \tag{8.21e}$$

The remaining array elements can be determined by invoking the property of symmetry $\mathcal{R}^{[S,n]}(t) = \mathcal{R}^{[n,S]}(t)$ applying to all $S = \{1, \ldots, N\}$ with $n = \{1, \ldots, N\}$ and all $t > 0$. Moreover, it can be observed that $\underline{\mathcal{R}}_k$ is a tridiagonal array for all $k = \{1, \ldots, M\}$, which can be employed for optimizing computer implementations of the numerical scheme.

Once the elements of the modified impeditivity array (8.20) are fully specified, the system of equations (8.19) can be solved via the marching-on-in-time technique, again. This way leads to (cf. (8.11))

$$\begin{aligned} \boldsymbol{J}_m = {} & \bar{\underline{\mathcal{Z}}}_1^{-1} \cdot \left[\boldsymbol{V}_m - \sum_{k=1}^{m-1} \left(\bar{\underline{\mathcal{Z}}}_{m-k+1} - 2\bar{\underline{\mathcal{Z}}}_{m-k} + \bar{\underline{\mathcal{Z}}}_{m-k-1}\right) \cdot \boldsymbol{J}_k\right] \\ & \text{for all } m = \{1, \ldots, M\}. \end{aligned} \tag{8.22}$$

As the form of time-domain (TD) functions $\Gamma^{A,B}(x, t)$ depends on the actual constitutive properties of the layer, we shall next discuss selected cases separately.

8.2.1 Screen with conductive properties

A thin highly conductive sheet can be characterized by its s-independent, equivalent layer admittance $\hat{Y}^{\mathrm{E}}(s) = 1/\hat{Z}^{\mathrm{E}}(s) = G^{\mathrm{E}}$ (see (3.5a)). Substituting this parameter in (D.5) and (D.8) and transforming the thus obtained s-domain expressions to TD, we arrive at

$$\Gamma^A(x,t) = \frac{c_0 t}{2G^{\mathrm{E}}} \frac{x^2}{\Delta c_0 \Delta t} \mathrm{H}(x)\mathrm{H}(t), \tag{8.23}$$

$$\Gamma^B(x,t) = \frac{c_0 t}{G^{\mathrm{E}}} \frac{x}{c_0 \Delta t} \mathrm{H}(x)\mathrm{H}(t), \tag{8.24}$$

respectively. Apparently, if $G^{\mathrm{E}} \to \infty$, then $\Gamma^{A,B} = 0$, which then yields the electric-current surface density on a PEC screen.

8.2.2 Screen with conductive and dielectric properties

A thin highly contrasting screen with conductive and dielectric properties is characterized by its equivalent layer conductance and capacitance via $\hat{Y}^{\mathrm{E}}(s) = 1/\hat{Z}^{\mathrm{E}}(s) = G^{\mathrm{E}} + sC^{\mathrm{E}}$ (see (3.5a) and (3.5b)). Again, substituting the corresponding layer impedance in (D.5) and (D.8), and transforming the result to the TD, we get

$$\Gamma^A(x,t) = \frac{1}{2G^{\mathrm{E}}} \frac{1}{\alpha^{\mathrm{E}} \Delta t} \frac{x^2}{\Delta} \left[\alpha^{\mathrm{E}} t - 1 + \exp(-\alpha^{\mathrm{E}} t)\right] \mathrm{H}(x)\mathrm{H}(t), \tag{8.25}$$

$$\Gamma^B(x,t) = \frac{1}{G^{\mathrm{E}}} \frac{x}{\alpha^{\mathrm{E}} \Delta t} \left[\alpha^{\mathrm{E}} t - 1 + \exp(-\alpha^{\mathrm{E}} t)\right] \mathrm{H}(x)\mathrm{H}(t), \tag{8.26}$$

respectively, where $\alpha^{\mathrm{E}} = G^{\mathrm{E}}/C^{\mathrm{E}} > 0$ (see (3.19)). As $C^{\mathrm{E}} \downarrow 0$, then the parameter $\alpha^{\mathrm{E}} \to \infty$, so that (8.25) and (8.26) approach (8.23) and (8.24), respectively. The same effect has the limit $t \to \infty$, which implies that the late-time behavior of the screen is governed by its conductive properties. For $G^{\mathrm{E}} \to \infty$ or/and $C^{\mathrm{E}} \to \infty$, the TD functions tend to the zero value, thus approaching the model of a PEC screen, again.

8.2.3 Screen with plasmonic properties

For a thin screen whose conductivity is described via the Drude-type conduction relaxation function (5.21), the equivalent layer admittance is given by $\hat{Y}^{\mathrm{E}}(s) = 1/\hat{Z}^{\mathrm{E}}(s) = \hat{G}^{\mathrm{E}}(s) = \delta\epsilon_0\omega_{\mathrm{pe}}^2/(s + \nu_{\mathrm{c}})$, where $\delta > 0$ is the thickness of the screen, with ω_{pe} and ν_{c} denoting the electron plasma angular frequency and the collision frequency, respectively. In such a case, the TD counterparts of (D.5) and (D.8) have the following form:

$$\Gamma^A(x,t) = \frac{1}{2\hat{G}^{\mathrm{E}}(0)} \frac{x^2}{\Delta \nu_{\mathrm{c}} \Delta t} (\nu_{\mathrm{c}} t + 1)\mathrm{H}(x)\mathrm{H}(t), \tag{8.27}$$

$$\Gamma^B(x,t) = \frac{1}{\hat{G}^{\mathrm{E}}(0)} \frac{x}{\nu_{\mathrm{c}} \Delta t} (\nu_{\mathrm{c}} t + 1)\mathrm{H}(x)\mathrm{H}(t), \tag{8.28}$$

respectively, where $\hat{G}^{\mathrm{E}}(0) = \delta\epsilon_0\omega_{\mathrm{pe}}^2/\nu_{\mathrm{c}}$ has the meaning of "static conductance". Again, in the late-time limit, (8.27) and (8.28) have the form of (8.23) and (8.24), respectively, thus resembling the behavior of a conductive screen.

8.2.4 Illustrative examples

To provide an illustrative example, we shall analyze the transient response of a bounded conductive screen whose constitutive properties are described by $G^{\mathrm{E}}/\eta_0 = 1.0$. Adopting the problem parameters from Section 8.1.3, the screen occupying a domain $\{a = -w/2 < x < w/2 = b, -\infty < y < \infty, z = 0\}$ is excited by the uniform EM plane wave that is defined by its angle of incidence, $\beta = 0$, and the pulse shape, $e^{\mathrm{i}}(t)$, specified by (8.15) with the unit amplitude, $e_{\mathrm{m}} = 1.0\,\mathrm{V/m}$, again (see Figure 8.3).

Again, the first screen under consideration is relatively narrow with respect to $c_0 t_{\mathrm{w}}$, so that the assumptions that led to the results presented in Section 7.2.1.1 are met. Referring to the corresponding response pertaining to the PEC screen (see Figure 8.4(a)), the induced current density shown in Figure 8.5(a) is relatively weaker. Moreover, its pulse shape still resembles the excitation bipolar triangle (see Figure 8.3) and not its time integral. Overall, the CdH-MoM solution with $N = 2$ and the corresponding narrow-strip approximation (see Section 7.2.1.1) correlate well.

In the second example, the relative width of the screen is increased such that $c_0 t_{\mathrm{w}}/w = 1.0$. Figure 8.5(b) shows the corresponding induced current distribution calculated via the CdH-MoM solution with $N = 11$ discretization nodes. For the sake of comparison, the induced electric-current surface density is expressed with the aid of Kirchhoff's approximation (see (6.132) with (6.135))

$$\partial J_y^{\mathrm{s}}(x,t) \overset{k}{\simeq} 2\{G^{\mathrm{E}}/[G^{\mathrm{E}} + 2Y_{\perp}(\kappa_0)]\}Y_{\perp}(\kappa_0)e^{\mathrm{i}}(t - \kappa_0 x). \tag{8.29}$$

Figure 8.5(b) demonstrates that the approximation provides satisfactory results for the problem under consideration. Comparing Figure 8.5(a) and (b), it can be concluded that the Kirchhoff approximation represented by (8.29) may be applicable even to relatively narrow conductive screens.

8.3 Pulsed EM scattering by a bounded metasurface

To account for the presence of magnetic currents supported by a metasurface, we shall pursue the approach proposed in [3]. Accordingly, the reciprocity relation (8.1) is extended to the following form (cf. (7.28)):

$$\begin{aligned}
&\int_{x=a}^{b} \langle \hat{E}_y^{\mathrm{T}}\rangle_{-}^{+}(x,s)\partial\hat{J}_y^{\mathrm{s}}(x,s)\mathrm{d}x - \int_{x=a}^{b} \langle \hat{H}_x^{\mathrm{T}}\rangle_{-}^{+}(x,s)\partial\hat{K}_x^{\mathrm{s}}(x,s)\mathrm{d}x \\
&= \int_{x=a}^{b} \langle \hat{E}_y^{\mathrm{s}}\rangle_{-}^{+}(x,s)\partial\hat{J}_y^{\mathrm{T}}(x,s)\mathrm{d}x - \int_{x=a}^{b} \langle \hat{H}_x^{\mathrm{s}}\rangle_{-}^{+}(x,s)\partial\hat{K}_x^{\mathrm{T}}(x,s)\mathrm{d}x,
\end{aligned} \tag{8.30}$$

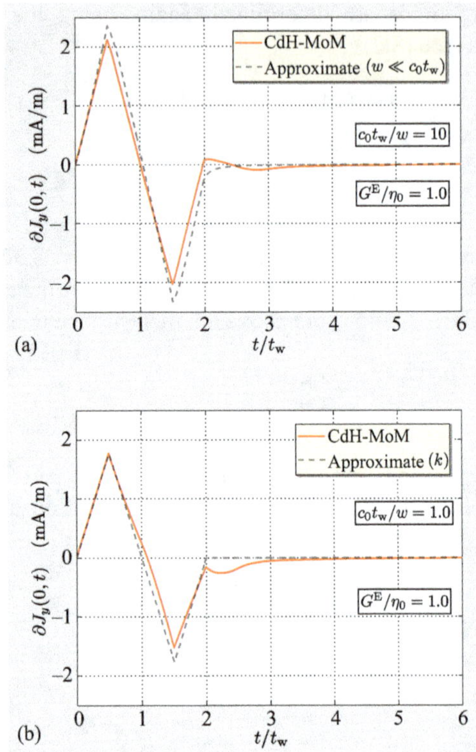

Figure 8.5 Plane-wave induced electric-current surface density at the center of the conductive screen. (a) The CdH-MoM solution (8.22) with (8.23) and (8.24) and the narrow-strip approximation (7.20) with (7.25) for $c_0 t_w / w = 10$. (b) The CdH-MoM solution (8.22) with (8.23) and (8.24) and the Kirchhoff approximation (8.29) for $c_0 t_w / w = 1.0$.

where the operator $\langle\rangle_-^+$ is defined by (3.11). Employing further the wave-slowness representation along the x-direction (see (4.33)) for the EM field strengths in (8.30), the reciprocity relation can be cast into the following form (cf. (8.3)):

$$
\begin{aligned}
&\frac{s}{2\pi i}\int_{\kappa=-i\infty}^{i\infty} \langle \tilde{E}_y^{\mathrm{T}} \rangle_-^+(\kappa, s)\partial \tilde{J}_y^{\mathrm{s}}(-\kappa, s)\mathrm{d}\kappa \\
&\qquad - \frac{s}{2\pi i}\int_{\kappa=-i\infty}^{i\infty} \langle \tilde{H}_x^{\mathrm{T}} \rangle_-^+(\kappa, s)\partial \tilde{K}_x^{\mathrm{s}}(-\kappa, s)\mathrm{d}\kappa \\
&= \frac{s}{2\pi i}\int_{\kappa=-i\infty}^{i\infty} \langle \tilde{E}_y^{\mathrm{s}} \rangle_-^+(\kappa, s)\partial \tilde{J}_y^{\mathrm{T}}(-\kappa, s)\mathrm{d}\kappa \\
&\qquad - \frac{s}{2\pi i}\int_{\kappa=-i\infty}^{i\infty} \langle \tilde{H}_x^{\mathrm{s}} \rangle_-^+(\kappa, s)\partial \tilde{K}_x^{\mathrm{T}}(-\kappa, s)\mathrm{d}\kappa,
\end{aligned}
\tag{8.31}
$$

in which the testing fields are determined, again, using the transform-domain excitation conditions (see (8.2))

$$\lim_{z\downarrow\delta/2}\tilde{H}_x^{\mathrm{T}}(\kappa,z,s)-\lim_{z\uparrow-\delta/2}\tilde{H}_x^{\mathrm{T}}(\kappa,z,s)=\partial\tilde{J}_y^{\mathrm{T}}(\kappa,s), \tag{8.32a}$$

$$\lim_{z\downarrow\delta/2}\tilde{E}_y^{\mathrm{T}}(\kappa,z,s)-\lim_{z\uparrow-\delta/2}\tilde{E}_y^{\mathrm{T}}(\kappa,z,s)=\partial\tilde{K}_x^{\mathrm{T}}(\kappa,s). \tag{8.32b}$$

Hence, the transform-domain testing field is represented as a 1-D wave motion propagating away from the screen and the excitation conditions (8.32a) and (8.32b) are enforced to determine its amplitude. The desired average of the horizontal EM field strengths as observed from below and above the metasurface then readily follow as

$$\langle\tilde{E}_y^{\mathrm{T}}\rangle_-^+(\kappa,s)=-[\mu_0/2p(\kappa)]\partial\tilde{J}_y^{\mathrm{T}}(\kappa,s), \tag{8.33}$$

$$\langle\tilde{H}_x^{\mathrm{T}}\rangle_-^+(\kappa,s)=-[p(\kappa)/2\mu_0]\partial\tilde{K}_x^{\mathrm{T}}(\kappa,s). \tag{8.34}$$

The property that $\langle\tilde{E}_y^{\mathrm{T}}\rangle_-^+$ and $\langle\tilde{H}_x^{\mathrm{T}}\rangle_-^+$ are generated by $\partial\tilde{J}_y^{\mathrm{T}}$ and $\partial\tilde{K}_x^{\mathrm{T}}$ alone, respectively, makes it possible to decouple the reciprocity relation to two independent relations, that is,

$$\begin{aligned}\frac{s}{2\pi\mathrm{i}}\int_{\kappa=-\mathrm{i}\infty}^{\mathrm{i}\infty}\frac{\mu_0}{2p(\kappa)}\partial\tilde{J}_y^{\mathrm{s}}(-\kappa,s)\partial\tilde{J}_y^{\mathrm{T}}(\kappa,s)\mathrm{d}\kappa&\\+\frac{s\hat{Z}^{\mathrm{E}}(s)}{2\pi\mathrm{i}}\int_{\kappa=-\mathrm{i}\infty}^{\mathrm{i}\infty}\partial\tilde{J}_y^{\mathrm{s}}(\kappa,s)\partial\tilde{J}_y^{\mathrm{T}}(-\kappa,s)\mathrm{d}\kappa&\\=\frac{s}{2\pi\mathrm{i}}\int_{\kappa=-\mathrm{i}\infty}^{\mathrm{i}\infty}\tilde{E}_y^{\mathrm{i}}(\kappa,0,s)\partial\tilde{J}_y^{\mathrm{T}}(-\kappa,s)\mathrm{d}\kappa,&\end{aligned} \tag{8.35a}$$

and

$$\begin{aligned}\frac{s}{2\pi\mathrm{i}}\int_{\kappa=-\mathrm{i}\infty}^{\mathrm{i}\infty}\frac{p(\kappa)}{2\mu_0}\partial\tilde{K}_x^{\mathrm{s}}(-\kappa,s)\partial\tilde{K}_x^{\mathrm{T}}(\kappa,s)\mathrm{d}\kappa&\\+\frac{s\hat{Y}^{\mathrm{M}}(s)}{2\pi\mathrm{i}}\int_{\kappa=-\mathrm{i}\infty}^{\mathrm{i}\infty}\partial\tilde{K}_x^{\mathrm{s}}(\kappa,s)\partial\tilde{K}_x^{\mathrm{T}}(-\kappa,s)\mathrm{d}\kappa&\\=\frac{s}{2\pi\mathrm{i}}\int_{\kappa=-\mathrm{i}\infty}^{\mathrm{i}\infty}\tilde{H}_x^{\mathrm{i}}(\kappa,0,s)\partial\tilde{K}_x^{\mathrm{T}}(-\kappa,s)\mathrm{d}\kappa,&\end{aligned} \tag{8.35b}$$

where we used the pertaining saltus-type conditions to express the scattered fields on the right-hand side of the reciprocity relation (8.31) (see (4.2a) and (4.2b)), viz

$$\langle\tilde{E}_y^{\mathrm{s}}\rangle_-^+(\kappa,s)=-\tilde{E}_y^{\mathrm{i}}(\kappa,0,s)+\hat{Z}^{\mathrm{E}}(s)\partial\tilde{J}_y^{\mathrm{s}}(\kappa,s), \tag{8.36a}$$

$$\langle\tilde{H}_x^{\mathrm{s}}\rangle_-^+(\kappa,s)=-\tilde{H}_x^{\mathrm{i}}(\kappa,0,s)+\hat{Y}^{\mathrm{M}}(s)\partial\tilde{K}_x^{\mathrm{s}}(\kappa,s). \tag{8.36b}$$

As a matter of fact, the first reciprocity relation (8.35a) has been handled before in Sections 7.2 and 8.2 to analyze the TD EM scattering by planar strips with conductive and dielectric properties. Therefore, the subsequent analysis is focused on the

(dual) magnetic-type reciprocity relation (8.35b) whose solution yields the induced magnetic-current surface density on a screen with highly contrasting magnetic constitutive properties.

8.3.1 Problem solution

To solve the reciprocity relation (8.35b) in the TD, we shall pursue the line of reasoning introduced in Section 8.1.2. Taking into the account the end conditions applying to the magnetic-current surface density, that is,

$$\partial K_x^{\mathrm{s}}(x,t) = 0 \text{ at } x = \{a,b\} \text{ and for all } t > 0, \tag{8.37}$$

we use the space–time piecewise linear expansion (see (8.6))

$$\partial K_x^{\mathrm{s}}(x,t) \simeq \sum_{n=2}^{N-1}\sum_{k=1}^{M} k_k^{[n]}\Lambda^{[n]}(x)\Lambda_k(t), \tag{8.38}$$

where $k_k^{[n]}$ are unknown coefficients (in V/m) to be determined and the triangular functions $\Lambda^{[n]}(x)$ and $\Lambda_k(t)$ are defined via (8.7a) and (7.19), respectively. Furthermore, the distribution of the testing magnetic-current density in space and time is chosen to be represented by the rectangular function and the Dirac-delta impulse, respectively. Hence, we write (see (8.9a))

$$\partial K_x^{\mathrm{T}}(x,t) = \Pi^{[S]}(x)\delta(t) \quad \text{for all } S = \{2,\dots,N-1\}, \tag{8.39}$$

where the rectangular function is defined by (8.9a). Since index S runs over the inner nodes only (see Figure 8.2), the end condition applying to the testing current-distribution, $\partial K_x^{\mathrm{T}}(x,t) = 0$ at $x = \{a,b\}$ for all $t > 0$, holds true.

The transform-domain counterparts of (8.38) and (8.39) are next substituted in the reciprocity relation (8.35b) and the thus obtained system of equations in the s-domain is transformed to the TD with the aid of the CdH technique. Following this way, we end up with (cf. (8.19))

$$\sum_{k=1}^{m}\left(\bar{\underline{\mathcal{Y}}}_{m-k+1} - 2\bar{\underline{\mathcal{Y}}}_{m-k} + \bar{\underline{\mathcal{Y}}}_{m-k-1}\right)\cdot \boldsymbol{K}_k = \boldsymbol{I}_m, \tag{8.40}$$

where $\bar{\underline{\mathcal{Y}}}_k = \bar{\underline{\mathcal{Y}}}(t_k)$ denotes a 2-D $[(N-2)\times(N-2)]$ "admittivity array" at $t = t_k = k\Delta t$. This array can be further expressed as a sum of two arrays

$$\bar{\underline{\mathcal{Y}}} = \underline{\mathcal{Y}} + \underline{\mathcal{G}}, \tag{8.41}$$

where the first array, $\underline{\mathcal{Y}}$, pertains to the perfectly magnetically conducting (PMC) screen (see Appendix E) and the elements of $\underline{\mathcal{G}}$ can be expressed in a similar way as the impeditivity array pertaining to a screen with conductive and dielectric properties (see (D.1)). Hence, we write

$$\hat{\mathcal{G}}^{[S,n]}(s) = \frac{\hat{Y}^{\mathrm{M}}(s)}{c_0\Delta t}\frac{c_0}{s}\frac{1}{2\pi\mathrm{i}}\int_{\kappa=-\mathrm{i}\infty}^{\mathrm{i}\infty}\tilde{\Pi}^{[S]}(-\kappa)\tilde{\Lambda}^{[n]}(\kappa)\mathrm{d}\kappa$$
$$\text{for all } S = \{2,\dots,N-1\} \text{ and } n = \{2,\dots,N-1\}, \tag{8.42}$$

with $\{s \in \mathbb{R}; s > 0\}$. Making use of the generic integral (D.3), (8.42) can be easily transformed to the TD, which yields

$$\mathcal{G}^{[S,n]}(t) = \Big[\Gamma^C(x_n - x_S + 3\Delta/2, t) - 3\Gamma^C(x_n - x_S + \Delta/2, t) + 3\Gamma^C(x_n - x_S - \Delta/2, t) - \Gamma^C(x_n - x_S - 3\Delta/2, t)\Big]. \tag{8.43}$$

Here, the TD function $\Gamma^C(x,t)$ has the form similar to (8.25), that is,

$$\Gamma^C(x,t) = \frac{1}{2R^{\mathrm{M}}}\frac{1}{\alpha^{\mathrm{M}}\Delta t}\frac{x^2}{\Delta}\left[\alpha^{\mathrm{M}}t - 1 + \exp(-\alpha^{\mathrm{M}}t)\right]\mathrm{H}(x)\mathrm{H}(t), \tag{8.44}$$

where $\alpha^{\mathrm{M}} = R^{\mathrm{M}}/L^{\mathrm{M}} > 0$ (see (3.21)). Apparently, if $R^{\mathrm{M}} \to \infty$ or/and $L^{\mathrm{M}} \to \infty$, then $\Gamma^C(x,t)$ and hence $\mathcal{G}^{[S,n]}(t)$ tend to zero for all $t > 0$, thus approaching the model of a PMC screen. Finally, the elements of the excitation array on the right-hand side of (8.40), $\boldsymbol{I}$, readily follow by using (8.13a) as

$$I^{[S]}(t) = \frac{Y_\perp(\kappa_0)}{\kappa_0}\left\{\int_{\tau=0}^{t-\kappa_0(x_S-\Delta/2)} e^{\mathrm{i}}(\tau)\mathrm{d}\tau - \int_{\tau=0}^{t-\kappa_0(x_S+\Delta/2)} e^{\mathrm{i}}(\tau)\mathrm{d}\tau\right\}$$
$$\text{for all } S = \{2,\ldots,N-1\}, \tag{8.45}$$

and recall that $Y_\perp(\kappa_0) = \eta_0\cos(\beta)$. Concerning the normal incidence, the limit of (8.45) as $\kappa_0 = 0$ yields $\eta_0 e^{\mathrm{i}}(t)\Delta$ applying to all inner nodes. Having defined all the necessary elements in system (8.40), the expansion coefficients representing the induced magnetic-current density can be expressed via the updating step-by-step scheme (cf. (8.22))

$$\boldsymbol{K}_m = \bar{\underline{\boldsymbol{\mathfrak{Y}}}}_1^{-1} \cdot \left[\boldsymbol{I}_m - \sum_{k=1}^{m-1}\left(\bar{\underline{\boldsymbol{\mathfrak{Y}}}}_{m-k+1} - 2\bar{\underline{\boldsymbol{\mathfrak{Y}}}}_{m-k} + \bar{\underline{\boldsymbol{\mathfrak{Y}}}}_{m-k-1}\right)\cdot \boldsymbol{K}_k\right]$$
$$\text{for all } S = \{2,\ldots,N-1\}, \tag{8.46}$$

where $\bar{\underline{\boldsymbol{\mathfrak{Y}}}}_1^{-1}$ denotes the inverse of the admittivity matrix at $t = \Delta t$.

8.3.2 *Illustrative examples*

To validate the numerical solution methodologies presented in Sections 7.3.1 and 8.3.1, we shall first calculate the induced magnetic-current surface density on a loss-free bounded screen whose constitutive properties are described by $L^{\mathrm{M}}\eta_0/t_{\mathrm{w}} = 1.0$ and $R^{\mathrm{M}} = 0$. Again, we analyze the screen occupying a bounded domain $\{a = -w/2 < x < w/2 = b, z = 0\}$ that is irradiated by the uniform EM plane wave with $\beta = 0$ and the plane-wave signature $e^{\mathrm{i}}(t)$ (see (8.15) and Figure 8.3).

In the first case, the screen is relatively narrow with respect to the spatial support of the exciting pulse such that the marching-on-in-time scheme (7.39) can be employed. Figure 8.6(a) shows that the pulse shapes of $\partial K_x^{\mathrm{s}}(0,t)$ as calculated

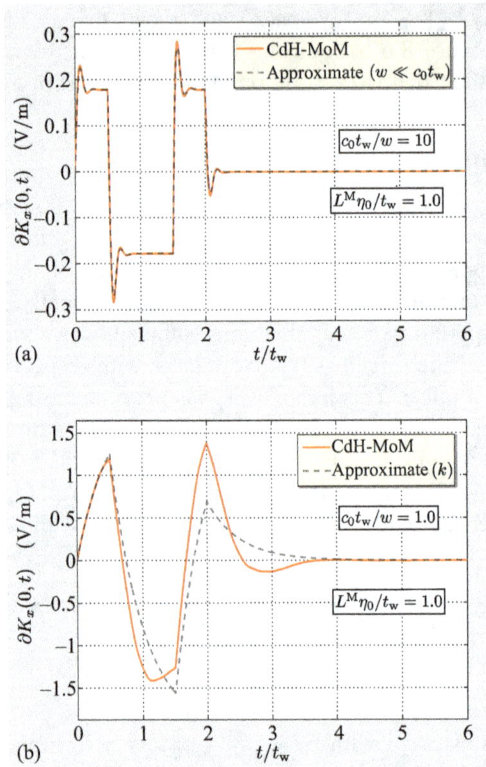

Figure 8.6 Plane-wave induced magnetic-current surface density at the center of the magnetic screen. (a) The CdH-MoM solution (8.46) and the narrow-strip approximation (7.39) for $c_0 t_\mathrm{w}/w = 10$. (b) The CdH-MoM solution (8.46) and the Kirchhoff approximation (8.47) for $c_0 t_\mathrm{w}/w = 1.0$.

via the narrow-strip approximation (7.39) and the CdH-MoM scheme (8.46) with $N = 3$ are almost identical, as expected. Finally, it is interesting to notice that the pulse shape of the magnetic-current surface density bears a resemblance to the time derivative of the triangular signature (see Figure 8.3).

Second, we take $c_0 t_\mathrm{w}/w = 1.0$ and calculate the TD response with the aid of the CdH-MoM solution (8.46) with $N = 11$ discretization nodes. Moreover, the magnetic-current distribution is calculated with the aid of Kirchhoff's approximation (cf. (6.136))

$$\partial K_x^\mathrm{s}(x,t) \overset{k}{\simeq} 2\left\{\delta(t) - \frac{2Z_\perp(\kappa_0)}{L^\mathrm{M}} \exp\left[-\frac{2Z_\perp(\kappa_0)}{L^\mathrm{M}} t\right] \mathrm{H}(t)\right\} \ast_t e^\mathrm{i}(t - \kappa_0 x), \tag{8.47}$$

where $Z_\perp(\kappa_0) = 1/Y_\perp(\kappa_0) = \zeta_0/\cos(\beta)$ and recall that $\zeta_0 = c_0\mu_0$ denotes the wave impedance. Figure 8.6(b) indicates good agreement between the Kirchhoff approximation and the CdH-MoM solution in the initial part of the response.

8.3.3 Concluding remarks

In this final section, we shall heuristically demonstrate the transparency of a metasurface that has been previously discussed in Chapter 4. To that end, we first note that the Kirchhoff approximation provides, as a matter of fact, the valid induced current distribution pertaining to the infinite screen. Consequently, it can be expected that the Kirchhoff approximation applied to a bounded screen will yield acceptable results if the width of the screen is relatively large with respect to the spatial support of the excitation pulse. This hypothesis was previously supported numerically by observing that the Kirchhoff approximation accurately estimates the early-time part of the current response (see Figures 8.4(b), 8.5(b) and 8.6(b)). Assuming now that the early-time limit is meaningful even for a narrow strip, (7.37) expressing the scattered electric-field strength at (x_0, y_0) can be replaced with

$$\hat{E}_y^{\mathrm{s}}(x_0, z_0, s) \simeq -\frac{1}{2\eta_0}\left(\frac{s}{2\pi c_0 r_0}\right)^{1/2}\left[\hat{I}^{\mathrm{s}}(s) - \hat{K}^{\mathrm{s}}(s)\eta_0\cos(\vartheta)\right]\exp(-sr_0/c_0), \tag{8.48}$$

as $|s| \to \infty$, where $\hat{I}^{\mathrm{s}}(s)$ and $\hat{K}^{\mathrm{s}}(s)$ denote the induced electric and magnetic currents and $\cos(\vartheta) = z_0/r_0$ with $r_0 = (x_0^2 + z_0^2)^{1/2} > 0$. Equation (8.48) clearly reveals that the fields radiated from the induced, equivalent electric and magnetic sources may cancel each other if

$$\hat{I}^{\mathrm{s}}(s) = \hat{K}^{\mathrm{s}}(s)\eta_0\cos(\vartheta). \tag{8.49}$$

To demonstrate when the zero-scattering condition (8.49) can be met, we shall consider a loss-free metasurface that supports both magnetic and electric currents. Since the former was estimated via (8.47), it remains to determine the induced electric currents on a dielectric strip. To this end, we may pursue the line of reasoning that led to the approximate electric-current distribution on an EM-penetrable semi-infinite screen (see Section 6.6.2) and get

$$\begin{aligned}\partial J_y^{\mathrm{s}}(x,t) &\overset{k}{\simeq} 2\left\{\delta(t) - \frac{2Y_\perp(\kappa_0)}{C^{\mathrm{E}}}\exp\left[-\frac{2Y_\perp(\kappa_0)}{C^{\mathrm{E}}}t\right]\mathrm{H}(t)\right\}\\ &\quad *_t\, Y_\perp(\kappa_0)e^{\mathrm{i}}(t - \kappa_0 x).\end{aligned} \tag{8.50}$$

Upon inspection of (8.47) with (8.50), it is concluded that

$$\partial J_y^{\mathrm{s}}(x,t) = \partial K_x^{\mathrm{s}}(x,t)Y_\perp(\kappa_0), \tag{8.51}$$

provided that

$$Y_\perp(\kappa_0)/C^{\mathrm{E}} = Z_\perp(\kappa_0)/L^{\mathrm{M}}, \tag{8.52}$$

or, eventually, if $Y_\perp(\kappa_0)t_\mathrm{w}/C^\mathrm{E}$ and $Z_\perp(\kappa_0)t_\mathrm{w}/L^\mathrm{M}$ are both sufficiently small such that the second terms in the braces of (8.47) and (8.50) can be neglected. The use of $Y_\perp(\kappa_0) = \eta_0 \cos(\beta)$ in (8.51) reveals that the zero-reflection condition (8.49) can be satisfied at points where $\cos(\vartheta) = \cos(\beta)$ (see Figure 8.7), which is in agreement with the conclusions drawn in Sections 4.1.3, 4.2, 4.2.3, 4.3 and 4.3.3. As a matter of fact, (8.52) can be viewed as a special case of the transparency condition (4.8) for $\hat{Y}^\mathrm{E} = sC^\mathrm{E}$ and $\hat{Y}^\mathrm{M} = 1/sL^\mathrm{M}$ applying to a loss-free metasurface. To validate the results, we next apply the CdH-MoM to calculate the electric- and magnetic-current surface densities at the center of such a screen that is described by $C^\mathrm{E}/\eta_0 t_\mathrm{w} = 0.10$ and $L^\mathrm{M} = C^\mathrm{E}/Y_\perp^2(\kappa_0)$ so that the transparency condition (8.52) is met. Again, the thin layer is normally irradiated by the impulsive plane wave with the triangular pulse shape $e^\mathrm{i}(t)$ with $c_0 t_\mathrm{w}/w = 1.0$ and the unit amplitude (see Figure 8.3). Figure 8.8 clearly shows that (8.51) indeed holds true at the initial part of the response, thus confirming our expectations numerically.

It is finally noted that condition (4.8) has been deduced from the reflection coefficient occurring in the solution of a 1-D wave propagation problem. The discussion given in this section indicates that the transparency of a metasurface can be explained by the counterbalance of induced electric and magnetic currents that virtually cancels the reflected EM field.

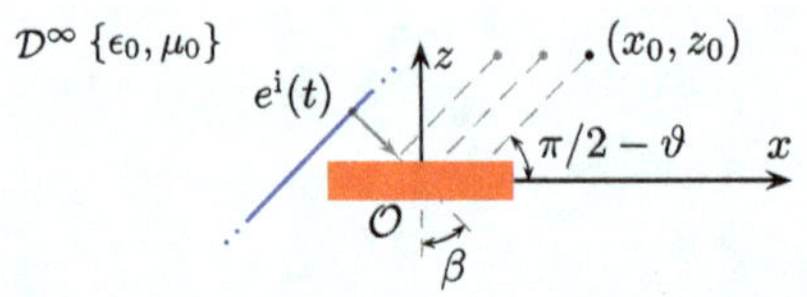

Figure 8.7 Plane-wave (non)-scattering by a bounded metasurface

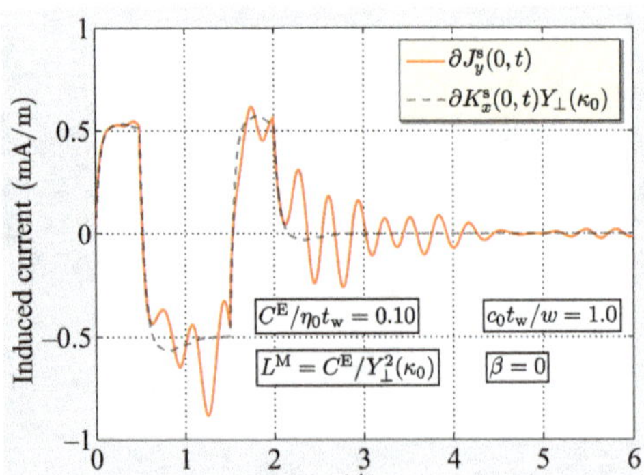

Figure 8.8 Plane-wave induced current surface densities at the center of a loss-free metasurface

References

[1] M. Štumpf. *Time-Domain Electromagnetic Reciprocity in Antenna Modeling*. Hoboken, NJ: John Wiley & Sons, Inc., 2019.
[2] J. G. Van Bladel. *Singular Electromagnetic Fields and Sources*. Piscataway, NJ: IEEE Press, 1991.
[3] M. Štumpf. Pulsed electromagnetic scattering by metasurfaces—a numerical solution based on the Cagniard-DeHoop method of moments. *IEEE Transactions on Antennas and Propagation*, 69(11):7761–7770, 2021.

Chapter 9

Pulsed EM-field scattering by bounded and narrow screens in a layered embedding

The need for efficient and controlled synthesis of thin-film structures that are encountered in various applications of electronics and semiconductor industries (e.g. microelectronic circuits, integrated optical devices) puts severe demands on electromagnetic (EM) solvers for achieving their EM scattering characteristics. Such structures typically consist of highly conductive layers located in a piece-wise homogeneous embedding, which cannot be directly analyzed with the aid of the Cagniard–DeHoop method of moments (CdH-MoM) formulation introduced in Chapter 8. Therefore, it is shown in this chapter how the CdH-MoM can be generalized, thus providing an efficient computational tool for analyzing the pulsed EM scattering by bounded, highly contrasting thin sheets on a layered background. For the sake of clarity, the methodology is illustrated based on the analysis of EM-penetrable sheets supporting electric currents only. The incorporation of combined effects of both electric and magnetic currents flowing on metasurfaces can be handled along similar lines.

9.1 Pulsed EM scattering by a bounded screen above the PEC ground

The problem configuration under consideration is shown in Figure 9.1. It consists of a thin highly contrasting screen that occupies $\{a < x < b, -\infty < y < \infty, -\delta/2 < z < \delta/2\}$, with $\delta > 0$ and $w = b - a > 0$ denoting its thickness and width, respectively. The screen is located in a homogeneous, isotropic and loss-free half-space

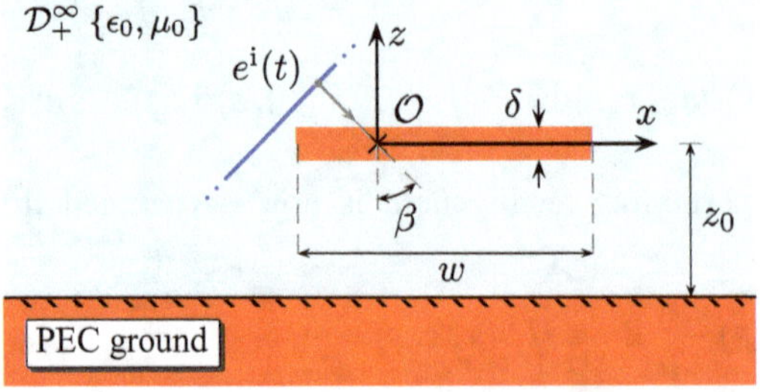

Figure 9.1 Pulsed EM plane wave impinging upon a grounded plane screen

$\mathcal{D}_{+}^{\infty}$ at height $z_0 > 0$ above the unbounded perfectly electrically conducting (PEC) ground plane extending over $\{-\infty < x < \infty, -\infty < y < \infty, z = -z_0\}$. The EM properties of the upper half-space are defined by its (real-valued, scalar and positive) electric permittivity ϵ_0 and magnetic permeability μ_0. Again, the screen is excited by the uniform, E-polarized plane wave

$$\hat{E}_y^{\mathrm{i}}(x, z, s) = \hat{e}^{\mathrm{i}}(s) \exp\{-s[\kappa_0 x - p(\kappa_0) z]\}, \tag{4.1 revisited}$$

where $\kappa_0 = \sin(\beta)/c_0$ and $p(\kappa_0) = \cos(\beta)/c_0$ are its slowness parameters in the x- and z-directions, respectively. Since the problem configuration as well as its excitation are y-independent, the non-vanishing field components, $\{E_y, H_x, H_z\}$, are governed by a 2-D system of y-independent EM field equations. To properly account for the presence of the PEC ground, the scattered field (denoted by superscript ^{s}) is defined as the difference between the total EM field in the configuration and the excitation field (denoted by superscript ^{e}):

$$\{\hat{E}_y^{\mathrm{s}}, \hat{H}_x^{\mathrm{s}}, \hat{H}_z^{\mathrm{s}}\}(x, z, s) = \{\hat{E}_y - \hat{E}_y^{\mathrm{e}}, \hat{H}_x - \hat{H}_x^{\mathrm{e}}, \hat{H}_z - \hat{H}_z^{\mathrm{e}}\}(x, z, s). \tag{9.1}$$

As a matter of fact, the excitation EM fields can be viewed as the total EM fields if the screen were absent and can be hence represented as the sum of the incident EM field and the field reflected from the perfect ground. Accordingly, we write

$$\begin{aligned}\hat{E}_y^{\mathrm{e}}(x, z, s) = {} & \hat{e}^{\mathrm{i}}(s) \exp\{-s[\kappa_0 x - p(\kappa_0) z]\} \\ & - \hat{e}^{\mathrm{i}}(s) \exp\{-s[\kappa_0 x + p(\kappa_0)(z + 2z_0)]\},\end{aligned} \tag{9.2}$$

thus satisfying $\hat{E}_y^{\mathrm{e}}(x, z, s) = 0$ as $z \downarrow -z_0$ for all $x \in \mathbb{R}$. Since the total field $\hat{E}_y$ must vanish on the PEC ground, combination of (9.1) and (9.2) implies that

$$\lim_{z \downarrow -z_0} \hat{E}_y^{\mathrm{s}}(x, z, s) = 0 \quad \text{for } x \in \mathbb{R}. \tag{9.3}$$

9.1.1 Problem formulation

Following the approach described in Section 7.1, the Lorentz reciprocity theorem [1, Section 1.4.1] is applied to $\mathcal{D}_{+}^{\infty}$ excluding the domain occupied by the screen and to the scattered- and testing-field states (denoted by superscript ^{T}). Taking into account the property of causality of the interrelated wave fields and the boundary condition on the PEC plane (see (9.3)), we arrive at

$$\int_{x=a}^{b} \hat{E}_y^{\mathrm{T}}(x, 0, s) \partial \hat{J}_y^{\mathrm{s}}(x, s) \mathrm{d}x = \int_{x=a}^{b} \hat{E}_y^{\mathrm{s}}(x, 0, s) \partial \hat{J}_y^{\mathrm{T}}(x, s) \mathrm{d}x, \tag{8.1 revisited}$$

again, where the pertaining testing field is now determined from the excitation condition

$$\begin{aligned}&\lim_{z \downarrow \delta/2} \hat{H}_x^{\mathrm{T}}(x, z, s) - \lim_{z \uparrow -\delta/2} \hat{H}_x^{\mathrm{T}}(x, z, s) \\ &= \partial \hat{J}_y^{\mathrm{T}}(x, s) \big[\mathrm{H}(x - a) - \mathrm{H}(x - b)\big] \quad \text{for all } x \in \mathbb{R},\end{aligned} \tag{8.2 revisited}$$

as $\delta \downarrow 0$, and the explicit-type boundary condition (cf. (9.3))

$$\lim_{z\downarrow -z_0} \hat{E}_y^{\mathrm{T}}(x,z,s) = 0 \quad \text{for all } x \in \mathbb{R}. \tag{9.4}$$

To solve the problem for the induced electric-current surface density in the time-domain (TD), we next apply the CdH-MoM, again. To that end, the reciprocity relation is cast to its slowness-domain form (8.3), in which (8.4) is replaced with

$$\tilde{E}_y^{\mathrm{T}}(\kappa,0,s) = -[\mu_0/2p(\kappa)]\{1-\exp[-2sp(\kappa)z_0]\}\partial\tilde{J}_y^{\mathrm{T}}(\kappa,s), \tag{9.5}$$

thus interrelating the testing-source density with the corresponding testing field in the transform domain. Upon comparison of (9.5) with (8.4), it can be concluded that the presence of the ground plane can be modeled by the additional (image) testing electric-current source distributed along $z = -2z_0$. Now, combining the slowness-domain reciprocity relation (8.3) with (9.5) and with the pertaining transition condition (cf. (8.17)), i.e.

$$\tilde{E}_y^{\mathrm{s}}(\kappa,0,s) = -\tilde{E}_y^{\mathrm{e}}(\kappa,0,s) + \hat{Z}^{\mathrm{E}}(s)\partial\tilde{J}_y^{\mathrm{s}}(\kappa,s), \tag{9.6}$$

we arrive at (cf. (8.18))

$$\begin{aligned}
&\frac{s\mu_0}{2\pi\mathrm{i}}\int_{\kappa=-\mathrm{i}\infty}^{\mathrm{i}\infty}\partial\tilde{J}_y^{\mathrm{T}}(\kappa,s)\partial\tilde{J}_y^{\mathrm{s}}(-\kappa,s)\frac{\mathrm{d}\kappa}{2p(\kappa)}\\
&\quad-\frac{s\mu_0}{2\pi\mathrm{i}}\int_{\kappa=-\mathrm{i}\infty}^{\mathrm{i}\infty}\partial\tilde{J}_y^{\mathrm{T}}(\kappa,s)\partial\tilde{J}_y^{\mathrm{s}}(-\kappa,s)\exp[-2sp(\kappa)z_0]\frac{\mathrm{d}\kappa}{2p(\kappa)}\\
&\quad+\frac{s\hat{Z}^{\mathrm{E}}(s)}{2\pi\mathrm{i}}\int_{\kappa=-\mathrm{i}\infty}^{\mathrm{i}\infty}\partial\tilde{J}_y^{\mathrm{s}}(\kappa,s)\partial\tilde{J}_y^{\mathrm{T}}(-\kappa,s)\mathrm{d}\kappa\\
&\quad=\frac{s}{2\pi\mathrm{i}}\int_{\kappa=-\mathrm{i}\infty}^{\mathrm{i}\infty}\tilde{E}_y^{\mathrm{e}}(\kappa,0,s)\partial\tilde{J}_y^{\mathrm{T}}(-\kappa,s)\mathrm{d}\kappa\\
&\quad\text{for } \{s\in\mathbb{R}; s>0\}.
\end{aligned} \tag{9.7}$$

The transform-domain reciprocity relation (9.7) is the point of departure for the numerical solution presented in the following section.

9.1.2 Problem solution

Upon introducing the (space–time) discretization in the reciprocity relation (9.7), the first and third interaction terms on its left-hand side can be cast into the form of impeditivity arrays $\underline{\mathbf{\mathcal{Z}}}$ and $\underline{\mathbf{\mathcal{R}}}$ as in Sections 8.1.2 and 8.2, respectively. In a similar fashion, the second term accounting for EM-field reflections against the ground plane can be represented by the "modified impeditivity array", say $\underline{\mathbf{\mathcal{K}}}$, so that we end up with (cf. (7.42) and (8.19))

$$\sum_{k=1}^{m}\left(\bar{\underline{\mathbf{\mathcal{Z}}}}_{m-k+1} - 2\bar{\underline{\mathbf{\mathcal{Z}}}}_{m-k} + \bar{\underline{\mathbf{\mathcal{Z}}}}_{m-k-1}\right)\cdot\boldsymbol{J}_k = \bar{\boldsymbol{V}}_m, \tag{9.8}$$

in which

$$\bar{\underline{\mathcal{Z}}} = \underline{\mathcal{Z}} - \underline{\mathcal{K}} + \underline{\mathcal{R}} \tag{9.9}$$

is the total impeditivity array and $\bar{\boldsymbol{V}}$ on the right-hand side of (9.8) denotes the corresponding excitation array. The elements of $\underline{\mathcal{Z}}$ and $\underline{\mathcal{R}}$ have been specified before and can be found in Appendix B and in Section 8.2, respectively. Furthermore, the elements of $\underline{\mathcal{K}}$ are derived in Appendix F and the elements of the excitation array, $\bar{V}^{[S]}$, can be determined by generalizing (8.13a)–(8.14) pertaining to the situation in the absence of the ground plane. Accordingly, upon expressing the (transform-domain) excitation-field distribution as a sum of the incident plane wave and its reflection from the PEC plane (see (9.2)), that is,

$$\tilde{E}^{\mathrm{e}}_y(\kappa, 0, s) = \tilde{E}^{\mathrm{i}}_y(\kappa, 0, s)\{1 - \exp[-2sp(\kappa_0)z_0]\}, \tag{9.10}$$

and substituting this expression in the (right-hand side of the) reciprocity relation (9.7), we may employ the results of Appendix C and get

$$\bar{V}^{[S]}(t) = V^{[S]}(t) - V^{[S]}[t - 2p(\kappa_0)z_0] \quad \text{for all } S = \{1, \ldots, N\}, \tag{9.11}$$

where the excitation-array elements, $V^{[S]}(t)$, are given via (8.13a)–(8.14).

Since the system of equations (9.8) has been fully determined, the marching-on-in-time technique can be employed to calculate the desired electric-current coefficients. In this way, we end up with (cf. (7.44) and (8.22))

$$\boldsymbol{J}_m = \bar{\underline{\mathcal{Z}}}_1^{-1} \cdot \left[\bar{\boldsymbol{V}}_m - \sum_{k=1}^{m-1} \left(\bar{\underline{\mathcal{Z}}}_{m-k+1} - 2\bar{\underline{\mathcal{Z}}}_{m-k} + \bar{\underline{\mathcal{Z}}}_{m-k-1}\right) \cdot \boldsymbol{J}_k\right]$$
$$\text{for all } m = \{1, \ldots, M\}. \tag{9.12}$$

9.1.3 Narrow-screen approximation

The concept of external impedance has been employed in Chapter 7 to calculate the transient current induced in a narrow conductive strip located on a homogeneous, isotropic and loss-free background. In this section, it is briefly illustrated how the presence of the PEC ground plane manifests itself in the expression for the external impedance. To that end, we start from the reciprocity relation (7.1), again, in which we substitute the testing field satisfying conditions (7.4a) and (7.4b) with the uniform-electric current distribution (see (7.2)) and the explicit-type boundary condition (9.4) applying on the surface of the PEC ground plane. The reciprocity relation then leads to (cf. (7.24))

$$\left[\hat{Z}_{\mathrm{ext}}(s) + \hat{Z}^{\mathrm{E}}(s)/w\right] \hat{I}^{\mathrm{s}}(s) = \hat{e}^{\mathrm{i}}(s)\{1 - \exp[-2sp(\kappa_0)z_0]\}, \tag{9.13}$$

as $w \downarrow 0$, where we use the following representation for the external impedance (cf. (7.8a)):

$$\hat{Z}_{\mathrm{ext}}(s) = \frac{s\mu_0}{2\pi\mathrm{i}} \int_{\kappa=-\mathrm{i}\infty}^{\mathrm{i}\infty} \mathrm{i}_0^2(s\kappa w/2)\{1 - \exp[-2sp(\kappa)z_0]\} \frac{\mathrm{d}\kappa}{2p(\kappa)}. \tag{9.14}$$

Following now the approach described in Section 7.1.1, the integral representation can be approximately evaluated to get (see (7.16a))

$$\hat{Z}_{\text{ext}}(s) \simeq -(s\mu_0/2\pi)[\log(sw/c_0) - 3/2 - \log(2) + \gamma] + (s\mu_0/2\pi)[\log(sz_0/c_0) + \gamma], \tag{9.15}$$

where $\gamma = 0.57721\ldots$ and the second term on the right-hand side follows from the small-argument expansion of $\mathrm{K}_0(2sz_0/c_0)$ (see [2, Eq. (9.6.13)]), which implies that z_0 was also assumed to be relatively small. The latter expression can be further cast into the form pertaining to a cylindrical wire of radius $a > 0$ above the infinite PEC plane (see [3, Section 11.1]), that is,

$$\hat{Z}_{\text{ext}}(s) \simeq (s\mu_0/2\pi)\log(2z_0/a), \tag{9.16}$$

with $a = w\exp(-3/2) \simeq 0.2231w$, thereby providing the relation for the "*equivalent radius*". In this respect, it is noted that the postulate of the electric-current distribution in the form of (7.3) yields $a = w/4$ as reported elsewhere [4]. Finally, substituting the external impedance given by (9.16) in (9.13) with $\hat{Z}^{\mathrm{E}}(s) = 0$, and transforming the thus obtained relation to the TD, we get an approximate expression for the induced electric current in a narrow, PEC grounded strip

$$I^{\mathrm{s}}(t) \simeq \frac{2\pi}{\mu_0}\frac{1}{\log(2z_0/a)}\left[\int_{\tau=0}^{t} e^{\mathrm{i}}(\tau)\mathrm{d}\tau - \int_{\tau=0}^{t-2p(\kappa_0)z_0} e^{\mathrm{i}}(\tau)\mathrm{d}\tau\right]. \tag{9.17}$$

Since z_0 is assumed to be relatively small, the latter can be further simplified to

$$I^{\mathrm{s}}(t) \simeq [4\pi h Y_{\perp}(\kappa_0)/\log(2z_0/a)]e^{\mathrm{i}}(t), \tag{9.18}$$

where $Y_{\perp}(\kappa_0) = \eta_0\cos(\beta)$. Equation (9.18) thus clearly indicates that the electric current induced on a narrow PEC strip grounded by a nearby PEC plane preserves the shape of the incident plane-wave signature, much like as in the case of a relatively wide conductive screen in the absence of the ground plane (see Figures 8.3 and 8.4(b), for instance). On the other hand, the electric-current pulse induced in a relatively narrow PEC strip in the homogeneous, isotropic and loss-free embedding resembles the shape of the time integral of the exciting signature (see Figures 8.3 and 8.4(a)).

9.1.4 *Illustrative examples*

In this section, we shall analyze the electric-current surface density induced on a bounded PEC screen located along $\{a = -w/2 < x < w/2 = b, z = 0\}$ at a height $z_0 = 1.50\,w$ above the PEC ground. As in previous examples (see Section 8.1.3, for example), conductive screens under consideration are irradiated by the impulsive EM plane wave with the triangular signature $e^{\mathrm{i}}(t)$ (see Figure 8.3) and $\beta = 0$.

In the first example, both parameters $w/c_0t_{\mathrm{w}} = 0.10$ and $z_0/c_0t_{\mathrm{w}} = 0.15$ are relatively small such that formulas (9.17) and (9.18) expressing the TD electric-current response can be employed. The calculated pulse shapes of $\partial J_y^{\mathrm{s}}$ at $x = 0$

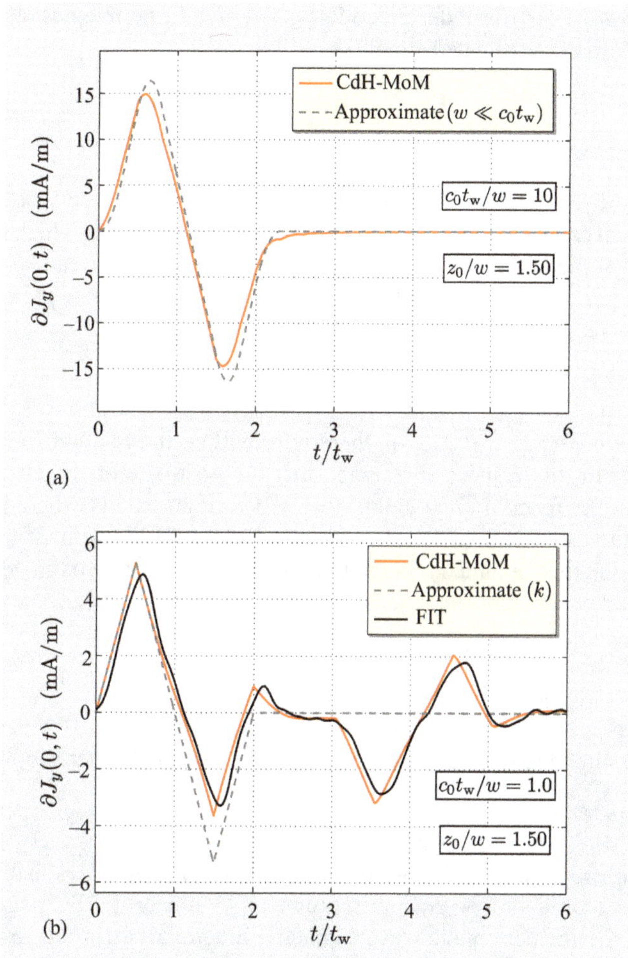

Figure 9.2 *Plane-wave induced electric-current surface density at the center of the grounded PEC screen. (a) The CdH-MoM solution (9.12) with $\hat{Z}^{\mathrm{E}}(s) = 0$ and the narrow-strip approximation (9.17) for $c_0 t_{\mathrm{w}}/w = 10$. (b) The CdH-MoM solution (9.12) with $\hat{Z}^{\mathrm{E}}(s) = 0$, the Kirchhoff approximation (8.16) and the FIT solution for $c_0 t_{\mathrm{w}}/w = 1.0$.*

and in $\{0 \leq t/t_{\mathrm{w}} \leq 6\}$ are shown in Figure 9.2(a). It can be seen that the CdH-MoM solution with $N = 2$ discretization nodes agrees well with the pulse shape predicted by (9.17). Indeed, it has been confirmed that the pulse shape of the induced current is similar to the one of the incident plane wave (see (9.18)).

In the second case, we increase the width of the screen such that $c_0 t_{\mathrm{w}}/w = 1.0$. Comparison between Figures 9.2(b) and 8.4(b) clearly reveals that the presence of the PEC ground does not manifest itself before the ground-reflected wave arrives back to

the screen at $t/t_{\mathrm{w}} = 3$. It is further seen that the Kirchhoff approximation accurately estimates the initial part of the response only. Apparently, the effect of the ground plane is not incorporated in the saltus-type model represented by (8.16). Finally, to validate the CdH-MoM formulation, the grounded strip has also been analyzed using the finite-integration technique (FIT) as implemented in CST Microwave Studio®. The corresponding pulse shapes shown in Figure 9.2(b) demonstrate the validity of the introduced CdH-MoM.

9.2 Pulsed EM scattering by a bounded screen on a dielectric half-space

The problem configuration analyzed in this section is shown in Figure 9.3. It consists of a bounded conductive screen located at the interface of two homogeneous, isotropic and loss-free half-spaces. The EM properties of the upper half-space, denoted by $\mathcal{D}_{+}^{\infty}$, are described by (scalar, real-valued and positive) parameters $\{\epsilon_0, \mu_0\}$. The medium in the lower half-space, denoted by $\mathcal{D}_{-}^{\infty}$, differ from the upper one in the electric permittivity only. Hence, its constitutive parameters are $\{\epsilon_1, \mu_0\}$. The corresponding EM wave speeds and admittances are then denoted by $c_{0,1} = (\epsilon_{0,1}\mu_0)^{-1/2} > 0$ and $\eta_{0,1} = 1/c_{0,1}\mu_0$, respectively. Again, the conductive screen is activated by the impulsive EM plane wave that can be represented in the s-domain via (4.1). The EM field scattered by the screen is, again, defined as the difference between the total field and the excitation field

$$\{\hat{E}_y^{\mathrm{s}}, \hat{H}_x^{\mathrm{s}}, \hat{H}_z^{\mathrm{s}}\}(x, z, s) = \{\hat{E}_y - \hat{E}_y^{\mathrm{e}}, \hat{H}_x - \hat{H}_x^{\mathrm{e}}, \hat{H}_z - \hat{H}_z^{\mathrm{e}}\}(x, z, s), \tag{9.1 revisited}$$

where the latter can be represented as a superposition of the incident and reflected EM plane waves in $\mathcal{D}_{+}^{\infty}$ and as the transmitted EM plane wave in the dielectric half-space occupying $\mathcal{D}_{-}^{\infty}$. Therefore, we may write

$$\hat{E}_y^{\mathrm{e}}(x, z, s) = \hat{e}^{\mathrm{i}}(s) \exp\{-s[\kappa_0 x - p_0(\kappa_0) z]\} + R_{\perp}\, \hat{e}^{\mathrm{i}}(s) \exp\{-s[\kappa_0 x + p_0(\kappa_0) z]\} \quad \text{in } \mathcal{D}_{+}^{\infty}, \tag{9.19}$$

$$\hat{E}_y^{\mathrm{e}}(x, z, s) = T_{\perp}\, \hat{e}^{\mathrm{i}}(s) \exp\{-s[\kappa_0 x - p_1(\kappa_0) z]\} \quad \text{in } \mathcal{D}_{-}^{\infty}, \tag{9.20}$$

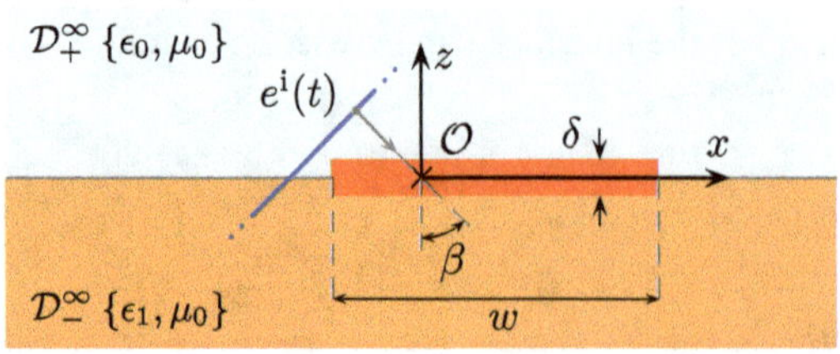

Figure 9.3 Pulsed EM plane wave impinging upon a bounded plane screen located on a dielectric half-space

where $p_1(\kappa_0) = \cos(\beta^{\rm t})/c_1$ with $\kappa_0 = \sin(\beta)/c_0$ and $p_0(\kappa_0) = \cos(\beta)/c_0$. Upon enforcing the continuity-type boundary conditions applying to $\hat{E}_y^{\rm e}$ and $\hat{H}_x^{\rm e}$ across the interface, the reflection and transmission coefficients follow

$$R_\perp = \frac{\eta_0 \cos(\beta) - \eta_1 \cos(\beta^{\rm t})}{\eta_0 \cos(\beta) + \eta_1 \cos(\beta^{\rm t})}, \tag{9.21}$$

$$T_\perp = \frac{2\eta_0 \cos(\beta)}{\eta_0 \cos(\beta) + \eta_1 \cos(\beta^{\rm t})}, \tag{9.22}$$

respectively, where $\cos(\beta^{\rm t}) = [1 - (\epsilon_0/\epsilon_1)\sin^2(\beta)]^{1/2}$ by virtue of Snell's law.

9.2.1 Problem formulation

The problem under consideration is tackled, again, with the aid of the EM reciprocity theorem of the time convolution type [1, Section 1.4.1]. Hence, the starting reciprocity relation (8.1) is arrived at upon applying the theorem to $\mathcal{D}_+^\infty \cup \mathcal{D}_-^\infty$ and to the scattered and testing-field states, again, while employing the continuity of $\{\hat{E}_y^{\rm s,T}, \hat{H}_x^{\rm s,T}\}$ across the interface. The pertinent testing field is determined from the excitation condition

$$\lim_{z\downarrow\delta/2} \hat{H}_x^{\rm T}(x,z,s) - \lim_{z\uparrow-\delta/2} \hat{H}_x^{\rm T}(x,z,s)$$
$$= \partial\hat{J}_y^{\rm T}(x,s)\big[{\rm H}(x-a) - {\rm H}(x-b)\big] \quad \text{for all } x \in \mathbb{R}, \tag{8.2 revisited}$$

as $\delta \downarrow 0$ and the continuity-type boundary condition (cf. (9.4))

$$\lim_{z\downarrow\delta/2} \hat{E}_y^{\rm T}(x,z,s) = \lim_{z\uparrow-\delta/2} \hat{E}_y^{\rm T}(x,z,s) \quad \text{for all } x \in \mathbb{R}, \tag{9.23}$$

as $\delta \downarrow 0$, again.

In order to solve the problem via the CdH-MoM, the s-domain reciprocity relation (8.1) is cast into its slowness-domain form (8.3), where we substitute the (transform-domain) testing field complying with conditions (8.2) and (9.23). Under the wave-slowness representation of the form specified by (4.33), the testing-field distribution at the level of the interface can be written as (cf. (8.4))

$$\tilde{E}_y^{\rm T}(\kappa,0,s) = -\frac{\mu_0}{p_0(\kappa)+p_1(\kappa)}\partial\tilde{J}_y^{\rm T}(\kappa,s), \tag{9.24}$$

in which (cf. (4.36))

$$p_{0,1}(\kappa) = (1/c_{0,1}^2 - \kappa^2)^{1/2}, \quad \text{with } {\rm Re}(p_{0,1}) \geq 0, \tag{9.25}$$

respectively. Hence, substituting (9.24) and (9.6) in the transform-domain reciprocity relation (8.3), we obtain (see (8.18) and (9.7))

$$\begin{aligned}\frac{s\mu_0}{2\pi\mathrm{i}}\int_{\kappa=-\mathrm{i}\infty}^{\mathrm{i}\infty}\partial\tilde{J}_y^{\mathrm{T}}(\kappa,s)\partial\tilde{J}_y^{\mathrm{s}}(-\kappa,s)\frac{\mathrm{d}\kappa}{p_0(\kappa)+p_1(\kappa)}\\+\frac{s\hat{Z}^{\mathrm{E}}(s)}{2\pi\mathrm{i}}\int_{\kappa=-\mathrm{i}\infty}^{\mathrm{i}\infty}\partial\tilde{J}_y^{\mathrm{s}}(\kappa,s)\partial\tilde{J}_y^{\mathrm{T}}(-\kappa,s)\mathrm{d}\kappa\\=\frac{s}{2\pi\mathrm{i}}\int_{\kappa=-\mathrm{i}\infty}^{\mathrm{i}\infty}\tilde{E}_y^{\mathrm{e}}(\kappa,0,s)\partial\tilde{J}_y^{\mathrm{T}}(-\kappa,s)\mathrm{d}\kappa\\\text{for } \{s\in\mathbb{R};s>0\},\end{aligned}\tag{9.26}$$

which is the point of departure for the numerical solution that follows.

9.2.2 *Problem solution*

To solve the starting reciprocity relation (9.26) numerically for the TD electric-current surface density induced on the screen, we pursue the line of reasoning described in Section 8.1.2. Accordingly, the solution domain is discretized uniformly both in space and time, and the transform-domain reciprocity relation is cast into the system of equations in the s-domain, the coefficients of which are transformed to back the TD via the standard CdH technique. In the way, we end up with the convolution-type system of equations

$$\sum_{k=1}^{m}\left(\bar{\underline{\mathcal{Z}}}_{m-k+1}-2\bar{\underline{\mathcal{Z}}}_{m-k}+\bar{\underline{\mathcal{Z}}}_{m-k-1}\right)\cdot\boldsymbol{J}_k=\bar{\boldsymbol{V}}_m, \qquad \text{(9.8 revisited)}$$

in which

$$\bar{\underline{\mathcal{Z}}}=\underline{\mathcal{X}}+\underline{\mathcal{R}}\tag{9.27}$$

denotes the modified impeditivity array. A detailed description of arrays $\underline{\mathcal{X}}$ and $\underline{\mathcal{R}}$ can be found in Appendix G and Section 8.2, respectively, and the elements of the excitation array $\bar{\boldsymbol{V}}$ on the right-hand side of (9.8) can be readily found by generalizing (8.13a)–(8.14). Thus, referring to (9.20), the excitation-field field distribution at the level of the interface is expressed as

$$\tilde{E}_y^{\mathrm{e}}(\kappa,0,s)=T_\perp\tilde{E}_y^{\mathrm{i}}(\kappa,0,s),\tag{9.28}$$

which yields

$$\bar{V}^{[S]}(t)=T_\perp V^{[S]}(t)\quad\text{for all } S=\{1,\ldots,N\},\tag{9.29}$$

where the expressions for $V^{[S]}(t)$ are given by (8.13a)–(8.14). Apparently, the vanishing contrast between the media in $\mathcal{D}_\pm^\infty$ implies $T_\perp=1$ and hence $\bar{V}^{[S]}(t)=V^{[S]}(t)$.

In the final step, the system of equations (9.8) is solved for the array of coefficients representing the induced electric-current surface density (see (8.6)). This can be accomplished via the marching-on-time scheme (9.12), again.

9.2.3 Narrow-screen approximation

In this subsection, we shall derive the external impedance of a narrow conductive strip located at a dielectric interface. The starting point is, again, the reciprocity relation (7.1), where we substitute the testing field satisfying (7.4a) and (7.4b). Postulating, again, the uniform-electric current spatial distribution, the reciprocity relation can be cast into the following form:

$$\left[\hat{Z}_{\text{ext}}(s) + \hat{Z}^{\text{E}}(s)/w\right] \hat{I}^{\text{s}}(s) = T_{\perp}\hat{e}^{\text{i}}(s), \tag{9.30}$$

as $w \downarrow 0$, where we use the following representation for the external impedance (cf. (7.8a)):

$$\hat{Z}_{\text{ext}}(s) = \frac{s\mu_0}{2\pi\text{i}} \int_{\kappa=-\text{i}\infty}^{\text{i}\infty} \text{i}_0^2(s\kappa w/2) \frac{\text{d}\kappa}{p_0(\kappa) + p_1(\kappa)}. \tag{9.31}$$

Recall here that the slowness parameters $p_{0,1}(\kappa)$ are given by (9.25). Adopting further the approach presented in Section 7.1.1 to evaluate the slowness representation (9.31) approximately, we end up with (cf. (7.16a) and (9.15))

$$\hat{Z}_{\text{ext}}(s) \simeq -(s\mu_0/2\pi)[\log(sw/c) - 3/2 - \log(2) + \gamma], \tag{9.32}$$

in which

$$c = \exp\left\{\frac{c_0^2[\log(c_1) + 1/2] - c_1^2[\log(c_0) + 1/2]}{c_0^2 - c_1^2}\right\}, \tag{9.33}$$

denotes the effective EM wave speed having the property $c_1 \leq c \leq c_0$. It is straightforward to verify that $c = c_0$ in the limit $c_1 \uparrow c_0$, thus getting the external impedance of a narrow strip in the homogeneous, isotropic and loss-free embedding (see (7.16a)). Consequently, upon replacing c_0 by its effective value c and $e^{\text{i}}(t)$ by its equivalent signature, $T_{\perp}\, e^{\text{i}}(t)$, reduced by the transmission coefficient $T_{\perp}$ (see (9.22)), the results of Section 7.1.2 can be employed. For instance, the electric current induced on a PEC strip located on a dielectric half-space can be then calculated via (cf. (7.20))

$$i_m = \ddot{\Phi}_1^{-1}\left[c_0\Delta t\eta_0 T_{\perp} e^{\text{i}}(t_m) - \sum_{k=1}^{m-1} i_k\left(\ddot{\Phi}_{m-k+1} - 2\ddot{\Phi}_{m-k} + \ddot{\Phi}_{m-k-1}\right)\right]$$
$$\text{for all } m = \{1, \ldots, M\}, \tag{9.34}$$

where $\ddot{\Phi}_k = \Phi(\ddot{t}_k)$ with $\ddot{t}_k = k\Delta\ddot{t} = k(c/c_0)\Delta t$, and $\Phi(t)$ is given by (7.21), again.

9.2.4 Illustrative examples

In this subsection, we shall calculate the electric-current surface density induced on a screen that is located along $\{a = -w/2 < x < w/2 = b, z = 0\}$ on a dielectric half-space. The relative permittivity of the dielectric half-space is $\epsilon_1/\epsilon_0 = 9.0$. We assume, again, that the exciting impulsive EM plane wave with the triangular

signature, $e^{\mathrm{i}}(t)$ (see Figure 8.3), impinges normally upon the screen. For the chosen parameters, the transmission coefficient is $T_{\perp} = 1/2$ (see (9.22)).

At first, the induced electric-current surface density on a relatively narrow screen with $c_0 t_{\mathrm{w}}/w = 10$ is examined. Its pulse shapes at $x = 0$ as calculated via the pertinent CdH-MoM solution with $N = 2$ (see Section 9.2.2) and the narrow-strip approximation (9.34) correlate very well (see Figure 9.4(a)). The calculated pulse shape shown in Figure 9.4(a) is similar to the time integral of the exciting plane-wave signature, much like the current response induced on a narrow strip located on the homogeneous background (see Figure 8.4(a)). With respect to the latter, however,

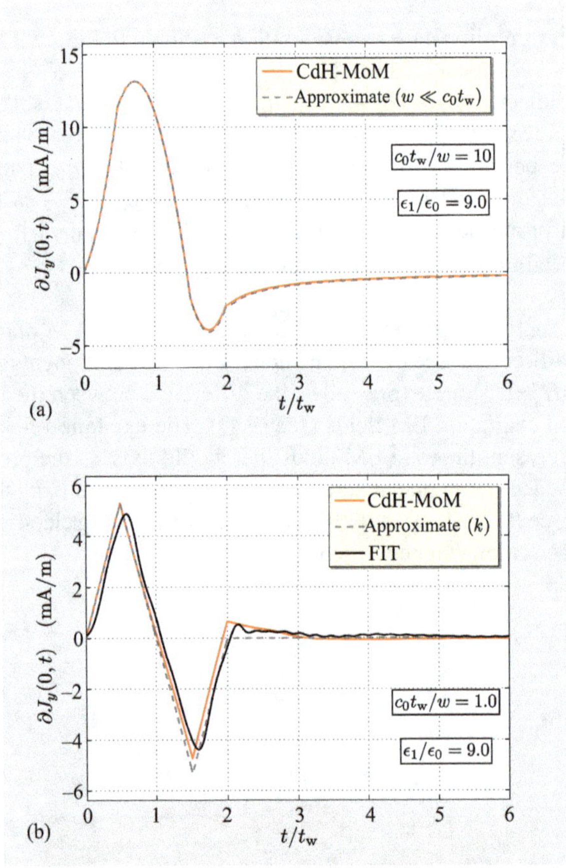

Figure 9.4 Plane-wave induced electric-current surface density at the center of the PEC screen placed on the dielectric half-space. (a) The CdH-MoM solution (9.12) with (9.27), (9.29) and $\hat{Z}^{\mathrm{E}}(s) = 0$ and the narrow-strip approximation (9.34) for $c_0 t_{\mathrm{w}}/w = 10$. (b) The CdH-MoM solution (9.12) with (9.27), (9.29) and $\hat{Z}^{\mathrm{E}}(s) = 0$, the Kirchhoff approximation (8.16) and the FIT solution for $c_0 t_{\mathrm{w}}/w = 1.0$.

the actual pulse shape is somewhat attenuated due to the presence of the dielectric half-space.

The width of the second screen under consideration is equal to the spatial support excitation pulse, that is, $c_0 t_w/w = 1.0$. Figure 9.4(b) demonstrates that the Kirchhoff approximation (8.16) estimates the CdH-MoM-based response very well, in particular, its initial part, again. As can be observed in Figure 9.4(b), this result has also been satisfactorily confirmed with the aid of the FIT as implemented in CST Microwave Studio®.

9.3 Pulsed EM scattering by a bounded screen on a grounded dielectric slab

The problem configuration under consideration is shown in Figure 9.5. It consists of a bounded thin screen located along $\{a < x < b, -\infty < y < \infty, -\delta/2 < z < \delta/2\}$ on a grounded dielectric slab. The slab occupies domain $\mathcal{D}_1$ that is filled by a homogeneous, isotropic and loss-free medium characterized by its (real-valued, scalar and positive) electric permittivity ϵ_1 and magnetic permeability μ_0. The ground plane is assumed to be modeled by a PEC wall located along $z = -z_0 < 0$ below the screen. The EM properties of the upper half-space $\mathcal{D}_+^\infty$ are $\{\epsilon_0, \mu_0\}$, again. The EM wave speeds pertaining to $\mathcal{D}_+^\infty$ and $\mathcal{D}_1$ are then given by $c_{0,1} = (\epsilon_{0,1}\mu_0)^{-1/2} > 0$, respectively.

Again, the analyzed structure is activated the impulsive EM plane wave polarized along the y-direction (see (4.1)). The non-vanishing components of the scattered EM field, $\{\hat{E}_y^s, \hat{H}_x^s, \hat{H}_z^s\}$ are expressed as the difference between the total EM fields and the so-called excitation EM fields (see (9.1)). The excitation fields, denoted by superscript ^{e}, represent the total EM fields that would exist in the problem configuration in the absence of the planar screen. Hence, upon generalizing the expressions derived before (see (9.2), (9.19) and (9.20)), the corresponding electric-field strength at the level of the screen can be written as

$$\hat{E}_y^e(x, 0, s) = \hat{E}_y^i(x, 0, s)\big[1 + \hat{R}_\perp(s)\big], \tag{9.35}$$

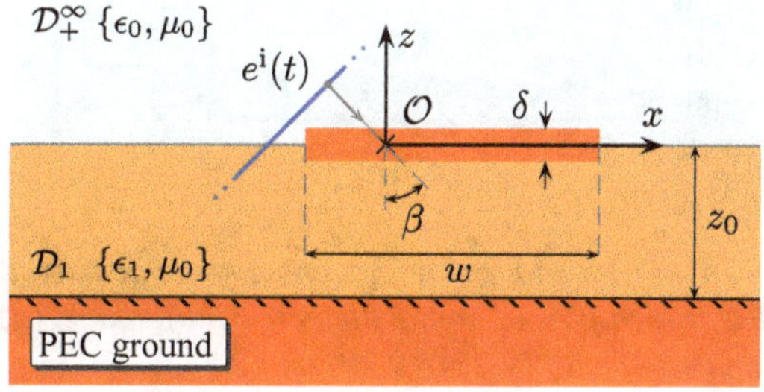

Figure 9.5 Pulsed EM plane wave impinging upon a bounded plane screen located on a grounded dielectric slab

where $\hat{R}_{\perp}(s)$ is the pertaining (generalized) reflection coefficient that can be expressed via

$$\hat{R}_{\perp}(s) = \frac{R_{\perp} - \exp[-2sp_1(\kappa_0)z_0]}{1 - R_{\perp}\exp[-2sp_1(\kappa_0)z_0]}. \tag{9.36}$$

Here, $p_1(\kappa_0) = [1 - (c_1/c_0)^2 \sin^2(\beta)]^{1/2}/c_1$ and the reflection coefficient, $R_{\perp}$, is given by (9.21). Upon expanding (9.36) in a (convergent) geometric series, we may write

$$\hat{E}^{\mathrm{e}}_y(x,0,s) = T_{\perp}\hat{E}^{\mathrm{i}}_y(x,0,s)\left\{1 - (1 - R_{\perp})\sum_{q=0}^{\infty} R_{\perp}^q \exp[-2s(q+1)p_1(\kappa_0)z_0]\right\}, \tag{9.37}$$

where the transmission coefficient, $T_{\perp}$, is defined by (9.22). The TD counterpart of (9.37) reveals that the excitation field at the level of the screen can be expressed as a superposition of delayed and attenuated copies of $e^{\mathrm{i}}(t - \kappa_0 x)$. This property makes it possible to obtain the elements of the actual excitation array just by summing the copies of $V^{[S]}(t)$ as given by (8.13a)–(8.14). Finally, we note that since $\hat{E}^{\mathrm{e}}_y = 0$ on the surface of the PEC ground, the explicit-type boundary condition (9.3) applying to the scattered electric field holds true.

9.3.1 Problem formulation

In an analogous manner to Sections 9.1.1 and 9.2.1, the starting point is the reciprocity relation (8.1), where we substitute the testing field satisfying the excitation condition

$$\begin{aligned}&\lim_{z\downarrow\delta/2} \hat{H}^{\mathrm{T}}_x(x,z,s) - \lim_{z\uparrow-\delta/2} \hat{H}^{\mathrm{T}}_x(x,z,s)\\ &= \partial\hat{J}^{\mathrm{T}}_y(x,s)\left[\mathrm{H}(x-a) - \mathrm{H}(x-b)\right] \quad \text{for all } x \in \mathbb{R},\end{aligned} \tag{8.2 revisited}$$

as $\delta \downarrow 0$, the continuity-type condition

$$\lim_{z\downarrow\delta/2} \hat{E}^{\mathrm{T}}_y(x,z,s) = \lim_{z\uparrow-\delta/2} \hat{E}^{\mathrm{T}}_y(x,z,s) \quad \text{for all } x \in \mathbb{R}, \tag{9.23 revisited}$$

as $\delta \downarrow 0$, and, finally, the explicit-type boundary condition on the perfect ground, that is,

$$\lim_{z\downarrow -z_0} \hat{E}^{\mathrm{T}}_y(x,z,s) = 0 \quad \text{for all } x \in \mathbb{R}. \tag{9.4 revisited}$$

In accordance with the boundary conditions, the resulting (transform-domain) electric-field distribution along $z = 0$ can be expressed as (cf. (9.24))

$$\tilde{E}^{\mathrm{T}}_y(\kappa,0,s) = -\frac{\mu_0}{p_0(\kappa)+p_1(\kappa)}\partial\tilde{J}^{\mathrm{T}}_y(\kappa,s)\frac{1-\exp[-2sp_1(\kappa)z_0]}{1-\tilde{R}_{\perp}(\kappa)\exp[-2sp_1(\kappa)z_0]}, \tag{9.38}$$

where the pertaining reflection coefficient has the following form (cf. (9.21)):

$$\tilde{R}_\perp(\kappa) = \frac{p_0(\kappa) - p_1(\kappa)}{p_0(\kappa) + p_1(\kappa)}. \tag{9.39}$$

Recall, again, that the vertical slowness coefficients, $p_{0,1}(\kappa)$, are defined by (9.25). Likewise the excitation field given by (9.37), the testing-field distribution (9.38) can be expressed via the geometric-series representation

$$\begin{aligned}\tilde{E}^{\mathrm{T}}_y(\kappa, 0, s) = &-\frac{\mu_0}{p_0(\kappa) + p_1(\kappa)} \partial \tilde{J}^{\mathrm{T}}_y(\kappa, s) \\ &\times \left\{1 - \left[1 - \tilde{R}_\perp(\kappa)\right] \sum_{q=0}^{\infty} \left[\tilde{R}_\perp(\kappa)\right]^q \exp[-2s(q+1)p_1(\kappa)z_0]\right\},\end{aligned} \tag{9.40}$$

which can be interpreted as a generalization of both (9.5) and (9.24). Indeed, if $c_1 = c_0$, then $\tilde{R}_\perp(\kappa) = 0$ and (9.40) reduces to (9.5) pertaining to the grounded screen (see Figure 9.1). Next, if $z_0 \to \infty$, the exponential terms in the braces of (9.40) are vanishingly small such that (9.40) simplifies to (9.24) applying to the screen located on the homogeneous dielectric half-space (see Figure 9.3).

Finally, making use of (9.40) and (9.6) in the transform-domain reciprocity relation (8.3), we arrive at (cf. (9.26))

$$\begin{aligned}&\frac{s\mu_0}{2\pi \mathrm{i}} \int_{\kappa=-\mathrm{i}\infty}^{\mathrm{i}\infty} \partial \tilde{J}^{\mathrm{T}}_y(\kappa, s) \partial \tilde{J}^{\mathrm{s}}_y(-\kappa, s) \frac{\mathrm{d}\kappa}{p_0(\kappa) + p_1(\kappa)} \\ &- \frac{s\mu_0}{2\pi \mathrm{i}} \sum_{q=0}^{\infty} \int_{\kappa=-\mathrm{i}\infty}^{\mathrm{i}\infty} \partial \tilde{J}^{\mathrm{T}}_y(\kappa, s) \partial \tilde{J}^{\mathrm{s}}_y(-\kappa, s) \tilde{F}^{[q]}[\kappa, 2(q+1)z_0, s] \frac{\mathrm{d}\kappa}{p_0(\kappa) + p_1(\kappa)} \\ &\qquad + \frac{s\hat{Z}^{\mathrm{E}}(s)}{2\pi \mathrm{i}} \int_{\kappa=-\mathrm{i}\infty}^{\mathrm{i}\infty} \partial \tilde{J}^{\mathrm{s}}_y(\kappa, s) \partial \tilde{J}^{\mathrm{T}}_y(-\kappa, s) \mathrm{d}\kappa \\ &\qquad = \frac{s}{2\pi \mathrm{i}} \int_{\kappa=-\mathrm{i}\infty}^{\mathrm{i}\infty} \tilde{E}^{\mathrm{e}}_y(\kappa, 0, s) \partial \tilde{J}^{\mathrm{T}}_y(-\kappa, s) \mathrm{d}\kappa \\ &\qquad \text{for } \{s \in \mathbb{R}; s > 0\},\end{aligned} \tag{9.41}$$

in which

$$\tilde{F}^{[q]}(\kappa, z, s) = \left[1 - \tilde{R}_\perp(\kappa)\right] \left[\tilde{R}_\perp(\kappa)\right]^q \exp[-sp_1(\kappa)z]. \tag{9.42}$$

Upon comparison of the derived reciprocity relation (9.41) with the one applying to the screen on a dielectric half-space (see (9.26)), it is concluded that the second term on the left-hand side of (9.41) accounts for the reflected-wave constituents propagating inside the dielectric slab. The impact of the new term on the induced electric-current distribution is the main subject of the next sections.

9.3.2 *Problem solution*

The reciprocity relation (9.41) is next handled numerically. To this end, we pursue the solution strategy introduced in Section 8.1.2, which yields the system of equations

representable by (9.8). The pertaining total impeditivity array, $\bar{\underline{\mathcal{Z}}}$, consists of three terms and can be expressed as (cf. (9.27))

$$\bar{\underline{\mathcal{Z}}} = \underline{\mathcal{X}} - \underline{\mathcal{S}} + \underline{\mathcal{R}}, \tag{9.43}$$

where $\underline{\mathcal{X}}$ and $\underline{\mathcal{R}}$ have been specified before (see Appendix G and Section 8.2) and a close description of generic integrals representing the elements of $\underline{\mathcal{S}}$ can be found in Appendix H. Furthermore, the elements of the pertinent excitation array on the right-hand side of (9.8) can be derived by generalizing (8.13a)–(8.14) using (9.37). In this way, we get at once (cf. (9.29))

$$\bar{V}^{[S]}(t) = T_\perp V^{[S]}(t) - T_\perp(1 - R_\perp)\sum_{q=0}^{\infty} R_\perp^q V^{[S]}[t - 2p_1(\kappa_0)(q+1)z_0]$$
$$\text{for all } S = \{1, \dots, N\}. \tag{9.44}$$

Owing to the property $e^{\mathrm{i}}(t) = 0$ and hence $V^{[S]}(t) = 0$ for $t < 0$, only a finite number of terms in (9.44) are needed in any bounded time window of observation. In the final step, the thus defined system of equations (9.8) is solved for the coefficients representing the desired electric-current distribution (see (8.6)). This can be accomplished via the marching-on-in-time technique, again (see (9.12)).

9.3.3 Narrow-screen approximation

The concept of external impedance is next used to provide an approximate expression for the TD electric current induced on a narrow strip located along $\{-w/2 < x < w/2, -\infty < y < \infty, z = 0\}$ on the grounded dielectric slab. To this end, we start from the reciprocity relation (7.1), again, where we substitute the testing field complying with conditions (8.2), (9.23) and (9.4) as revisited in Section 9.3.1. Under the assumption of the uniform electric-current spatial distribution (see (7.2)), the starting reciprocity relation yields (cf. (9.30))

$$\left[\hat{Z}_{\text{ext}}(s) + \hat{Z}^{\text{E}}(s)/w\right]\hat{I}^{\text{s}}(s) = T_\perp \hat{e}^{\text{i}}(s)$$
$$\times \left\{1 - (1 - R_\perp)\sum_{q=0}^{\infty} R_\perp^q \exp[-2s(q+1)p_1(\kappa_0)z_0]\right\}, \tag{9.45}$$

as $w \downarrow 0$, and the corresponding external impedance can be represented via (cf. (9.31))

$$\hat{Z}_{\text{ext}}(s) = \frac{s\mu_0}{2\pi\mathrm{i}}\int_{\kappa=-\mathrm{i}\infty}^{\mathrm{i}\infty} \mathrm{i}_0^2(s\kappa w/2)$$
$$\times \left\{1 - \frac{[1 - \tilde{R}_\perp(\kappa)]\exp[-2sp_1(\kappa)z_0]}{1 - \tilde{R}_\perp(\kappa)\exp[-2sp_1(\kappa)z_0]}\right\}\frac{\mathrm{d}\kappa}{p_0(\kappa) + p_1(\kappa)}. \tag{9.46}$$

The (transform-domain) reflection coefficient, $\tilde{R}_\perp(\kappa)$, as occurs in (9.46) is given by (9.39). Upon carrying out the limiting procedure introduced in Section 7.1.1, the slowness integral representation (9.46) can be approximated by (cf. (9.16))

$$\hat{Z}_{\text{ext}}(s) \simeq (s\mu_0/2\pi)\log[2(c/c_1)z_0/a], \tag{9.47}$$

provided that both w and z_0 are relatively small. Recall that the "*equivalent radius*" in (9.47) is $a = w\exp(-3/2) \simeq 0.2231w$ and that the effective wave speed c was described by (9.33). Comparison of the thus obtained approximation (9.47) with the one pertaining to the grounded slab (see (9.16)) reveals that the presence of the dielectric slab manifests itself in a relative extension of the distance between the screen and the ground, $z_0 > 0$, by factor $c/c_1 > 1$. Owing to the fact that $c = c_0$ as $c_1 \uparrow c_0$, (9.16) can be viewed as the limit of (9.47).

With the closed-form formula for the external impedance at our disposal, we may finally express the TD electric current induced in the strip. The desired response follows upon substituting (9.47) in (9.45) and transforming the resulting expression for $\hat{I}^{\text{s}}(s)$ back to the TD. Assuming, for the sake of simplicity, the PEC strip characterized by $\hat{Z}^{\text{E}}(s) = 0$, we arrive at (cf. (9.17))

$$I^{\text{s}}(t) \simeq \frac{2\pi}{\mu_0} \frac{T_\perp}{\log[2(c/c_1)z_0/a]} \left[\int_{\tau=0}^{t} e^{\text{i}}(\tau)\text{d}\tau \right. \\ \left. - (1 - R_\perp) \sum_{q=0}^{\infty} R_\perp^q \int_{\tau=0}^{t-2p_1(\kappa_0)(q+1)z_0} e^{\text{i}}(\tau)\text{d}\tau \right]. \tag{9.48}$$

Clearly, under the limit $c_1 \uparrow c_0$, (9.48) boils down to (9.17), as expected.

9.3.4 *Illustrative examples*

In this section, we shall calculate the electric-current surface density induced on a PEC screen located along $\{a = -w/2 < x < w/2 = b, -\infty < y < \infty, z = 0\}$ on a dielectric slab. Its relative permittivity is $\epsilon_1/\epsilon_0 = 9.0$ and the thickness of the dielectric slab is $z_0 = 1.50\,w$.

In the first case, the width of the screen, $w > 0$, and the thickness of the slab, $z_0 > 0$, are assumed to be small with respect to the spatial support of the excitation pulse in the dielectric slab. Therefore, we take $w/c_1 t_{\text{w}} = 0.10$ and hence $z_0/c_1 t_{\text{w}} = 0.15$. For the sake of validation, the pulse shape of $\partial J_y^{\text{s}}(0,t)$ is calculated with the aid of the corresponding CdH-MoM procedure (see Section 9.3.2) with $N = 2$ discretization nodes and the narrow-strip approximation represented by (9.48). As can be observed in Figure 9.6(a), the resulting pulse shapes correlate very well. Much as the pulse shape induced on a grounded screen, or on a relatively wide screen in the homogeneous background (see Figures 9.2(a) and 8.4(b), for example), the calculated pulses have the shape similar to the excitation (triangular) one (see Figure 8.3).

In the next example, the screen on the grounded slab is irradiated by the impulsive plane wave, the signature of which is ten times shorter than the one used in

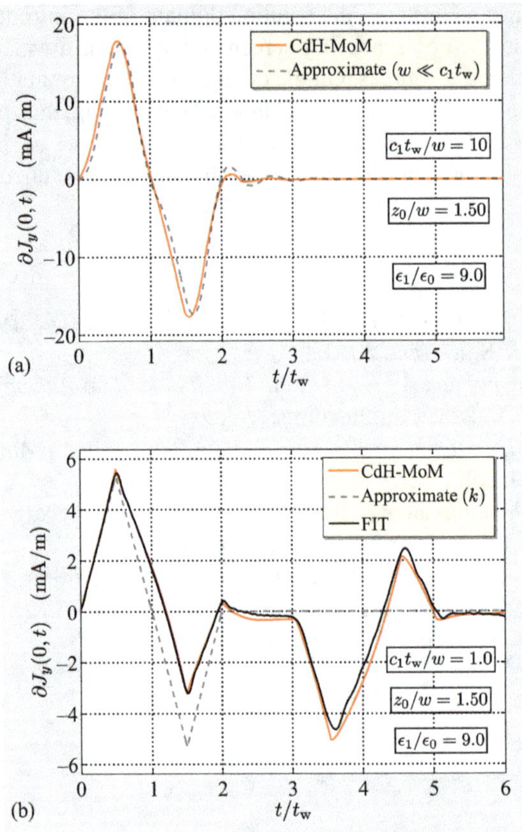

Figure 9.6 *Plane-wave induced electric-current surface density at the center of the PEC screen placed on the grounded slab. (a) The CdH-MoM solution (9.12) with (9.43), (9.44) and $\hat{Z}^{\mathrm{E}}(s) = 0$ and the narrow-strip approximation (9.48) for $c_1 t_w / w = 10$. (b) The CdH-MoM solution (9.12) with (9.43), (9.44) and $\hat{Z}^{\mathrm{E}}(s) = 0$, the Kirchhoff approximation (8.16) and the FIT solution for $c_1 t_w / w = 1.0$.*

the previous example. Figure 9.6(b) demonstrates that the induced current responses as calculated via the CdH-MoM with $N = 11$ and the FIT as implemented in CST Microwave Studio® agree well. Again, the saltus-type model represented by (8.16) estimates well the early-time part of the response only.

9.4 Concluding remarks

On a few selected examples, we have demonstrated the use of CdH-MoM for analyzing pulsed EM scattering by conducting plane sheets embedded in a planarly layered background medium. In contrast to standard brute-force computational approaches

(e.g. finite-difference technique), the solution domain of the introduced CdH-MoM is limited to the surface of scattering screens only, which drastically reduces the required computational resources. On the other hand, the computational scheme has to be adapted to the structure of the actual layered embedding, thus making the CdH-MoM less universal. This is, however, a common feature of all integral-equation techniques employing the concept of stratified-media Green's function [5].

References

[1] M. Štumpf. *Electromagnetic Reciprocity in Antenna Theory*. Hoboken, NJ: John Wiley & Sons, Inc., 2018.

[2] M. Abramowitz and I. A. Stegun. *Handbook of Mathematical Functions*. New York, NY: Dover Publications, 1972.

[3] M. Štumpf. *Time-Domain Electromagnetic Reciprocity in Antenna Modeling*. Hoboken, NJ: John Wiley & Sons, Inc., 2019.

[4] C. Butler. The equivalent radius of a narrow conducting strip. *IEEE Transactions on Antennas and Propagation*, 30(4):755–758, 1982.

[5] J. R. Mosig. Integral equation technique. In Itoh T., editor, *Numerical Techniques for Microwave and Millimeter-Wave Passive Structures*, chapter 3, pp. 133–213. New York, NY: John Wiley & Sons, 1989.

Chapter 10

Pulsed EM-field coupling between bounded and narrow conductive screens

In order to evaluate electromagnetic (EM) phenomena such as cross talk or substrate noise coupling in thin-film multilayered structures, the pulsed EM field interaction between conductive strips has to be properly incorporated in a computational model. This is exactly the main purpose of this chapter, where it is demonstrated how mutual-coupling effects can be included in the Cagniard–DeHoop method of moments (CdH-MoM). Again, the main mathematical tool for the analysis is the EM reciprocity theorem of the time-convolution type. The reciprocity-based formulation yields a reciprocity relation that is upon discretizing the (space–time) solution domain cast into its matrix form, elements of which represent local and remote interactions involved. A few illustrative examples, including the EM coupling between two plane screens on the dielectric half-space and on the grounded dielectric slab, are discussed and analyzed in detail. More complex problems can be analyzed along the same lines.

10.1 Problem formulation

Without loss of generality, let us consider the EM interaction between two perfectly electrically conducting (PEC) plane screens embedded in a (piecewise) homogeneous medium. The two screens occupy domains $\mathcal{S}_A = \{a < x < b, -\infty < y < \infty, -\delta/2 < z < \delta/2\}$ and $\mathcal{S}_B = \{c < x < d, -\infty < y < \infty, z_B - \delta/2 < z < z_B + \delta/2\}$, where $w_A = b - a > 0$ and $w_B = d - c > 0$ are their widths, and $\delta > 0$ denotes their (relatively) small thickness. Again, the non-vanishing y-independent EM fields in the problem configuration are excited by the uniform EM plane wave that is described by

$$\hat{E}_y^{\mathrm{i}}(x, z, s) = \hat{e}^{\mathrm{i}}(s) \exp\{-s[\kappa_0 x - p(\kappa_0) z]\}, \qquad \text{(4.1 revisited)}$$

with $\kappa_0 = \sin(\beta)/c_0$ and $p(\kappa_0) = \cos(\beta)/c_0$ (see Figure 10.1). Application of the EM reciprocity theorem of the time-convolution type [1, Section 1.4.1] to the actual

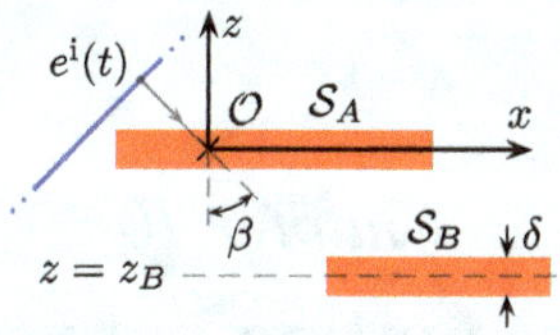

Figure 10.1 Pulsed EM plane wave impinging upon two conductive planar screens

scattered-field state (denoted by superscript ^{s}) and to the testing-field state (denoted by superscript ^{T}) yields

$$\int_{x=a}^{b} \left[\hat{E}_y^{\mathrm{T};A}(x,0,s) + \hat{E}_y^{\mathrm{T};B}(x,0,s)\right]\partial\hat{J}_y^{\mathrm{s};A}(x,s)\mathrm{d}x$$
$$+ \int_{x=c}^{d} \left[\hat{E}_y^{\mathrm{T};B}(x,z_B,s) + \hat{E}_y^{\mathrm{T};A}(x,z_B,s)\right]\partial\hat{J}_y^{\mathrm{s};B}(x,s)\mathrm{d}x$$
$$= \int_{x=a}^{b} \hat{E}_y^{\mathrm{s}}(x,0,s)\partial\hat{J}_y^{\mathrm{T};A}(x,s)\mathrm{d}x + \int_{x=c}^{d} \hat{E}_y^{\mathrm{s}}(x,z_B,s)\partial\hat{J}_y^{\mathrm{T};B}(x,s)\mathrm{d}x, \quad (10.1)$$

which can be viewed as a straightforward extension of the reciprocity relation (8.1) pertaining to a single plane screen. In (10.1), $\partial\hat{J}_y^{\mathrm{s};A}(x,s)$ and $\partial\hat{J}_y^{\mathrm{s};B}(x,s)$ denote the induced electric-current surface density on screen $\mathcal{S}_A$ and screen $\mathcal{S}_B$, respectively. A similar meaning has the EM-field quantities pertaining to the testing-field state. In particular, $\hat{E}_y^{\mathrm{T};A}(x,z,s)$ denotes the testing electric-field strength at (x,z) due to $\partial\hat{J}_y^{\mathrm{T};A}(x,s)$ distributed on screen $\mathcal{S}_A$ along $z=0$. A similar definition applies to $\hat{E}_y^{\mathrm{T};B}(x,z,s)$ that is hence radiated by $\partial\hat{J}_y^{\mathrm{T};B}(x,s)$ located along $z=z_B$ on screen $\mathcal{S}_B$.

In order to solve (10.1) numerically using the CdH-MoM, the wave-slowness representation (4.33) is used in the s-domain reciprocity relation (10.1) to get its equivalent form

$$\frac{s}{2\pi\mathrm{i}}\int_{\kappa=-\mathrm{i}\infty}^{\mathrm{i}\infty} \left[\tilde{E}_y^{\mathrm{T};A}(\kappa,0,s) + \tilde{E}_y^{\mathrm{T};B}(\kappa,0,s)\right]\partial\tilde{J}_y^{\mathrm{s};A}(-\kappa,s)\mathrm{d}\kappa$$
$$+ \frac{s}{2\pi\mathrm{i}}\int_{\kappa=-\mathrm{i}\infty}^{\mathrm{i}\infty} \left[\tilde{E}_y^{\mathrm{T};B}(\kappa,z_B,s) + \tilde{E}_y^{\mathrm{T};A}(\kappa,z_B,s)\right]\partial\tilde{J}_y^{\mathrm{s};B}(-\kappa,s)\mathrm{d}\kappa$$
$$= \frac{s}{2\pi\mathrm{i}}\int_{\kappa=-\mathrm{i}\infty}^{\mathrm{i}\infty} \tilde{E}_y^{\mathrm{s}}(\kappa,0,s)\partial\tilde{J}_y^{\mathrm{T};A}(-\kappa,s)\mathrm{d}\kappa$$
$$+ \frac{s}{2\pi\mathrm{i}}\int_{\kappa=-\mathrm{i}\infty}^{\mathrm{i}\infty} \tilde{E}_y^{\mathrm{s}}(\kappa,z_B,s)\partial\tilde{J}_y^{\mathrm{T};B}(-\kappa,s)\mathrm{d}\kappa$$
$$\text{for } \{s\in\mathbb{R}; s>0\}. \quad (10.2)$$

On the left-hand side of the latter reciprocity relation, we can recognize the self-interaction terms pertaining to screen $\mathcal{S}_A$

$$\frac{s}{2\pi\mathrm{i}}\int_{\kappa=-\mathrm{i}\infty}^{\mathrm{i}\infty}\tilde{E}_y^{\mathrm{T};A}(\kappa,0,s)\partial\tilde{J}_y^{\mathrm{s};A}(-\kappa,s)\mathrm{d}\kappa, \tag{10.3a}$$

and to screen $\mathcal{S}_B$,

$$\frac{s}{2\pi\mathrm{i}}\int_{\kappa=-\mathrm{i}\infty}^{\mathrm{i}\infty}\tilde{E}_y^{\mathrm{T};B}(\kappa,z_B,s)\partial\tilde{J}_y^{\mathrm{s};B}(-\kappa,s)\mathrm{d}\kappa. \tag{10.3b}$$

Further, the mutual EM-field coupling between the screens is captured by the remaining two interaction terms

$$\frac{s}{2\pi\mathrm{i}}\int_{\kappa=-\mathrm{i}\infty}^{\mathrm{i}\infty}\tilde{E}_y^{\mathrm{T};B}(\kappa,0,s)\partial\tilde{J}_y^{\mathrm{s};A}(-\kappa,s)\mathrm{d}\kappa, \tag{10.3c}$$

and

$$\frac{s}{2\pi\mathrm{i}}\int_{\kappa=-\mathrm{i}\infty}^{\mathrm{i}\infty}\tilde{E}_y^{\mathrm{T};A}(\kappa,z_B,s)\partial\tilde{J}_y^{\mathrm{s};B}(-\kappa,s)\mathrm{d}\kappa. \tag{10.3d}$$

To solve the starting reciprocity relation numerically, the space–time solution domain has to be discretized. The spatial solution domains, $\mathcal{S}_A$ and $\mathcal{S}_B$, as well as the time axis $\{t \in \mathbb{R}; t > 0\}$ are discretized uniformly. The uniform discretization is not mandatory, but it greatly simplifies the resulting solution scheme and hence its code implementation. Accordingly, the spatial domain $\mathcal{S}_A$ occupying by the first screen is divided into segments with the constant length $\Delta_A = w_A/(N_A - 1)$, so that

$$\Delta\mathcal{S}_A = \left\{x_n = a + (n-1)\Delta_A, z = 0\right\} \quad \text{for } n = \{1,\ldots,N_A\}, \tag{10.4a}$$

where $\Delta\mathcal{S}_A$ denotes the partition of $\mathcal{S}_A$ and N_A is the number of nodes along the uniform grid. In a similar manner, the spatial domain $\mathcal{S}_B$ occupied by the second screen is represented by the partition $\Delta\mathcal{S}_B \subset \mathcal{S}_B$ that consists of N_B nodes according to

$$\Delta\mathcal{S}_B = \left\{x_n = c + (n - N_A - 1)\Delta_B, z = z_B\right\}$$
$$\text{for } n = \{N_A+1,\ldots,N_A+N_B\}, \tag{10.4b}$$

where $\Delta_B = w_B/(N_B - 1)$. Furthermore, the time axis is discretized as $\{t_k = k\Delta t; k = 1,2,\ldots,M\}$, where $\Delta t > 0$ has the meaning of time step. Expanding next the induced electric-current surface density in a piecewise linear manner over the discretized space–time solution domain, while assuming the piecewise constant testing-source distribution (see (8.6) and (8.8)), the reciprocity relation can be cast into the matrix form as shown, symbolically, in Figure 10.2. Here, the impeditivity $[(N_A + N_B) \times (N_A + N_B)]$ array, $\hat{\boldsymbol{\mathcal{Z}}}$, interrelates the electric-current $[(N_A + N_B) \times 1]$ array $\hat{\boldsymbol{J}}$ of unknown electric-current coefficients (in A/m) with the voltage excitation $[(N_A + N_B) \times 1]$ array, denoted by $\hat{\boldsymbol{V}}$. The subarrays of the

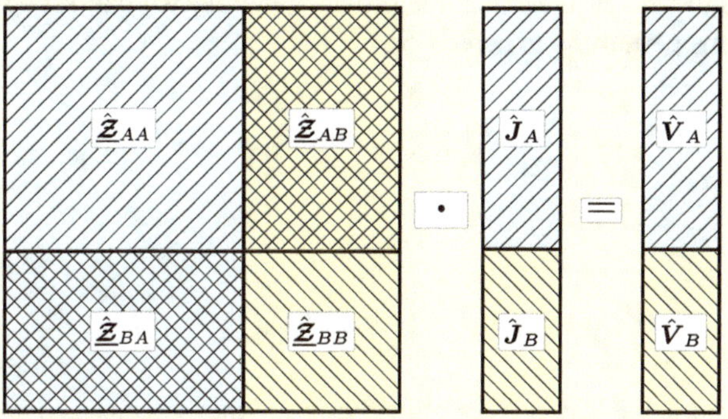

Figure 10.2 Symbolical partitioned matrix description of the EM coupling problem consisting of two conductive screens

impeditivity array $\hat{\underline{\mathcal{Z}}}$ can be associated with the interactions listed above via (10.3a)–(10.3d). Indeed, they can be interpreted as follows:

- $\hat{\underline{\mathcal{Z}}}_{AA}$ is the $[N_A \times N_A]$ impeditivity subarray representing the local interactions on screen $\mathcal{S}_A$ pertaining to (10.3a);
- $\hat{\underline{\mathcal{Z}}}_{BB}$ is the $[N_B \times N_B]$ impeditivity subarray representing the local interactions on screen $\mathcal{S}_B$ pertaining to (10.3b);
- $\hat{\underline{\mathcal{Z}}}_{BA}$ is the $[N_B \times N_A]$ impeditivity subarray representing the remote interactions from screen $\mathcal{S}_B$ to screen $\mathcal{S}_A$ pertaining to (10.3c);
- $\hat{\underline{\mathcal{Z}}}_{AB}$ is the $[N_A \times N_B]$ impeditivity subarray representing the remote interactions from screen $\mathcal{S}_A$ to screen $\mathcal{S}_B$ pertaining to (10.3d).

Whenever the assumed surrounding medium is self-adjoint in its EM properties [1, Section 1.4.1], the mutual-coupling impeditivity subarrays have the property

$$\hat{\mathcal{Z}}_{AB}^{[S,n]}(s) = \hat{\mathcal{Z}}_{BA}^{[n,S]}(s) \quad \text{for all } S = \{1, \ldots, N_A\} \text{ and } n = \{N_A + 1, \ldots, N_A + N_B\}, \tag{10.5}$$

by virtue of EM reciprocity (see [1, Section 7.2]). Further, the array of electric-current coefficients consists of two subarrays, $\hat{\boldsymbol{J}}_A$ and $\hat{\boldsymbol{J}}_B$, of dimensions $[N_A \times 1]$ and $[N_B \times 1]$, respectively. The former one contains the coefficients representing the induced electric-current surface density on $\mathcal{S}_A$ and $\hat{\boldsymbol{J}}_B$ has the similar meaning but pertains to $\mathcal{S}_B$. Likewise, the excitation array consists of two subarrays $\hat{\boldsymbol{V}}_A$ and $\hat{\boldsymbol{V}}_B$ that represent the first and second interaction terms on the right-hand side of (10.2), respectively.

Since the way how the testing field is related to its source depends on the properties of the background medium, we shall analyze a few illustrative problem configurations in the following sections.

10.2 Problem solution

The CdH inversion of self-coupling interactions (10.3a) and (10.3b) pertaining to a single bounded screen has been thoroughly analyzed in Chapters 8 and 9 and the corresponding Appendices. Therefore, we shall next be concerned with the remote-interaction terms (10.3c) and (10.3d) describing the mutual coupling between two screens in various settings.

10.2.1 EM coupling between screens in a homogeneous embedding

To begin with a relatively simple problem configuration, we shall study the EM coupling between a pair of coplanar screens located in an unbounded, homogeneous, isotropic and loss-free medium (see Figure 10.3). To that end, we may combine (10.3d) with $z_B = 0$ and (8.4) to get a slowness representation of (the elements of) the mutual-coupling array $\hat{\underline{\mathcal{Z}}}_{AB}$. Hence, we may write (cf. (B.1))

$$\hat{\mathcal{Z}}_{AB}^{[S,n]}(s) = \frac{\zeta_0}{sc_0\Delta t}\frac{1}{2\pi\mathrm{i}}\int_{\kappa=-\mathrm{i}\infty}^{\mathrm{i}\infty}\tilde{\Pi}^{[S]}(\kappa)\tilde{\Lambda}^{[n]}(-\kappa)\frac{\mathrm{d}\kappa}{2p(\kappa)}$$
$$\text{for all } S = \{1,\dots,N_A\} \text{ and } n = \{N_A+1,\dots,N_A+N_B\}, \tag{10.6}$$

assuming the indexes of the total impeditivity array (see Figure 10.2). We next follow the line of reasoning pursued in Appendix B and express the elements of the coupling array in terms of generic integral representations $\hat{\Upsilon}^{A,B}(x,s)$, the time-domain (TD) counterparts of which were derived in Sections B.1 and B.2. Using these results, we arrive at

$$\begin{aligned}\mathcal{Z}_{AB}^{[S,n]}(t) &= \frac{\zeta_0}{c_0\Delta t}\frac{1}{\Delta_B}\Big[\Upsilon^A(x_S - x_n + \Delta_B + \Delta_A/2, t)\\ &- \Upsilon^A(x_S - x_n + \Delta_B - \Delta_A/2, t) - 2\Upsilon^A(x_S - x_n + \Delta_A/2, t)\\ &+ 2\Upsilon^A(x_S - x_n - \Delta_A/2, t) + \Upsilon^A(x_S - x_n - \Delta_B + \Delta_A/2, t)\\ &- \Upsilon^A(x_S - x_n - \Delta_B - \Delta_A/2, t)\Big]\end{aligned}$$
$$\text{for all } S = \{2,\dots,N_A-1\} \text{ with } n = \{N_A+2,\dots,N_A+N_B-1\}, \tag{10.7a}$$

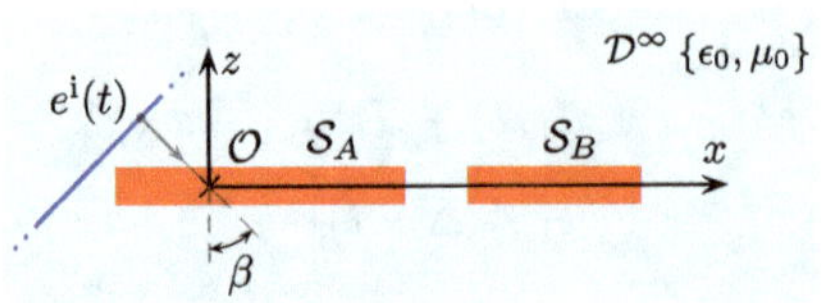

Figure 10.3 Pulsed EM plane wave impinging upon two coplanar strips embedded in the unbounded, homogeneous and isotropic medium

$$\mathcal{Z}_{AB}^{[S,N_A+1]}(t) = \frac{\zeta_0}{c_0\Delta t}\frac{1}{\Delta_B}\Big[-\Upsilon^A(x_S - x_{N_A+1} + \Delta_A/2, t) + \Upsilon^A(x_S - x_{N_A+1} - \Delta_A/2, t) + \Upsilon^A(x_S - x_{N_A+1} - \Delta_B + \Delta_A/2, t) - \Upsilon^A(x_S - x_{N_A+1} - \Delta_B - \Delta_A/2, t)\Big] + \frac{\zeta_0}{c_0\Delta t}\Big[\Upsilon^B(x_S - x_{N_A+1} + \Delta_A/2, t) - \Upsilon^B(x_S - x_{N_A+1} - \Delta_A/2, t)\Big] \quad \text{for all } S = \{2, \ldots, N_A - 1\}, \tag{10.7b}$$

$$\mathcal{Z}_{AB}^{[S,N_A+N_B]}(t) = \frac{\zeta_0}{c_0\Delta t}\frac{1}{\Delta_B}\Big[\Upsilon^A(x_S - x_{N_A+N_B} + \Delta_B + \Delta_A/2, t) - \Upsilon^A(x_S - x_{N_A+N_B} + \Delta_A/2, t) - \Upsilon^A(x_S - x_{N_A+N_B} + \Delta_B - \Delta_A/2, t) + \Upsilon^A(x_S - x_{N_A+N_B} - \Delta_A/2, t)\Big] - \frac{\zeta_0}{c_0\Delta t}\Big[\Upsilon^B(x_S - x_{N_A+N_B} + \Delta_A/2, t) - \Upsilon^B(x_S - x_{N_A+N_B} - \Delta_A/2, t)\Big] \quad \text{for all } S = \{2, \ldots, N_A - 1\}, \tag{10.7c}$$

$$\begin{aligned}\mathcal{Z}_{AB}^{[1,n]}(t) = \frac{\zeta_0}{c_0\Delta t}\frac{1}{\Delta_B}\Big[&\Upsilon^A(x_1 - x_n + \Delta_B + \Delta_A/2, t) - \Upsilon^A(x_1 - x_n + \Delta_B, t)\\ &- 2\Upsilon^A(x_1 - x_n + \Delta_A/2, t) + 2\Upsilon^A(x_1 - x_n, t)\\ &+ \Upsilon^A(x_1 - x_n - \Delta_B + \Delta_A/2, t) - \Upsilon^A(x_1 - x_n - \Delta_B, t)\Big]\\ &\text{for all } n = \{N_A + 2, \ldots, N_A + N_B - 1\},\end{aligned} \tag{10.7d}$$

and

$$\begin{aligned}\mathcal{Z}_{AB}^{[N_A,n]}(t) = \frac{\zeta_0}{c_0\Delta t}\frac{1}{\Delta_B}\Big[&\Upsilon^A(x_{N_A} - x_n + \Delta_B, t)\\ &- \Upsilon^A(x_{N_A} - x_n + \Delta_B - \Delta_A/2, t) - 2\Upsilon^A(x_{N_A} - x_n, t)\\ &+ 2\Upsilon^A(x_{N_A} - x_n - \Delta_A/2, t) + \Upsilon^A(x_{N_A} - x_n - \Delta_B, t)\\ &- \Upsilon^A(x_{N_A} - x_n - \Delta_B - \Delta_A/2, t)\Big]\\ &\text{for all } n = \{N_A + 2, \ldots, N_A + N_B - 1\},\end{aligned} \tag{10.7e}$$

with

$$\begin{aligned}\mathcal{Z}_{AB}^{[1,N_A+1]}(t) = \frac{\zeta_0}{c_0\Delta t}\frac{1}{\Delta_B}\Big[&-\Upsilon^A(x_1 - x_{N_A+1} + \Delta_A/2, t)\\ &+ \Upsilon^A(x_1 - x_{N_A+1}, t) + \Upsilon^A(x_1 - x_{N_A+1} - \Delta_B + \Delta_A/2, t)\\ &- \Upsilon^A(x_1 - x_{N_A+1} - \Delta_B, t)\Big]\\ &+ \frac{\zeta_0}{c_0\Delta t}\Big[\Upsilon^B(x_1 - x_{N_A+1} + \Delta_A/2, t) - \Upsilon^B(x_1 - x_{N_A+1}, t)\Big],\end{aligned} \tag{10.7f}$$

$$\mathcal{Z}_{AB}^{[N_A,N_A+1]}(t) = \frac{\zeta_0}{c_0\Delta t}\frac{1}{\Delta_B}\Big[-\Upsilon^A(x_{N_A}-x_{N_A+1},t)$$
$$+\Upsilon^A(x_{N_A}-x_{N_A+1}-\Delta_A/2,t)+\Upsilon^A(x_{N_A}-x_{N_A+1}-\Delta_B,t)$$
$$-\Upsilon^A(x_{N_A}-x_{N_A+1}-\Delta_B-\Delta_A/2,t)\Big]+\frac{\zeta_0}{c_0\Delta t}\Big[\Upsilon^B(x_{N_A}-x_{N_A+1},t)$$
$$-\Upsilon^B(x_{N_A}-x_{N_A+1}-\Delta_A/2,t)\Big], \tag{10.7g}$$

$$\mathcal{Z}_{AB}^{[1,N_A+N_B]}(t) = \frac{\zeta_0}{c_0\Delta t}\frac{1}{\Delta_B}\Big[\Upsilon^A(x_1-x_{N_A+N_B}+\Delta_B+\Delta_A/2,t)$$
$$-\Upsilon^A(x_1-x_{N_A+N_B}+\Delta_B,t)-\Upsilon^A(x_1-x_{N_A+N_B}+\Delta_A/2,t)$$
$$+\Upsilon^A(x_1-x_{N_A+N_B},t)\Big]-\frac{\zeta_0}{c_0\Delta t}\Big[\Upsilon^B(x_1-x_{N_A+N_B}+\Delta_A/2,t)$$
$$-\Upsilon^B(x_1-x_{N_A+N_B},t)\Big], \tag{10.7h}$$

and

$$\mathcal{Z}_{AB}^{[N_A,N_A+N_B]}(t) = \frac{\zeta_0}{c_0\Delta t}\frac{1}{\Delta_B}\Big[\Upsilon^A(x_{N_A}-x_{N_A+N_B}+\Delta_B,t)$$
$$-\Upsilon^A(x_{N_A}-x_{N_A+N_B}+\Delta_B-\Delta_A/2,t)-\Upsilon^A(x_{N_A}-x_{N_A+N_B},t)$$
$$+\Upsilon^A(x_{N_A}-x_{N_A+N_B}-\Delta_A/2,t)\Big]-\frac{\zeta_0}{c_0\Delta t}\Big[\Upsilon^B(x_{N_A}-x_{N_A+N_B},t)$$
$$-\Upsilon^B(x_{N_A}-x_{N_A+N_B}-\Delta_A/2,t)\Big]. \tag{10.7i}$$

Having determined the elements of $\underline{\mathcal{Z}}_{AB}$ via (10.7a)–(10.7i), the elements of $\underline{\mathcal{Z}}_{BA}$ immediately follow upon applying (10.5). Recall that the elements of self-coupling impeditivity arrays, $\underline{\mathcal{Z}}_{AA}$ and $\underline{\mathcal{Z}}_{BB}$, pertaining to isolated screens $\mathcal{S}_A$ and $\mathcal{S}_B$, respectively, can be calculated using the results introduced in Appendix B. Finally, to fully specify the system depicted in Figure 10.2, it remains to define the elements of the excitation array. This can be accomplished via a straightforward extension of (8.13a)–(8.14) applying to a single screen. Indeed, the elements of $\boldsymbol{V}_A$ are given by

$$V_A^{[S]}(t) = \frac{1}{\kappa_0}\left\{\int_{\tau=0}^{t-\kappa_0(x_S-\Delta_A/2)} e^{\mathrm{i}}(\tau)\mathrm{d}\tau - \int_{\tau=0}^{t-\kappa_0(x_S+\Delta_A/2)} e^{\mathrm{i}}(\tau)\mathrm{d}\tau\right\}$$
$$\text{for all } S=\{2,\ldots,N_A-1\}, \tag{10.8a}$$

with

$$V_A^{[1]}(t) = \frac{1}{\kappa_0}\left\{\int_{\tau=0}^{t-\kappa_0 x_1} e^{\mathrm{i}}(\tau)\mathrm{d}\tau - \int_{\tau=0}^{t-\kappa_0(x_1+\Delta_A/2)} e^{\mathrm{i}}(\tau)\mathrm{d}\tau\right\}, \tag{10.8b}$$

$$V_A^{[N_A]}(t) = \frac{1}{\kappa_0}\left\{\int_{\tau=0}^{t-\kappa_0(x_{N_A}-\Delta_A/2)} e^{\mathrm{i}}(\tau)\mathrm{d}\tau - \int_{\tau=0}^{t-\kappa_0 x_{N_A}} e^{\mathrm{i}}(\tau)\mathrm{d}\tau\right\}. \tag{10.8c}$$

In a similar manner, the elements of $\boldsymbol{V}_B$ can be written as

$$V_B^{[S]}(t) = \frac{1}{\kappa_0}\left\{\int_{\tau=0}^{t-\kappa_0(x_S-\Delta_B/2)} e^{\mathrm{i}}(\tau)\mathrm{d}\tau - \int_{\tau=0}^{t-\kappa_0(x_S+\Delta_B/2)} e^{\mathrm{i}}(\tau)\mathrm{d}\tau\right\}$$
$$\text{for all } S=\{N_A+2,\ldots,N_A+N_B-1\}, \tag{10.9a}$$

with

$$V_B^{[N_A+1]}(t) = \frac{1}{\kappa_0}\left\{\int_{\tau=0}^{t-\kappa_0 x_{N_A+1}} e^{\mathrm{i}}(\tau)\mathrm{d}\tau - \int_{\tau=0}^{t-\kappa_0(x_{N_A+1}+\Delta_B/2)} e^{\mathrm{i}}(\tau)\mathrm{d}\tau\right\}, \tag{10.9b}$$

$$V_B^{[N_A+N_B]}(t) = \frac{1}{\kappa_0}\left\{\int_{\tau=0}^{t-\kappa_0(x_{N_A+N_B}-\Delta_B/2)} e^{\mathrm{i}}(\tau)\mathrm{d}\tau - \int_{\tau=0}^{t-\kappa_0 x_{N_A+N_B}} e^{\mathrm{i}}(\tau)\mathrm{d}\tau\right\}. \tag{10.9c}$$

for the edge nodes. The limits corresponding to the normal incidence, characterized by $\kappa_0 = 0$, directly follow as (cf. (8.14))

$$V_A^{[S]}(t) = \begin{cases} e^{\mathrm{i}}(t)\Delta_A & \text{for } S=\{2,\ldots,N_A-1\}, \\ e^{\mathrm{i}}(t)\Delta_A/2 & \text{for } S=\{1,N_A\}, \end{cases} \tag{10.10}$$

and

$$V_B^{[S]}(t) = \begin{cases} e^{\mathrm{i}}(t)\Delta_B & \text{for } S=\{N_A+2,\ldots,N_A+N_B-1\}, \\ e^{\mathrm{i}}(t)\Delta_B/2 & \text{for } S=\{N_A+1,N_A+N_B\}. \end{cases} \tag{10.11}$$

The thus fully determined system of equations in the TD has the form of (8.10), again, which can be solved using the marching-on-in-time technique. In this manner, we end up with electric-current coefficients describing the desired electric-current surface density on both $\mathcal{S}_A$ and $\mathcal{S}_B$ (see (8.11)).

The problem configuration with a nonzero vertical distance between the screens, $|z_B| > 0$ (see Figure 10.1), can be readily handled with the aid of TD functions derived in Appendix F. Indeed, employing the following relation between the testing field and its source (cf. (8.4)), viz

$$\tilde{E}_y^{\mathrm{T};A}(\kappa, z_B, s) = -[\mu_0/2p(\kappa)]\partial\tilde{J}_y^{\mathrm{T};A}(\kappa, s)\exp[-sp(\kappa)|z_B|], \tag{10.12}$$

in (10.3d), the (elements of the) mutual-coupling impeditivity array can be represented via (cf. (10.6))

$$\hat{\mathcal{Z}}_{AB}^{[S,n]}(s) = \frac{\zeta_0}{sc_0\Delta t}\frac{1}{2\pi\mathrm{i}}\int_{\kappa=-\mathrm{i}\infty}^{\mathrm{i}\infty} \tilde{\Pi}^{[S]}(\kappa)\tilde{\Lambda}^{[n]}(-\kappa)\exp[-sp(\kappa)|z_B|]\frac{\mathrm{d}\kappa}{2p(\kappa)}$$
$$\text{for all } S=\{1,\ldots,N_A\} \text{ and } n=\{N_A+1,\ldots,N_A+N_B\}. \tag{10.13}$$

Consequently, the pertinent expressions for defining the mutual-coupling impeditivity array can be derived from (10.7a) to (10.7i), upon replacing $\Upsilon^{A,B}(x,t)$ with $\Pi^{A,B}(x,|z_B|,t)$, respectively (see Sections F.1 and F.2). The shift of $\mathcal{S}_B$ must also be accounted for in the corresponding excitation subarray. Therefore, we substitute $t + p(\kappa_0)z_B$ for t on the right-hand side of (10.9a)–(10.9c) and (10.11), thereby incorporating the propagation time of the incident plane wave to traverse the vertical offset.

10.2.1.1 Narrow-screen approximation

We shall next consider the problem configuration consisting of two thin PEC screens whose width $w > 0$ is supposed to be relatively small with respect to the spatial support of the exciting plane-wave pulse shape. The center of the first screen is located at the origin so that $\mathcal{S}_A = \{-w/2 < x < w/2, -\infty < y < \infty, -\delta/2 < z < \delta/2\}$. The second screen is located at (x_B, z_B) and hence occupies $\mathcal{S}_B = \{x_B - w/2 < x < x_B + w/2, -\infty < y < \infty, z_B - \delta/2 < z < z_B + \delta/2\}$ (see Figure 10.4).

The problem is formulated with the aid of the EM reciprocity theorem, again. Accordingly, the starting reciprocity relation (10.1) is modified accordingly:

$$\begin{aligned}&\int_{x=-w/2}^{w/2}\left[\hat{E}_y^{\mathrm{T};A}(x,0,s) + \hat{E}_y^{\mathrm{T};B}(x,0,s)\right]\partial\hat{J}_y^{\mathrm{s};A}(x,s)\mathrm{d}x\\&+\int_{x=x_B-w/2}^{x_B+w/2}\left[\hat{E}_y^{\mathrm{T};B}(x,z_B,s) + \hat{E}_y^{\mathrm{T};A}(x,z_B,s)\right]\partial\hat{J}_y^{\mathrm{s};B}(x,s)\mathrm{d}x\\&=\int_{x=-w/2}^{w/2}\hat{E}_y^{\mathrm{s}}(x,0,s)\partial\hat{J}_y^{\mathrm{T};A}(x,s)\mathrm{d}x + \int_{x=x_B-w/2}^{x_B+w/2}\hat{E}_y^{\mathrm{s}}(x,z_B,s)\partial\hat{J}_y^{\mathrm{T};B}(x,s)\mathrm{d}x.\end{aligned} \tag{10.14}$$

The testing fields on the left-hand side of the reciprocity relation are next determined through the wave-slowness representation (see (4.33)), again. Hence, assuming the uniform spatial testing-source distribution over $\mathcal{S}_A$ and $\mathcal{S}_B$ (cf. (7.2)), i.e.

$$\partial\hat{J}_y^{\mathrm{T};A}(x,s) = \frac{\hat{I}^{\mathrm{T};A}(s)}{w}\left[\mathrm{H}(x+w/2) - \mathrm{H}(x-w/2)\right], \tag{10.15a}$$

$$\partial\hat{J}_y^{\mathrm{T};B}(x,s) = \frac{\hat{I}^{\mathrm{T};B}(s)}{w}\left[\mathrm{H}(x-x_B+w/2) - \mathrm{H}(x-x_B-w/2)\right], \tag{10.15b}$$

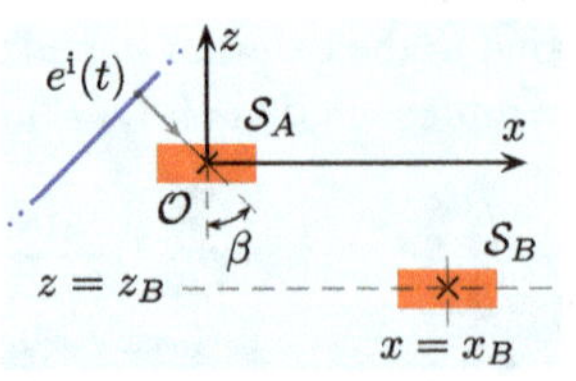

Figure 10.4 Pulsed EM plane wave impinging upon two thin and narrow strips

we may follow the procedure described in Section 7.1 to arrive at the slowness representations of the corresponding testing fields (cf. (7.5))

$$\hat{E}_y^{\mathrm{T};A}(x,z,s) = -\frac{s\mu_0 \hat{I}^{\mathrm{T};A}(s)}{2\pi \mathrm{i}} \int_{\kappa=-\mathrm{i}\infty}^{\mathrm{i}\infty} \exp\{-s[\kappa x + p(\kappa)|z|]\} \times \mathrm{i}_0(s\kappa w/2)\frac{\mathrm{d}\kappa}{2p(\kappa)}, \tag{10.16a}$$

$$\hat{E}_y^{\mathrm{T};B}(x,z,s) = -\frac{s\mu_0 \hat{I}^{\mathrm{T};B}(s)}{2\pi \mathrm{i}} \int_{\kappa=-\mathrm{i}\infty}^{\mathrm{i}\infty} \exp\{-s[\kappa (x-x_B) + p(\kappa)|z-z_B|]\} \times \mathrm{i}_0(s\kappa w/2)\frac{\mathrm{d}\kappa}{2p(\kappa)} \quad \text{for } \{s \in \mathbb{R}; s > 0\}, \tag{10.16b}$$

where $\mathrm{i}_0(x)$ denotes the modified spherical Bessel function of the first kind. The slowness representations of testing fields are further substituted in the left-hand side of the reciprocity relation (10.14), where we change the order of the integrations with respect to κ and x and, subsequently, evaluate the inner integral with respect to x. Since the induced electric-current surface density is assumed to be uniformly distributed much as the testing source (see (10.15a) and (10.15b)), the integration can be carried out in closed form. In the next step, we substitute (10.15a) and (10.15b) in the right-hand side of (10.14) and employ the explicit-type boundary conditions applying to PEC surfaces $\mathcal{S}_A$ and $\mathcal{S}_B$. In this way, assuming the uniform EM plane wave specified by (4.1), we get

$$\int_{x=-w/2}^{w/2} \hat{E}_y^{\mathrm{s}}(x,0,s)\partial \hat{J}_y^{\mathrm{T};A}(x,s)\mathrm{d}x = -\hat{E}_y^{\mathrm{i}}(0,0,s)\,\mathrm{i}_0(s\kappa_0 w/2), \tag{10.17a}$$

$$\int_{x=x_B-w/2}^{x_B+w/2} \hat{E}_y^{\mathrm{s}}(x,z_B,s)\partial \hat{J}_y^{\mathrm{T};B}(x,s)\mathrm{d}x = -\hat{E}_y^{\mathrm{i}}(x_B,z_B,s)\,\mathrm{i}_0(s\kappa_0 w/2). \tag{10.17b}$$

Finally, upon collecting the results, we end up with the following system of equations in the s-domain (cf. Figure 10.2):

$$\begin{bmatrix} \hat{Z}_{AA}(s) & \hat{Z}_{AB}(s) \\ \hat{Z}_{BA}(s) & \hat{Z}_{BB}(s) \end{bmatrix} \cdot \begin{bmatrix} \hat{I}^{\mathrm{s};A}(s) \\ \hat{I}^{\mathrm{s};B}(s) \end{bmatrix} = \begin{bmatrix} \hat{E}_y^{\mathrm{i}}(0,0,s) \\ \hat{E}_y^{\mathrm{i}}(x_B,z_B,s) \end{bmatrix}, \tag{10.18}$$

where we used $\mathrm{i}_0(x) = 1 + O(x^2)$ as $x \downarrow 0$ to approximate the right-hand sides of (10.17a) and (10.17b) for $\kappa_0 \neq 0$. In system (10.18), $\hat{Z}_{AA}(s)$ and $\hat{Z}_{BB}(s)$ have the meaning of self-coupling impedances per unit length. As a matter of fact, impedances $\hat{Z}_{AA}(s)$ and $\hat{Z}_{BB}(s)$ correspond to the external impedances of screens $\mathcal{S}_A$ and $\mathcal{S}_B$, respectively, and can be hence represented via (cf. (7.8a))

$$\hat{Z}_{AA}(s) = \hat{Z}_{BB}(s) = \frac{s\mu_0}{2\pi \mathrm{i}} \int_{\kappa=-\mathrm{i}\infty}^{\mathrm{i}\infty} \mathrm{i}_0^2(s\kappa w/2)\frac{\mathrm{d}\kappa}{2p(\kappa)}, \tag{10.19}$$

where $p(\kappa) = (1/c_0^2 - \kappa^2)^{1/2}$ with $\mathrm{Re}(p) \geq 0$. Furthermore, the EM interaction between the narrow screens is characterized by the mutual-coupling impedances,

$\hat{Z}_{AB}(s)$ and $\hat{Z}_{BA}(s)$, that are, in virtue of EM reciprocity, equal to each other. Hence, their integral representations have the following form:

$$\hat{Z}_{AB}(s) = \frac{s\mu_0}{2\pi\mathrm{i}}\int_{\kappa=-\mathrm{i}\infty}^{\mathrm{i}\infty} \exp\{-s[\kappa x_B + p(\kappa)|z_B|]\}\,\mathrm{i}_0^2(s\kappa w/2)\frac{\mathrm{d}\kappa}{2p(\kappa)}. \tag{10.20}$$

Since the transformation of $\hat{Z}_{AA}(s)\hat{I}^{\mathrm{s};A}(s)$ and $\hat{Z}_{BB}(s)\hat{I}^{\mathrm{s};B}(s)$ back to the TD has been, in fact, solved in Section 7.1.2, we shall further focus on the mutual-coupling terms $\hat{Z}_{AB}(s)\hat{I}^{\mathrm{s};B}(s)$ and $\hat{Z}_{BA}(s)\hat{I}^{\mathrm{s};A}(s)$ (see (10.18)). To that end, we shall first approximate the integration in (10.20). Supposing that screen $\mathcal{S}_B$ is located not too close to $\mathcal{S}_A$, we may use $\mathrm{i}_0^2(x) = 1 + O(x^2)$ as $x \downarrow 0$ in (10.20) and get a slowness representation that can be handled via the standard CdH technique (see Section 2.1.1). Subsequently, the integrand of the thus obtained approximate integral representation is analytically continued away from $\mathrm{Re}(\kappa) = 0$ and the integration contour is replaced with the CdH path defined via

$$\kappa x_B + p(\kappa)|z_B| = \tau, \tag{10.21}$$

where τ is the real-valued and positive time parameter. Mapping further the variable of integration from κ to τ, the mutual impedance can be expressed through the modified Bessel function of the second kind, that is,

$$\hat{Z}_{AB}(s) \simeq \frac{s\mu_0}{2\pi}\int_{\tau=T_B}^{\infty} \exp(-s\tau)\frac{\mathrm{d}\tau}{(\tau^2 - T_B^2)^{1/2}} = \frac{s\mu_0}{2\pi}\mathrm{K}_0(sT_B), \tag{10.22}$$

where we used $T_B = (x_B^2 + z_B^2)^{1/2}/c_0 > 0$ and the integral formula [2, Eq. (29.3.119)]. With the analytical expression for the s-domain mutual impedance at our disposal, we shall proceed with the solution of (10.18) in the TD. For this purpose, the induced electric current induced on $\mathcal{S}_A$ and $\mathcal{S}_B$ is expanded in a set of triangular functions, which can be expressed as (cf. (7.18))

$$\hat{I}^{\mathrm{s};A}(s) \simeq \sum_{k=1}^{M} i_k^A \hat{\Lambda}_k(s), \tag{10.23a}$$

$$\hat{I}^{\mathrm{s};B}(s) \simeq \sum_{k=1}^{M} i_k^B \hat{\Lambda}_k(s), \tag{10.23b}$$

where i_k^A and i_k^B are yet unknown electric-current coefficients (in A). Finally, making use of (10.19) and (10.22)–(10.23b) in (10.18) and transforming the result to the TD, we obtain a convolution-type system of equations (cf. (7.42) and (8.10))

$$\sum_{k=1}^{m} \left(\underline{\boldsymbol{Z}}_{m-k+1} - 2\underline{\boldsymbol{Z}}_{m-k} + \underline{\boldsymbol{Z}}_{m-k-1}\right) \cdot \boldsymbol{I}_k = \boldsymbol{E}_m, \tag{10.24}$$

where $\underline{\boldsymbol{Z}}_k = \underline{\boldsymbol{Z}}(t_k)$ is a $[2 \times 2]$ impedance array that is given by

$$\underline{\boldsymbol{Z}}_k = \frac{\zeta_0}{c_0\Delta t}\begin{bmatrix} \Phi(t_k) & \Psi(t_k) \\ \Psi(t_k) & \Phi(t_k) \end{bmatrix}. \tag{10.25}$$

The TD functions can be expressed as

$$\Phi(t) = [\Upsilon^B(w,t) - 2\Upsilon^B(0,t) + \Upsilon^B(-w,t)]/w^2, \tag{7.21 revisited}$$
$$\Psi(t) = \cosh^{-1}(t/T_B)\mathrm{H}(t - T_B)/2\pi, \tag{10.26}$$

and recall that $\Upsilon^B(x,t)$ is given by (B.9). Further, $\boldsymbol{I}_k$ is a $[2 \times 1]$ electric-current array containing the desired current coefficients

$$\boldsymbol{I}_k = \begin{bmatrix} i_k^A \\ i_k^B \end{bmatrix}, \tag{10.27}$$

and finally,

$$\boldsymbol{E}_m = \begin{bmatrix} e^{\mathrm{i}}(t_m) \\ e^{\mathrm{i}}[t_m - \kappa_0 x_B + p(\kappa_0) z_B] \end{bmatrix} \tag{10.28}$$

denotes the corresponding excitation array. Thanks to the universal property of causality, the system of TD equations (10.24) can be solved for the unknown coefficients with the aid of the marching-on-in-time technique. Pursuing this approach, we end up with (cf. (7.44) and (8.11))

$$\boldsymbol{I}_m = \underline{\boldsymbol{Z}}_1^{-1} \cdot \left[\boldsymbol{E}_m - \sum_{k=1}^{m-1} \left(\underline{\boldsymbol{Z}}_{m-k+1} - 2\underline{\boldsymbol{Z}}_{m-k} + \underline{\boldsymbol{Z}}_{m-k-1}\right) \cdot \boldsymbol{I}_k\right]$$
$$\text{for all } m = \{1, \ldots, M\}, \tag{10.29}$$

where $\underline{\boldsymbol{Z}}_1^{-1}$ denotes the inverse of the impedance matrix at $t = \Delta t$.

10.2.1.2 Numerical examples

The results presented in Section 10.2.1 are first used to calculate the induced electric-current response on two PEC screens located along $\mathcal{S}_A = \{0 < x < w, -\infty < y < \infty, z = 0\}$ and $\mathcal{S}_B = \{w/3 < x < 4w/3, -\infty < y < \infty, z = z_B = -1.50w\}$ (see Figure 10.1). The screens are irradiated by the impulsive plane EM wave with the triangular signature, $e^{\mathrm{i}}(t)$ (see Figure 8.3), and the angle of incidence is $\beta = \pi/4$. The excitation pulse-time width is taken such that $c_0 t_{\mathrm{w}}/w = 1.0$. The pertaining CdH-MoM solution accounting for the EM coupling between the screens is shown in Figure 10.5. In addition, we calculate the corresponding electric-current response on $\mathcal{S}_A$ in (a) the presence an infinite PEC plane lying at the level of $\mathcal{S}_B$ (grounded $\mathcal{S}_A$) and (b) in the absence of $\mathcal{S}_B$ ($\mathcal{S}_A$ only) (see Sections 8.1.2 and 9.1.2). As can be observed in Figure 10.5, the early-time parts of the resulting responses are identical. The differences with respect to the referential configurations appear at $t/t_{\mathrm{w}} = (7/2)\sin(\pi/4) \simeq 2.48$ and at $t/t_{\mathrm{w}} = (11/6)\sin(\pi/4) + \sqrt{41/18} \simeq 2.81$ when the wave reflected against the infinite ground plane and the cylindrical diffracted wave emanating from the edge at $(w/3, -1.50w)$ arrive at the field point $(w/2, 0)$, respectively.

Finally, in order to validate the narrow-screen approximation (see Section 10.2.1.1), we calculate the induced electric-current on two narrow screens

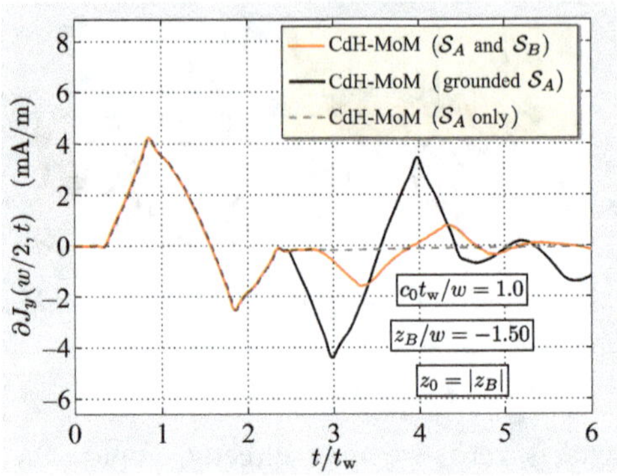

Figure 10.5 Plane-wave induced electric-current surface density at the center of screen $\mathcal{S}_A$

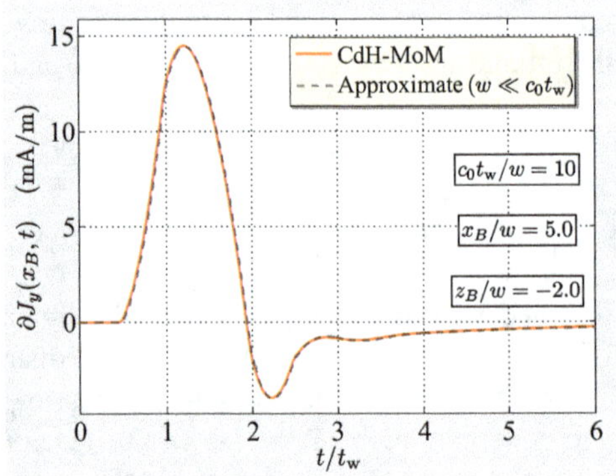

Figure 10.6 Plane-wave induced electric-current surface density at the center of narrow screen $\mathcal{S}_B$

occupying $\mathcal{S}_A = \{-w/2 < x < w/2, z = 0\}$ and $\mathcal{S}_B = \{x_B - w/2 < x < x_B + w/2, z = z_B\}$ with $(x_B, z_B) = (5.0w, -2.0w)$. The width of the screens, $w > 0$, is supposed to be a tenth of $c_0 t_w$, that is, $c_0 t_w/w = 10$. The incident plane wave has the shape of bipolar triangle, again, and $\beta = \pi/4$. Figure 10.6 shows the calculated electric-current surface density at the center of screen $\mathcal{S}_B$. Apparently, the narrow-strip approximation correlates with the relevant CdH-MoM solution (with $N_A = N_B = 2$) very well.

10.2.2 EM coupling between screens on a dielectric half-space

In this section, we shall describe the EM coupling between two screens lying on the surface of a dielectric half-space (see Figure 10.7). To that end, we employ the mutual interaction term (10.3d), again, to describe the elements of the corresponding mutual-coupling impeditivity subarray $\underline{\mathbf{Z}}_{AB}$. Since the vertical offset $|z_B|$ between

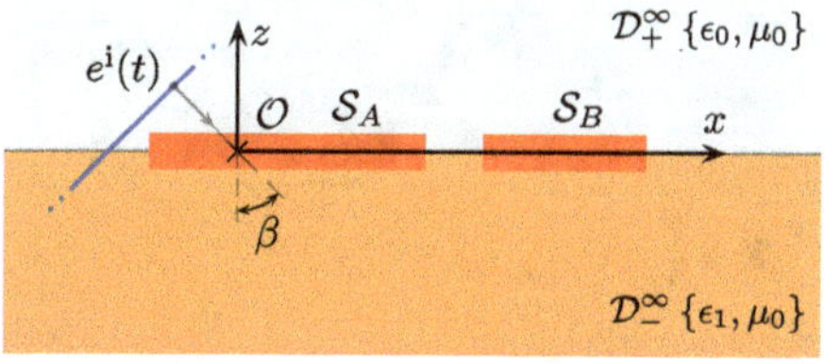

Figure 10.7 Pulsed EM plane wave impinging upon two coplanar strips located on a dielectric half-space

the interacting screens is zero, we may directly employ the transform-domain relation (9.24) and write

$$\tilde{E}_y^{\mathrm{T};A}(\kappa, 0, s) = -\frac{\mu_0}{p_0(\kappa) + p_1(\kappa)} \partial \tilde{J}_y^{\mathrm{T};A}(\kappa, s). \tag{10.30}$$

This relation is next substituted in (10.3d), which yields the slowness representation of (the elements of) the mutual-impeditivity array (cf. (10.6) and (G.1))

$$\hat{\mathcal{Z}}_{AB}^{[S,n]}(s) = \frac{\zeta_0}{sc_0\Delta t} \frac{1}{2\pi\mathrm{i}} \int_{\kappa=-\mathrm{i}\infty}^{\mathrm{i}\infty} \tilde{\Pi}^{[S]}(\kappa)\tilde{\Lambda}^{[n]}(-\kappa) \frac{\mathrm{d}\kappa}{p_0(\kappa) + p_1(\kappa)}$$
$$\text{for all } S = \{1, \ldots, N_A\} \text{ and } n = \{N_A + 1, \ldots, N_A + N_B\}, \tag{10.31}$$

where we assumed the piecewise-linear expansion of the induced electric-current surface density on $\Delta\mathcal{S}_B$ and the piecewise-constant spatial distribution of the testing source over $\Delta\mathcal{S}_A$. The elements of the mutual-coupling impeditivity array can be expanded in terms of generic integrals $\hat{\Lambda}^{A,B}(x, s)$, the TD counterparts of which can be found in Sections G.1 and G.2. Following these lines, we arrive at expressions similar to (10.7a)–(10.7i). As a matter of fact, the relevant expressions for all $\mathcal{Z}_{AB}^{[S,n]}(t)$ can be deduced directly from (10.7a)–(10.7i), again, upon replacing $\Upsilon^{A,B}(x, t)$ with $\Lambda^{A,B}(x, t)$, respectively. The effect of the dielectric half-space is further accounted for in the excitation array. The modified excitation array, say $\bar{\boldsymbol{V}}(t)$, consists, again, of $\bar{\boldsymbol{V}}_A(t)$ and $\bar{\boldsymbol{V}}_B(t)$ pertaining to $\mathcal{S}_A$ and $\mathcal{S}_B$, respectively, and can be evaluated via (cf. (9.29))

$$\bar{V}^{[S]}(t) = T_\perp V^{[S]}(t) \quad \text{for all } S = \{1, \ldots, N_A + N_B\}, \tag{9.29 revisited}$$

where elements $V^{[S]}(t)$ were specified before via (10.8a)–(10.11). Recall here that $T_\perp$ is the transmission coefficient given by (9.22). Employing these results, we finally end up with the convolution-type system of equations (8.10) whose solution yields the (time-dependent) arrays of coefficients representing the desired induced electric-current surface density on both $\mathcal{S}_A$ and $\mathcal{S}_B$.

10.2.2.1 Narrow-screen approximation

In this subsection, we shall analyze the EM mutual coupling between two coplanar narrow screens lying on the dielectric half-space. The strips occupy $\mathcal{S}_A = \{-w/2 <$

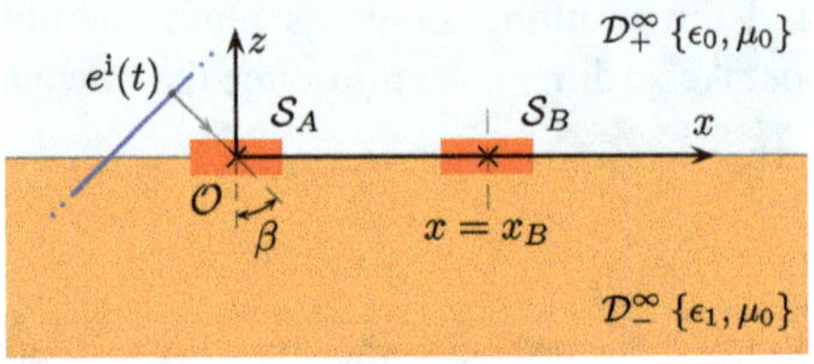

Figure 10.8 Pulsed EM plane wave impinging upon two coplanar narrow strips located on a dielectric half-space

$x < w/2, -\infty < y < \infty, -\delta/2 < z < \delta/2\}$ and $\mathcal{S}_B = \{x_B - w/2 < x < x_B + w/2, -\infty < y < \infty, -\delta/2 < z < \delta/2\}$, where $w > 0$ and $\delta > 0$ denote their (relatively small) width and thickness, respectively, and x_B is the shift of screen $\mathcal{S}_B$ with respect to the origin (see Figure 10.8).

The point of departure for the following analysis is the reciprocity relation (10.14), again, where we let $z_B = 0$. Under the assumption of uniform spatial distribution of both testing and induced electric-current densities on the screens (e.g. (10.15a) and (10.15b)), we may pursue the approach introduced in Section 10.2.1.1 and cast the reciprocity relation into its matrix form (cf. (10.18))

$$\begin{bmatrix} \hat{Z}_{AA}(s) & \hat{Z}_{AB}(s) \\ \hat{Z}_{BA}(s) & \hat{Z}_{BB}(s) \end{bmatrix} \cdot \begin{bmatrix} \hat{I}^{s;A}(s) \\ \hat{I}^{s;B}(s) \end{bmatrix} = \begin{bmatrix} T_\perp \hat{E}^{i}_y(0,0,s) \\ T_\perp \hat{E}^{i}_y(x_B,0,s) \end{bmatrix}. \tag{10.32}$$

Here, the self-coupling impedances $\hat{Z}_{AA}(s)$ and $\hat{Z}_{BB}(s)$ are, in fact, identical with the external impedance of a narrow strip on the dielectric half-space (see (9.31)) and can be hence represented as follows (cf. (10.19)):

$$\hat{Z}_{AA}(s) = \hat{Z}_{BB}(s) = \frac{s\mu_0}{2\pi \mathrm{i}} \int_{\kappa=-\mathrm{i}\infty}^{\mathrm{i}\infty} \mathrm{i}_0^2(s\kappa w/2) \frac{\mathrm{d}\kappa}{\mathrm{p}_0(\kappa) + \mathrm{p}_1(\kappa)}, \tag{10.33}$$

where the slowness parameters $\mathrm{p}_{0,1}(\kappa)$ were defined via (9.25). Furthermore, the horizontal offset between the screens, $|x_B| > 0$, is incorporated in the mutual-coupling impedances as (cf. (10.20))

$$\hat{Z}_{AB}(s) = \hat{Z}_{BA}(s) = \frac{s\mu_0}{2\pi \mathrm{i}} \int_{\kappa=-\mathrm{i}\infty}^{\mathrm{i}\infty} \exp(-s\kappa x_B)\, \mathrm{i}_0^2(s\kappa w/2) \frac{\mathrm{d}\kappa}{\mathrm{p}_0(\kappa) + \mathrm{p}_1(\kappa)}. \tag{10.34}$$

Finally, to specify the right-hand side of (10.32), we used the explicit-type boundary condition, $\hat{E}^{s}_y(x,0,s) = -\hat{E}^{e}_y(x,0,s)$, on $\mathcal{S}_A$ and $\mathcal{S}_B$ (see (9.1)) with (9.20) and the limit $w \downarrow 0$ to get rid of $\mathrm{i}_0(s\kappa_0 w/2)$ for all angles of incidence.

As the inversion of $\hat{Z}_{AA}(s)\hat{I}^{s;A}(s)$ and $\hat{Z}_{BB}(s)\hat{I}^{s;B}(s)$ to the TD can be found in Section 9.2.3, we next limit our analysis to the transformation of mutual-coupling terms $\hat{Z}_{AB}(s)\hat{I}^{s;B}(s)$ and $\hat{Z}_{BA}(s)\hat{I}^{s;A}(s)$. To that end, we next pursue the approach applied in Section 10.2.1.1. Hence, under the assumption that both screens are relatively narrow, the integrand in (10.34) is approximated via $\mathrm{i}_0^2(x) =$

$1 + O(x^2)$ as $x \downarrow 0$ and the resulting slowness representation is handled via the standard CdH technique. Accordingly, we introduce the new variable of integration, $\{\tau \in \mathbb{R}; \tau >, 0\}$, via

$$\kappa x_B = \tau, \tag{10.35}$$

which is satisfied along a loop encircling the branch cuts along $\{\mathrm{Im}(\kappa) = 0, 1/c_{0,1} \leq |\mathrm{Re}(\kappa)| < \infty\}$, and transform the integral to (cf. (10.22))

$$\hat{Z}_{AB}(s) \simeq \frac{s\mu_0}{\pi} \int_{\tau=T_B}^{\infty} \exp(-s\tau)\mathrm{Re}\left\{\frac{p_0[\kappa(\tau)]}{p_0[\kappa(\tau)] + p_1[\kappa(\tau)]}\right\} \frac{\mathrm{d}\tau}{(\tau^2 - T_B^2)^{1/2}}, \tag{10.36}$$

where we take the values of $p_{0,1}[\kappa(\tau)]$ just above the branch cuts and $T_B = |x_B|/c_0$. In the next step, the induced electric currents in $\mathcal{S}_A$ and $\mathcal{S}_B$ are expanded in time via the triangular functions (see (10.23a) and (10.23b)), which makes it possible to transform the system of equations (10.32) to the TD. The result of the transformation is the convolution-type system of equations that can be represented by (10.24), again, where (cf. (10.25))

$$\underline{\boldsymbol{Z}}_k = \frac{\zeta_0}{c_0 \Delta t} \begin{bmatrix} \Phi(\ddot{t}_k) & \Pi(t_k) \\ \Pi(t_k) & \Phi(\ddot{t}_k) \end{bmatrix}, \tag{10.37}$$

where $t_k = k\Delta t$ and $\ddot{t}_k = k\Delta\ddot{t} = k(c/c_0)\Delta t$ for $k = \{1, \ldots, M\}$ with the effective EM wave speed c given by (9.33). The TD function $\Phi(t)$ is defined via (7.21), again, and $\Pi(t)$ follows in closed form as (cf. (10.26))

$$\begin{aligned} \Pi(t) = {} & \frac{1}{2\pi} \frac{c_1^2}{c_0^2 - c_1^2} \left[\frac{c_0 t}{|x_B|} \left(\frac{c_0^2 t^2}{x_B^2} - 1 \right)^{1/2} - \cosh^{-1}\left(\frac{c_0 t}{|x_B|} \right) \right] \mathrm{H}\left(t - \frac{|x_B|}{c_0} \right) \\ & - \frac{1}{2\pi} \frac{c_0^2}{c_0^2 - c_1^2} \left[\frac{c_1 t}{|x_B|} \left(\frac{c_1^2 t^2}{x_B^2} - 1 \right)^{1/2} - \cosh^{-1}\left(\frac{c_1 t}{|x_B|} \right) \right] \mathrm{H}\left(t - \frac{|x_B|}{c_1} \right). \end{aligned} \tag{10.38}$$

The electric-current array, $\boldsymbol{I}_k$, is given by (10.27), again, and the excitation array is found upon transforming the right-hand side of (10.32) to the TD, that is (cf. (10.28)),

$$\boldsymbol{E}_m = \begin{bmatrix} T_\perp e^{\mathrm{i}}(t_m) \\ T_\perp e^{\mathrm{i}}(t_m - \kappa_0 x_B) \end{bmatrix}. \tag{10.39}$$

The thus defined system of equations can be solved with the aid of the updating, step-by-step procedure (10.29) that readily yields the pertaining electric-current coefficients, i_m^A and i_m^B, for all $m = \{1, \ldots, M\}$.

10.2.2.2 Numerical examples

The computational model proposed in Section 10.2.2 is implemented to calculate the induced electric-current surface density on two coplanar screens $\mathcal{S}_A$ and $\mathcal{S}_B$

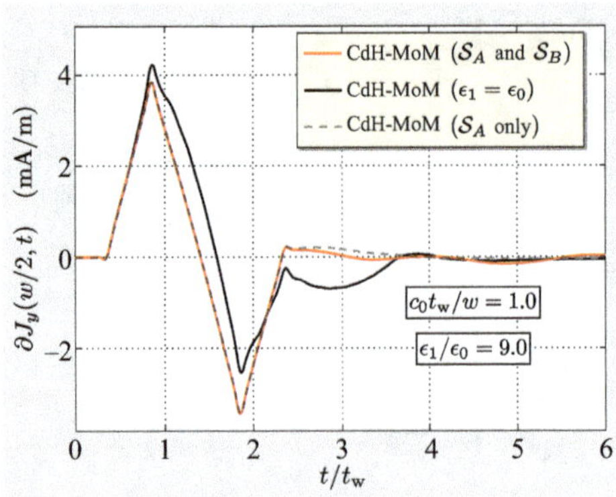

Figure 10.9 Plane-wave induced electric-current surface density at the center of screen $\mathcal{S}_A$ lying on the dielectric half-space

located on a dielectric half-space. In this example, we consider screens lying along $\mathcal{S}_A = \{0 < x < w, z = 0\}$ and $\mathcal{S}_B = \{3w/2 < x < 2w, z = 0\}$ on the half-space characterized by its electric permittivity $\epsilon_1 = 9.0\epsilon_0$. The structure is irradiated by the impulsive plane wave that is defined by its pulse shape, $e^{\mathrm{i}}(t)$, as shown in Figure 8.3 and by $\beta = \pi/4$, again. The excitation pulse-time width is taken such that $c_0 t_\mathrm{w}/w = 1.0$.

Figure 10.9 shows the resulting pulse shape at the mid-point of screen $\mathcal{S}_A$. To demonstrate the impact of the dielectric half-space and of the second screen $\mathcal{S}_B$ on the calculated current response $\partial J_y(w/2, t)$, we have also analyzed (a) the coplanar screens in the vacuum background ($\epsilon_1 = \epsilon_0$) and (b) the case in the absence of $\mathcal{S}_B$ ($\mathcal{S}_A$ only). It can be seen in Figure 10.9 that the rising edge of the calculated response is influenced neither by the contrasting half-space nor the second screen. In fact, it is observed that the coupling effect could be in this example essentially neglected in the entire time window of observation.

As the final example, we shall analyze the electric-current response of two narrow strips on the dielectric half-space. The coplanar strips occupy domains $\mathcal{S}_A = \{-w/2 < x < w/2, -\infty < y < \infty, z = 0\}$ and $\mathcal{S}_B = \{x_B - w/2 < x < x_B + w/2, -\infty < y < \infty, z = 0\}$ with $x_B = 5.0w$. The width of the screens, denoted by $w > 0$, is assumed to be relatively small with respect to $c_0 t_\mathrm{w}$ of the triangular plane-wave signature (see Figure 8.3). Therefore, we take $c_0 t_\mathrm{w}/w = 10$ and $\beta = \pi/4$, again. Figure 10.10 shows the calculated pulse shape of the electric-current surface density at the center of screen $\mathcal{S}_B$. It can be observed in Figure 10.10 that the pulse shape predicted by the approximate computational procedure as introduced in Section 10.2.2.1 agrees very well with the one obtained with the aid of CdH-MoM for $N_A = N_B = 2$ (see Section 10.2.2).

10.2.3 EM coupling between screens on a grounded dielectric slab

The planar screens under consideration in this subsection are located on a grounded dielectric slab of thickness $z_0 > 0$ (see Figure 10.11). To describe the elements of

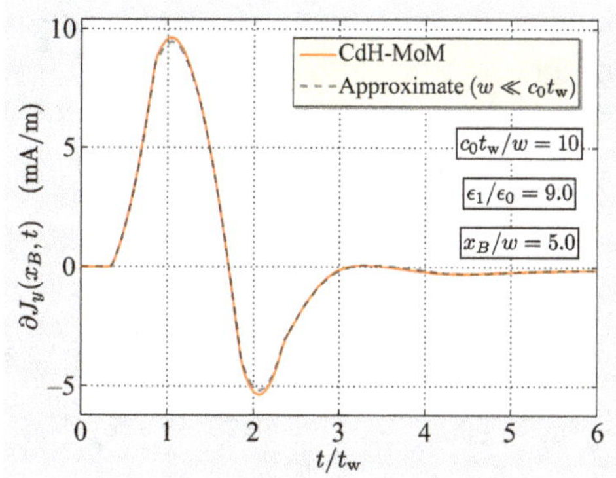

Figure 10.10 Plane-wave induced electric-current surface density at the center of narrow screen $\mathcal{S}_B$ on the dielectric half-space

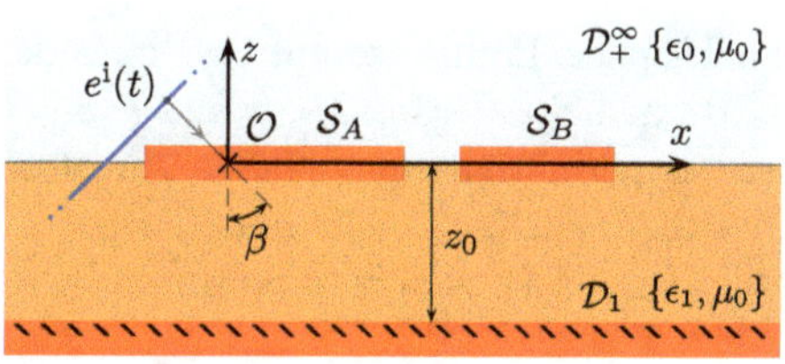

Figure 10.11 Pulsed EM plane wave impinging upon two coplanar strips located on a grounded dielectric slab

$\mathcal{Z}_{AB}$, we shall first specify the relevant relation for the testing (transform-domain) electric-field strength on $\mathcal{S}_B$ due to the testing source on $\mathcal{S}_A$. For this purpose, we may use (9.40) to write

$$\tilde{E}_y^{\mathrm{T};A}(\kappa, 0, s) = -\frac{\mu_0}{p_0(\kappa) + p_1(\kappa)} \partial \tilde{J}_y^{\mathrm{T};A}(\kappa, s) \times \left\{ 1 - [1 - \tilde{R}_\perp(\kappa)] \sum_{q=0}^{\infty} [\tilde{R}_\perp(\kappa)]^q \exp[-2s(q+1)p_1(\kappa)z_0] \right\}, \quad (10.40)$$

where $\tilde{R}_\perp(\kappa)$ is the (transform-domain) reflection coefficient as given by (9.39). Choosing, again, the piecewise-linear expansion of the induced current density on $\Delta\mathcal{S}_B$ and the piecewise-constant distribution of the testing source on $\Delta\mathcal{S}_A$, the mutual-interaction quantity (10.3d) with (10.40) leads to (cf. (10.31) and (H.1))

$$\hat{\mathcal{Z}}_{AB}^{[S,n]}(s) = \frac{\zeta_0}{s c_0 \Delta t} \frac{1}{2\pi \mathrm{i}} \int_{\kappa=-\mathrm{i}\infty}^{\mathrm{i}\infty} \tilde{\Pi}^{[S]}(\kappa) \tilde{\Lambda}^{[n]}(-\kappa) \frac{\mathrm{d}\kappa}{p_0(\kappa) + p_1(\kappa)} - \frac{\zeta_0}{s c_0 \Delta t} \sum_{q=0}^{\infty} \frac{1}{2\pi \mathrm{i}} \int_{\kappa=-\mathrm{i}\infty}^{\mathrm{i}\infty} \tilde{\Pi}^{[S]}(\kappa) \tilde{\Lambda}^{[n]}(-\kappa) [1 - \tilde{R}_\perp(\kappa)] [\tilde{R}_\perp(\kappa)]^q$$

$$\times \exp\left[-sp_1(\kappa)Z^{[q]}\right]\frac{\mathrm{d}\kappa}{p_0(\kappa)+p_1(\kappa)}$$
$$\text{for all } S = \{1,\ldots,N_A\} \text{ and } n = \{N_A+1,\ldots,N_A+N_B\}, \tag{10.41}$$

where $Z^{[q]} = 2(q+1)z_0$ denotes the vertical propagation path of the qth wave constituent. Upon comparison of (10.41) with (G.1) and (H.1), it is concluded that $\hat{\mathcal{Z}}_{AB}^{[S,n]}(s)$ can be expanded in terms of generic integrals $\hat{\Lambda}^{A,B}(x,s)$ and $\hat{\Psi}^{A,B}(x,z,s)$, the TD counterparts of which can be found in Appendices G and H, respectively. In this way, we finally arrive at the following expressions (cf. (10.7a)–(10.7i)):

$$\mathcal{Z}_{AB}^{[S,n]}(t) = \frac{\zeta_0}{c_0\Delta t}\frac{1}{\Delta_B}\Big[\Lambda^A(x_S - x_n + \Delta_B + \Delta_A/2, t)$$
$$- \Lambda^A(x_S - x_n + \Delta_B - \Delta_A/2, t) - 2\Lambda^A(x_S - x_n + \Delta_A/2, t)$$
$$+ 2\Lambda^A(x_S - x_n - \Delta_A/2, t) + \Lambda^A(x_S - x_n - \Delta_B + \Delta_A/2, t)$$
$$- \Lambda^A(x_S - x_n - \Delta_B - \Delta_A/2, t)\Big]$$
$$- \frac{\zeta_0}{c_0\Delta t}\frac{1}{\Delta_B}\sum_{q=0}^{\infty}\Big[\Psi^A(x_S - x_n + \Delta_B + \Delta_A/2, Z^{[q]}, t)$$
$$- \Psi^A(x_S - x_n + \Delta_B - \Delta_A/2, Z^{[q]}, t) - 2\Psi^A(x_S - x_n + \Delta_A/2, Z^{[q]}, t)$$
$$+ 2\Psi^A(x_S - x_n - \Delta_A/2, Z^{[q]}, t) + \Psi^A(x_S - x_n - \Delta_B + \Delta_A/2, Z^{[q]}, t)$$
$$- \Psi^A(x_S - x_n - \Delta_B - \Delta_A/2, Z^{[q]}, t)\Big]$$
$$\text{for all } S = \{2,\ldots,N_A-1\}$$
$$\text{and } n = \{N_A+2,\ldots,N_A+N_B-1\}, \tag{10.42a}$$

$$\mathcal{Z}_{AB}^{[S,N_A+1]}(t) = \frac{\zeta_0}{c_0\Delta t}\frac{1}{\Delta_B}\Big[-\Lambda^A(x_S - x_{N_A+1} + \Delta_A/2, t)$$
$$+ \Lambda^A(x_S - x_{N_A+1} - \Delta_A/2, t) + \Lambda^A(x_S - x_{N_A+1} - \Delta_B + \Delta_A/2, t)$$
$$- \Lambda^A(x_S - x_{N_A+1} - \Delta_B - \Delta_A/2, t)\Big] + \frac{\zeta_0}{c_0\Delta t}\Big[\Lambda^B(x_S - x_{N_A+1} + \Delta_A/2, t)$$
$$- \Lambda^B(x_S - x_{N_A+1} - \Delta_A/2, t)\Big]$$
$$- \frac{\zeta_0}{c_0\Delta t}\frac{1}{\Delta_B}\sum_{q=0}^{\infty}\Big[-\Psi^A(x_S - x_{N_A+1} + \Delta_A/2, Z^{[q]}, t)$$
$$+ \Psi^A(x_S - x_{N_A+1} - \Delta_A/2, Z^{[q]}, t)$$
$$+ \Psi^A(x_S - x_{N_A+1} - \Delta_B + \Delta_A/2, Z^{[q]}, t)$$
$$- \Psi^A(x_S - x_{N_A+1} - \Delta_B - \Delta_A/2, Z^{[q]}, t)\Big]$$

$$
\begin{aligned}
&- \frac{\zeta_0}{c_0 \Delta t} \sum_{q=0}^{\infty} \Big[\Psi^B(x_S - x_{N_A+1} + \Delta_A/2, Z^{[q]}, t) \\
&- \Psi^B(x_S - x_{N_A+1} - \Delta_A/2, Z^{[q]}, t) \Big] \\
&\text{for all } S = \{2, \ldots, N_A - 1\},
\end{aligned} \tag{10.42b}
$$

$$
\begin{aligned}
&\mathcal{Z}_{AB}^{[S,N_A+N_B]}(t) = \frac{\zeta_0}{c_0 \Delta t} \frac{1}{\Delta_B} \Big[\Lambda^A(x_S - x_{N_A+N_B} + \Delta_B + \Delta_A/2, t) \\
&- \Lambda^A(x_S - x_{N_A+N_B} + \Delta_A/2, t) - \Lambda^A(x_S - x_{N_A+N_B} + \Delta_B - \Delta_A/2, t) \\
&+ \Lambda^A(x_S - x_{N_A+N_B} - \Delta_A/2, t) \Big] - \frac{\zeta_0}{c_0 \Delta t} \Big[\Lambda^B(x_S - x_{N_A+N_B} + \Delta_A/2, t) \\
&- \Lambda^B(x_S - x_{N_A+N_B} - \Delta_A/2, t) \Big] \\
&- \frac{\zeta_0}{c_0 \Delta t} \frac{1}{\Delta_B} \sum_{q=0}^{\infty} \Big[\Psi^A(x_S - x_{N_A+N_B} + \Delta_B + \Delta_A/2, Z^{[q]}, t) \\
&- \Psi^A(x_S - x_{N_A+N_B} + \Delta_A/2, Z^{[q]}, t) \\
&- \Psi^A(x_S - x_{N_A+N_B} + \Delta_B - \Delta_A/2, Z^{[q]}, t) \\
&+ \Psi^A(x_S - x_{N_A+N_B} - \Delta_A/2, Z^{[q]}, t) \Big] \\
&+ \frac{\zeta_0}{c_0 \Delta t} \sum_{q=0}^{\infty} \Big[\Psi^B(x_S - x_{N_A+N_B} + \Delta_A/2, Z^{[q]}, t) \\
&- \Psi^B(x_S - x_{N_A+N_B} - \Delta_A/2, Z^{[q]}, t) \Big] \\
&\text{for all } S = \{2, \ldots, N_A - 1\},
\end{aligned} \tag{10.42c}
$$

$$
\begin{aligned}
\mathcal{Z}_{AB}^{[1,n]}(t) = &\frac{\zeta_0}{c_0 \Delta t} \frac{1}{\Delta_B} \Big[\Lambda^A(x_1 - x_n + \Delta_B + \Delta_A/2, t) - \Lambda^A(x_1 - x_n + \Delta_B, t) \\
&- 2\Lambda^A(x_1 - x_n + \Delta_A/2, t) + 2\Lambda^A(x_1 - x_n, t) \\
&+ \Lambda^A(x_1 - x_n - \Delta_B + \Delta_A/2, t) - \Lambda^A(x_1 - x_n - \Delta_B, t) \Big] \\
&- \frac{\zeta_0}{c_0 \Delta t} \frac{1}{\Delta_B} \sum_{q=0}^{\infty} \Big[\Psi^A(x_1 - x_n + \Delta_B + \Delta_A/2, Z^{[q]}, t) \\
&- \Psi^A(x_1 - x_n + \Delta_B, Z^{[q]}, t) - 2\Psi^A(x_1 - x_n + \Delta_A/2, Z^{[q]}, t) \\
&+ 2\Psi^A(x_1 - x_n, Z^{[q]}, t) + \Psi^A(x_1 - x_n - \Delta_B + \Delta_A/2, Z^{[q]}, t) \\
&- \Psi^A(x_1 - x_n - \Delta_B, Z^{[q]}, t) \Big] \\
&\text{for all } n = \{N_A + 2, \ldots, N_A + N_B - 1\},
\end{aligned} \tag{10.42d}
$$

and

$$\mathcal{Z}_{AB}^{[N_A,n]}(t) = \frac{\zeta_0}{c_0\Delta t}\frac{1}{\Delta_B}\Big[\Lambda^A(x_{N_A} - x_n + \Delta_B, t)$$
$$- \Lambda^A(x_{N_A} - x_n + \Delta_B - \Delta_A/2, t) - 2\Lambda^A(x_{N_A} - x_n, t)$$
$$+ 2\Lambda^A(x_{N_A} - x_n - \Delta_A/2, t) + \Lambda^A(x_{N_A} - x_n - \Delta_B, t)$$
$$- \Lambda^A(x_{N_A} - x_n - \Delta_B - \Delta_A/2, t)\Big]$$
$$- \frac{\zeta_0}{c_0\Delta t}\frac{1}{\Delta_B}\sum_{q=0}^{\infty}\Big[\Psi^A\big(x_{N_A} - x_n + \Delta_B, Z^{[q]}, t\big)$$
$$- \Psi^A\big(x_{N_A} - x_n + \Delta_B - \Delta_A/2, Z^{[q]}, t\big) - 2\Psi^A\big(x_{N_A} - x_n, Z^{[q]}, t\big)$$
$$+ 2\Psi^A\big(x_{N_A} - x_n - \Delta_A/2, Z^{[q]}, t\big) + \Psi^A\big(x_{N_A} - x_n - \Delta_B, Z^{[q]}, t\big)$$
$$- \Psi^A\big(x_{N_A} - x_n - \Delta_B - \Delta_A/2, Z^{[q]}, t\big)\Big]$$
$$\text{for all } n = \{N_A + 2, \ldots, N_A + N_B - 1\}, \tag{10.42e}$$

with

$$\mathcal{Z}_{AB}^{[1,N_A+1]}(t) = \frac{\zeta_0}{c_0\Delta t}\frac{1}{\Delta_B}\Big[- \Lambda^A(x_1 - x_{N_A+1} + \Delta_A/2, t)$$
$$+ \Lambda^A(x_1 - x_{N_A+1}, t) + \Lambda^A(x_1 - x_{N_A+1} - \Delta_B + \Delta_A/2, t)$$
$$- \Lambda^A(x_1 - x_{N_A+1} - \Delta_B, t)\Big] + \frac{\zeta_0}{c_0\Delta t}\Big[\Lambda^B(x_1 - x_{N_A+1} + \Delta_A/2, t)$$
$$- \Lambda^B(x_1 - x_{N_A+1}, t)\Big] - \frac{\zeta_0}{c_0\Delta t}\frac{1}{\Delta_B}\sum_{q=0}^{\infty}\Big[- \Psi^A\big(x_1 - x_{N_A+1} + \Delta_A/2, Z^{[q]}, t\big)$$
$$+ \Psi^A\big(x_1 - x_{N_A+1}, Z^{[q]}, t\big) + \Psi^A\big(x_1 - x_{N_A+1} - \Delta_B + \Delta_A/2, Z^{[q]}, t\big)$$
$$- \Psi^A\big(x_1 - x_{N_A+1} - \Delta_B, Z^{[q]}, t\big)\Big]$$
$$- \frac{\zeta_0}{c_0\Delta t}\sum_{q=0}^{\infty}\Big[\Psi^B\big(x_1 - x_{N_A+1} + \Delta_A/2, Z^{[q]}, t\big) - \Psi^B\big(x_1 - x_{N_A+1}, Z^{[q]}, t\big)\Big], \tag{10.42f}$$

$$\mathcal{Z}_{AB}^{[N_A,N_A+1]}(t) = \frac{\zeta_0}{c_0\Delta t}\frac{1}{\Delta_B}\Big[- \Lambda^A(x_{N_A} - x_{N_A+1}, t)$$
$$+ \Lambda^A(x_{N_A} - x_{N_A+1} - \Delta_A/2, t) + \Lambda^A(x_{N_A} - x_{N_A+1} - \Delta_B, t)$$
$$- \Lambda^A(x_{N_A} - x_{N_A+1} - \Delta_B - \Delta_A/2, t)\Big] + \frac{\zeta_0}{c_0\Delta t}\Big[\Lambda^B(x_{N_A} - x_{N_A+1}, t)$$
$$- \Lambda^B(x_{N_A} - x_{N_A+1} - \Delta_A/2, t)\Big]$$
$$- \frac{\zeta_0}{c_0\Delta t}\frac{1}{\Delta_B}\sum_{q=0}^{\infty}\Big[- \Psi^A\big(x_{N_A} - x_{N_A+1}, Z^{[q]}, t\big)$$

$$+ \Psi^A\big(x_{N_A} - x_{N_A+1} - \Delta_A/2, Z^{[q]}, t\big) + \Psi^A\big(x_{N_A} - x_{N_A+1} - \Delta_B, Z^{[q]}, t\big)$$
$$- \Psi^A\big(x_{N_A} - x_{N_A+1} - \Delta_B - \Delta_A/2, Z^{[q]}, t\big)\Big]$$
$$- \frac{\zeta_0}{c_0 \Delta t} \sum_{q=0}^{\infty} \Big[\Psi^B\big(x_{N_A} - x_{N_A+1}, Z^{[q]}, t\big)$$
$$- \Psi^B\big(x_{N_A} - x_{N_A+1} - \Delta_A/2, Z^{[q]}, t\big)\Big], \tag{10.42g}$$

$$\mathcal{Z}_{AB}^{[1, N_A+N_B]}(t) = \frac{\zeta_0}{c_0 \Delta t} \frac{1}{\Delta_B} \Big[\Lambda^A(x_1 - x_{N_A+N_B} + \Delta_B + \Delta_A/2, t)$$
$$- \Lambda^A(x_1 - x_{N_A+N_B} + \Delta_B, t) - \Lambda^A(x_1 - x_{N_A+N_B} + \Delta_A/2, t)$$
$$+ \Lambda^A(x_1 - x_{N_A+N_B}, t)\Big] - \frac{\zeta_0}{c_0 \Delta t} \Big[\Lambda^B(x_1 - x_{N_A+N_B} + \Delta_A/2, t)$$
$$- \Lambda^B(x_1 - x_{N_A+N_B}, t)\Big]$$
$$- \frac{\zeta_0}{c_0 \Delta t} \frac{1}{\Delta_B} \sum_{q=0}^{\infty} \Big[\Psi^A\big(x_1 - x_{N_A+N_B} + \Delta_B + \Delta_A/2, Z^{[q]}, t\big)$$
$$- \Psi^A\big(x_1 - x_{N_A+N_B} + \Delta_B, Z^{[q]}, t\big) - \Psi^A\big(x_1 - x_{N_A+N_B} + \Delta_A/2, Z^{[q]}, t\big)$$
$$+ \Psi^A\big(x_1 - x_{N_A+N_B}, Z^{[q]}, t\big)\Big]$$
$$+ \frac{\zeta_0}{c_0 \Delta t} \sum_{q=0}^{\infty} \Big[\Psi^B\big(x_1 - x_{N_A+N_B} + \Delta_A/2, Z^{[q]}, t\big)$$
$$- \Psi^B\big(x_1 - x_{N_A+N_B}, Z^{[q]}, t\big)\Big], \tag{10.42h}$$

and

$$\mathcal{Z}_{AB}^{[N_A, N_A+N_B]}(t) = \frac{\zeta_0}{c_0 \Delta t} \frac{1}{\Delta_B} \Big[\Lambda^A(x_{N_A} - x_{N_A+N_B} + \Delta_B, t)$$
$$- \Lambda^A(x_{N_A} - x_{N_A+N_B} + \Delta_B - \Delta_A/2, t) - \Lambda^A(x_{N_A} - x_{N_A+N_B}, t)$$
$$+ \Lambda^A(x_{N_A} - x_{N_A+N_B} - \Delta_A/2, t)\Big] - \frac{\zeta_0}{c_0 \Delta t} \Big[\Lambda^B(x_{N_A} - x_{N_A+N_B}, t)$$
$$- \Lambda^B(x_{N_A} - x_{N_A+N_B} - \Delta_A/2, t)\Big]$$
$$- \frac{\zeta_0}{c_0 \Delta t} \frac{1}{\Delta_B} \sum_{q=0}^{\infty} \Big[\Psi^A\big(x_{N_A} - x_{N_A+N_B} + \Delta_B, Z^{[q]}, t\big)$$
$$- \Psi^A\big(x_{N_A} - x_{N_A+N_B} + \Delta_B - \Delta_A/2, Z^{[q]}, t\big)$$
$$- \Psi^A\big(x_{N_A} - x_{N_A+N_B}, Z^{[q]}, t\big) + \Psi^A\big(x_{N_A} - x_{N_A+N_B} - \Delta_A/2, Z^{[q]}, t\big)\Big]$$
$$+ \frac{\zeta_0}{c_0 \Delta t} \sum_{q=0}^{\infty} \Big[\Psi^B\big(x_{N_A} - x_{N_A+N_B}, Z^{[q]}, t\big)$$
$$- \Psi^B\big(x_{N_A} - x_{N_A+N_B} - \Delta_A/2, Z^{[q]}, t\big)\Big]. \tag{10.42i}$$

It remains to specify the items of the (modified) excitation array $\bar{V}^{[S]}(t)$. This can be achieved via

$$\bar{V}^{[S]}(t) = T_{\perp} V^{[S]}(t) - T_{\perp}(1 - R_{\perp}) \sum_{q=0}^{\infty} R_{\perp}^{q} V^{[S]}[t - 2p_1(\kappa_0)(q+1)z_0] \quad \text{for all } S = \{1, \ldots, N_A + N_B\}, \tag{9.44 revisited}$$

where $V^{[S]}(t)$ denotes the excitation array elements pertaining to the unbounded, homogeneous and isotropic background (see (10.8a)–(10.11)). Further recall that the reflection coefficient, $R_{\perp}$, and the transmission coefficient, $T_{\perp}$, are defined via (9.21) and (9.22), respectively. Since the self-coupling impeditivity arrays can be readily filled according to Section 9.3.2, the system of equations (8.10) is fully determined. Its solution via the marching-on-in-time technique (see (8.11)) yields the induced electric current on $\mathcal{S}_A$ and $\mathcal{S}_B$.

10.2.3.1 Narrow-screen approximation

We shall analyze the electric current induced in two coplanar narrow screens located on the grounded dielectric slab of a finite thickness $z_0 > 0$. The screens are placed along $\mathcal{S}_A = \{-w/2 < x < w/2, -\infty < y < \infty, -\delta/2 < z < \delta/2\}$ and $\mathcal{S}_B = \{x_B - w/2 < x < x_B + w/2, -\infty < y < \infty, -\delta/2 < z < \delta/2\}$, where $w > 0$ and $\delta > 0$ denote their (relatively small) width and thickness, respectively, and x_B is the horizontal shift of screen $\mathcal{S}_B$ with respect to the origin (see Figure 10.12).

The analysis that follows starts from the reciprocity relation (10.14) with $z_B = 0$, again. Supposing next that both testing and actual current density has the uniform spatial distribution over the strips, the interaction terms in the reciprocity relation can be cast into the matrix form (cf. (10.32) and Figure 10.2), viz

$$\begin{bmatrix} \hat{Z}_{AA}(s) & \hat{Z}_{AB}(s) \\ \hat{Z}_{BA}(s) & \hat{Z}_{BB}(s) \end{bmatrix} \cdot \begin{bmatrix} \hat{I}^{s;A}(s) \\ \hat{I}^{s;B}(s) \end{bmatrix} = \begin{bmatrix} \hat{E}_A(s) \\ \hat{E}_B(s) \end{bmatrix}, \tag{10.43}$$

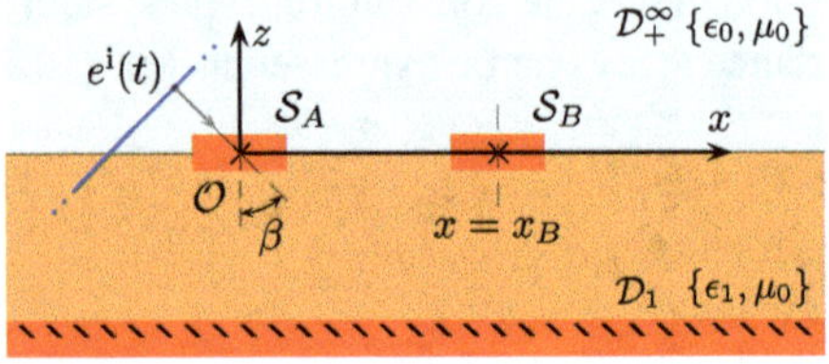

Figure 10.12 Pulsed EM plane wave impinging upon two coplanar narrow strips located on a grounded dielectric slab

where the self-coupling impedances $\hat{Z}_{AA}(s)$ and $\hat{Z}_{BB}(s)$ follow from (cf. (9.46))

$$\hat{Z}_{AA}(s) = \hat{Z}_{BB}(s) = \frac{s\mu_0}{2\pi\mathrm{i}} \int_{\kappa=-\mathrm{i}\infty}^{\mathrm{i}\infty} \mathrm{i}_0^2(s\kappa w/2) \frac{\mathrm{d}\kappa}{p_0(\kappa)+p_1(\kappa)}$$
$$- \frac{s\mu_0}{2\pi\mathrm{i}} \int_{\kappa=-\mathrm{i}\infty}^{\mathrm{i}\infty} \mathrm{i}_0^2(s\kappa w/2) \frac{[1-\tilde{R}_\perp(\kappa)]\exp[-2sp_1(\kappa)z_0]}{1-\tilde{R}_\perp(\kappa)\exp[-2sp_1(\kappa)z_0]} \frac{\mathrm{d}\kappa}{p_0(\kappa)+p_1(\kappa)}. \tag{10.44}$$

Recall that the transform-domain reflection coefficient, $\tilde{R}_\perp(\kappa)$, was defined via (9.39). In a similar fashion, we may represent the mutual-coupling impedances

$$\hat{Z}_{AB}(s) = \hat{Z}_{BA}(s) = \frac{s\mu_0}{2\pi\mathrm{i}} \int_{\kappa=-\mathrm{i}\infty}^{\mathrm{i}\infty} \exp(-s\kappa x_B)\,\mathrm{i}_0^2(s\kappa w/2) \frac{\mathrm{d}\kappa}{p_0(\kappa)+p_1(\kappa)}$$
$$- \frac{s\mu_0}{2\pi\mathrm{i}} \int_{\kappa=-\mathrm{i}\infty}^{\mathrm{i}\infty} \exp(-s\kappa x_B)\,\mathrm{i}_0^2(s\kappa w/2) \frac{[1-\tilde{R}_\perp(\kappa)]\exp[-2sp_1(\kappa)z_0]}{1-\tilde{R}_\perp(\kappa)\exp[-2sp_1(\kappa)z_0]}$$
$$\times \frac{\mathrm{d}\kappa}{p_0(\kappa)+p_1(\kappa)}. \tag{10.45}$$

Since the first integrals on the right-hand side of (10.44) and (10.45) are identical with the ones pertaining to the dielectric half-space (see (10.33) and (10.34)), it is concluded that the second integrals incorporate the effect of wave constituents reflecting between the ground plane and the dielectric interface. The elements of the excitation array in (10.43) follow upon employing the explicit-type boundary condition with (9.37) in the right-hand side the reciprocity relation (10.14). Making use of these results, we arrive at

$$\hat{E}_A(s) = T_\perp \hat{e}^{\mathrm{i}}(s) \left\{ 1 - (1-R_\perp) \sum_{q=0}^{\infty} R_\perp^q \exp[-2s(q+1)p_1(\kappa_0)z_0] \right\}, \tag{10.46a}$$

$$\hat{E}_B(s) = \hat{E}_A(s)\exp(-sp_0 x_B), \tag{10.46b}$$

where we replaced $\mathrm{i}_0(s\kappa_0 w/2)$ with the number one for all the angles of incidence, β, and recall that $R_\perp$ and $T_\perp$ are given by (9.21) and (9.22), respectively. The system of equations (10.43) in the s-domain is subsequently transformed to the TD. The inverse transformation yields the convolution-type system of equations (10.24), again, where the impedance array can be expressed as (cf. (10.37))

$$\underline{\boldsymbol{Z}}_k = \frac{\zeta_0}{c_0\Delta t} \begin{bmatrix} \Phi(\ddot{t}_k) & \Pi(t_k) \\ \Pi(t_k) & \Phi(\ddot{t}_k) \end{bmatrix} - \frac{\zeta_0}{c_0\Delta t} \sum_{q=0}^{\infty} \begin{bmatrix} \Lambda^{[q]}(t_k) & \Omega^{[q]}(t_k) \\ \Omega^{[q]}(t_k) & \Lambda^{[q]}(t_k) \end{bmatrix}. \tag{10.47}$$

Here, the first impedance array on the right-hand side corresponds to the half-space configuration (see (10.37)) and the additional terms account for the upgoing and downgoing wave constituents propagating in the slab. The diagonal elements in the

summation can be then expressed through the TD counterpart of the generic integral representation $\hat{\Psi}^B$, the TD counterpart of which is specified in Section H.2. Accordingly, we may write

$$\Lambda^{[q]}(t) = [\Psi^B(w, Z^{[q]}, t) - 2\Psi^B(0, Z^{[q]}, t) + \Psi^B(-w, Z^{[q]}, t)]/w^2, \qquad (10.48)$$

where $Z^{[q]} = 2(q+1)z_0 > 0$ denotes the vertical propagation path. It is noted that if the approximate formula (9.47) were taken as the starting point, we would simply get $\Lambda^{[q]}(t) = (\mu_0/2\pi)\log[2(c/c_1)z_0/a]\mathrm{H}(t)$. The expression given by (10.48) is, however, not limited to (relatively) thin slabs and is hence more general. Next, the off-diagonal terms in the summation can be expressed via the original of $\hat{\Omega}^{[q]}(s)$ that can be found upon converting the integrand of the second integral in (10.45) into the geometric series, while employing $\mathrm{i}_0^2(x) = 1 + O(x^2)$ as $x \downarrow 0$ under the narrow-screen assumption. In this manner, we end up with (see (H.4) and (H.21))

$$\hat{\Omega}^{[q]}(s) = \frac{1}{s}\frac{1}{2\pi\mathrm{i}}\int_{p=-\mathrm{i}\infty}^{\mathrm{i}\infty} \exp\big\{-s\big[\kappa x_B + p_1(\kappa)Z^{[q]}\big]\big\}\frac{\tilde{G}^{[q]}(\kappa)\mathrm{d}\kappa}{p_0(\kappa)+p_1(\kappa)}$$
$$\text{for } \{s \in \mathbb{R}; s > 0\}, \qquad (10.49)$$

where $\tilde{G}^{[q]}$ is given by (H.2). The transformation of (10.49) to the TD can be carried out via the CdH technique (see Section 2.1.1), again. Pursuing approach described in Appendix H, the desired TD function can be expressed as a sum of body- and head-wave terms, that is,

$$\Omega^{[q]}(t) = \Omega_{\mathrm{B}}^{[q]}(t) + \Omega_{\mathrm{H}}^{[q]}(t), \qquad (10.50)$$

where the body-wave contribution reads

$$\Omega_{\mathrm{B}}^{[q]}(t) = \frac{1}{\pi}\int_{T_{\mathrm{B}}^{[q]}}^{t} \mathrm{Re}\left\{\frac{\tilde{G}^{[q]}[\kappa(\tau)]p_1[\kappa(\tau)]}{p_0[\kappa(\tau)]+p_1[\kappa(\tau)]}\right\}_{\mathcal{G}_{\mathrm{B}}} \frac{\mathrm{d}\tau}{(\tau^2 - T_{\mathrm{B}}^{[q]2})^{1/2}}, \qquad (10.51)$$

with $T_{\mathrm{B}}^{[q]} = (x_B^2 + Z^{[q]2})^{1/2}/c_1$. The non-vanishing head-wave contribution appears when $|x_B| > (c_1/c_0)(x_B^2 + Z^{[q]2})^{1/2}$ and can be expressed in a similar form as

$$\Omega_{\mathrm{H}}^{[q]}(t) = \frac{1}{\pi}\int_{T_{\mathrm{H}}^{[q]}}^{\min(t, T_{\mathrm{B}}^{[q]})} \mathrm{Im}\left\{\frac{\tilde{G}^{[q]}[\kappa(\tau)]p_1[\kappa(\tau)]}{p_0[\kappa(\tau)]+p_1[\kappa(\tau)]}\right\}_{\mathcal{G}_{\mathrm{H}}} \frac{\mathrm{d}\tau}{(T_{\mathrm{B}}^{[q]2} - \tau^2)^{1/2}}, \qquad (10.52)$$

with $T_{\mathrm{H}}^{[q]} = |x_B|/c_0 + Z^{[q]}(1/c_1^2 - 1/c_0^2)^{1/2}$ (cf. (2.90)). The integrals in (10.51) and (10.52) are evaluated along the body- and head-wave parts of the CdH path located in the upper half of the complex κ-plane. The body-wave path, say $\mathcal{G}_{\mathrm{B}}$, has the shape of a hyperbolic arc that can be parametrized via

$$\mathcal{G}_{\mathrm{B}} = \left\{\kappa(\tau) = \frac{x_B}{x_B^2 + Z^{[q]2}}\tau + \mathrm{i}\frac{Z^{[q]}}{x_B^2 + Z^{[q]2}}(\tau^2 - T_{\mathrm{B}}^{[q]2})^{1/2}\right\}$$
$$\text{for } \tau \geq T_{\mathrm{B}}^{[q]}. \qquad (10.53)$$

The head-wave path, denoted by $\mathcal{G}_{\mathrm{H}}$, is a straight line running just above the branch cut of $p_0(\kappa)$ and can be described by

$$\mathcal{G}_{\mathrm{H}} = \left\{ \kappa(\tau) = \frac{x_B}{x_B^2 + Z^{[q]2}}\tau - \frac{Z^{[q]}}{x_B^2 + Z^{[q]2}}(T_{\mathrm{B}}^{[q]2} - \tau^2)^{1/2} + \mathrm{i}0 \right\}$$
$$\text{for } \{T_{\mathrm{H}}^{[q]} \leq \tau \leq T_{\mathrm{B}}^{[q]}\}. \tag{10.54}$$

The inverse square-root singularities in (10.51) and (10.52) can be readily handled via $\tau = T_{\mathrm{B}}^{[q]} \cosh(u)$ in $\{0 \leq u < \infty\}$ and $\tau = T_{\mathrm{B}}^{[q]} \cos(v)$ in $\{0 \leq v \leq \pi/2\}$, respectively, which leads to well-behaved integrals that can be evaluated via standard numerical routines.

Having fully defined the impedance array, it remains to specify the excitation array on the right-hand side of (10.24). The pertinent excitation array can be found upon including the additional reflected-wave constituents from (10.46a) in (10.39). This way leads to

$$\boldsymbol{E}_m = \begin{bmatrix} T_\perp \left\{ e^{\mathrm{i}}(t_m) - (1 - R_\perp) \sum_{q=0}^{\infty} R_\perp^q e^{\mathrm{i}}\left[t_m - p_1(\kappa_0) Z^{[q]}\right] \right\} \\ T_\perp \Big\{ e^{\mathrm{i}}(t_m - \kappa_0 x_B) \\ -(1 - R_\perp) \sum_{q=0}^{\infty} R_\perp^q e^{\mathrm{i}}\left[t_m - \kappa_0 x_B - p_1(\kappa_0) Z^{[q]}\right] \Big\} \end{bmatrix}. \tag{10.55}$$

Owing to the property $e^{\mathrm{i}}(t) = 0$ for $t < 0$, the number of terms in the summations in (10.55) is finite in any bounded time window of observation. In the final step, the thus defined convolution-type system of equations (10.24) is solved for coefficients i_m^A and i_m^B for all $m = \{1, \ldots, M\}$. These coefficients represent the electric currents induced in the narrow screens via the TD originals of (10.23a) and (10.23b). The solution can be achieved via the marching-on-in-time technique (10.29), again.

10.2.3.2 Numerical examples

We shall next demonstrate the calculation of the electric-current surface density induced along two coplanar PEC screens placed on a grounded dielectric slab. To that end, we shall employ the results presented in Section 10.2.3. The screens under consideration are located along $\mathcal{S}_A = \{0 < x < w, -\infty < y < \infty, z = 0\}$ and $\mathcal{S}_B = \{3w/2 < x < 2w, -\infty < y < \infty, z = 0\}$ on the slab that is characterized by its electric permittivity $\epsilon_1 = 9.0\epsilon_0$ and thickness $z_0 = 1.50\,w$. The structure is irradiated by the impulsive plane-wave that is specified by its unit triangular signature, $e^{\mathrm{i}}(t)$, and by the angle of incidence, $\beta = \pi/4$. The pulse-time width is taken such that $c_1 t_{\mathrm{w}}/w = 1.0$.

The calculated pulse shape of the induced electric-current surface density at the center of $\mathcal{S}_A$ is shown in Figure 10.13. To evaluate the effect of the perfect ground plane and of the second screen $\mathcal{S}_B$, we have also analyzed the response of (a) the coplanar screens in the absence of the ground plane (half-space) and (b) the case in the absence of $\mathcal{S}_B$ ($\mathcal{S}_A$ only). Again, the early-time responses are identical. The effect of the ground plane becomes apparent just after $t/t_{\mathrm{w}} = \sqrt{2}/12 + 3\sqrt{17/18} \simeq 3.03$ when the wave reflected from the ground plane arrives back to the field point. The

impact of EM coupling between the screen is noticeable much earlier, shortly after the initial peak.

Finally, we shall employ the approximate methodology introduced in Section 10.2.3.1 to calculate the electric-current response of two narrow strips on the grounded slab. The coplanar strips occupy domains $\mathcal{S}_A = \{-w/2 < x < w/2, -\infty < y < \infty, z = 0\}$ and $\mathcal{S}_B = \{x_B - w/2 < x < x_B + w/2, -\infty < y < \infty, z = 0\}$ with $x_B = 5.0w$. For the narrow-screen approximation to apply, the width of the strips is taken to be $c_1 t_w / w = 10$. The remaining configurational and excitation parameters remain the same as in the previous example. Figure 10.14 shows the resulting pulse shapes as found using the approximate procedure and the pertaining CdH-MoM solution for $N_A = N_B = 2$. Apparently, the narrow-screen approximation is prone to late-time instabilities [3, Chapter 3], which limits its applicability to achieving the early-time solution only. Therefore, to get a stable solution

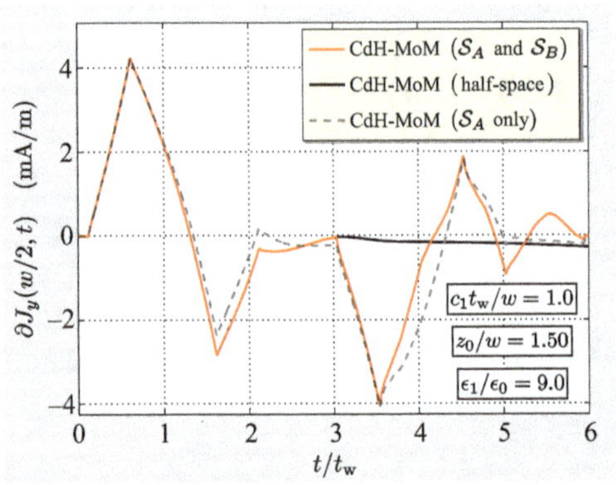

Figure 10.13 Plane-wave induced electric-current surface density at the center of screen $\mathcal{S}_A$ lying on the dielectric slab

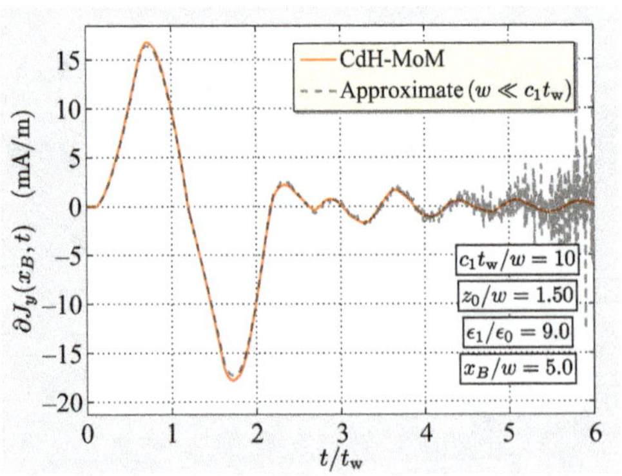

Figure 10.14 Plane-wave induced electric-current surface density at the center of narrow screen $\mathcal{S}_B$ on the grounded dielectric slab

in the entire time window, the use of CdH-MoM with the piecewise linear space–time electric-current distribution is recommended.

References

[1] M. Štumpf. *Electromagnetic Reciprocity in Antenna Theory*. Hoboken, NJ: John Wiley & Sons, Inc., 2018.
[2] M. Abramowitz and I. A. Stegun. *Handbook of Mathematical Functions*. New York, NY: Dover Publications, 1972.
[3] A. G. Tijhuis. *Electromagnetic Inverse Profiling: Theory and Numerical Implementation*. Utrecht, the Netherlands: VNU Science Press, 1987.

Chapter 11

Pulsed EM-field coupling between bounded and narrow metasurfaces

We have demonstrated in Chapter 10 that the Cagniard–DeHoop method of moments (CdH-MoM) can serve for evaluating the electromagnetic (EM) coupling between planar strips supporting electric currents. In this chapter, the EM reciprocity theorem of the time-convolution type is applied, again, to put forward an efficient computational methodology that is capable of calculating the pulsed EM scattering by multiple metasurfaces exhibiting both magnetic and dielectric constitutive properties. In particular, it is shown that under the assumption of piecewise linear space–time distribution of unknown currents, the pertaining EM reciprocity relation can be cast into a system of time-convolution type that is amenable to its solution via the marching-on-in-time technique. In addition, a special case concerning the EM coupling between relatively narrow metastrips is analyzed in detail.

11.1 Problem formulation

Referring to Figure 11.1, we shall analyze the pulsed EM scattering by two metasurfaces occupying domains $\mathcal{S}_A = \{a < x < b, -\infty < y < \infty, -\delta/2 < z < \delta/2\}$ and $\mathcal{S}_B = \{c < x < d, -\infty < y < \infty, z_B - \delta/2 < z < z_B + \delta/2\}$, with $\delta > 0$ denoting their (relatively) small thickness. As far as their EM properties are concerned, the metasurfaces $\mathcal{S}_{A,B}$ under consideration are characterized by their (scalar, real-valued and positive) electric permittivities and conductivities $\{\epsilon_{A,B}, \sigma_{A,B}\}$ and by $\{\mu_{A,B}, \Gamma_{A,B}\}$ representing their magnetic permeability and magnetic hysteresis losses, respectively. It is assumed that both metasurfaces show high contrasts in their EM properties with respect to the parameters of the homogeneous, isotropic and loss-free embedding, $\mathcal{D}^\infty$, such that their interaction with the incident EM field can be described using their equivalent layer impedances and admittances (see (3.15a) and (3.15b)). Thus, the EM parameters pertaining to metasurfaces $\mathcal{S}_A$ and $\mathcal{S}_B$ are described by $\{\hat{Z}_A^{\mathrm{E}}, \hat{Y}_A^{\mathrm{M}}\}$ and $\{\hat{Z}_B^{\mathrm{E}}, \hat{Y}_B^{\mathrm{M}}\}$, respectively. As in Section 10.1, for instance, the metasurfaces are activated by an impulsive plane EM wave, the electric-field strength of which is described by (4.1).

The problem will be next approached via the EM reciprocity theorem of the time-convolution type [1, Section 1.4.1], again. Through the theorem, we interrelate the (actual) scattered-field state (denoted by superscript ^{s}) and the (auxiliary) testing-field state (denoted by superscript ^{T}) in the unbounded domain exterior to

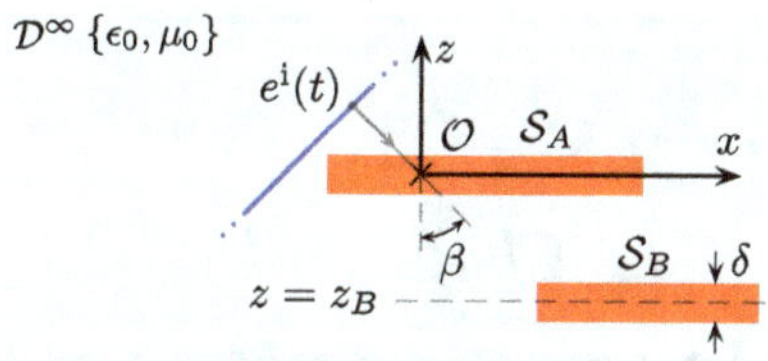

Figure 11.1 Pulsed EM plane wave impinging upon two metasurfaces

the metasurfaces. Accounting for the self-adjointness of the embedding and the property of causality of the interacting field states, we arrive at an s-domain reciprocity relation that is subsequently rewritten in terms of complex slowness-domain integrals (see (4.33)). The thus obtained reciprocity relation has then the following form (cf. (8.31) and (10.2)):

$$\begin{aligned}
&\frac{s}{2\pi \mathrm{i}}\int_{\kappa=-\mathrm{i}\infty}^{\mathrm{i}\infty}\big[\langle\tilde{E}_y^{\mathrm{T};A}\rangle_-^+(\kappa,0,s)+\langle\tilde{E}_y^{\mathrm{T};B}\rangle_-^+(\kappa,0,s)\big]\partial\tilde{J}_y^{\mathrm{s};A}(-\kappa,s)\mathrm{d}\kappa\\
&-\frac{s}{2\pi \mathrm{i}}\int_{\kappa=-\mathrm{i}\infty}^{\mathrm{i}\infty}\big[\langle\tilde{H}_x^{\mathrm{T};A}\rangle_-^+(\kappa,0,s)+\langle\tilde{H}_x^{\mathrm{T};B}\rangle_-^+(\kappa,0,s)\big]\partial\tilde{K}_x^{\mathrm{s};A}(-\kappa,s)\mathrm{d}\kappa\\
&+\frac{s}{2\pi \mathrm{i}}\int_{\kappa=-\mathrm{i}\infty}^{\mathrm{i}\infty}\big[\langle\tilde{E}_y^{\mathrm{T};B}\rangle_-^+(\kappa,z_B,s)+\langle\tilde{E}_y^{\mathrm{T};A}\rangle_-^+(\kappa,z_B,s)\big]\partial\tilde{J}_y^{\mathrm{s};B}(-\kappa,s)\mathrm{d}\kappa\\
&-\frac{s}{2\pi \mathrm{i}}\int_{\kappa=-\mathrm{i}\infty}^{\mathrm{i}\infty}\big[\langle\tilde{H}_x^{\mathrm{T};B}\rangle_-^+(\kappa,z_B,s)+\langle\tilde{H}_x^{\mathrm{T};A}\rangle_-^+(\kappa,z_B,s)\big]\partial\tilde{K}_x^{\mathrm{s};B}(-\kappa,s)\mathrm{d}\kappa\\
&=\frac{s}{2\pi \mathrm{i}}\int_{\kappa=-\mathrm{i}\infty}^{\mathrm{i}\infty}\langle\tilde{E}_y^{\mathrm{s}}\rangle_-^+(\kappa,0,s)\partial\tilde{J}_y^{\mathrm{T};A}(-\kappa,s)\mathrm{d}\kappa\\
&-\frac{s}{2\pi \mathrm{i}}\int_{\kappa=-\mathrm{i}\infty}^{\mathrm{i}\infty}\langle\tilde{H}_x^{\mathrm{s}}\rangle_-^+(\kappa,0,s)\partial\tilde{K}_x^{\mathrm{T};A}(-\kappa,s)\mathrm{d}\kappa\\
&+\frac{s}{2\pi \mathrm{i}}\int_{\kappa=-\mathrm{i}\infty}^{\mathrm{i}\infty}\langle\tilde{E}_y^{\mathrm{s}}\rangle_-^+(\kappa,z_B,s)\partial\tilde{J}_y^{\mathrm{T};B}(-\kappa,s)\mathrm{d}\kappa\\
&-\frac{s}{2\pi \mathrm{i}}\int_{\kappa=-\mathrm{i}\infty}^{\mathrm{i}\infty}\langle\tilde{H}_x^{\mathrm{s}}\rangle_-^+(\kappa,z_B,s)\partial\tilde{K}_x^{\mathrm{T};B}(-\kappa,s)\mathrm{d}\kappa \quad \text{for } \{s\in\mathbb{R}; s>0\}. \quad (11.1)
\end{aligned}$$

Here, $\{\partial\tilde{J}_y^{\mathrm{s};A}, \partial\tilde{J}_y^{\mathrm{s};B}\}$ and $\{\partial\tilde{K}_x^{\mathrm{s};A}, \partial\tilde{K}_x^{\mathrm{s};B}\}$ denote the (transform-domain) electric- and magnetic-current surface densities induced on metasurface $\mathcal{S}_A$ and metasurface $\mathcal{S}_B$, respectively. A similar definition applies to the testing sources, $\{\partial\tilde{J}_y^{\mathrm{T};A}, \partial\tilde{J}_y^{\mathrm{T};B}\}$ and $\{\partial\tilde{K}_x^{\mathrm{T};A}, \partial\tilde{K}_x^{\mathrm{T};B}\}$, that occur on the right-hand side of the reciprocity relation. Furthermore, we use $\{\tilde{E}_y^{\mathrm{T};A}, \tilde{H}_x^{\mathrm{T};A}\}$ to denote the (transform-domain) electric- and magnetic-field strength generated by the testing current sources on metasurface $\mathcal{S}_A$ and a similar meaning is attributed to $\{\tilde{E}_y^{\mathrm{T};B}, \tilde{H}_x^{\mathrm{T};B}\}$ representing the testing fields excited by the testing sources located on $\mathcal{S}_B$. Finally, $\{\tilde{E}_y^{\mathrm{s}}, \tilde{H}_x^{\mathrm{s}}\}$ are the transform-domain scattered fields and the meaning of operator $\langle\rangle_-^+$ was defined by (3.11).

The way how the testing-field quantities are related to their sources can be readily determined using the relevant transform-domain excitation conditions. Starting with the testing fields excited by the sources on $\mathcal{S}_A$, we have (cf. (8.32a) and (8.32b))

$$\lim_{z\downarrow\delta/2}\tilde{H}_x^{\mathrm{T};A}(\kappa,z,s)-\lim_{z\uparrow-\delta/2}\tilde{H}_x^{\mathrm{T};A}(\kappa,z,s)=\partial\tilde{J}_y^{\mathrm{T};A}(\kappa,s), \tag{11.2a}$$

$$\lim_{z\downarrow\delta/2}\tilde{E}_y^{\mathrm{T};A}(\kappa,z,s)-\lim_{z\uparrow-\delta/2}\tilde{E}_y^{\mathrm{T};A}(\kappa,z,s)=\partial\tilde{K}_x^{\mathrm{T};A}(\kappa,s). \tag{11.2b}$$

Expressing the pertinent transform-domain testing field as a one-dimensional (1-D) wave motion propagating away from metasurface $\mathcal{S}_A$, the excitation conditions then yield (the average of) the horizontal EM field strengths as observed from below and above the metasurface (cf. (8.33) and (8.34)), i.e.

$$\langle\tilde{E}_y^{\mathrm{T};A}\rangle_-^+(\kappa,0,s)=-[\mu_0/2p(\kappa)]\partial\tilde{J}_y^{\mathrm{T};A}(\kappa,s), \tag{11.3}$$

$$\langle\tilde{H}_x^{\mathrm{T};A}\rangle_-^+(\kappa,0,s)=-[p(\kappa)/2\mu_0]\partial\tilde{K}_x^{\mathrm{T};A}(\kappa,s). \tag{11.4}$$

At the level of metasurface $\mathcal{S}_B$, we then obtain

$$\langle\tilde{E}_y^{\mathrm{T};A}\rangle_-^+(\kappa,z_B,s)=\left[-\frac{\mu_0}{2p(\kappa)}\partial\tilde{J}_y^{\mathrm{T};A}(\kappa,s)\pm\frac{1}{2}\partial\tilde{K}_x^{\mathrm{T};A}(\kappa,s)\right]\exp[\mp sp(\kappa)z_B], \tag{11.5}$$

$$\langle\tilde{H}_x^{\mathrm{T};A}\rangle_-^+(\kappa,z_B,s)=\left[-\frac{p(\kappa)}{2\mu_0}\partial\tilde{K}_x^{\mathrm{T};A}(\kappa,s)\pm\frac{1}{2}\partial\tilde{J}_y^{\mathrm{T};A}(\kappa,s)\right]\exp[\mp sp(\kappa)z_B],$$

$$\text{for } z_B\gtrless 0, \tag{11.6}$$

respectively. In a similar manner, we shall determine the transform-domain testing fields due to the sources located on $\mathcal{S}_B$. The pertaining excitation conditions, that is,

$$\lim_{z\downarrow z_B+\delta/2}\tilde{H}_x^{\mathrm{T};B}(\kappa,z,s)-\lim_{z\uparrow z_B-\delta/2}\tilde{H}_x^{\mathrm{T};B}(\kappa,z,s)=\partial\tilde{J}_y^{\mathrm{T};B}(\kappa,s), \tag{11.7a}$$

$$\lim_{z\downarrow z_B+\delta/2}\tilde{E}_y^{\mathrm{T};B}(\kappa,z,s)-\lim_{z\uparrow z_B-\delta/2}\tilde{E}_y^{\mathrm{T};B}(\kappa,z,s)=\partial\tilde{K}_x^{\mathrm{T};B}(\kappa,s), \tag{11.7b}$$

yield the following testing fields at the source level:

$$\langle\tilde{E}_y^{\mathrm{T};B}\rangle_-^+(\kappa,z_B,s)=-[\mu_0/2p(\kappa)]\partial\tilde{J}_y^{\mathrm{T};B}(\kappa,s), \tag{11.8}$$

$$\langle\tilde{H}_x^{\mathrm{T};B}\rangle_-^+(\kappa,z_B,s)=-[p(\kappa)/2\mu_0]\partial\tilde{K}_x^{\mathrm{T};B}(\kappa,s). \tag{11.9}$$

Furthermore, at the level of metasurface $\mathcal{S}_A$, we get

$$\langle\tilde{E}_y^{\mathrm{T};B}\rangle_-^+(\kappa,0,s)=\left[-\frac{\mu_0}{2p(\kappa)}\partial\tilde{J}_y^{\mathrm{T};B}(\kappa,s)\mp\frac{1}{2}\partial\tilde{K}_x^{\mathrm{T};B}(\kappa,s)\right]\exp[\mp sp(\kappa)z_B], \tag{11.10}$$

$$\langle\tilde{H}_x^{\mathrm{T};B}\rangle_-^+(\kappa,0,s)=\left[-\frac{p(\kappa)}{2\mu_0}\partial\tilde{K}_x^{\mathrm{T};B}(\kappa,s)\mp\frac{1}{2}\partial\tilde{J}_y^{\mathrm{T};B}(\kappa,s)\right]\exp[\mp sp(\kappa)z_B],$$

$$\text{for } z_B\gtrless 0, \tag{11.11}$$

respectively. The right-hand side of the reciprocity relation is finally rewritten with the aid of the pertinent saltus-type conditions (4.2a) and (4.2b). Since the incident field is continuous across the metasurfaces, we may write

$$\langle \tilde{E}_y^{\mathrm{s}} \rangle_-^+(\kappa, 0, s) = -\tilde{E}_y^{\mathrm{i}}(\kappa, 0, s) + \hat{Z}_A^{\mathrm{E}}(s)\partial \tilde{J}_y^{\mathrm{s};A}(\kappa, s), \tag{11.12a}$$

$$\langle \tilde{H}_x^{\mathrm{s}} \rangle_-^+(\kappa, 0, s) = -\tilde{H}_x^{\mathrm{i}}(\kappa, 0, s) + \hat{Y}_A^{\mathrm{M}}(s)\partial \tilde{K}_x^{\mathrm{s};A}(\kappa, s), \tag{11.12b}$$

when crossing metasurface $\mathcal{S}_A$, and

$$\langle \tilde{E}_y^{\mathrm{s}} \rangle_-^+(\kappa, z_B, s) = -\tilde{E}_y^{\mathrm{i}}(\kappa, z_B, s) + \hat{Z}_B^{\mathrm{E}}(s)\partial \tilde{J}_y^{\mathrm{s};B}(\kappa, s), \tag{11.12c}$$

$$\langle \tilde{H}_x^{\mathrm{s}} \rangle_-^+(\kappa, z_B, s) = -\tilde{H}_x^{\mathrm{i}}(\kappa, z_B, s) + \hat{Y}_B^{\mathrm{M}}(s)\partial \tilde{K}_x^{\mathrm{s};B}(\kappa, s), \tag{11.12d}$$

when crossing metasurface $\mathcal{S}_B$.

11.2 Problem solution

The reciprocity relation (11.1) is next solved numerically via the CdH-MoM. To that end, we follow in part the approach pursued in Section 10.1 and discretize the solution space–time domain uniformly. In this process, the continuous time axis is replaced by $\{t_k = k\Delta t; \Delta t > 0, k = 1, 2, \ldots, M\}$, and the spatial domains $\mathcal{S}_A$ and $\mathcal{S}_B$ are represented by their partitions $\Delta\mathcal{S}_A$ and $\Delta\mathcal{S}_B$, respectively (see (10.4a) and (10.4b)). The induced electric- and magnetic-current densities are further expanded in triangular functions both in space and time. The unknown current densities induced on metasurface $\mathcal{S}_A$ can be then represented in the transform domain via (cf. (8.6) and (8.38))

$$\partial \tilde{J}_y^{\mathrm{s};A}(\kappa, s) \simeq \sum_{n=1}^{N_A} \sum_{k=1}^{M} j_k^{[n]} \tilde{\Lambda}_A^{[n]}(\kappa) \hat{\Lambda}_k(s), \tag{11.13a}$$

$$\partial \tilde{K}_x^{\mathrm{s};A}(\kappa, s) \simeq \sum_{n=2}^{N_A-1} \sum_{k=1}^{M} k_k^{[n]} \tilde{\Lambda}_A^{[n]}(\kappa) \hat{\Lambda}_k(s), \tag{11.13b}$$

where the original of $\tilde{\Lambda}_A^{[n]}(\kappa)$ can be defined via (8.7a)–(8.7c) with the spatial step $\Delta = \Delta_A$ pertaining to $\Delta\mathcal{S}_A$. In a similar fashion, we may write

$$\partial \tilde{J}_y^{\mathrm{s};B}(\kappa, s) \simeq \sum_{n=N_A+1}^{N_A+N_B} \sum_{k=1}^{M} j_k^{[n]} \tilde{\Lambda}_B^{[n]}(\kappa) \hat{\Lambda}_k(s), \tag{11.13c}$$

$$\partial \tilde{K}_x^{\mathrm{s};B}(\kappa, s) \simeq \sum_{n=N_A+2}^{N_A+N_B-1} \sum_{k=1}^{M} k_k^{[n]} \tilde{\Lambda}_B^{[n]}(\kappa) \hat{\Lambda}_k(s), \tag{11.13d}$$

for the induced currents over metasurface $\mathcal{S}_B$. Furthermore, the testing current densities are chosen to behave as the Dirac-impulse distribution in time with the constant

distribution over a spatial discretization segment (see (8.8)). In the transform domain, the testing functions can be hence represented via

$$\partial \tilde{J}_y^{\mathrm{T};A}(\kappa, s) = \tilde{\Pi}_A^{[S]}(\kappa) \text{ for } S = \{1, \ldots, N_A\}, \tag{11.14a}$$

$$\partial \tilde{K}_x^{\mathrm{T};A}(\kappa, s) = \tilde{\Pi}_A^{[S]}(\kappa) \text{ for } S = \{2, \ldots, N_A - 1\}, \tag{11.14b}$$

$$\partial \tilde{J}_y^{\mathrm{T};B}(\kappa, s) = \tilde{\Pi}_B^{[S]}(\kappa) \text{ for } S = \{N_A + 1, \ldots, N_A + N_B\}, \tag{11.14c}$$

$$\partial \tilde{K}_x^{\mathrm{T};B}(\kappa, s) = \tilde{\Pi}_B^{[S]}(\kappa) \text{ for } S = \{N_A + 2, \ldots, N_A + N_B - 1\}, \tag{11.14d}$$

where $\tilde{\Pi}_{A,B}^{[S]}(\kappa)$ denotes the slowness-domain counterpart of the rectangular functions, $\Pi_{A,B}^{[S]}(x)$, whose definition can be deduced from (8.9a)–(8.9c) with $\Delta = \Delta_{A,B}$, respectively. Finally, making use of (11.3)–(11.6) with (11.8)–(11.14d) in the starting reciprocity relation (11.1), we end up with a system of equations in the s-domain whose block-matrix form is depicted in Figure 11.2. The square $[2(N_A + N_B - 2) \times 2(N_A + N_B - 2)]$ two-dimensional (2-D) array on the left-hand side of the system of equations is partitioned into an impeditivity $[(N_A + N_B) \times (N_A + N_B)]$ array, $\hat{\underline{\mathcal{Z}}}$, an admittivity $[(N_A + N_B - 4) \times (N_A + N_B - 4)]$ array, $\hat{\underline{\mathcal{Y}}}$, and two additional (magneto-dielectric) coupling arrays, $\hat{\underline{\mathcal{F}}}$ and $\hat{\underline{\mathcal{Q}}}$, of dimensions $[(N_A + N_B) \times (N_A + N_B - 4)]$ and $[(N_A + N_B - 4) \times (N_A + N_B)]$, respectively. Furthermore, the $[(N_A + N_B) \times 1]$ array of unknown coefficients representing the induced electric-current surface density on $\mathcal{S}_A$ and $\mathcal{S}_B$ consists of two parts, $\hat{\boldsymbol{J}}_A$ and $\hat{\boldsymbol{J}}_B$, respectively. Owing to the end conditions applying

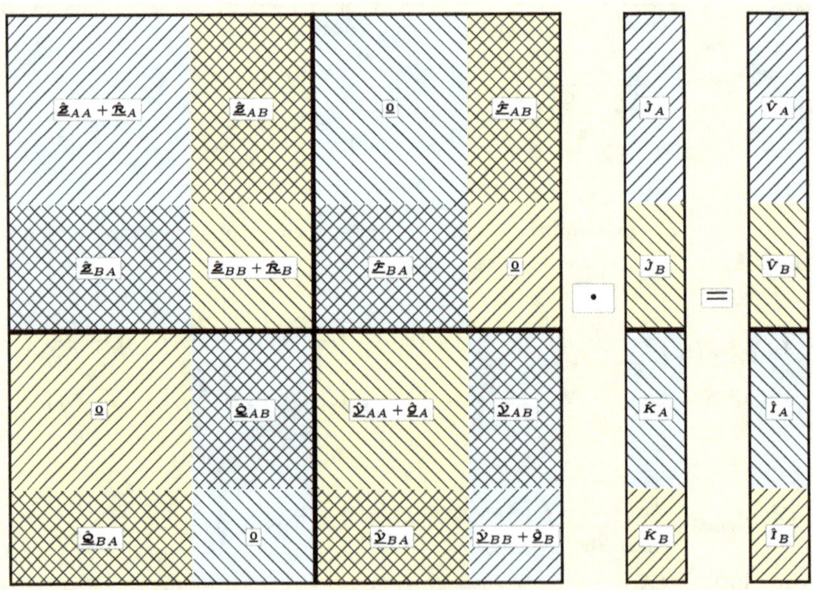

Figure 11.2 Symbolical partitioned matrix description of the EM coupling problem consisting of two metasurfaces

to the induced magnetic-current surface density on the two metasurfaces (see (8.37)), the magnetic current is represented by $\hat{\boldsymbol{K}}_A$ and $\hat{\boldsymbol{K}}_B$ via a $[(N_A + N_B - 4) \times 1]$ array. In a similar way, the excitation array on the right-hand side of the system is composed of two parts, $\hat{\boldsymbol{V}}_{A,B}$ and $\hat{\boldsymbol{I}}_{A,B}$, pertaining to the action of the tangential components of incident electric- and magnetic-field strengths along $\mathcal{S}_{A,B}$, respectively. The elements of the subarrays are specified separately in the following subsections.

11.2.1 Entries of $\hat{\underline{\mathcal{Z}}}$ and $\hat{\underline{\mathcal{R}}}$

The entries of the impeditivity array can be represented via the following integrals (cf. (B.1)):

$$\hat{\mathcal{Z}}_{AA}^{[S,n]}(s) = \frac{\zeta_0}{sc_0\Delta t}\frac{1}{2\pi \mathrm{i}}\int_{\kappa=-\mathrm{i}\infty}^{\mathrm{i}\infty} \tilde{\Pi}_A^{[S]}(\kappa)\tilde{\Lambda}_A^{[n]}(-\kappa)\frac{\mathrm{d}\kappa}{2p(\kappa)}, \tag{11.15}$$

$$\hat{\mathcal{R}}_{A}^{[S,n]}(s) = \frac{\hat{Z}_A^{\mathrm{E}}(s)}{c_0\Delta t}\frac{c_0}{s}\frac{1}{2\pi \mathrm{i}}\int_{\kappa=-\mathrm{i}\infty}^{\mathrm{i}\infty} \tilde{\Pi}_A^{[S]}(-\kappa)\tilde{\Lambda}_A^{[n]}(\kappa)\mathrm{d}\kappa,$$

$$\text{for all } S = \{1,\ldots,N_A\} \text{ and } n = \{1,\ldots,N_A\}, \tag{11.16}$$

thus representing the local interactions on metasurface $\mathcal{S}_A$ whose EM constitutive properties are specified by the equivalent layer impedance $\hat{Z}_A^{\mathrm{E}}$. The local interactions on metasurface $\mathcal{S}_B$ can be described in a similar manner via

$$\hat{\mathcal{Z}}_{BB}^{[S,n]}(s) = \frac{\zeta_0}{sc_0\Delta t}\frac{1}{2\pi \mathrm{i}}\int_{\kappa=-\mathrm{i}\infty}^{\mathrm{i}\infty} \tilde{\Pi}_B^{[S]}(\kappa)\tilde{\Lambda}_B^{[n]}(-\kappa)\frac{\mathrm{d}\kappa}{2p(\kappa)}, \tag{11.17}$$

$$\hat{\mathcal{R}}_{B}^{[S,n]}(s) = \frac{\hat{Z}_B^{\mathrm{E}}(s)}{c_0\Delta t}\frac{c_0}{s}\frac{1}{2\pi \mathrm{i}}\int_{\kappa=-\mathrm{i}\infty}^{\mathrm{i}\infty} \tilde{\Pi}_B^{[S]}(-\kappa)\tilde{\Lambda}_B^{[n]}(\kappa)\mathrm{d}\kappa,$$

$$\text{for all } S = \{N_A+1,\ldots,N_A+N_B\} \text{ and } n = \{N_A+1,\ldots,N_A+N_B\}. \tag{11.18}$$

The elements of the mutual-coupling impeditivity subarrays are given by (cf. (10.13))

$$\hat{\mathcal{Z}}_{AB}^{[S,n]}(s) = \frac{\zeta_0}{sc_0\Delta t}\frac{1}{2\pi \mathrm{i}}\int_{\kappa=-\mathrm{i}\infty}^{\mathrm{i}\infty} \tilde{\Pi}_A^{[S]}(\kappa)\tilde{\Lambda}_B^{[n]}(-\kappa)\exp[-sp(\kappa)|z_B|]\frac{\mathrm{d}\kappa}{2p(\kappa)}$$

$$\text{for all } S = \{1,\ldots,N_A\} \text{ and } n = \{N_A+1,\ldots,N_A+N_B\}, \tag{11.19a}$$

and

$$\hat{\mathcal{Z}}_{BA}^{[S,n]}(s) = \frac{\zeta_0}{sc_0\Delta t}\frac{1}{2\pi \mathrm{i}}\int_{\kappa=-\mathrm{i}\infty}^{\mathrm{i}\infty} \tilde{\Pi}_B^{[S]}(\kappa)\tilde{\Lambda}_A^{[n]}(-\kappa)\exp[-sp(\kappa)|z_B|]\frac{\mathrm{d}\kappa}{2p(\kappa)}$$

$$\text{for all } S = \{N_A+1,\ldots,N_A+N_B\} \text{ and } n = \{1,\ldots,N_A\}. \tag{11.19b}$$

By virtue of EM reciprocity, the elements of the mutual-coupling impeditivity subarrays are interrelated via condition (10.5).

Finally, the time-domain (TD) counterparts of self-impeditivity entries, $\hat{\mathcal{Z}}_{AA}^{[S,n]}(s)$ and $\hat{\mathcal{Z}}_{BB}^{[S,n]}(s)$, can be found using the results of Appendix B. The inclusion of bounded electric constitutive parameters via $\hat{\mathcal{R}}_{A,B}^{[S,n]}(s)$ is described in Section 8.2 and the TD original of the mutual-impeditivity array has been derived in Section 10.2.1 using the results of Appendix B or F.

11.2.2 Entries of $\underline{\hat{\mathcal{Y}}}$ *and* $\underline{\hat{\mathcal{G}}}$

The entries of the admittivity array can be represented via the following integrals (cf. (E.1) and (8.42)):

$$\hat{\mathcal{Y}}_{AA}^{[S,n]}(s) = \frac{\eta_0}{sc_0\Delta t}\frac{c_0^2}{2\pi\mathrm{i}}\int_{\kappa=-\mathrm{i}\infty}^{\mathrm{i}\infty}\tilde{\Pi}_A^{[S]}(\kappa)\tilde{\Lambda}_A^{[n]}(-\kappa)\frac{p(\kappa)\mathrm{d}\kappa}{2}, \tag{11.20}$$

$$\hat{\mathcal{G}}_{A}^{[S,n]}(s) = \frac{\hat{Y}_A^{\mathrm{M}}(s)}{c_0\Delta t}\frac{c_0}{s}\frac{1}{2\pi\mathrm{i}}\int_{\kappa=-\mathrm{i}\infty}^{\mathrm{i}\infty}\tilde{\Pi}_A^{[S]}(-\kappa)\tilde{\Lambda}_A^{[n]}(\kappa)\mathrm{d}\kappa,$$

$$\text{for all } S = \{2,\ldots,N_A-1\} \text{ and } n = \{2,\ldots,N_A-1\}, \tag{11.21}$$

thereby representing the local interactions on metasurface $\mathcal{S}_A$ whose EM constitutive properties are specified by the equivalent layer admittance $\hat{Y}_A^{\mathrm{M}}$. In a similar fashion, we may describe the local interactions on metasurface $\mathcal{S}_B$, that is,

$$\hat{\mathcal{Y}}_{BB}^{[S,n]}(s) = \frac{\eta_0}{sc_0\Delta t}\frac{c_0^2}{2\pi\mathrm{i}}\int_{\kappa=-\mathrm{i}\infty}^{\mathrm{i}\infty}\tilde{\Pi}_B^{[S]}(\kappa)\tilde{\Lambda}_B^{[n]}(-\kappa)\frac{p(\kappa)\mathrm{d}\kappa}{2}, \tag{11.22}$$

$$\hat{\mathcal{G}}_{B}^{[S,n]}(s) = \frac{\hat{Y}_B^{\mathrm{M}}(s)}{c_0\Delta t}\frac{c_0}{s}\frac{1}{2\pi\mathrm{i}}\int_{\kappa=-\mathrm{i}\infty}^{\mathrm{i}\infty}\tilde{\Pi}_B^{[S]}(-\kappa)\tilde{\Lambda}_B^{[n]}(\kappa)\mathrm{d}\kappa,$$

$$\text{for all } S = \{N_A+2,\ldots,N_A+N_B-1\}$$

$$\text{and } n = \{N_A+2,\ldots,N_A+N_B-1\}. \tag{11.23}$$

Finally, the entries of the mutual-admittance subarrays are given by

$$\hat{\mathcal{Y}}_{AB}^{[S,n]}(s) = \frac{\eta_0}{sc_0\Delta t}\frac{c_0^2}{2\pi\mathrm{i}}\int_{\kappa=-\mathrm{i}\infty}^{\mathrm{i}\infty}\tilde{\Pi}_A^{[S]}(\kappa)\tilde{\Lambda}_B^{[n]}(-\kappa)\exp[-sp(\kappa)|z_B|]\frac{p(\kappa)\mathrm{d}\kappa}{2}$$

$$\text{for all } S = \{2,\ldots,N_A-1\} \text{ and } n = \{N_A+2,\ldots,N_A+N_B-1\}, \tag{11.24}$$

and

$$\hat{\mathcal{Y}}_{BA}^{[S,n]}(s) = \frac{\eta_0}{sc_0\Delta t}\frac{c_0^2}{2\pi\mathrm{i}}\int_{\kappa=-\mathrm{i}\infty}^{\mathrm{i}\infty}\tilde{\Pi}_B^{[S]}(\kappa)\tilde{\Lambda}_A^{[n]}(-\kappa)\exp[-sp(\kappa)|z_B|]\frac{p(\kappa)\mathrm{d}\kappa}{2}$$

$$\text{for all } S = \{N_A+2,\ldots,N_A+N_B-1\} \text{ and } n = \{2,\ldots,N_A-1\}. \tag{11.25}$$

Owing to the property of EM reciprocity, the mutual-admittance subarrays are each other's transpose.

Finally, the TD counterparts of self-admittivity entries, $\hat{\mathcal{Y}}_{AA}^{[S,n]}(s)$ and $\hat{\mathcal{Y}}_{BB}^{[S,n]}(s)$, can be found using the results of Appendix E. The inclusion of bounded magnetic constitutive parameters via $\hat{\mathcal{G}}_{A,B}^{[S,n]}(s)$ is described in Section 8.3.1 and the TD original of the mutual-admittivity array can be readily found with the aid of Appendix I.

11.2.3 Entries of $\hat{\underline{\mathcal{F}}}$ and $\hat{\underline{\mathcal{Q}}}$

The nonzero elements of $\hat{\underline{\mathcal{F}}}$ can be expressed via

$$\hat{\mathcal{F}}_{AB}^{[S,n]}(s) = \frac{\operatorname{sign}(z_B)}{c_0\Delta t}\frac{c_0}{2s}\frac{1}{2\pi\mathrm{i}}\int_{\kappa=-\mathrm{i}\infty}^{\mathrm{i}\infty}\tilde{\Pi}_A^{[S]}(\kappa)\tilde{\Lambda}_B^{[n]}(-\kappa)\exp[-sp(\kappa)|z_B|]\mathrm{d}\kappa$$
$$\text{for all } S=\{1,\ldots,N_A\} \text{ and } n=\{N_A+2,\ldots,N_A+N_B-1\}, \quad (11.26\text{a})$$

and

$$\hat{\mathcal{F}}_{BA}^{[S,n]}(s) = -\frac{\operatorname{sign}(z_B)}{c_0\Delta t}\frac{c_0}{2s}\frac{1}{2\pi\mathrm{i}}\int_{\kappa=-\mathrm{i}\infty}^{\mathrm{i}\infty}\tilde{\Pi}_B^{[S]}(\kappa)\tilde{\Lambda}_A^{[n]}(-\kappa)\exp[-sp(\kappa)|z_B|]\mathrm{d}\kappa$$
$$\text{for all } S=\{N_A+1,\ldots,N_A+N_B\} \text{ and } n=\{2,\ldots,N_A-1\}. \quad (11.26\text{b})$$

The entries of the remaining mutual-coupling array, $\hat{\underline{\mathcal{Q}}}$, can be expressed in a much similar fashion as

$$\hat{\mathcal{Q}}_{AB}^{[S,n]}(s) = \frac{\operatorname{sign}(z_B)}{c_0\Delta t}\frac{c_0}{2s}\frac{1}{2\pi\mathrm{i}}\int_{\kappa=-\mathrm{i}\infty}^{\mathrm{i}\infty}\tilde{\Pi}_A^{[S]}(\kappa)\tilde{\Lambda}_B^{[n]}(-\kappa)\exp[-sp(\kappa)|z_B|]\mathrm{d}\kappa$$
$$\text{for all } S=\{2,\ldots,N_A-1\} \text{ and } n=\{N_A+1,\ldots,N_A+N_B\}, \quad (11.27\text{a})$$

and

$$\hat{\mathcal{Q}}_{BA}^{[S,n]}(s) = -\frac{\operatorname{sign}(z_B)}{c_0\Delta t}\frac{c_0}{2s}\frac{1}{2\pi\mathrm{i}}\int_{\kappa=-\mathrm{i}\infty}^{\mathrm{i}\infty}\tilde{\Pi}_B^{[S]}(\kappa)\tilde{\Lambda}_A^{[n]}(-\kappa)\exp[-sp(\kappa)|z_B|]\mathrm{d}\kappa$$
$$\text{for all } S=\{N_A+2,\ldots,N_A+N_B-1\} \text{ and } n=\{1,\ldots,N_A\}. \quad (11.27\text{b})$$

The TD originals of the coupling terms described by (11.26a)–(11.27b) can be found with the aid of the generic integral representation closely analyzed in Appendix J. It is finally noted that the coupling arrays, $\hat{\underline{\mathcal{F}}}$ and $\hat{\underline{\mathcal{Q}}}$, are identically zero when the metasurfaces lie in a plane, i.e. when $z_B = 0$. In such a case, the system of equations can be decoupled into two independent systems that can be solved, separately, for the induced electric and magnetic currents (see Figure (11.2)).

11.2.4 Entries of the excitation array

Since the elements of the voltage-type excitation arrays $\boldsymbol{V}_{A,B}$ have been specified in Section 10.2.1, we shall further focus on the electric-current-type excitation arrays

$\boldsymbol{I}_{A,B}$ in the TD. Starting with the elements pertaining to metasurface $\mathcal{S}_A$, we may write (cf. (8.45))

$$I_A^{[S]}(t) = \frac{Y_\perp(\kappa_0)}{\kappa_0} \left\{ \int_{\tau=0}^{t-\kappa_0(x_S-\Delta_A/2)} e^{\mathrm{i}}(\tau)\mathrm{d}\tau - \int_{\tau=0}^{t-\kappa_0(x_S+\Delta_A/2)} e^{\mathrm{i}}(\tau)\mathrm{d}\tau \right\}$$
$$\text{for all } S = \{2, \ldots, N_A - 1\}, \tag{11.28}$$

if $\kappa_0 \neq 0$, and

$$I_A^{[S]}(t) = \eta_0 e^{\mathrm{i}}(t)\Delta_A \quad \text{for all } S = \{2, \ldots, N_A - 1\}, \tag{11.29}$$

for the normal incidence characterized by $\beta = 0$ and hence $\kappa_0 = 0$ (see Figure 11.1). In a similar manner, we shall proceed with the entries corresponding to metasurface $\mathcal{S}_B$. Hence, we write

$$I_B^{[S]}(t) = \frac{Y_\perp(\kappa_0)}{\kappa_0} \left\{ \int_{\tau=0}^{t+p(\kappa_0)z_B-\kappa_0(x_S-\Delta_B/2)} e^{\mathrm{i}}(\tau)\mathrm{d}\tau \right.$$
$$\left. - \int_{\tau=0}^{t+p(\kappa_0)z_B-\kappa_0(x_S+\Delta_B/2)} e^{\mathrm{i}}(\tau)\mathrm{d}\tau \right\}$$
$$\text{for all } S = \{N_A + 2, \ldots, N_A + N_B - 1\}, \tag{11.30}$$

if $\kappa_0 \neq 0$, and

$$I_B^{[S]}(t) = \eta_0 e^{\mathrm{i}}(t + z_B/c_0)\Delta_B$$
$$\text{for all } S = \{N_A + 2, \ldots, N_A + N_B - 1\}, \tag{11.31}$$

applying to the limit $\beta = 0$.

11.2.5 Solving the TD system of equations

In the final step, the subarrays forming the system of equations are cast into the block-matrix form (see Figure 11.2) and transformed to the TD. In this way, we arrive at the following convolution-type system of equations (cf. (7.42)):

$$\sum_{k=1}^{m} \left(\underline{\boldsymbol{A}}_{m-k+1} - 2\underline{\boldsymbol{A}}_{m-k} + \underline{\boldsymbol{A}}_{m-k-1} \right) \cdot \boldsymbol{X}_k = \boldsymbol{F}_m \tag{11.32}$$

in which $\underline{\boldsymbol{A}}_k = \underline{\boldsymbol{A}}(t_k)$ represents a $[2(N_A + N_B - 2) \times 2(N_A + N_B - 2)]$ 2-D array that interrelates the 1-D $[(N_A + N_B - 4) \times 1]$ array of unknown electric- and magnetic-current coefficients, $\boldsymbol{X}_k$, with the 1-D $[(N_A + N_B - 4) \times 1]$ excitation array. If the metasurfaces under consideration are coplanar, (11.32) decouples into (8.10) and (8.40) that can be solved independently via (8.11) and (8.46) for the current coefficients. In general, however, the induced electric and magnetic currents

are not independent, so that the full system (11.32) must be solved. This can be accomplished, again, via the stable marching-on-in-time procedure as (cf. (8.11))

$$\boldsymbol{X}_m = \underline{\boldsymbol{A}}_1^{-1} \cdot \left[\boldsymbol{F}_m - \sum_{k=1}^{m-1} \left(\underline{\boldsymbol{A}}_{m-k+1} - 2\underline{\boldsymbol{A}}_{m-k} + \underline{\boldsymbol{A}}_{m-k-1} \right) \cdot \boldsymbol{X}_k \right]$$

$$\text{for all } m = \{1, \ldots, M\}, \tag{11.33}$$

thereby achieving $j_m^{[n]}$ for all $n = \{1, \ldots, N_A + N_B\}$ and $k_m^{[n]}$ for all $n = \{2, ...N_A - 1\} \cup \{N_A + 2, \ldots, N_A + N_B - 1\}$ (see (11.13a)–(11.13d)).

11.3 Narrow-screen approximation

The problem configuration under consideration consists of two metasurfaces whose width $w > 0$ is relatively small with respect to the spatial support of the excitation EM plane-wave signature. In accordance with Figure 10.4, it is assumed that such metastrips occupy domains $\mathcal{S}_A = \{-w/2 < x < w/2, -\infty < y < \infty, -\delta/2 < z < \delta/2\}$ and $\mathcal{S}_B = \{x_B - w/2 < x < x_B + w/2, -\infty < y < \infty, z_B - \delta/2 < z < z_B + \delta/2\}$.

To solve the problem, we employ the transform-domain reciprocity relation (11.1), where we use relations (11.3)–(11.6) with (11.8)–(11.12d), again. The testing-current distribution is assumed to be uniform (see (7.29a) and (7.29b)), again, which yields

$$\partial \tilde{J}_y^{\mathrm{T};A}(\kappa, s) = \hat{I}^{\mathrm{T};A}(s)\mathrm{i}_0(s\kappa w/2), \tag{11.34a}$$

$$\partial \tilde{K}_x^{\mathrm{T};A}(\kappa, s) = \hat{K}^{\mathrm{T};A}(s)\mathrm{i}_0(s\kappa w/4), \tag{11.34b}$$

on metastrip $\mathcal{S}_A$, and

$$\partial \tilde{J}_y^{\mathrm{T};B}(\kappa, s) = \hat{I}^{\mathrm{T};B}(s)\mathrm{i}_0(s\kappa w/2)\exp(s\kappa x_B), \tag{11.34c}$$

$$\partial \tilde{K}_x^{\mathrm{T};B}(\kappa, s) = \hat{K}^{\mathrm{T};B}(s)\mathrm{i}_0(s\kappa w/4)\exp(s\kappa x_B), \tag{11.34d}$$

on metastrip $\mathcal{S}_B$, where x_B denotes the horizontal shift of $\mathcal{S}_B$ with respect to the origin. With the narrow-screen assumption in mind, the induced electric-current surface density is supposed to be uniformly distributed over the metastrips. To meet the pertinent edge condition, we further postulate the triangular spatial distribution of the induced magnetic-current surface density (see (7.32b)). Accordingly, in the transform domain, we get

$$\partial \tilde{J}_y^{\mathrm{s};A}(\kappa, s) = \hat{I}^{\mathrm{s};A}(s)\mathrm{i}_0(s\kappa w/2), \tag{11.35a}$$

$$\partial \tilde{K}_x^{\mathrm{s};A}(\kappa, s) = \hat{K}^{\mathrm{s};A}(s)\mathrm{i}_0^2(s\kappa w/4), \tag{11.35b}$$

for the electric- and magnetic-current surface densities induced in metastrip $\mathcal{S}_A$, respectively. In a similar fashion, we find the expressions for the transform-domain electric- and magnetic-current surface densities induced in metastrip $\mathcal{S}_B$, that is,

$$\partial \tilde{J}_y^{\mathrm{s};B}(\kappa, s) = \hat{I}^{\mathrm{s};B}(s)\mathrm{i}_0(s\kappa w/2)\exp(s\kappa x_B), \tag{11.35c}$$

$$\partial \tilde{K}_x^{\mathrm{s};B}(\kappa, s) = \hat{K}^{\mathrm{s};B}(s)\mathrm{i}_0^2(s\kappa w/4)\exp(s\kappa x_B), \tag{11.35d}$$

respectively. Then, making use of (11.34a)–(11.35d) in the starting reciprocity relation, we end up with the following system of equations (cf. Figures 11.2 and (10.18)):

$$\begin{bmatrix} \hat{Z}_{AA} + \hat{Z}_A^{\mathrm{E}}/w & \hat{Z}_{AB} & 0 & \hat{F}_{AB} \\ \hat{Z}_{BA} & \hat{Z}_{BB} + \hat{Z}_B^{\mathrm{E}}/w & \hat{F}_{BA} & 0 \\ 0 & \hat{Q}_{AB} & \hat{Y}_{AA} + \frac{3}{2}\hat{Y}_A^{\mathrm{M}}/w & \hat{Y}_{AB} \\ \hat{Q}_{BA} & 0 & \hat{Y}_{BA} & \hat{Y}_{BB} + \frac{3}{2}\hat{Y}_B^{\mathrm{M}}/w \end{bmatrix}$$

$$\cdot \begin{bmatrix} \hat{I}^{\mathrm{s};A} \\ \hat{I}^{\mathrm{s};B} \\ \hat{K}^{\mathrm{s};A} \\ \hat{K}^{\mathrm{s};B} \end{bmatrix} = \begin{bmatrix} \hat{E}_y^{\mathrm{i}}(0,0,s) \\ \hat{E}_y^{\mathrm{i}}(x_B, z_B, s) \\ \hat{H}_x^{\mathrm{i}}(0,0,s) \\ \hat{H}_x^{\mathrm{i}}(x_B, z_B, s) \end{bmatrix}, \tag{11.36}$$

where $\hat{Z}_{AA}$ and $\hat{Z}_{BB}$ represent the external (per-unit-length) impedances of perfectly electrically conducting (PEC) strips pertaining to $\mathcal{S}_A$ and $\mathcal{S}_B$, respectively, whose slowness representation is given by

$$\hat{Z}_{AA}(s) = \hat{Z}_{BB}(s) = \frac{s}{2\pi \mathrm{i}} \int_{\kappa=-\mathrm{i}\infty}^{\mathrm{i}\infty} \mathrm{i}_0^2(s\kappa w/2)\frac{\mu_0}{2p(\kappa)}\mathrm{d}\kappa. \qquad \text{(10.19 revisited)}$$

The mutual-coupling impedances, $\hat{Z}_{AB}$ and $\hat{Z}_{BA}$, can be represented via

$$\hat{Z}_{AB}(s) = \frac{s}{2\pi \mathrm{i}} \int_{\kappa=-\mathrm{i}\infty}^{\mathrm{i}\infty} \exp\{-s[\kappa x_B + p(\kappa)|z_B|]\}\, \mathrm{i}_0^2(s\kappa w/2)\frac{\mu_0}{2p(\kappa)}\mathrm{d}\kappa, \qquad \text{(10.20 revisited)}$$

which has been approximated by (10.22) as $(s\mu_0/2\pi)\mathrm{K}_0(sT_B)$, with $T_B > 0$ representing the travel time of the pulse to traverse the distance between the centers of metastrips. Furthermore, the self-coupling admittances, $\hat{Y}_{AA}$ and $\hat{Y}_{BB}$, correspond to the external admittances of the pertinent perfectly magnetically conducting (PMC) screens. Consequently, they are given via (cf. (7.35))

$$\hat{Y}_{AA}(s) = \hat{Y}_{BB}(s) = \frac{s}{2\pi \mathrm{i}} \int_{\kappa=-\mathrm{i}\infty}^{\mathrm{i}\infty} \mathrm{i}_0^3(s\kappa w/4)\frac{p(\kappa)}{2\mu_0}\mathrm{d}\kappa. \tag{11.37}$$

The mutual-coupling admittances, as denoted by $\hat{Y}_{AB}$ and $\hat{Y}_{BA}$, are, again, equal to each other. Their slowness representation have the following form:

$$\hat{Y}_{AB}(s) = \frac{s}{2\pi \mathrm{i}} \int_{\kappa=-\mathrm{i}\infty}^{\mathrm{i}\infty} \exp\{-s[\kappa x_B + p(\kappa)|z_B|]\}\, \mathrm{i}_0^3(s\kappa w/4)\frac{p(\kappa)}{2\mu_0}\mathrm{d}\kappa. \tag{11.38}$$

Finally, the remaining coupling entries in the system equations represented by (11.36), that is, $\hat{F}_{AB}$ with $\hat{F}_{BA}$ and $\hat{Q}_{AB}$, $\hat{Q}_{BA}$, are represented in a similar manner:

$$\hat{F}_{AB}(s) = -\hat{F}_{BA}(s) = \frac{\mathrm{sign}(z_B)}{2}\frac{s}{2\pi\mathrm{i}}\int_{\kappa=-\mathrm{i}\infty}^{\mathrm{i}\infty} \exp\{-s[\kappa x_B + p(\kappa)|z_B|]\} \times \mathrm{i}_0(s\kappa w/2)\mathrm{i}_0^2(s\kappa w/4)\mathrm{d}\kappa, \tag{11.39}$$

$$\hat{Q}_{AB}(s) = -\hat{Q}_{BA}(s) = \frac{\mathrm{sign}(z_B)}{2}\frac{s}{2\pi\mathrm{i}}\int_{\kappa=-\mathrm{i}\infty}^{\mathrm{i}\infty} \exp\{-s[\kappa x_B + p(\kappa)|z_B|]\} \times \mathrm{i}_0(s\kappa w/2)\mathrm{i}_0(s\kappa w/4)\mathrm{d}\kappa. \tag{11.40}$$

To derive approximate expressions for the mutual-coupling terms applying to $w \ll r_B = (x_B^2 + z_B^2)^{1/2}$, we use the small-argument expansions of the spherical modified Bessel function along with the standard CdH technique (see Sections 2.1.1 and 10.2.1.1). In this manner, after a few straightforward steps, we arrive at (cf. (10.22))

$$\hat{Y}_{AB}(s) \simeq \frac{s}{2\pi\mu_0}\frac{z_B^2}{r_B^4}\frac{\mathrm{d}^2}{\mathrm{d}s^2}\int_{\tau=T_B}^{\infty} \exp(-s\tau)\frac{\mathrm{d}\tau}{(\tau^2 - T_B^2)^{1/2}} - \frac{s}{2\pi\mu_0}\frac{x_B^2}{r_B^4}\frac{\mathrm{d}^2}{\mathrm{d}s^2}\int_{\tau=T_B}^{\infty} \exp(-s\tau)(\tau^2 - T_B^2)^{1/2}\frac{\mathrm{d}\tau}{\tau^2} \quad \text{for } \{s \in \mathbb{R}; s > 0\}, \tag{11.41}$$

and

$$\hat{F}_{AB}(s) \simeq \hat{Q}_{AB}(s) = -\frac{s}{2\pi}\frac{z_B}{r_B^2}\frac{\mathrm{d}}{\mathrm{d}s}\int_{\tau=T_B}^{\infty} \exp(-s\tau)\frac{\mathrm{d}\tau}{(\tau^2 - T_B^2)^{1/2}} \quad \text{for } \{s \in \mathbb{R}; s > 0\}. \tag{11.42}$$

With the aid of [2, Eq. (29.3.119)], (11.41) and (11.42) are rewritten in terms of the modified Bessel functions of the second kind, that is,

$$\hat{Y}_{AB}(s) \simeq \frac{\eta_0}{2\pi}\frac{z_B^2 - x_B^2}{r_B^3}\mathrm{K}_1(sT_B) + \frac{\eta_0}{2\pi}\frac{z_B^2}{r_B^2}\frac{s}{c_0}\mathrm{K}_0(sT_B), \tag{11.43}$$

$$\hat{F}_{AB}(s) \simeq \hat{Q}_{AB}(s) \simeq \frac{1}{2\pi}\frac{z_B}{r_B}\frac{s}{c_0}\mathrm{K}_1(sT_B). \tag{11.44}$$

In the following step, the induced currents are expanded in a piecewise linear manner in time, which can be represented by

$$\hat{I}^{s;A}(s) \simeq \sum_{k=1}^{M} i_k^A \hat{\Lambda}_k(s), \tag{10.23a revisited}$$

$$\hat{I}^{s;B}(s) \simeq \sum_{k=1}^{M} i_k^B \hat{\Lambda}_k(s), \tag{10.23b revisited}$$

$$\hat{K}^{s;A}(s) \simeq \sum_{k=1}^{M} v_k^A \hat{\Lambda}_k(s), \tag{11.45a}$$

$$\hat{K}^{s;B}(s) \simeq \sum_{k=1}^{M} v_k^B \hat{\Lambda}_k(s), \tag{11.45b}$$

where $i_k^{A,B}$ and $v_k^{A,B}$ denote the unknown coefficients representing the induced electric and magnetic currents induced in the metastrips. Further recall that $\Lambda_k(t)$ denotes the triangular function, an expression for which is given by (7.19).

Having fully specified the array elements in (11.36), we may finally transform the system to the TD. The transformation leads to a convolution-type system of equations, again, that is written as (cf. (10.24))

$$\sum_{k=1}^{m} \left(\underline{\boldsymbol{M}}_{m-k+1} - 2\underline{\boldsymbol{M}}_{m-k} + \underline{\boldsymbol{M}}_{m-k-1} \right) \cdot \boldsymbol{C}_k = \boldsymbol{G}_m, \tag{11.46}$$

where $\underline{\boldsymbol{M}}_k = \underline{\boldsymbol{M}}(t_k)$ is a $[4 \times 4]$ array that is composed of four $[2 \times 2]$ subarrays as

$$\underline{\boldsymbol{M}}_k = \begin{bmatrix} \underline{\boldsymbol{Z}}_k & \underline{\boldsymbol{F}}_k \\ \underline{\boldsymbol{Q}}_k & \underline{\boldsymbol{Y}}_k \end{bmatrix}. \tag{11.47}$$

Its entries are next specified in detail. At first, $\underline{\boldsymbol{Z}}_k$ is a $[2 \times 2]$ impedance array (cf. (10.25))

$$\underline{\boldsymbol{Z}}_k = \frac{\zeta_0}{c_0 \Delta t} \begin{bmatrix} \Phi_A(t_k) & \Psi(t_k) \\ \Psi(t_k) & \Phi_B(t_k) \end{bmatrix}, \tag{11.48}$$

with TD functions given by (cf. (7.26))

$$\begin{aligned} \Phi_A(t) &= [\Upsilon^B(w,t) - 2\Upsilon^B(0,t) + \Upsilon^B(-w,t)]/w^2 \\ &\quad + (c_0/w\alpha_A^{\mathrm{E}})(\eta_0/G_A^{\mathrm{E}})\left[\alpha_A^{\mathrm{E}} t - 1 + \exp(-\alpha_A^{\mathrm{E}} t)\right]\mathrm{H}(t), \end{aligned} \tag{11.49}$$

$$\begin{aligned} \Phi_B(t) &= [\Upsilon^B(w,t) - 2\Upsilon^B(0,t) + \Upsilon^B(-w,t)]/w^2 \\ &\quad + (c_0/w\alpha_B^{\mathrm{E}})(\eta_0/G_B^{\mathrm{E}})\left[\alpha_B^{\mathrm{E}} t - 1 + \exp(-\alpha_B^{\mathrm{E}} t)\right]\mathrm{H}(t), \end{aligned} \tag{11.50}$$

$$\Psi(t) = \cosh^{-1}(t/T_B)\mathrm{H}(t - T_B)/2\pi. \tag{10.26 revisited}$$

Here, $\alpha_{A,B}^{\mathrm{E}} = G_{A,B}^{\mathrm{E}}/C_{A,B}^{\mathrm{E}} > 0$ (see (3.19)) and $\Upsilon^B(x,t)$ is given in Appendix B. Next, $\underline{\boldsymbol{Y}}_k$ denotes a $[2 \times 2]$ admittance array

$$\underline{\boldsymbol{Y}}_k = \frac{\eta_0}{c_0 \Delta t} \begin{bmatrix} \Theta_A(t_k) & \Omega(t_k) \\ \Omega(t_k) & \Theta_B(t_k) \end{bmatrix}, \tag{11.51}$$

where (cf. (7.40))

$$\Theta_A(t) = 8[\Xi(3w/4,t) - 3\Xi(w/4,t) + 3\Xi(-w/4,t) - \Xi(-3w/4,t)]/w^3 + (3c_0/2w\alpha_A^{\mathrm{M}})(\zeta_0/R_A^{\mathrm{M}})\big[\alpha_A^{\mathrm{M}}t - 1 + \exp(-\alpha_A^{\mathrm{M}}t)\big]\mathrm{H}(t), \tag{11.52}$$

$$\Theta_B(t) = 8[\Xi(3w/4,t) - 3\Xi(w/4,t) + 3\Xi(-w/4,t) - \Xi(-3w/4,t)]/w^3 + (3c_0/2w\alpha_B^{\mathrm{M}})(\zeta_0/R_B^{\mathrm{M}})\big[\alpha_B^{\mathrm{M}}t - 1 + \exp(-\alpha_B^{\mathrm{M}}t)\big]\mathrm{H}(t), \tag{11.53}$$

$$\Omega(t) = (z_B^2/4\pi r_B^2)[(t/T_B)(t^2/T_B^2-1)^{1/2} + \cosh^{-1}(t/T_B)]\mathrm{H}(t-T_B) - (x_B^2/4\pi r_B^2)[(t/T_B)(t^2/T_B^2-1)^{1/2} - \cosh^{-1}(t/T_B)]\mathrm{H}(t-T_B). \tag{11.54}$$

Here, $\alpha_{A,B}^{\mathrm{M}} = R_{A,B}^{\mathrm{M}}/L_{A,B}^{\mathrm{M}} > 0$ (see (3.21)) and $\Xi(x,t)$ is given in Appendix E. Finally, the remaining coupling arrays are given as

$$\underline{\boldsymbol{F}}_k = \underline{\boldsymbol{Q}}_k = \frac{1}{c_0\Delta t}\begin{bmatrix} 0 & \Gamma(t_k) \\ \Gamma(t_k) & 0 \end{bmatrix}, \tag{11.55}$$

where

$$\Gamma(t) = (z_B/2\pi r_B)(t^2/T_B^2-1)^{1/2}\mathrm{H}(t-T_B). \tag{11.56}$$

Apparently, if the metastrips lie in one plane, $z_B = 0$, and hence $\Gamma(t) = 0$ for all $t > 0$. In such a circumstance, the convolution-type system (11.46) can be decomposed into two independent systems whose solution yields the coefficients representing the induced electric and magnetic currents. These coefficients are included in a $[4 \times 1]$ 1-D array as follows (cf. (10.27)):

$$\boldsymbol{C}_k = [i_k^A \; i_k^B \; v_k^A \; v_k^B]^{\mathcal{T}}, \tag{11.57}$$

where $^{\mathcal{T}}$ denotes the transpose operator, and the excitation array on the right-hand side of (11.46) is given by (cf. (10.28))

$$\boldsymbol{G}_m = \begin{bmatrix} e^{\mathrm{i}}(t_m) \\ e^{\mathrm{i}}[t_m - \kappa_0 x_B + p(\kappa_0)z_B] \\ Y_\perp(\kappa_0)e^{\mathrm{i}}(t_m) \\ Y_\perp(\kappa_0)e^{\mathrm{i}}[t_m - \kappa_0 x_B + p(\kappa_0)z_B] \end{bmatrix}. \tag{11.58}$$

Recall here that $Y_\perp(\kappa_0) = \eta_0\cos(\beta)$. With all the components at our disposal, the system of equations can be solved for the current coefficients using the following step-by-step updating procedure (cf. (8.11)):

$$\boldsymbol{C}_m = \underline{\boldsymbol{M}}_1^{-1} \cdot \left[\boldsymbol{G}_m - \sum_{k=1}^{m-1}\left(\underline{\boldsymbol{M}}_{m-k+1} - 2\underline{\boldsymbol{M}}_{m-k} + \underline{\boldsymbol{M}}_{m-k-1}\right)\cdot \boldsymbol{C}_k\right] \quad \text{for all } m = \{1,\ldots,M\}. \tag{11.59}$$

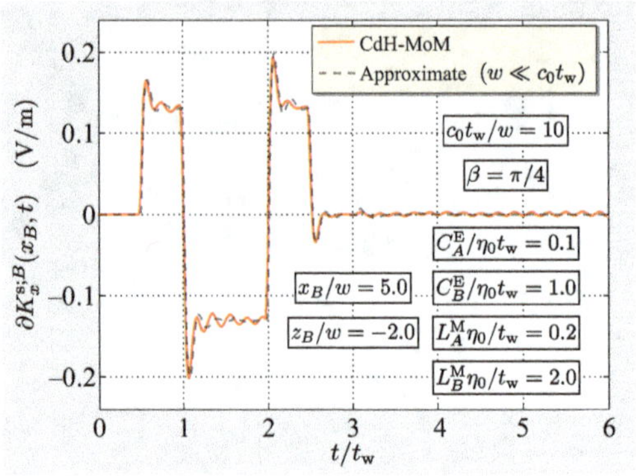

Figure 11.3 Plane-wave induced magnetic-current surface density at the center of metastrip $\mathcal{S}_B$

11.4 Illustrative examples

In this section, we shall analyze EM scattering by two noncoplanar relatively narrow metasurfaces. Their width, $w > 0$, is taken to be a tenth of $c_0 t_w$ of the triangular signature, $e^i(t)$, as defined by (8.15) (see Figure 8.3). We further take the unit amplitude $e_m = 1.0\,\text{V/m}$ and the angle of incidence is $\beta = \pi/4$ (see Figure 11.1). The metastrips occupy $\mathcal{S}_A = \{-w/2 < x < w/2, -\infty < y < \infty, z = 0\}$ and $\mathcal{S}_B = \{x_B - w/2 < x < x_B + w/2, -\infty < y < \infty, z = z_B\}$, where $(x_B, z_B) = (5.0w, -2.0w)$ defines the spatial shift of $\mathcal{S}_B$ with respect to the origin. As far as their EM constitutive properties are concerned, we consider loss-free metastrips described by $C_A^E/\eta_0 t_w = 0.1$ and $L_A^M \eta_0/t_w = 0.2$ with $C_B^E/\eta_0 t_w = 1.0$ and $L_B^M \eta_0/t_w = 2.0$, thus meeting the pertinent transparency conditions $C_{A,B}^E/L_{A,B}^M = Y_\perp^2(\kappa_0)$ (see (4.8)). Owing to the relatively small width, $c_0 t_w/w = 10$, the CdH-MoM solution proposed in Section 11.2 was carried out with $N_{A,B} = 3$ nodes only. Figure 11.3 shows the magnetic-current surface density on $\mathcal{S}_B$ as computed with the help of the CdH-MoM and its narrow-screen approximation (see Section 11.3). In conclusion, a satisfactory agreement between the numerical results is observed.

References

[1] M. Štumpf. *Electromagnetic Reciprocity in Antenna Theory*. Hoboken, NJ: John Wiley & Sons, Inc., 2018.

[2] M. Abramowitz and I. A. Stegun. *Handbook of Mathematical Functions*. New York, NY: Dover Publications, 1972.

Chapter 12

Pulsed EM-field scattering by 3-D bounded metasurfaces

If the analyzed structure and its excitation do not vary along the y-direction, the electromagnetic (EM) scattering problem can be solved through a set of y-independent Maxwell's equations supplemented with the pertaining boundary conditions. A variety of such two-dimensional (2-D) EM scattering problems have been analyzed throughout the previous chapters. If the problem dimensionality reduction is no longer appropriate, one has to resort to a general, fully three-dimensional (3-D) methodology. Such a computational approach is the subject of this chapter, where the pulsed EM scattering by thin, highly contrasting layers is analyzed with the aid of the Cagniard–DeHoop method of moments (CdH-MoM) [1]. Special attention is paid to the incorporation of combined magneto-dielectric EM constitutive properties of metasurfaces.

12.1 Pulsed EM scattering by a bounded 3-D screen with dielectric and conductive properties

In this section, we shall analyze the pulsed EM scattering by a 3-D screen whose electric permittivity and conductivity show a relatively high contrast with respect to the EM parameters of the surrounding medium. The screen occupies the bounded domain $\mathcal{V} = \{(x, y) \in \Sigma \subset \mathbb{R}^2, -\delta/2 < z < \delta/2\}$, where $\delta > 0$ denotes its (vanishing) thickness (see Figure 12.1).

The thin layer is supposed to be irradiated by an impulsive E-polarized plane wave that can be in the s-domain represented via (cf. (4.1))

$$\hat{\boldsymbol{E}}^{\mathrm{i}}(x, y, z, s) = \boldsymbol{\alpha}\, \hat{e}^{\mathrm{i}}(s) \exp\{-s[\kappa_0 x + \sigma_0 y - p(\kappa_0, \sigma_0) z]\}, \tag{12.1}$$

where $\boldsymbol{\alpha} = [\sin(\phi), -\cos(\phi), 0]$ denotes a unit vector in the direction of polarization and

$$\kappa_0 = \cos(\phi)\sin(\theta)/c_0, \tag{12.2a}$$

$$\sigma_0 = \sin(\phi)\sin(\theta)/c_0 \tag{12.2b}$$

have the meaning of slowness parameters in the x- and y-directions, respectively. Here, ϕ and θ denote the azimuthal and polar angles, respectively. Consequently,

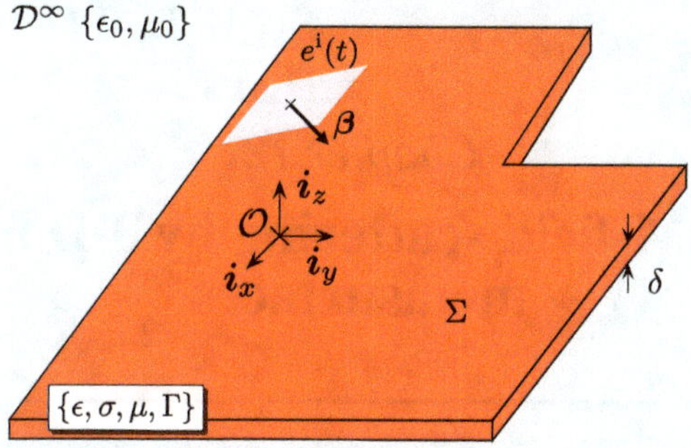

Figure 12.1 Pulsed EM plane wave impinging upon a bounded 3-D metasurface

by virtue of (4.77), the vertical slowness follows as $p(\kappa_0, \sigma_0) = \cos(\theta)/c_0$. The main goal of the following sections is to calculate the electric-current surface density induced in the layer.

12.1.1 Problem formulation

The EM scattering problem is tackled via the EM reciprocity theorem of the time-convolution type [2, Section 28.4] that is applied to the unbounded domain exterior to the layer and to the scattered (denoted by superscript ^{s}) and testing (denoted by superscript ^{T}) EM-field states, again. Representing the jump of (the tangential component of) the magnetic-field strength across the layer via the electric-current surface density (see (3.7b), for instance), we may then readily arrive at (cf. (8.1))

$$\int_{\boldsymbol{r}\in\Sigma} \hat{\boldsymbol{E}}^{\mathrm{T}}(x, y, 0, s) \cdot \partial\hat{\boldsymbol{J}}^{\mathrm{s}}(x, y, s)\mathrm{d}A = \int_{\boldsymbol{r}\in\Sigma} \hat{\boldsymbol{E}}^{\mathrm{s}}(x, y, 0, s) \cdot \partial\hat{\boldsymbol{J}}^{\mathrm{T}}(x, y, s)\mathrm{d}A. \tag{12.3}$$

In the reciprocity relation, the testing electric-field strength, $\hat{\boldsymbol{E}}^{\mathrm{T}}(x, y, 0, s)$, is related to the testing-source density, $\partial\hat{\boldsymbol{J}}^{\mathrm{T}}(x, y, s)$, via the standard source-type representations [2, Eqs. (26.3-1) and (26.3-3)]. Further, the scattered electric-field strength on its right-hand side is expressed via (3.6) in the following form:

$$\hat{\boldsymbol{E}}^{\mathrm{s}}(x, y, 0, s) = -\hat{\boldsymbol{E}}^{\mathrm{i}}(x, y, 0, s) + \hat{Z}^{\mathrm{E}}(s)\partial\hat{\boldsymbol{J}}^{\mathrm{s}}(x, y, s) \quad \text{for all } (x, y) \in \Sigma. \tag{12.4}$$

Note that for $\hat{Z}^{\mathrm{E}}(s) \downarrow 0$, the tangential electric field on screen's surface vanishes identically, which is characteristic for a perfectly electrically conducting (PEC)

surface. Upon substituting (12.4) in (12.3), we get the s-domain reciprocity relation pertaining to a screen with dielectric and conductive properties, viz

$$\int_{\boldsymbol{r}\in\Sigma}\left[\hat{\boldsymbol{E}}^{\mathrm{T}}(x,y,0,s)-\hat{Z}^{\mathrm{E}}(s)\partial\hat{\boldsymbol{J}}^{\mathrm{T}}(x,y,s)\right]\cdot\partial\hat{\boldsymbol{J}}^{\mathrm{s}}(x,y,s)\mathrm{d}A$$
$$=-\int_{\boldsymbol{r}\in\Sigma}\hat{\boldsymbol{E}}^{\mathrm{i}}(x,y,0,s)\cdot\partial\hat{\boldsymbol{J}}^{\mathrm{T}}(x,y,s)\mathrm{d}A. \tag{12.5}$$

The reciprocity relation (12.5) is further solved for the time-domain (TD) induced electric-current surface density using the CdH-MoM [1, Section 14.5]. To that end, we employ the 2-D wave-slowness representation in the plane parallel to the layer (see (4.75)) to cast the reciprocity relation into the following form:

$$\begin{aligned}
&\left(\frac{s}{2\mathrm{i}\pi}\right)^2\int_{\kappa=-\mathrm{i}\infty}^{\mathrm{i}\infty}\mathrm{d}\kappa\int_{\sigma=-\mathrm{i}\infty}^{\mathrm{i}\infty}\Big[\tilde{E}_x^{\mathrm{T}}(\kappa,\sigma,0,s)\partial\tilde{J}_x^{\mathrm{s}}(-\kappa,-\sigma,s)\\
&\qquad+\tilde{E}_y^{\mathrm{T}}(\kappa,\sigma,0,s)\partial\tilde{J}_y^{\mathrm{s}}(-\kappa,-\sigma,s)\\
&-\hat{Z}^{\mathrm{E}}(s)\partial\tilde{J}_x^{\mathrm{T}}(-\kappa,-\sigma,s)\partial\tilde{J}_x^{\mathrm{s}}(\kappa,\sigma,s)-\hat{Z}^{\mathrm{E}}(s)\partial\tilde{J}_y^{\mathrm{T}}(-\kappa,-\sigma,s)\partial\tilde{J}_y^{\mathrm{s}}(\kappa,\sigma,s)\Big]\mathrm{d}\sigma\\
&=-\left(\frac{s}{2\mathrm{i}\pi}\right)^2\int_{\kappa=-\mathrm{i}\infty}^{\mathrm{i}\infty}\mathrm{d}\kappa\int_{\sigma=-\mathrm{i}\infty}^{\mathrm{i}\infty}\Big[\tilde{E}_x^{\mathrm{i}}(\kappa,\sigma,0,s)\partial\tilde{J}_x^{\mathrm{T}}(-\kappa,-\sigma,s)\\
&\qquad+\tilde{E}_y^{\mathrm{i}}(\kappa,\sigma,0,s)\partial\tilde{J}_y^{\mathrm{T}}(-\kappa,-\sigma,s)\Big]\mathrm{d}\sigma\\
&\qquad\text{for } \{s\in\mathbb{R};s>0\},
\end{aligned}\tag{12.6}$$

where we use (cf. [1, Eqs. (14.33) and (14.34)])

$$\begin{aligned}
\tilde{E}_x^{\mathrm{T}}(\kappa,\sigma,0,s)=&-(s/\epsilon_0)\Omega_0^2(\kappa)\partial\tilde{J}_x^{\mathrm{T}}(\kappa,\sigma,s)/2sp(\kappa,\sigma)\\
&+(s/\epsilon_0)\kappa\sigma\,\partial\tilde{J}_y^{\mathrm{T}}(\kappa,\sigma,s)/2sp(\kappa,\sigma),
\end{aligned}\tag{12.7a}$$

$$\begin{aligned}
\tilde{E}_y^{\mathrm{T}}(\kappa,\sigma,0,s)=&-(s/\epsilon_0)\Omega_0^2(\sigma)\partial\tilde{J}_y^{\mathrm{T}}(\kappa,\sigma,s)/2sp(\kappa,\sigma)\\
&+(s/\epsilon_0)\sigma\kappa\,\partial\tilde{J}_x^{\mathrm{T}}(\kappa,\sigma,s)/2sp(\kappa,\sigma).
\end{aligned}\tag{12.7b}$$

In (12.7a) and (12.7b), we used $\Omega_0(\kappa)=(1/c_0^2-\kappa^2)^{1/2}$ and $p(\kappa,\sigma)$ is defined by (4.77). The slowness-domain reciprocity relation (12.6) with the transform-domain source-type representations (12.7a) and (12.7b) will next be solved with the help of the CdH technique (see Chapter 1).

12.1.2 Problem solution

The reciprocity-based formulation described in the previous section is further employed to develop a computationally efficient scheme that yields the space–time distribution of the induced electric-current density induced in the layer. For this purpose, the spatial solution domain, Σ, is uniformly divided into rectangular cells of dimensions $\Delta_x>0$ and $\Delta_y>0$ along the x- and y-directions, respectively. It is emphasized that the uniform spatial grid is not mandatory, but

it greatly simplifies the numerical solution and hence its code implementation. Likewise in Section 8.1.2, the continuous time axis, $\{t \in \mathbb{R}; t > 0\}$, is replaced with its discretized version, $\{t_k = k\Delta t; \Delta t > 0, k = 1, 2, \ldots, M\}$, where $\Delta t > 0$ is the time step. Once the solution domain is discretized, the unknown electric-current surface density can be expanded in a piecewise linear manner according to (cf. (8.6))

$$\partial J_x^{\mathrm{s}}(x, y, t) \simeq \frac{1}{\Delta_y} \sum_{u=1}^{U} \sum_{k=1}^{M} i_k^{[u]} \Lambda^{[u]}(x) \Pi^{[u]}(y) \Lambda_k(t), \tag{12.8}$$

where $i_k^{[u]}$ represents a (yet unknown) coefficient (in A) pertaining to the uth spatial node and to the temporal point $t_k = k\Delta t$. The definition of triangular functions $\Lambda^{[u]}(x)$ and $\Lambda_k(t)$ follows from (8.7a) and (7.19), respectively, and $\Pi^{[u]}(y)$ represents the rectangular function that is described by (cf. (8.9a))

$$\Pi^{[u]}(y) = \begin{cases} 1 & \text{for } y \in [y_u - \Delta_y/2, y_u + \Delta_y/2], \\ 0 & \text{elsewhere,} \end{cases}$$
$$\text{for all } u = \{1, \ldots, U\}. \tag{12.9}$$

The y-component of the induced electric-current surface density can be represented in a similar fashion as

$$\partial J_y^{\mathrm{s}}(x, y, t) \simeq \frac{1}{\Delta_x} \sum_{v=1}^{V} \sum_{k=1}^{M} i_k^{[v]} \Lambda^{[v]}(y) \Pi^{[v]}(x) \Lambda_k(t), \tag{12.10}$$

where $i_k^{[v]}$ denotes an unknown coefficient (in A) representing the yth component of the induced electric-current surface density at the vth spatial node and at $t_k = k\Delta t$, again. For the reader's convenience, the spatial expansion functions are depicted in Figure 12.2 for $u = 3$ and $v = 12$. In the next step, the testing electric-current distribution is specified. Following the solution approach pursued in [1, Section 14.5], the "razor-type" testing current is chosen, that is,

$$\partial J_x^{\mathrm{T}}(x, y, t) = \Pi^{[P]}(x)\delta(y - y_P)\delta(t), \tag{12.11a}$$
$$\partial J_y^{\mathrm{T}}(x, y, t) = \Pi^{[Q]}(y)\delta(x - x_Q)\delta(t),$$
$$\text{for all } P = \{1, \ldots, U\} \text{ and } Q = \{1, \ldots, V\}, \tag{12.11b}$$

where the meaning of rectangular functions, $\Pi^{[P]}(x)$ and $\Pi^{[Q]}(y)$, immediately follows from (12.9). In the final step, the transform-domain images of (12.8) and (12.10)–(12.11b) are substituted in slowness-domain reciprocity relation (12.6), which yields a system of equations in the s-domain that can be, symbolically, cast

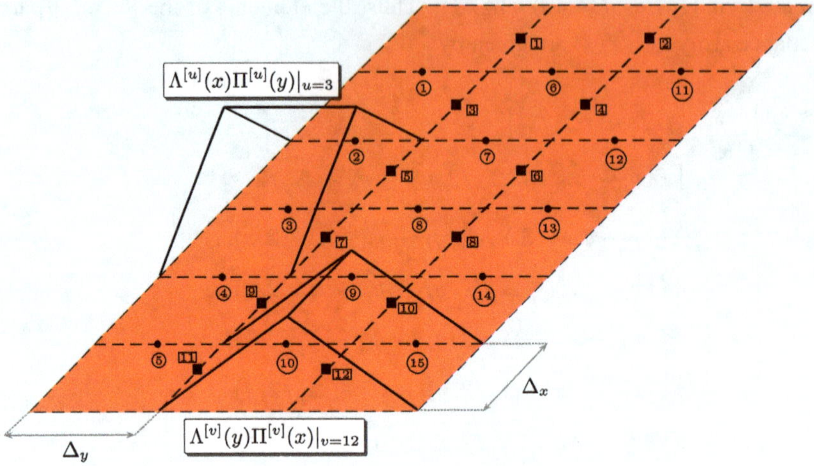

Figure 12.2 Uniform discretization of the spatial solution domain

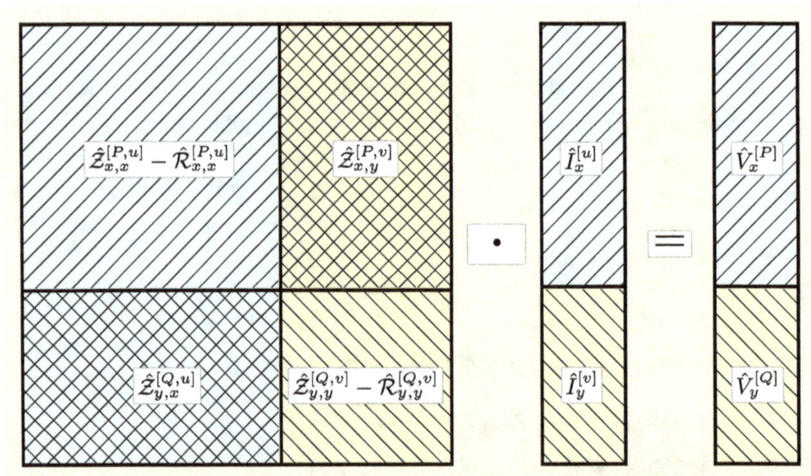

Figure 12.3 Symbolical partitioned matrix description of the EM scattering problem consisting of a thin layer with dielectric and conductive properties

into the form shown in Figure 12.3. Here, the one-dimensional (1-D) arrays $\hat{I}^{[u]}_x$ and $\hat{I}^{[v]}_y$ of dimensions $[U \times 1]$ and $[V \times 1]$ correspond to the induced electric-current surface densities $\partial \hat{J}^{\rm s}_x(x, y, s)$ and $\partial \hat{J}^{\rm s}_y(x, y, s)$ through the (unknown) coefficients $i^{[u]}_k$ and $i^{[v]}_k$, respectively. Further, the subarrays of the impedance array can be

expressed via [1, Eqs. (14.40)–(14.43)]. Thus, the elements of the $[U \times U]$ partial impedance array $\hat{\mathcal{Z}}_{x,x}^{[P,u]}(s)$ are given by

$$\begin{aligned}\hat{\mathcal{Z}}_{x,x}^{[P,u]}(s) = {} & \left[\zeta_0/c_0\Delta t\Delta_x\Delta_y\right] \\ & \times \Big[\hat{\Upsilon}(x_P - x_u + 3\Delta_x/2, y_P - y_u + \Delta_y/2, s) \\ & - \hat{\Upsilon}(x_P - x_u + 3\Delta_x/2, y_P - y_u - \Delta_y/2, s) \\ & - 3\hat{\Upsilon}(x_P - x_u + \Delta_x/2, y_P - y_u + \Delta_y/2, s) \\ & + 3\hat{\Upsilon}(x_P - x_u + \Delta_x/2, y_P - y_u - \Delta_y/2, s) \\ & + 3\hat{\Upsilon}(x_P - x_u - \Delta_x/2, y_P - y_u + \Delta_y/2, s) \\ & - 3\hat{\Upsilon}(x_P - x_u - \Delta_x/2, y_P - y_u - \Delta_y/2, s) \\ & - \hat{\Upsilon}(x_P - x_u - 3\Delta_x/2, y_P - y_u + \Delta_y/2, s) \\ & + \hat{\Upsilon}(x_P - x_u - 3\Delta_x/2, y_P - y_u - \Delta_y/2, s)\Big] \\ & \text{for all } P = \{1, \ldots, U\} \text{ and } u = \{1, \ldots, U\}, \end{aligned} \tag{12.12a}$$

where $\hat{\Upsilon}(x, y, s)$ with its TD original are given in Section K.1. The elements of the $[V \times V]$ partial impedance array $\hat{\mathcal{Z}}_{y,y}^{[Q,v]}(s)$ follow in a similar form:

$$\begin{aligned}\hat{\mathcal{Z}}_{y,y}^{[Q,v]}(s) = {} & \left[\zeta_0/c_0\Delta t\Delta_x\Delta_y\right] \\ & \times \Big[\hat{\Upsilon}(y_Q - y_v + 3\Delta_y/2, x_Q - x_v + \Delta_x/2, s) \\ & - \hat{\Upsilon}(y_Q - y_v + 3\Delta_y/2, x_Q - x_v - \Delta_x/2, s) \\ & - 3\hat{\Upsilon}(y_Q - y_v + \Delta_y/2, x_Q - x_v + \Delta_x/2, s) \\ & + 3\hat{\Upsilon}(y_Q - y_v + \Delta_y/2, x_Q - x_v - \Delta_x/2, s) \\ & + 3\hat{\Upsilon}(y_Q - y_v - \Delta_y/2, x_Q - x_v + \Delta_x/2, s) \\ & - 3\hat{\Upsilon}(y_Q - y_v - \Delta_y/2, x_Q - x_v - \Delta_x/2, s) \\ & - \hat{\Upsilon}(y_Q - y_v - 3\Delta_y/2, x_Q - x_v + \Delta_x/2, s) \\ & + \hat{\Upsilon}(y_Q - y_v - 3\Delta_y/2, x_Q - x_v - \Delta_x/2, s)\Big] \\ & \text{for all } Q = \{1, \ldots, V\} \text{ and } v = \{1, \ldots, V\}. \end{aligned} \tag{12.12b}$$

The off-diagonal blocks effectuating the coupling of $\hat{I}_x^{[u]}$ with $\hat{V}_y^{[Q]}$ and of $\hat{I}_y^{[v]}$ with $\hat{V}_x^{[P]}$ are described via $\hat{\mathcal{Z}}_{y,x}^{[Q,u]}(s)$ and $\hat{\mathcal{Z}}_{x,y}^{[P,v]}(s)$, respectively, that can be expressed via

$$
\begin{aligned}
&\hat{\mathcal{Z}}_{y,x}^{[Q,u]}(s) = [\zeta_0/c_0\Delta t\Delta_x\Delta_y] \\
&\times \Big[\hat{\Pi}(x_Q - x_u + \Delta_x, y_Q - y_u + \Delta_y, s) - 2\hat{\Pi}(x_Q - x_u + \Delta_x, y_Q - y_u, s) \\
&+ \hat{\Pi}(x_Q - x_u + \Delta_x, y_Q - y_u - \Delta_y, s) - 2\hat{\Pi}(x_Q - x_u, y_Q - y_u + \Delta_y, s) \\
&+ 4\hat{\Pi}(x_Q - x_u, y_Q - y_u, s) - 2\hat{\Pi}(x_Q - x_u, y_Q - y_u - \Delta_y, s) \\
&+ \hat{\Pi}(x_Q - x_u - \Delta_x, y_Q - y_u + \Delta_y, s) - 2\hat{\Pi}(x_Q - x_u - \Delta_x, y_Q - y_u, s) \\
&+ \hat{\Pi}(x_Q - x_u - \Delta_x, y_Q - y_u - \Delta_y, s)\Big] \\
&\quad \text{for all } Q = \{1, \ldots, V\} \text{ and } u = \{1, \ldots, U\},
\end{aligned}
\tag{12.12c}
$$

where $\hat{\Pi}(x, y, s)$ with its TD original are given in Section K.2. Furthermore, in a similar fashion, we write

$$
\begin{aligned}
&\hat{\mathcal{Z}}_{x,y}^{[P,v]}(s) = [\zeta_0/c_0\Delta t\Delta_x\Delta_y] \\
&\times \Big[\hat{\Pi}(y_P - y_v + \Delta_y, x_P - x_v + \Delta_x, s) - 2\hat{\Pi}(y_P - y_v + \Delta_y, x_P - x_v, s) \\
&+ \hat{\Pi}(y_P - y_v + \Delta_y, x_P - x_v - \Delta_x, s) - 2\hat{\Pi}(y_P - y_v, x_P - x_v + \Delta_x, s) \\
&+ 4\hat{\Pi}(y_P - y_v, x_P - x_v, s) - 2\hat{\Pi}(y_P - y_v, x_P - x_v - \Delta_x, s) \\
&+ \hat{\Pi}(y_P - y_v - \Delta_y, x_P - x_v + \Delta_x, s) - 2\hat{\Pi}(y_P - y_v - \Delta_y, x_P - x_v, s) \\
&+ \hat{\Pi}(y_P - y_v - \Delta_y, x_P - x_v - \Delta_x, s)\Big] \\
&\quad \text{for all } P = \{1, \ldots, U\} \text{ and } v = \{1, \ldots, V\}.
\end{aligned}
\tag{12.12d}
$$

The effect of bounded electric permittivity and conductivity is incorporated via two diagonal subarrays, $\hat{\mathcal{R}}_{x,x}^{[P,u]}(s)$ and $\hat{\mathcal{R}}_{y,y}^{[Q,v]}(s)$, whose elements are given as follows:

$$
\begin{aligned}
\hat{\mathcal{R}}_{x,x}^{[P,u]}(s) = {} & [\hat{Z}^{\mathrm{E}}(s)/c_0\Delta t\Delta_x\Delta_y] \\
& \times \Big[\hat{\Phi}(x_u - x_P + 3\Delta_x/2, y_u - y_P + \Delta_y/2, s) \\
& - \hat{\Phi}(x_u - x_P + 3\Delta_x/2, y_u - y_P - \Delta_y/2, s) \\
& - 3\hat{\Phi}(x_u - x_P + \Delta_x/2, y_u - y_P + \Delta_y/2, s) \\
& + 3\hat{\Phi}(x_u - x_P + \Delta_x/2, y_u - y_P - \Delta_y/2, s) \\
& + 3\hat{\Phi}(x_u - x_P - \Delta_x/2, y_u - y_P + \Delta_y/2, s) \\
& - 3\hat{\Phi}(x_u - x_P - \Delta_x/2, y_u - y_P - \Delta_y/2, s) \\
& - \hat{\Phi}(x_u - x_P - 3\Delta_x/2, y_u - y_P + \Delta_y/2, s) \\
& + \hat{\Phi}(x_u - x_P - 3\Delta_x/2, y_u - y_P - \Delta_y/2, s)\Big] \\
& \quad \text{for all } P = \{1, \ldots, U\} \text{ and } u = \{1, \ldots, U\},
\end{aligned}
\tag{12.13a}
$$

and

$$\begin{aligned}\hat{\mathcal{R}}_{y,y}^{[Q,v]}(s) = {} & \left[\hat{Z}^{\mathrm{E}}(s)/c_0\Delta t\Delta_x\Delta_y\right] \\ & \times \Big[\hat{\Phi}(y_v - y_Q + 3\Delta_y/2, x_v - x_Q + \Delta_x/2, s) \\ & - \hat{\Phi}(y_v - y_Q + 3\Delta_y/2, x_v - x_Q - \Delta_x/2, s) \\ & - 3\hat{\Phi}(y_v - y_Q + \Delta_y/2, x_v - x_Q + \Delta_x/2, s) \\ & + 3\hat{\Phi}(y_v - y_Q + \Delta_y/2, x_v - x_Q - \Delta_x/2, s) \\ & + 3\hat{\Phi}(y_v - y_Q - \Delta_y/2, x_v - x_Q + \Delta_x/2, s) \\ & - 3\hat{\Phi}(y_v - y_Q - \Delta_y/2, x_v - x_Q - \Delta_x/2, s) \\ & - \hat{\Phi}(y_v - y_Q - 3\Delta_y/2, x_v - x_Q + \Delta_x/2, s) \\ & + \hat{\Phi}(y_v - y_Q - 3\Delta_y/2, x_v - x_Q - \Delta_x/2, s)\Big] \\ & \text{for all } Q = \{1, \ldots, V\} \text{ and } v = \{1, \ldots, V\}, \end{aligned} \tag{12.13b}$$

where $\hat{\Phi}(x, y, s)$ is expressed in Section K.3 as $\hat{\Phi}(x, y, s) = (c_0 x^2/2s^2)\mathrm{H}(x)\mathrm{H}(y)$. Consequently, the TD original of (cf. (D.5) and (D.8))

$$\hat{\Gamma}^{\mathrm{E}}(x, y, s) = \left[\hat{Z}^{\mathrm{E}}(s)/\Delta_x\Delta_y c_0\Delta t\right]\hat{\Phi}(x, y, s) \tag{12.14}$$

can be used to get the elements of TD subarrays $\mathcal{R}_{x,x}^{[P,u]}(t)$ and $\mathcal{R}_{y,y}^{[Q,v]}(t)$. For a thin, highly conductive sheet that is characterized by its s-independent, equivalent layer admittance $\hat{Y}^{\mathrm{E}}(s) = 1/\hat{Z}^{\mathrm{E}}(s) = G^{\mathrm{E}}$, we get (cf. (8.23))

$$\Gamma^{\mathrm{E}}(x, y, t) = \frac{c_0 t}{2G^{\mathrm{E}}}\frac{x^2}{\Delta_x\Delta_y c_0\Delta t}\mathrm{H}(x)\mathrm{H}(y)\mathrm{H}(t). \tag{12.15}$$

Apparently, $\Gamma^{\mathrm{E}}(x, y, t) = 0$ for a PEC surface characterized by $G^{\mathrm{E}} \to \infty$. More generally, for a thin layer showing a high contrast in both dielectric and conductive properties, the layer admittance is given by $\hat{Y}^{\mathrm{E}}(s) = 1/\hat{Z}^{\mathrm{E}}(s) = G^{\mathrm{E}} + sC^{\mathrm{E}}$, which yields at once (cf. (8.25))

$$\Gamma^{\mathrm{E}}(x, y, t) = \frac{c_0^2 t^2}{2c_0 C^{\mathrm{E}}}\left[\frac{1}{\alpha^{\mathrm{E}}t} - \frac{1 - \exp(-\alpha^{\mathrm{E}}t)}{(\alpha^{\mathrm{E}}t)^2}\right]\frac{x^2}{\Delta_x\Delta_y c_0\Delta t}\mathrm{H}(x)\mathrm{H}(y)\mathrm{H}(t), \tag{12.16}$$

where $\alpha^{\mathrm{E}} = G^{\mathrm{E}}/C^{\mathrm{E}} > 0$ (see (3.19)). It is noted that for a loss-free highly dielectric layer, we may use $1/\alpha t - [1 - \exp(-\alpha t)]/\alpha^2 t^2 = 1/2 + O(\alpha t)$ as $\alpha t \downarrow 0$.

To fully specify the system depicted in Figure 12.3, we shall next evaluate the elements of the excitation array on its right-hand side. For the sake of illustration, the excitation array shall be determined through the evaluation of the spatial integral

on the right-hand side of (12.5). Hence, making use of (12.1) with (the s-domain counterpart of) (12.11a) in the right-hand side of the reciprocity relation, we get

$$\begin{aligned}\hat{V}_x^{[P]}(s) &= -\int_{x\in\mathbb{R}} \mathrm{d}x \int_{y\in\mathbb{R}} \hat{E}_x^{\mathrm{i}}(x,y,0,s)\partial\hat{J}_x^{\mathrm{T}}(x,y,s)\mathrm{d}y \\ &= -\hat{e}^{\mathrm{i}}(s)\Delta_x \sin(\phi)\mathrm{i}_0(s\kappa_0\Delta_x/2)\exp[-s(\kappa_0 x_P + \sigma_0 y_P)] \\ &\quad \text{for all } P = \{1,\ldots,U\},\end{aligned} \tag{12.17}$$

where $\mathrm{i}_0(x)$ denotes the modified spherical Bessel function of the first kind. The latter s-domain expression can be transformed to the TD at once and we get (cf. (8.13a))

$$\begin{aligned}V_x^{[P]}(t) = -\frac{\sin(\phi)}{\kappa_0}\Bigg\{ &\int_{\tau=0}^{t-\kappa_0(x_P-\Delta_x/2)-\sigma_0 y_P} e^{\mathrm{i}}(\tau)\mathrm{d}\tau \\ &- \int_{\tau=0}^{t-\kappa_0(x_P+\Delta_x/2)-\sigma_0 y_P} e^{\mathrm{i}}(\tau)\mathrm{d}\tau \Bigg\} \quad \text{for all } P = \{1,\ldots,U\}.\end{aligned} \tag{12.18}$$

If $s\kappa_0\Delta_x$ is relatively small, then $\mathrm{i}_0(s\kappa_0\Delta_x/2)$ in (12.17) can be replaced by the value one, which leads to the following approximation:

$$V_x^{[P]}(t) \simeq -e^{\mathrm{i}}(t-\kappa_0 x_P - \sigma_0 y_P)\Delta_x \sin(\phi) \quad \text{for all } P = \{1,\ldots,U\}. \tag{12.19}$$

In particular, for the normal incidence, $\theta = 0$ and hence $\kappa_0 = 0$ (see (12.2a)), the limit of (12.18) follows as [1, Eq. (14.49)]

$$V_x^{[P]}(t) = -e^{\mathrm{i}}(t)\Delta_x \sin(\phi) \quad \text{for all } P = \{1,\ldots,U\}. \tag{12.20}$$

Similar lines of reasoning lead to

$$\begin{aligned}\hat{V}_y^{[Q]}(s) &= -\int_{x\in\mathbb{R}} \mathrm{d}x \int_{y\in\mathbb{R}} \hat{E}_y^{\mathrm{i}}(x,y,0,s)\partial\hat{J}_y^{\mathrm{T}}(x,y,s)\mathrm{d}y \\ &= \hat{e}^{\mathrm{i}}(s)\Delta_y \cos(\phi)\mathrm{i}_0(s\sigma_0\Delta_y/2)\exp[-s(\kappa_0 x_Q + \sigma_0 y_Q)] \\ &\quad \text{for all } Q = \{1,\ldots,V\}.\end{aligned} \tag{12.21}$$

Transforming the latter to the TD, we obtain

$$\begin{aligned}V_y^{[Q]}(t) = \frac{\cos(\phi)}{\sigma_0}\Bigg\{ &\int_{\tau=0}^{t-\kappa_0 x_Q-\sigma_0(y_Q-\Delta_y/2)} e^{\mathrm{i}}(\tau)\mathrm{d}\tau \\ &- \int_{\tau=0}^{t-\kappa_0 x_Q-\sigma_0(y_Q+\Delta_y/2)} e^{\mathrm{i}}(\tau)\mathrm{d}\tau \Bigg\} \quad \text{for all } Q = \{1,\ldots,V\}.\end{aligned} \tag{12.22}$$

Again, if $s\sigma_0\Delta_y \ll 1$, we may use

$$V_y^{[Q]}(t) \simeq e^{\mathrm{i}}(t-\kappa_0 x_Q - \sigma_0 y_Q)\Delta_y \cos(\phi) \quad \text{for all } Q = \{1,\ldots,V\}. \tag{12.23}$$

For the normal incidence, $\theta = 0$ and hence $\sigma_0 = 0$ (see (12.2b)), we finally get [1, Eq. (14.50)]

$$V_y^{[Q]}(t) = e^{\mathrm{i}}(t)\Delta_y \cos(\phi) \quad \text{for all } Q = \{1, \ldots, V\}. \tag{12.24}$$

Upon transforming the s-domain system of equations to the TD, we finally end up with (cf. (7.42) and (8.19))

$$\sum_{k=1}^{m} \left(\underline{\boldsymbol{Z}}_{m-k+1} - 2\underline{\boldsymbol{Z}}_{m-k} + \underline{\boldsymbol{Z}}_{m-k-1}\right) \cdot \boldsymbol{I}_k = \boldsymbol{V}_m, \tag{12.25}$$

where

$$\underline{\boldsymbol{Z}} = \underline{\boldsymbol{\mathcal{Z}}} - \boldsymbol{\mathcal{R}}. \tag{12.26}$$

Here, $\underline{\boldsymbol{Z}}_k = \underline{\boldsymbol{Z}}(t_k)$ denotes a 2-D $[(U+V) \times (U+V)]$ TD impedance array evaluated at $t = t_k = k\Delta t$. It consists of (the TD originals of) impedance elements (12.12a)–(12.13b) according to Figure 12.3. Further, $\boldsymbol{I}_k$ is a 1-D $[(U+V) \times 1]$ array that is filled by the electric-current coefficients, $i_k^{[u]}$ and $i_k^{[v]}$, that through (12.8) and (12.9) represent the x- and y-components of the induced electric-current surface density, respectively. Finally, the excitation 1-D $[(U+V) \times 1]$ array is composed of the excitation voltage elements as given by (12.18)–(12.20) and (12.22)–(12.24). In the final step, the convolution-type system of equation (12.25) is solved for the array of current coefficients, $\boldsymbol{I}_m$, via the marching-on-in-time technique, that is (cf. (7.44) and (8.11)),

$$\boldsymbol{I}_m = \underline{\boldsymbol{Z}}_1^{-1} \cdot \left[\boldsymbol{V}_m - \sum_{k=1}^{m-1} \left(\underline{\boldsymbol{Z}}_{m-k+1} - 2\underline{\boldsymbol{Z}}_{m-k} + \underline{\boldsymbol{Z}}_{m-k-1}\right) \cdot \boldsymbol{I}_k\right]$$
$$\text{for all } m = \{1, \ldots, M\}, \tag{12.27}$$

where $\underline{\boldsymbol{Z}}_1^{-1}$ denotes the inverse of the impedance matrix at $t = \Delta t$.

12.2 Pulsed EM scattering by a bounded 3-D metasurface

If the thin layer under consideration exhibits the (relatively high) contrast in both electric permittivity and magnetic permeability, then all tangential EM-field components experience a jump proportional to the contrast sheet electric- and magnetic-current densities (see (3.10a) and (3.10b)). Accounting for these jumps in the EM

reciprocity relation interrelating the scattered and testing EM-field states in the exterior domain, we arrive at an extension of (12.3) that reads

$$\begin{aligned}\int_{\boldsymbol{r}\in\Sigma}&\langle\hat{\boldsymbol{E}}^{\mathrm{T}}\rangle_{-}^{+}(x,y,s)\cdot\partial\hat{\boldsymbol{J}}^{\mathrm{s}}(x,y,s)\mathrm{d}A\\&\qquad-\int_{\boldsymbol{r}\in\Sigma}\langle\hat{\boldsymbol{H}}^{\mathrm{T}}\rangle_{-}^{+}(x,y,s)\cdot\partial\hat{\boldsymbol{K}}^{\mathrm{s}}(x,y,s)\mathrm{d}A\\=\int_{\boldsymbol{r}\in\Sigma}&\langle\hat{\boldsymbol{E}}^{\mathrm{s}}\rangle_{-}^{+}(x,y,s)\cdot\partial\hat{\boldsymbol{J}}^{\mathrm{T}}(x,y,s)\mathrm{d}A\\&\qquad-\int_{\boldsymbol{r}\in\Sigma}\langle\hat{\boldsymbol{H}}^{\mathrm{s}}\rangle_{-}^{+}(x,y,s)\cdot\partial\hat{\boldsymbol{K}}^{\mathrm{T}}(x,y,s)\mathrm{d}A,\end{aligned}\tag{12.28}$$

where we used the operator $\langle\rangle_{-}^{+}$ as defined by (3.11). In the reciprocity relation, the testing electric- and magnetic-field strengths, $\hat{\boldsymbol{E}}^{\mathrm{T}}(x,y,0^{\pm},s)$ and $\hat{\boldsymbol{H}}^{\mathrm{T}}(x,y,0^{\pm},s)$, are related to the testing-source densities, $\partial\hat{\boldsymbol{J}}^{\mathrm{T}}(x,y,s)$ and $\partial\hat{\boldsymbol{K}}^{\mathrm{T}}(x,y,s)$, via the standard source-type representations [2, Eqs. (26.3-1) and (26.3-3)]. Next, the (average values of the) scattered electric- and magnetic-field strengths on its right-hand side are expressed in the following form (cf. (12.4)):

$$\langle\hat{\boldsymbol{E}}^{\mathrm{s}}\rangle_{-}^{+}(x,y,s)=-\hat{\boldsymbol{E}}^{\mathrm{i}}(x,y,0,s)+\hat{Z}^{\mathrm{E}}(s)\partial\hat{\boldsymbol{J}}^{\mathrm{s}}(x,y,s),\tag{12.29a}$$

$$\begin{aligned}&\langle\hat{\boldsymbol{H}}^{\mathrm{s}}\rangle_{-}^{+}(x,y,s)=-\hat{\boldsymbol{H}}^{\mathrm{i}}(x,y,0,s)+\hat{Y}^{\mathrm{M}}(s)\partial\hat{\boldsymbol{K}}^{\mathrm{s}}(x,y,s),\\&\quad\text{for all }(x,y)\in\Sigma,\end{aligned}\tag{12.29b}$$

and recall that the definitions of the equivalent layer impedance and admittance are given by (3.15a) and (3.15b). Substituting next the cross-layer conditions (12.29a) and (12.29b) in (12.28), the reciprocity relation can be rewritten as (cf. (12.5))

$$\begin{aligned}&\int_{\boldsymbol{r}\in\Sigma}\left[\langle\hat{\boldsymbol{E}}^{\mathrm{T}}\rangle_{-}^{+}(x,y,s)-\hat{Z}^{\mathrm{E}}(s)\partial\hat{\boldsymbol{J}}^{\mathrm{T}}(x,y,s)\right]\cdot\partial\hat{\boldsymbol{J}}^{\mathrm{s}}(x,y,s)\mathrm{d}A\\&-\int_{\boldsymbol{r}\in\Sigma}\left[\langle\hat{\boldsymbol{H}}^{\mathrm{T}}\rangle_{-}^{+}(x,y,s)-\hat{Y}^{\mathrm{M}}(s)\partial\hat{\boldsymbol{K}}^{\mathrm{T}}(x,y,s)\right]\cdot\partial\hat{\boldsymbol{K}}^{\mathrm{s}}(x,y,s)\mathrm{d}A\\&=-\int_{\boldsymbol{r}\in\Sigma}\left[\hat{\boldsymbol{E}}^{\mathrm{i}}(x,y,0,s)\cdot\partial\hat{\boldsymbol{J}}^{\mathrm{T}}(x,y,s)-\hat{\boldsymbol{H}}^{\mathrm{i}}(x,y,0,s)\cdot\partial\hat{\boldsymbol{K}}^{\mathrm{T}}(x,y,s)\right]\mathrm{d}A.\end{aligned}\tag{12.30}$$

The reciprocity relation (12.30) is next solved for the induced currents using the CdH-MoM. To that end, the 2-D wave-slowness representation in the plane parallel to the layer (see (4.75)) is employed to rewrite the s-domain reciprocity relation in terms of slowness-domain integrals. Under the wave-slowness representation, it is observed that (cf. (12.7a) and (12.7b))

$$\begin{aligned}\langle\tilde{E}_x^{\mathrm{T}}\rangle_{-}^{+}(\kappa,\sigma,s)=&-(s/\epsilon_0)\Omega_0^2(\kappa)\partial\tilde{J}_x^{\mathrm{T}}(\kappa,\sigma,s)/2sp(\kappa,\sigma)\\&+(s/\epsilon_0)\kappa\sigma\,\partial\tilde{J}_y^{\mathrm{T}}(\kappa,\sigma,s)/2sp(\kappa,\sigma),\end{aligned}\tag{12.31a}$$

$$\begin{aligned}\langle\tilde{E}_y^{\mathrm{T}}\rangle_{-}^{+}(\kappa,\sigma,s)=&-(s/\epsilon_0)\Omega_0^2(\sigma)\partial\tilde{J}_y^{\mathrm{T}}(\kappa,\sigma,s)/2sp(\kappa,\sigma)\\&+(s/\epsilon_0)\sigma\kappa\,\partial\tilde{J}_x^{\mathrm{T}}(\kappa,\sigma,s)/2sp(\kappa,\sigma),\end{aligned}\tag{12.31b}$$

and similar relations apply to (the average of) the testing magnetic-field strength

$$\langle \tilde{H}_x^{\mathrm{T}} \rangle_-^+(\kappa,\sigma,s) = -\,(s/\mu_0)\Omega_0^2(\kappa)\partial\tilde{K}_x^{\mathrm{T}}(\kappa,\sigma,s)/2sp(\kappa,\sigma) + (s/\mu_0)\kappa\sigma\,\partial\tilde{K}_y^{\mathrm{T}}(\kappa,\sigma,s)/2sp(\kappa,\sigma), \tag{12.31c}$$

$$\langle \tilde{H}_y^{\mathrm{T}} \rangle_-^+(\kappa,\sigma,s) = -\,(s/\mu_0)\Omega_0^2(\sigma)\partial\tilde{K}_y^{\mathrm{T}}(\kappa,\sigma,s)/2sp(\kappa,\sigma) + (s/\mu_0)\sigma\kappa\,\partial\tilde{K}_x^{\mathrm{T}}(\kappa,\sigma,s)/2sp(\kappa,\sigma), \tag{12.31d}$$

The property that $\langle \hat{\boldsymbol{E}}^{\mathrm{T}} \rangle_-^+(x,y,s)$ and $\langle \hat{\boldsymbol{H}}^{\mathrm{T}} \rangle_-^+(x,y,s)$ are generated by $\partial\hat{\boldsymbol{J}}^{\mathrm{T}}(x,y,s)$ and $\partial\hat{\boldsymbol{K}}^{\mathrm{T}}(x,y,s)$ alone, respectively, makes it possible to decouple the reciprocity relation to two independent relations, that is (cf. (8.35a) and (12.6)),

$$\begin{aligned}
&\left(\frac{s}{2\mathrm{i}\pi}\right)^2 \int_{\kappa=-\mathrm{i}\infty}^{\mathrm{i}\infty} \mathrm{d}\kappa \int_{\sigma=-\mathrm{i}\infty}^{\mathrm{i}\infty} \Big[\langle \tilde{E}_x^{\mathrm{T}} \rangle_-^+(\kappa,\sigma,s)\partial\tilde{J}_x^{\mathrm{s}}(-\kappa,-\sigma,s) + \langle \tilde{E}_y^{\mathrm{T}} \rangle_-^+(\kappa,\sigma,s)\partial\tilde{J}_y^{\mathrm{s}}(-\kappa,-\sigma,s) \\
&\quad - \hat{Z}^{\mathrm{E}}(s)\partial\tilde{J}_x^{\mathrm{T}}(-\kappa,-\sigma,s)\partial\tilde{J}_x^{\mathrm{s}}(\kappa,\sigma,s) - \hat{Z}^{\mathrm{E}}(s)\partial\tilde{J}_y^{\mathrm{T}}(-\kappa,-\sigma,s)\partial\tilde{J}_y^{\mathrm{s}}(\kappa,\sigma,s)\Big]\mathrm{d}\sigma \\
&= -\left(\frac{s}{2\mathrm{i}\pi}\right)^2 \int_{\kappa=-\mathrm{i}\infty}^{\mathrm{i}\infty} \mathrm{d}\kappa \int_{\sigma=-\mathrm{i}\infty}^{\mathrm{i}\infty} \Big[\tilde{E}_x^{\mathrm{i}}(\kappa,\sigma,0,s)\partial\tilde{J}_x^{\mathrm{T}}(-\kappa,-\sigma,s) + \tilde{E}_y^{\mathrm{i}}(\kappa,\sigma,0,s)\partial\tilde{J}_y^{\mathrm{T}}(-\kappa,-\sigma,s)\Big]\mathrm{d}\sigma,
\end{aligned} \tag{12.32a}$$

and (cf. (8.35b))

$$\begin{aligned}
&\left(\frac{s}{2\mathrm{i}\pi}\right)^2 \int_{\kappa=-\mathrm{i}\infty}^{\mathrm{i}\infty} \mathrm{d}\kappa \int_{\sigma=-\mathrm{i}\infty}^{\mathrm{i}\infty} \Big[\langle \tilde{H}_x^{\mathrm{T}} \rangle_-^+(\kappa,\sigma,s)\partial\tilde{K}_x^{\mathrm{s}}(-\kappa,-\sigma,s) + \langle \tilde{H}_y^{\mathrm{T}} \rangle_-^+(\kappa,\sigma,s)\partial\tilde{K}_y^{\mathrm{s}}(-\kappa,-\sigma,s) \\
&\quad - \hat{Y}^{\mathrm{M}}(s)\partial\tilde{K}_x^{\mathrm{T}}(-\kappa,-\sigma,s)\partial\tilde{K}_x^{\mathrm{s}}(\kappa,\sigma,s) - \hat{Y}^{\mathrm{M}}(s)\partial\tilde{K}_y^{\mathrm{T}}(-\kappa,-\sigma,s)\partial\tilde{K}_y^{\mathrm{s}}(\kappa,\sigma,s)\Big]\mathrm{d}\sigma \\
&= -\left(\frac{s}{2\mathrm{i}\pi}\right)^2 \int_{\kappa=-\mathrm{i}\infty}^{\mathrm{i}\infty} \mathrm{d}\kappa \int_{\sigma=-\mathrm{i}\infty}^{\mathrm{i}\infty} \Big[\tilde{H}_x^{\mathrm{i}}(\kappa,\sigma,0,s)\partial\tilde{K}_x^{\mathrm{T}}(-\kappa,-\sigma,s) + \tilde{H}_y^{\mathrm{i}}(\kappa,\sigma,0,s)\partial\tilde{K}_y^{\mathrm{T}}(-\kappa,-\sigma,s)\Big]\mathrm{d}\sigma.
\end{aligned} \tag{12.32b}$$

Upon comparison of (12.6)–(12.7b) with (12.32a), (12.31a) and (12.31b), it is concluded that the solution methodology introduced in Section 12.1.2 is directly applicable to calculating the electric-current distribution induced in a metasurface. Hence, the subsequent analysis deals with the (dual) magnetic-type reciprocity relation (12.32b) only. Its solution that is described in the following section yields the corresponding magnetic-current surface density.

12.2.1 Problem solution

To solve the reciprocity relation (12.32b) numerically, we shall largely pursue the approach described in Section 12.1.2. Accordingly, the space–time solution domain is first discretized uniformly. Subsequently, the unknown current densities are expanded in terms of triangular functions (cf. (12.8) and (12.10))

$$\partial K_x^{\mathrm{s}}(x,y,t) \simeq \frac{1}{\Delta_y}\sum_{u=1}^{U}\sum_{k=1}^{M} v_k^{[u]}\Lambda^{[u]}(x)\Pi^{[u]}(y)\Lambda_k(t), \tag{12.33a}$$

$$\partial K_y^{\mathrm{s}}(x,y,t) \simeq \frac{1}{\Delta_x}\sum_{v=1}^{V}\sum_{k=1}^{M} v_k^{[v]}\Lambda^{[v]}(y)\Pi^{[v]}(x)\Lambda_k(t), \tag{12.33b}$$

where $v_k^{[u]}$ and $v_k^{[v]}$ denote magnetic-current coefficients (in V) pertaining to the uth and vth nodes of the spatial uniform grid, respectively (see Figure 12.2). For the magnetic-current testing sources, we again choose the "razor-type" spatial distribution and the impulsive Dirac-delta behavior with respect to time (cf. (12.11a) and (12.11b))

$$\partial K_x^{\mathrm{T}}(x,y,t) = \Pi^{[P]}(x)\delta(y-y_P)\delta(t), \tag{12.34a}$$

$$\begin{aligned}&\partial K_y^{\mathrm{T}}(x,y,t) = \Pi^{[Q]}(y)\delta(x-x_Q)\delta(t),\\ &\quad \text{for all } P=\{1,\ldots,U\} \text{ and } Q=\{1,\ldots,V\},\end{aligned} \tag{12.34b}$$

and recall that the rectangular functions $\Pi^{[P]}(x)$ and $\Pi^{[Q]}(y)$ are defined before via (12.9). Substitution of the transform-domain counterparts of (12.33a)–(12.34b) in the reciprocity relation (12.32b) yields a system of equations in the s-domain that can be, symbolically, represented as depicted in Figure 12.4. Here, the 1-D arrays $\hat{V}_x^{[u]}$ and $\hat{V}_y^{[v]}$ of dimensions $[U\times 1]$ and $[V\times 1]$ correspond to the induced magnetic-current surface densities $\partial\hat{K}_x^{\mathrm{s}}(x,y,s)$ and $\partial\hat{K}_y^{\mathrm{s}}(x,y,s)$ through the (unknown) coefficients $v_k^{[u]}$ and $v_k^{[v]}$, respectively. The blocks of the 2-D $[(U+V)\times(U+V)]$ admittance array $\hat{\boldsymbol{\mathcal{Y}}}$ can be, again, expressed through the generic integrals $\hat{\Upsilon}(x,y,s)$ and $\hat{\Pi}(x,y,s)$ (see Sections K.1 and K.2). In fact, since the electric- and magnetic-current quantities were expanded with respect to the same basis, we may simply use

$$\boldsymbol{\mathcal{Y}}(t) = \eta_0^2\boldsymbol{\mathcal{Z}}(t) \quad \text{for all } t>0, \tag{12.35}$$

where the elements of the impedance array are given by (12.12a)–(12.12d) in the s-domain. Further, the effect of bounded magnetic permeability is incorporated via

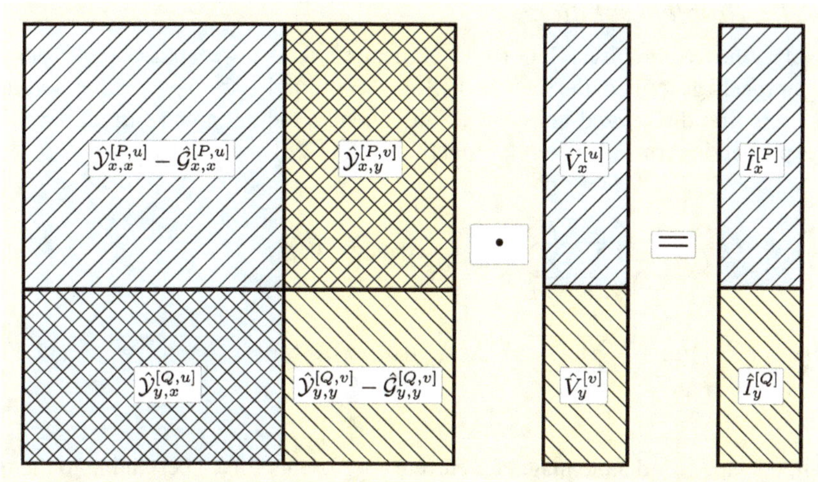

Figure 12.4 Symbolical partitioned matrix description of the EM scattering problem consisting a thin layer with magnetic properties

two diagonal subarrays $\hat{\mathcal{G}}^{[P,u]}_{x,x}(s)$ and $\hat{\mathcal{G}}^{[Q,v]}_{y,y}(s)$ whose elements are given as follows (cf. (12.13a)):

$$\begin{aligned}\hat{\mathcal{G}}^{[P,u]}_{x,x}(s) = {} & \left[\hat{Y}^{\mathrm{M}}(s)/c_0\Delta t\Delta_x\Delta_y\right] \\ & \times \Big[\hat{\Phi}(x_u - x_P + 3\Delta_x/2, y_u - y_P + \Delta_y/2, s) \\ & - \hat{\Phi}(x_u - x_P + 3\Delta_x/2, y_u - y_P - \Delta_y/2, s) \\ & - 3\hat{\Phi}(x_u - x_P + \Delta_x/2, y_u - y_P + \Delta_y/2, s) \\ & + 3\hat{\Phi}(x_u - x_P + \Delta_x/2, y_u - y_P - \Delta_y/2, s) \\ & + 3\hat{\Phi}(x_u - x_P - \Delta_x/2, y_u - y_P + \Delta_y/2, s) \\ & - 3\hat{\Phi}(x_u - x_P - \Delta_x/2, y_u - y_P - \Delta_y/2, s) \\ & - \hat{\Phi}(x_u - x_P - 3\Delta_x/2, y_u - y_P + \Delta_y/2, s) \\ & + \hat{\Phi}(x_u - x_P - 3\Delta_x/2, y_u - y_P - \Delta_y/2, s)\Big] \\ & \quad \text{for all } P = \{1, \ldots, U\} \text{ and } u = \{1, \ldots, U\},\end{aligned} \tag{12.36a}$$

and (cf. (12.13b))

$$\begin{aligned}\hat{\mathcal{G}}^{[Q,v]}_{y,y}(s) = {} & \left[\hat{Y}^{\mathrm{M}}(s)/c_0\Delta t\Delta_x\Delta_y\right] \\ & \times \Big[\hat{\Phi}(y_v - y_Q + 3\Delta_y/2, x_v - x_Q + \Delta_x/2, s) \\ & - \hat{\Phi}(y_v - y_Q + 3\Delta_y/2, x_v - x_Q - \Delta_x/2, s) \\ & - 3\hat{\Phi}(y_v - y_Q + \Delta_y/2, x_v - x_Q + \Delta_x/2, s)\end{aligned}$$

$$
\begin{aligned}
&+ 3\hat{\Phi}(y_v - y_Q + \Delta_y/2, x_v - x_Q - \Delta_x/2, s)\\
&+ 3\hat{\Phi}(y_v - y_Q - \Delta_y/2, x_v - x_Q + \Delta_x/2, s)\\
&- 3\hat{\Phi}(y_v - y_Q - \Delta_y/2, x_v - x_Q - \Delta_x/2, s)\\
&- \hat{\Phi}(y_v - y_Q - 3\Delta_y/2, x_v - x_Q + \Delta_x/2, s)\\
&+ \hat{\Phi}(y_v - y_Q - 3\Delta_y/2, x_v - x_Q - \Delta_x/2, s)\Big]\\
&\text{for all } Q = \{1, \dots, V\} \text{ and } v = \{1, \dots, V\}, \qquad (12.36\text{b})
\end{aligned}
$$

where $\hat{\Phi}(x, y, s)$ is given in Section K.3. Making use of this result, it follows that the TD originals of (12.36a) and (12.36b) can be found via the inverse of (cf. (12.14))

$$\hat{\Gamma}^{\mathrm{M}}(x, y, s) = \left[\hat{Y}^{\mathrm{M}}(s)/\Delta_x\Delta_y c_0\Delta t\right]\hat{\Phi}(x, y, s). \qquad (12.37)$$

For a loss-free magnetic layer that is characterized by its equivalent layer impedance $\hat{Z}^{\mathrm{M}}(s) = 1/\hat{Y}^{\mathrm{M}}(s) = sL^{\mathrm{M}}$, we get at once (cf. (12.16))

$$\Gamma^{\mathrm{M}}(x, y, t) = \frac{c_0^2 t^2}{4c_0 L^{\mathrm{M}}}\frac{x^2}{\Delta_x\Delta_y c_0\Delta t}\mathrm{H}(x)\mathrm{H}(y)\mathrm{H}(t). \qquad (12.38)$$

Clearly, $\Gamma^{\mathrm{M}}(x, y, t) = 0$ for a perfectly magnetically conducting (PMC) surface that is characterized by $L^{\mathrm{M}} \to \infty$.

In the next step, we shall specify the excitation (electric-current) array on the right-hand side of the system shown in Figure 12.4. Their explicit form can be achieved through the spatial integrals that follow (cf. (12.17))

$$
\begin{aligned}
\hat{I}_x^{[P]}(s) &= -\int_{x\in\mathbb{R}}\mathrm{d}x\int_{y\in\mathbb{R}}\hat{H}_x^{\mathrm{i}}(x, y, 0, s)\partial\hat{K}_x^{\mathrm{T}}(x, y, s)\mathrm{d}y\\
&= \eta_0\hat{e}^{\mathrm{i}}(s)\Delta_x\cos(\phi)\cos(\theta)\mathrm{i}_0(s\kappa_0\Delta_x/2)\exp[-s(\kappa_0 x_P + \sigma_0 y_P)]\\
&\quad\text{for all } P = \{1, \dots, U\}, \qquad (12.39)
\end{aligned}
$$

and (cf. (12.21))

$$
\begin{aligned}
\hat{I}_y^{[Q]}(s) &= -\int_{x\in\mathbb{R}}\mathrm{d}x\int_{y\in\mathbb{R}}\hat{H}_y^{\mathrm{i}}(x, y, 0, s)\partial\hat{K}_y^{\mathrm{T}}(x, y, s)\mathrm{d}y\\
&= \eta_0\hat{e}^{\mathrm{i}}(s)\Delta_y\sin(\phi)\cos(\theta)\mathrm{i}_0(s\sigma_0\Delta_y/2)\exp[-s(\kappa_0 x_Q + \sigma_0 y_Q)]\\
&\quad\text{for all } Q = \{1, \dots, V\}. \qquad (12.40)
\end{aligned}
$$

The inverse transform of excitation elements (12.39) and (12.40) yields (cf. (12.18))

$$
\begin{aligned}
I_x^{[P]}(t) =& \frac{\eta_0\cos(\phi)\cos(\theta)}{\kappa_0}\left\{\int_{\tau=0}^{t-\kappa_0(x_P-\Delta_x/2)-\sigma_0 y_P}e^{\mathrm{i}}(\tau)\mathrm{d}\tau\right.\\
&\left.-\int_{\tau=0}^{t-\kappa_0(x_P+\Delta_x/2)-\sigma_0 y_P}e^{\mathrm{i}}(\tau)\mathrm{d}\tau\right\}\quad\text{for all } P = \{1, \dots, U\},\\
&\qquad (12.41)
\end{aligned}
$$

and (cf. (12.22))

$$I_y^{[Q]}(t) = \frac{\eta_0 \sin(\phi)\cos(\theta)}{\sigma_0} \left\{ \int_{\tau=0}^{t-\kappa_0 x_Q - \sigma_0(y_Q - \Delta_y/2)} e^{\mathrm{i}}(\tau)\mathrm{d}\tau \right.$$
$$\left. - \int_{\tau=0}^{t-\kappa_0 x_Q - \sigma_0(y_Q + \Delta_y/2)} e^{\mathrm{i}}(\tau)\mathrm{d}\tau \right\} \quad \text{for all } Q = \{1, \ldots, V\}. \tag{12.42}$$

For the normal incidence, $\theta = 0$ and hence $\kappa_0 = \sigma_0 = 0$ (see (12.2a) and (12.2b)), the limits of (12.41) and (12.42) take the simple form (cf. (12.20) and (12.24))

$$I_x^{[P]}(t) = \eta_0 e^{\mathrm{i}}(t)\Delta_x \cos(\phi) \quad \text{for all } P = \{1, \ldots, U\}, \tag{12.43}$$
$$I_y^{[Q]}(t) = \eta_0 e^{\mathrm{i}}(t)\Delta_y \sin(\phi) \quad \text{for all } Q = \{1, \ldots, V\}. \tag{12.44}$$

Having thus transformed all the necessary elements, the system of s-domain equations can be transformed to the TD. This leads to the following (dual) system of convolution-type equations (cf. (12.25)):

$$\sum_{k=1}^{m} \left(\underline{\boldsymbol{Y}}_{m-k+1} - 2\underline{\boldsymbol{Y}}_{m-k} + \underline{\boldsymbol{Y}}_{m-k-1} \right) \cdot \boldsymbol{V}_k = \boldsymbol{I}_m, \tag{12.45}$$

where (cf. (12.26))

$$\underline{\boldsymbol{Y}} = \underline{\boldsymbol{\mathcal{Y}}} - \underline{\boldsymbol{\mathcal{G}}}. \tag{12.46}$$

Here, $\underline{\boldsymbol{Y}}_k = \underline{\boldsymbol{Y}}(t_k)$ denotes a 2-D $[(U+V) \times (U+V)]$ TD admittance array at $t = t_k = k\Delta t$. It consists of (the TD originals of) admittance elements in the form depicted in Figure 12.4. Next, $\boldsymbol{V}_k$ is a 1-D $[(U+V) \times 1]$ array that is filled by the magnetic-current coefficients, $v_k^{[u]}$ and $v_k^{[v]}$, that through (12.33a) and (12.33b) represent the x- and y-components of the induced magnetic-current surface density, respectively. Finally, the excitation 1-D $[(U+V) \times 1]$ array is composed of the excitation current elements as given by (12.41)–(12.44). In the final step, the system of equations (12.45) is solved for the array of current coefficients, $\boldsymbol{V}_m$, via the marching-on-in-time technique (cf. (12.27))

$$\boldsymbol{V}_m = \underline{\boldsymbol{Y}}_1^{-1} \cdot \left[\boldsymbol{I}_m - \sum_{k=1}^{m-1} \left(\underline{\boldsymbol{Y}}_{m-k+1} - 2\underline{\boldsymbol{Y}}_{m-k} + \underline{\boldsymbol{Y}}_{m-k-1} \right) \cdot \boldsymbol{V}_k \right]$$
$$\text{for all } m = \{1, \ldots, M\}, \tag{12.47}$$

where $\underline{\boldsymbol{Y}}_1^{-1}$ denotes the inverse of the admittance matrix at $t = \Delta t$.

12.3 Pulsed EM scattering by a small rectangular screen

A special form of the TD solution introduced in the previous sections can be arrived at for a (relatively) small planar screen with rectangular faces (see Figure 12.5). The CdH-MoM analysis of the pulsed EM-field interaction with such a

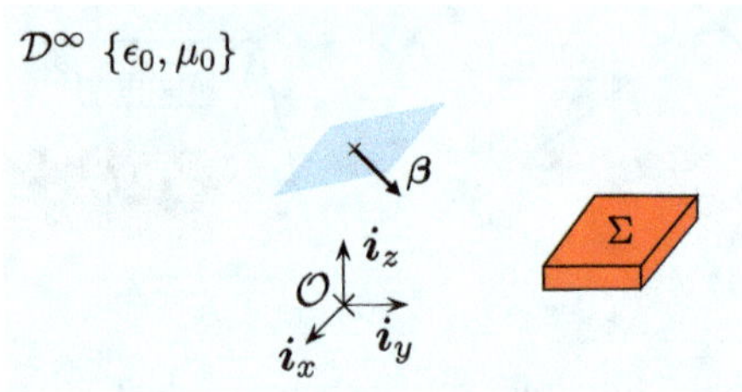

Figure 12.5 Pulsed EM plane wave impinging upon a small rectangular screen

small scatterer is the main subject of this section. The 3-D screen under consideration occupies a bounded volume $\mathcal{V} = \{(x,y) \in \Sigma \subset \mathbb{R}^2, -\delta/2 < z < \delta/2\}$ with $\Sigma = \{-\Delta_x < x < \Delta_x, \Delta_y < y < \Delta_y\}$, where $2\Delta_x > 0$, $2\Delta_y > 0$ and $\delta > 0$ are the (relatively small) dimensions of the rectangular screen. It is assumed that the screen is located in the homogeneous, isotropic and loss-free embedding, $\mathcal{D}^\infty$, whose EM characteristics are described by $\{\epsilon_0, \mu_0\}$. For the sake of simplicity, the screen is supposed to be irradiated by the uniform, E-polarized plane wave, again (see (12.1)). But, a more complex EM source can be readily incorporated in a similar way.

12.3.1 A small rectangular screen with dielectric and conductive properties

EM scattering by a small screen with dielectric and conductive properties can be analyzed with the aid of (12.6), again. In order to solve the reciprocity relation numerically, the induced electric-current surface density is first expanded in terms of (space–time) basis functions. Owing to the assumption that the dimensions of the screen are relatively small, we shall further postulate that the in-plane components of the unknown current density have the following simple form (cf. (12.8) and (12.10)):

$$\partial J_x^{\mathrm{s}}(x,y,t) \simeq \frac{I_x(t)}{2\Delta_y} \Lambda(x)\Pi(y), \tag{12.48a}$$

$$\partial J_y^{\mathrm{s}}(x,y,t) \simeq \frac{I_y(t)}{2\Delta_x} \Lambda(y)\Pi(x), \tag{12.48b}$$

where $I_x(t)$ and $I_y(t)$ represent the desired TD electric-current pulses. Furthermore, in (12.48a), we used (cf. (8.7a))

$$\Lambda(x) = \begin{cases} 1 + x/\Delta_x & \text{for } x \in [-\Delta_x, 0], \\ 1 - x/\Delta_x & \text{for } x \in [0, \Delta_x], \end{cases} \tag{12.49}$$

and (cf. (8.9a))

$$\Pi(y) = \begin{cases} 1 & \text{for } y \in [-\Delta_y, \Delta_y], \\ 0 & \text{elsewhere.} \end{cases} \tag{12.50}$$

The functions of spatial coordinates, $\Lambda(y)$ with $\Pi(x)$, as occur in (12.48b), have a similar meaning. The spatial distributions of $\partial J_x^{\mathrm{s}}(x,y,t)$ and $\partial J_y^{\mathrm{s}}(x,y,t)$ are

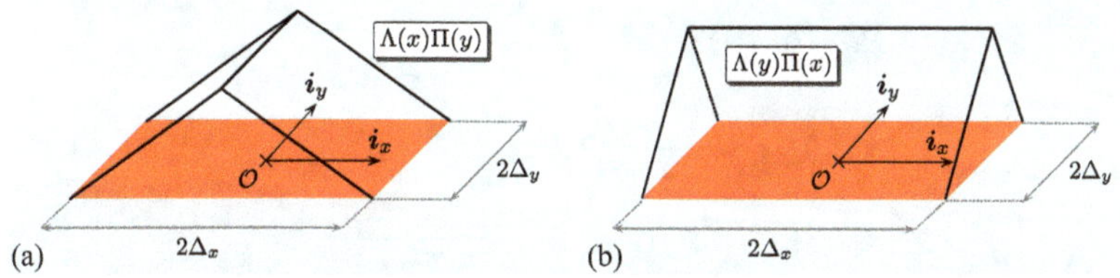

Figure 12.6 Spatial basis functions of the current density. (a) x-component; (b) y-component.

depicted in Figure 12.6(a) and (b), respectively. As can be seen, the chosen expansion meets the edge conditions $\partial J_x^s(\pm\Delta_x, y, t) = \partial J_y^s(x, \pm\Delta_y, t) = 0$ for all $t > 0$. The induced electric-current pulses are further expanded along the uniformly discretized time axis, $\{t_k = k\Delta t; \Delta t > 0, k = 1, 2, \ldots, M\}$, in terms of triangular functions (cf. (7.18) with (7.19))

$$I_x(t) \simeq \sum_{k=1}^{M} i_{k;x}\Lambda_k(t), \tag{12.51a}$$

$$I_y(t) \simeq \sum_{k=1}^{M} i_{k;y}\Lambda_k(t), \tag{12.51b}$$

where $i_{k;x}$ and $i_{k;y}$ are the electric-current coefficients (in A) to be found in the following solution. As for the testing currents, we choose the "razor-type" functions, again (cf. (12.11a) and (12.11b))

$$\partial J_x^{\mathrm{T}}(x, y, t) = \Pi(2x)\delta(y)\delta(t), \tag{12.52a}$$

$$\partial J_y^{\mathrm{T}}(x, y, t) = \Pi(2y)\delta(x)\delta(t), \tag{12.52b}$$

where we used the rectangular function as defined by (12.50). Substitution of the transform-domain counterparts of (12.48a), (12.48b) with (12.51a), (12.51b) and (12.52a), (12.52b) in the reciprocity relation (12.6) yields two (uncoupled) equations in the s-domain that interrelate the induced electric currents with the co-polarized components of the incident EM field (see Figure 12.3). Upon transforming the equations back to the original domain, we end up with the convolution-type system of equations (12.25) with (12.26), again, where

$$\underline{\mathcal{Z}}_k = \frac{\zeta_0}{2c_0\Delta t\Delta_x\Delta_y}\begin{bmatrix} \Phi_{x,x}(t_k) & 0 \\ 0 & \Phi_{y,y}(t_k) \end{bmatrix}, \tag{12.53}$$

with

$$\begin{aligned}\Phi_{x,x}(t) = {} & \Upsilon(3\Delta_x/2, \Delta_y, t) - \Upsilon(3\Delta_x/2, -\Delta_y, t) - 3\Upsilon(\Delta_x/2, \Delta_y, t) \\ & + 3\Upsilon(\Delta_x/2, -\Delta_y, t) + 3\Upsilon(-\Delta_x/2, \Delta_y, t) - 3\Upsilon(-\Delta_x/2, -\Delta_y, t) \\ & - \Upsilon(-3\Delta_x/2, \Delta_y, t) + \Upsilon(-3\Delta_x/2, -\Delta_y, t).\end{aligned} \tag{12.54}$$

The remaining TD function, $\Phi_{y,y}(t)$, then follows by replacing $\Delta_{x,y}$ with $\Delta_{y,x}$ in (12.54), respectively. Recall that the TD function, $\Upsilon(x, y, t)$, is specified in Section K.1 in terms of elementary functions. The effect of bounded electric conductivity or/and permittivity can be, again, described through $\Gamma^{\mathrm{E}}(x, y, t)$ as given by (12.15) and (12.16). Accordingly, we write

$$\underline{\boldsymbol{\mathcal{R}}}_k = \frac{1}{2} \begin{bmatrix} \Gamma^{\mathrm{E}}_{x,x}(t_k) & 0 \\ 0 & \Gamma^{\mathrm{E}}_{y,y}(t_k) \end{bmatrix}, \tag{12.55}$$

where

$$\begin{aligned} \Gamma^{\mathrm{E}}_{x,x}(t) &= \Gamma^{\mathrm{E}}(3\Delta_x/2, \Delta_y, t) - \Gamma^{\mathrm{E}}(3\Delta_x/2, -\Delta_y, t) - 3\Gamma^{\mathrm{E}}(\Delta_x/2, \Delta_y, t) \\ &\quad + 3\Gamma^{\mathrm{E}}(\Delta_x/2, -\Delta_y, t) + 3\Gamma^{\mathrm{E}}(-\Delta_x/2, \Delta_y, t) - 3\Gamma^{\mathrm{E}}(-\Delta_x/2, -\Delta_y, t) \\ &\quad - \Gamma^{\mathrm{E}}(-3\Delta_x/2, \Delta_y, t) + \Gamma^{\mathrm{E}}(-3\Delta_x/2, -\Delta_y, t). \end{aligned} \tag{12.56}$$

The second TD function, $\Gamma^{\mathrm{E}}_{y,y}(t)$, can be expressed in a similar manner with $\Delta_{x,y}$ replaced by $\Delta_{y,x}$, respectively. The convolution-type system of equations (12.25) is finally fully determined by specifying the excitation array on its right-hand side, that is,

$$\boldsymbol{V}_m = \begin{bmatrix} V_x(t_m) \\ V_y(t_m) \end{bmatrix}. \tag{12.57}$$

Its components pertaining to the E-polarized plane-wave excitation immediately follow as (cf. (12.18) and (12.22))

$$V_x(t) = -\frac{\sin(\phi)}{\kappa_0} \left\{ \int_{\tau=0}^{t+\kappa_0\Delta_x/2} e^{\mathrm{i}}(\tau)\mathrm{d}\tau - \int_{\tau=0}^{t-\kappa_0\Delta_x/2} e^{\mathrm{i}}(\tau)\mathrm{d}\tau \right\}, \tag{12.58a}$$

$$V_y(t) = \frac{\cos(\phi)}{\sigma_0} \left\{ \int_{\tau=0}^{t+\sigma_0\Delta_y/2} e^{\mathrm{i}}(\tau)\mathrm{d}\tau - \int_{\tau=0}^{t-\sigma_0\Delta_y/2} e^{\mathrm{i}}(\tau)\mathrm{d}\tau \right\}. \tag{12.58b}$$

Since the rectangular screen is assumed to be relatively small, we may use the TD expressions that are exact for the normal incidence characterized by $\theta = 0$, that is,

$$V_x(t) = -e^{\mathrm{i}}(t)\Delta_x \sin(\phi), \tag{12.59a}$$

$$V_y(t) = e^{\mathrm{i}}(t)\Delta_y \cos(\phi), \tag{12.59b}$$

again (see (12.20) and (12.24)). Finally, the array of unknown coefficients (see (12.51a) and (12.51b)), i.e.

$$\boldsymbol{I}_m = \begin{bmatrix} i_{m;x} \\ i_{m;y} \end{bmatrix} \quad \text{for all } m = \{1, \dots, M\}, \tag{12.60}$$

can be obtained via the updating step-by-step scheme (12.27).

12.3.2 A small rectangular metasurface

The pulsed EM scattering by a metasurface can be analyzed via the reciprocity relations (12.32a) and (12.32b). Owing to the similarity of (12.6) with (12.32a), the solution given in Section 12.3.1 can serve for calculating the electric-current distribution induced on a small rectangular metasurface. Consequently, it is sufficient to calculate the magnetic-current density from the second reciprocity relation (12.32b). To that end, we shall follow similar lines of reasoning used before. Hence, on account of the small-rectangle assumption, the unknown magnetic-current surface density is expanded as (cf. (12.48a), (12.48b) and (12.51a), (12.51b))

$$\partial K_x^{\mathrm{s}}(x,y,t) \simeq \frac{K_x(t)}{2\Delta_y}\Lambda(x)\Pi(y) = \frac{\Lambda(x)\Pi(y)}{2\Delta_y}\sum_{k=1}^{M} v_{k;x}\Lambda_k(t), \tag{12.61a}$$

$$\partial K_y^{\mathrm{s}}(x,y,t) \simeq \frac{K_y(t)}{2\Delta_x}\Lambda(y)\Pi(x) = \frac{\Lambda(y)\Pi(x)}{2\Delta_x}\sum_{k=1}^{M} v_{k;y}\Lambda_k(t), \tag{12.61b}$$

where $v_{k;x}$ and $v_{k;y}$ denote the magnetic-current coefficients (in V) and the definition of basis function is given by (12.49), (12.50) and (7.19). The spatial distribution of the induced currents is shown in Figure 12.6. Furthermore, as for the testing sources we take the "razor-type" functions, again (cf. (12.52a) and (12.52b))

$$\partial K_x^{\mathrm{T}}(x,y,t) = \Pi(2x)\delta(y)\delta(t), \tag{12.62a}$$

$$\partial K_y^{\mathrm{T}}(x,y,t) = \Pi(2y)\delta(x)\delta(t). \tag{12.62b}$$

Using (the transform-domain images of) (12.61a)–(12.62b) in the reciprocity relation (12.32b), we arrive at two mutually independent equations in the s-domain that interrelate the induced magnetic currents with the corresponding components of the incident magnetic-field strength (see Figure 12.4). Their inverse to the TD leads to the convolution-type system of (uncoupled equations) (12.45), the admittance array in which directly follows from (12.53) via (cf. (12.35))

$$\underline{\mathcal{Y}}_k = \eta_0^2 \underline{\mathcal{Z}}_k \quad \text{for all } k = \{1,\ldots,M\}. \tag{12.63}$$

The effect of bounded magnetic permeability is further incorporated via the following admittance array (cf. (12.55)):

$$\underline{\mathcal{G}}_k = \frac{1}{2}\begin{bmatrix} \Gamma^{\mathrm{M}}_{x,x}(t_k) & 0 \\ 0 & \Gamma^{\mathrm{M}}_{y,y}(t_k) \end{bmatrix}, \tag{12.64}$$

where (cf. (12.56))

$$\begin{aligned}\Gamma^{\mathrm{M}}_{x,x}(t) = {}& \Gamma^{\mathrm{M}}(3\Delta_x/2,\Delta_y,t) - \Gamma^{\mathrm{M}}(3\Delta_x/2,-\Delta_y,t) - 3\Gamma^{\mathrm{M}}(\Delta_x/2,\Delta_y,t)\\ &+ 3\Gamma^{\mathrm{M}}(\Delta_x/2,-\Delta_y,t) + 3\Gamma^{\mathrm{M}}(-\Delta_x/2,\Delta_y,t) - 3\Gamma^{\mathrm{M}}(-\Delta_x/2,-\Delta_y,t)\\ &- \Gamma^{\mathrm{M}}(-3\Delta_x/2,\Delta_y,t) + \Gamma^{\mathrm{M}}(-3\Delta_x/2,-\Delta_y,t).\end{aligned} \tag{12.65}$$

Recall that the TD function, $\Gamma^{\mathrm{M}}(t)$, applying to a loss-free high-permeability layer is defined by (12.38). The second TD function in (12.64), $\Gamma^{\mathrm{M}}_{y,y}(t)$, can be readily

obtained from (12.65) by replacing $\Delta_{x,y}$ with $\Delta_{y,x}$, respectively. The excitation array on the right-hand side of the convolution-type equation (12.45) consists of two terms

$$\boldsymbol{I}_m = \begin{bmatrix} I_x(t_m) \\ I_y(t_m) \end{bmatrix}. \tag{12.66}$$

For the E-polarized plane-wave excitation, they can be deduced from (12.41) and (12.42) as

$$I_x(t) = \frac{\eta_0 \cos(\phi)\cos(\theta)}{\kappa_0}\left\{ \int_{\tau=0}^{t+\kappa_0\Delta_x/2} \mathrm{e}^{\mathrm{i}}(\tau)\mathrm{d}\tau - \int_{\tau=0}^{t-\kappa_0\Delta_x/2} \mathrm{e}^{\mathrm{i}}(\tau)\mathrm{d}\tau \right\}, \tag{12.67a}$$

$$I_y(t) = \frac{\eta_0 \sin(\phi)\cos(\theta)}{\sigma_0}\left\{ \int_{\tau=0}^{t+\sigma_0\Delta_y/2} \mathrm{e}^{\mathrm{i}}(\tau)\mathrm{d}\tau - \int_{\tau=0}^{t-\sigma_0\Delta_y/2} \mathrm{e}^{\mathrm{i}}(\tau)\mathrm{d}\tau \right\}. \tag{12.67b}$$

Their small-rectangle approximations, which are exact for the normal incidence, $\theta = 0$, immediately follow as (cf. (12.43) and (12.44))

$$I_x(t) = \eta_0 \mathrm{e}^{\mathrm{i}}(t)\Delta_x \cos(\phi), \tag{12.68a}$$

$$I_y(t) = \eta_0 \mathrm{e}^{\mathrm{i}}(t)\Delta_y \sin(\phi). \tag{12.68b}$$

In the final step, the equations are solved, using the marching-on-in-time scheme (12.47), for example. This way finally leads to the array of the magnetic-current coefficients, that is,

$$\boldsymbol{V}_m = \begin{bmatrix} v_{m;x} \\ v_{m;y} \end{bmatrix} \quad \text{for all } m = \{1, \ldots, M\}. \tag{12.69}$$

Once the coefficients are known, they can be substituted in (12.61a) and (12.61b), thereby achieving the desired (space–time) magnetic-current distribution on a small metasurface.

12.4 Illustrative examples

To provide an illustrative numerical example, we shall analyze the electric- and magnetic-current surface distribution induced on a loss-free metasurface. The metasurface occupies a rectangular domain $\Sigma = \{-w/2 < x < w/2, -w/4 < y < w/4, z = 0\}$ and is excited by the E-polarized plane wave whose polarization and direction of propagation is specified by $\phi = \pi/4$ and $\theta = 0$ (see (12.1)). Its pulse shape, $\mathrm{e}^{\mathrm{i}}(t)$, is defined via (8.15) (see Figure 8.3), again, with the unit amplitude, $e_{\mathrm{m}} = 1.0\,\mathrm{V/m}$ and $c_0 t_{\mathrm{w}}/w = 0.50$. The EM properties of the metasurface are described via the normalized layer's capacitance and inductance as $C^{\mathrm{E}}/\eta_0 t_{\mathrm{w}} = 1.0$ and $L^{\mathrm{M}}\eta_0/t_{\mathrm{w}} = 2.0$, respectively. The surface of the metasurface is discretized uniformly with the spatial steps $\Delta_x = w/(N_x + 1)$ and $\Delta_y = 2w/(N_y + 1)$ in the

x- and y-directions, respectively, where N_x and N_y denote the number of inner nodes in the pertaining directions. In the numerical that follows, we take $N_x = N_y = 6$, which leads to a regular Cartesian grid with (inner) nodes at $(x_u, y_u, 0)$ and $(x_v, y_v, 0)$ for all $u = \{1, \ldots, U\}$ and $v = \{1, \ldots, V\}$, respectively, with $U = 42$ and $V = 42$. In a similar fashion, the time axis is discretized uniformly with the constant time step as $\{t_k = k\Delta t; \Delta t > 0, k = 1, 2, \ldots, M\}$. The induced current responses are observed within a bounded time window that is limited by $c_0 t_M / w = 6.0$.

Once the discretization of the space–time solution domain is set, the marching-on-in-time scheme (12.27) yields the coefficients $i_k^{[u]}$ and $i_k^{[v]}$ pertaining to the x- and y-components of the induced electric-current surface density (see (12.8) and (12.10)), respectively. Likewise, the x- and y-components of the induced magnetic-current surface density are specified by coefficients $v_k^{[u]}$ and $v_k^{[v]}$ (see (12.33a) and (12.33b)), respectively, that follow from (12.47) in a step-by-step manner. The thus calculated TD response is compared with the corresponding Kirchhoff's approximation (see Section 6.6). Assuming the current densities that would be excited in an unbounded metasurface, we arrive at the following TD expressions for the electric-current response (cf. (8.29)):

$$\partial J_x^{\mathrm{s}}(x,y,t) \overset{k}{\simeq} -2\left\{\delta(t) - \frac{2Y_\perp(\kappa_0,\sigma_0)}{C^{\mathrm{E}}}\exp\left[-\frac{2Y_\perp(\kappa_0,\sigma_0)}{C^{\mathrm{E}}}t\right]\mathrm{H}(t)\right\} *_t H_y^{\mathrm{i}}(x,y,0,t), \tag{12.70a}$$

$$\partial J_y^{\mathrm{s}}(x,y,t) \overset{k}{\simeq} 2\left\{\delta(t) - \frac{2Y_\perp(\kappa_0,\sigma_0)}{C^{\mathrm{E}}}\exp\left[-\frac{2Y_\perp(\kappa_0,\sigma_0)}{C^{\mathrm{E}}}t\right]\mathrm{H}(t)\right\} *_t H_x^{\mathrm{i}}(x,y,0,t) \quad \text{for all } (x,y)\in\Sigma \text{ and } t>0, \tag{12.70b}$$

where $Y_\perp(\kappa_0, \sigma_0) = p(\kappa_0, \sigma_0)/\mu_0 = \eta_0 \cos(\theta)$. Correspondingly, the Kirchhoff approximation of the induced magnetic-current surface density can be expressed as (cf. (8.47))

$$\partial K_x^{\mathrm{s}}(x,y,t) \overset{k}{\simeq} 2\left\{\delta(t) - \frac{2Z_\perp(\kappa_0,\sigma_0)}{L^{\mathrm{M}}}\exp\left[-\frac{2Z_\perp(\kappa_0,\sigma_0)}{L^{\mathrm{M}}}t\right]\mathrm{H}(t)\right\} *_t E_y^{\mathrm{i}}(x,y,0,t), \tag{12.70c}$$

$$\partial K_y^{\mathrm{s}}(x,y,t) \overset{k}{\simeq} -2\left\{\delta(t) - \frac{2Z_\perp(\kappa_0,\sigma_0)}{L^{\mathrm{M}}}\exp\left[-\frac{2Z_\perp(\kappa_0,\sigma_0)}{L^{\mathrm{M}}}t\right]\mathrm{H}(t)\right\} *_t E_x^{\mathrm{i}}(x,y,0,t) \quad \text{for all } (x,y)\in\Sigma \text{ and } t>0, \tag{12.70d}$$

where $Z_\perp(\kappa_0, \sigma_0) = 1/Y_\perp(\kappa_0, \sigma_0)$.

The pulse shapes of ∂J_x and ∂K_y as evaluated at the center of the metasurface in $\{0 \le t/t_{\mathrm{w}} \le 6.0\}$ are shown in Figure 12.7. Since the dimensions of the sheet are comparable to the spatial support of the exciting plane wave, a good correlation between the corresponding pulses is observed in their early-time parts only where

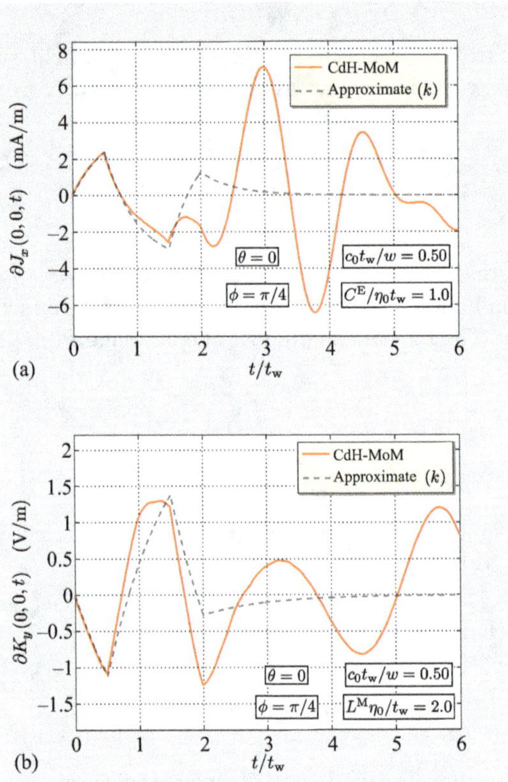

Figure 12.7 Plane-wave induced electric- and magnetic-current surface densities at the center of the metasurface. (a) The CdH-MoM solution (12.27) with the Kirchhoff approximation (12.70a). (b) The CdH-MoM solution (12.47) with the Kirchhoff approximation (12.70d).

the effects of boundaries do not occur. A similar observation has already been made before (see Figures 8.4(b), 8.5(b) and 8.6(b)).

12.5 Concluding remarks

The computational methodologies introduced in this chapter lead to the electric-or/and magnetic-current distributions induced in bounded 3-D screens. In this section, the relation of the calculated current densities to the EM scattering characteristics of such screens is discussed. We shall start by describing some scattering properties of a metasurface. Subsequently, we shall closely analyze the far-field scattered wave amplitudes pertaining to a small, electrically conducting rectangular screen.

12.5.1 EM scattering by a metasurface

At a large distance from the metasurface, the scattered EM field can be described by a radially propagating spherical wave as [2, Eq. (26.11-10)]

$$\{\hat{\boldsymbol{E}}^{\mathrm{s}}, \hat{\boldsymbol{H}}^{\mathrm{s}}\}(\boldsymbol{r}, s) = \{\hat{\boldsymbol{E}}^{\mathrm{s};\infty}, \hat{\boldsymbol{H}}^{\mathrm{s};\infty}\}(\boldsymbol{\xi}, s)\frac{\exp(-s|\boldsymbol{r}|/c_0)}{4\pi|\boldsymbol{r}|}\left[1 + O(|\boldsymbol{r}|^{-1})\right], \tag{12.71}$$

as $|\boldsymbol{r}| \to \infty$, where $\hat{\boldsymbol{E}}^{\mathrm{s};\infty}$ and $\hat{\boldsymbol{H}}^{\mathrm{s};\infty}$ denote the electric- and magnetic-field spherical-wave amplitudes, respectively, and $\boldsymbol{\xi} = \boldsymbol{r}/|\boldsymbol{r}|$ is the unit vector in the direction of observation. The far-field amplitudes can be found from [2, (26.11-11) and (26.11-12)]

$$\hat{\boldsymbol{E}}^{\mathrm{s};\infty}(\boldsymbol{\xi}, s) = s\mu_0(\boldsymbol{\xi}\boldsymbol{\xi} - \underline{\boldsymbol{I}}) \cdot \hat{\boldsymbol{\phi}}^{J;\infty}(\boldsymbol{\xi}, s) + sc_0^{-1}\boldsymbol{\xi} \times \hat{\boldsymbol{\phi}}^{K;\infty}(\boldsymbol{\xi}, s), \tag{12.72a}$$

$$\hat{\boldsymbol{H}}^{\mathrm{s};\infty}(\boldsymbol{\xi}, s) = s\epsilon_0(\boldsymbol{\xi}\boldsymbol{\xi} - \underline{\boldsymbol{I}}) \cdot \hat{\boldsymbol{\phi}}^{K;\infty}(\boldsymbol{\xi}, s) - sc_0^{-1}\boldsymbol{\xi} \times \hat{\boldsymbol{\phi}}^{J;\infty}(\boldsymbol{\xi}, s), \tag{12.72b}$$

with (the far-field amplitudes of) the electric- and magnetic-current source vector potentials [2, Eq. (26.11-9)]

$$\hat{\boldsymbol{\phi}}^{J;\infty}(\boldsymbol{\xi}, s) = \int_{\boldsymbol{r}'\in\Sigma} \exp(s\boldsymbol{\xi} \cdot \boldsymbol{r}'/c_0)\partial\hat{\boldsymbol{J}}^{\mathrm{s}}(\boldsymbol{r}', s)\mathrm{d}A(\boldsymbol{r}'), \tag{12.73a}$$

$$\hat{\boldsymbol{\phi}}^{K;\infty}(\boldsymbol{\xi}, s) = \int_{\boldsymbol{r}'\in\Sigma} \exp(s\boldsymbol{\xi} \cdot \boldsymbol{r}'/c_0)\partial\hat{\boldsymbol{K}}^{\mathrm{s}}(\boldsymbol{r}', s)\mathrm{d}A(\boldsymbol{r}'). \tag{12.73b}$$

Recall that the space–time distributions of the induced electric- and magnetic-current distributions, $\partial\boldsymbol{J}^{\mathrm{s}}(\boldsymbol{r}, t)$ and $\partial\boldsymbol{K}^{\mathrm{s}}(\boldsymbol{r}, t)$, are assumed to be attainable for all $\boldsymbol{r} \in \Sigma$ and $t > 0$ via the CdH-MoM procedures as described in Sections 12.1.2 and 12.2.1.

Now, upon inspection of the Kirchhoff approximations (12.70a)–(12.70d), it is concluded that (the mutually orthogonal components of) the electric- and magnetic-current surface densities are interrelated via (cf. (8.51))

$$\partial J_y^{\mathrm{s}}(x, y, t) = \partial K_x^{\mathrm{s}}(x, y, t)Y_\perp(\kappa_0, \sigma_0), \tag{12.74a}$$

$$\partial J_x^{\mathrm{s}}(x, y, t) = -\partial K_y^{\mathrm{s}}(x, y, t)Y_\perp(\kappa_0, \sigma_0), \tag{12.74b}$$

provided that

$$Y_\perp(\kappa_0, \sigma_0)/C^{\mathrm{E}} = Z_\perp(\kappa_0, \sigma_0)/L^{\mathrm{M}}. \tag{12.75}$$

Note that the "2-D counterpart" of (12.75) as given by (8.52) has been previously associated with the transparency condition (4.8). Consequently, under this condition, the magnetic-field far-field amplitudes can be expressed via

$$\hat{H}_x^{\mathrm{s};\infty}(\boldsymbol{\xi}, s) = -sc_0^{-1}(\xi_y/\xi_z)\left[\boldsymbol{\xi} \cdot \hat{\boldsymbol{\phi}}^{J;\infty}(\boldsymbol{\xi}, s)\right], \tag{12.76a}$$

$$\hat{H}_y^{\mathrm{s};\infty}(\boldsymbol{\xi}, s) = sc_0^{-1}(\xi_x/\xi_z)\left[\boldsymbol{\xi} \cdot \hat{\boldsymbol{\phi}}^{J;\infty}(\boldsymbol{\xi}, s)\right], \tag{12.76b}$$

$$\hat{H}_z^{\mathrm{s};\infty}(\boldsymbol{\xi}, s) = 0, \tag{12.76c}$$

where we may use $\boldsymbol{\xi} = \cos(\phi)\sin(\theta)\boldsymbol{i}_x + \sin(\phi)\sin(\theta)\boldsymbol{i}_y + \cos(\theta)\boldsymbol{i}_z$ and note that $\boldsymbol{i}_z \cdot \hat{\boldsymbol{\phi}}^{J;\infty}(\boldsymbol{\xi}, s) = 0$ (see (12.73a)). Once $\hat{H}_x^{s;\infty}(\boldsymbol{\xi}, s)$ and $\hat{H}_y^{s;\infty}(\boldsymbol{\xi}, s)$ are determined, the electric-field amplitudes immediately follow from [2, Eq. (26.11-13)] as

$$\hat{E}_x^{s;\infty}(\boldsymbol{\xi}, s) = \zeta_0 \xi_z \hat{H}_y^{s;\infty}(\boldsymbol{\xi}, s), \tag{12.77a}$$

$$\hat{E}_y^{s;\infty}(\boldsymbol{\xi}, s) = -\zeta_0 \xi_z \hat{H}_x^{s;\infty}(\boldsymbol{\xi}, s), \tag{12.77b}$$

$$\hat{E}_z^{s;\infty}(\boldsymbol{\xi}, s) = -\zeta_0 \left[\xi_x \hat{H}_y^{s;\infty}(\boldsymbol{\xi}, s) - \xi_y \hat{H}_x^{s;\infty}(\boldsymbol{\xi}, s)\right]. \tag{12.77c}$$

The relation to the 2-D problem configuration analyzed in Section 4.1.1 can be seen by using $\phi = 0$ and hence $\boldsymbol{\xi} = \sin(\theta)\boldsymbol{i}_x + \cos(\theta)\boldsymbol{i}_z$, and $\hat{\boldsymbol{\phi}}^{J;\infty}(\boldsymbol{\xi}, s) = \hat{\phi}_y^{J;\infty}\boldsymbol{i}_y$. As a consequence, all the scattered-wave amplitudes (12.76a)–(12.77c) vanish identically, which is consistent with the transparency properties discussed in Section 4.1.1.

Finally, to demonstrate the limits of validity of relations (12.74a) and (12.74b) for a bounded metasurface, the electric- and magnetic-current space–time distributions on $\Sigma = \{-w/2 < x < w/2, -w/4 < y < w/4, z = 0\}$ are calculated. The layer under consideration is excited by the E-polarized plane wave that is defined by its (bipolar) triangular signature with the unit amplitude $e_{\mathrm{m}} = 1.0\,\mathrm{V/m}$ and with $\phi = \pi/4$ and $\theta = 0$ (see (12.1)), again. Its relative spatial support is now taken as $c_0 t_{\mathrm{w}}/w = 0.10$; that is, the sheet is now relatively large with respect to the spatial support of the exciting pulse. The EM properties of the metasurface are described by $C^{\mathrm{E}}/\eta_0 t_{\mathrm{w}} = L^{\mathrm{M}}\eta_0/t_{\mathrm{w}} = 1.0$ such that the "transparency" condition (12.75) is met. Figure 12.8 clearly shows that relations (12.74a) and (12.74b) are for such a metasurface satisfied over a substantial part of the time window.

12.5.2 EM scattering by a small rectangular sheet

We next analyze EM scattering characteristics of a small, loss-free rectangular sheet supporting electric currents. To that end, we shall utilize expressions (12.71)–(12.73b) from the previous section. Hence, employing the postulated electric-current distributions (12.48a) and (12.48b) in (12.73a), the (non-vanishing components of the) pertaining potential function can be expressed in closed form as

$$\hat{\phi}_x^{J;\infty}(\boldsymbol{\xi}, s) = \hat{I}_x(s)\Delta_x\, \mathrm{i}_0^2(s\xi_x\Delta_x/2c_0)\mathrm{i}_0(s\xi_y\Delta_y/c_0), \tag{12.78a}$$

$$\hat{\phi}_y^{J;\infty}(\boldsymbol{\xi}, s) = \hat{I}_y(s)\Delta_y\, \mathrm{i}_0^2(s\xi_y\Delta_y/2c_0)\mathrm{i}_0(s\xi_x\Delta_x/c_0), \tag{12.78b}$$

where $\mathrm{i}_0(x)$ represents the modified spherical Bessel function of the first kind. Under the assumption that the sheet's dimensions are relatively small, we may use $\mathrm{i}_0(x) = 1 + O(x^2)$ as $x \downarrow 0$ in (12.78a) and (12.78b) to get their approximations, viz

$$\hat{\phi}_x^{J;\infty}(\boldsymbol{\xi}, s) \simeq \hat{I}_x(s)\Delta_x, \tag{12.79a}$$

$$\hat{\phi}_y^{J;\infty}(\boldsymbol{\xi}, s) \simeq \hat{I}_y(s)\Delta_y. \tag{12.79b}$$

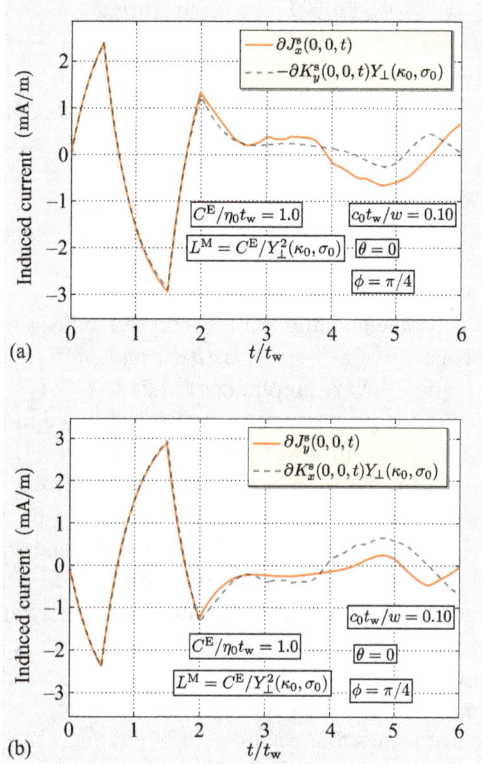

Figure 12.8 Plane-wave induced surface current densities at the center of the metasurface. (a) Validity of (12.74a). (b) Validity of (12.74b).

As a matter of fact, the latter expressions are exact along the directions normal to the sheet's plane, where $\xi_x = \xi_y = 0$. Consequently, the use of (12.79a) and (12.79b) in (12.72a) leads to the following far-field amplitudes of the electric type:

$$\hat{E}^{s;\infty}_x(\boldsymbol{\xi}, s) \simeq -s\mu_0(1-\xi_x^2)\Delta_x \hat{I}_x(s) + s\mu_0\xi_x\xi_z\Delta_y\hat{I}_y(s), \tag{12.80a}$$
$$\hat{E}^{s;\infty}_y(\boldsymbol{\xi}, s) \simeq -s\mu_0(1-\xi_y^2)\Delta_y \hat{I}_y(s) + s\mu_0\xi_x\xi_z\Delta_x\hat{I}_x(s), \tag{12.80b}$$
$$\hat{E}^{s;\infty}_z(\boldsymbol{\xi}, s) \simeq s\mu_0\xi_z\xi_x\Delta_x \hat{I}_x(s) + s\mu_0\xi_z\xi_y\Delta_y\hat{I}_y(s). \tag{12.80c}$$

In a similar manner, (12.72b) yields

$$\hat{H}^{s;\infty}_x(\boldsymbol{\xi}, s) \simeq sc_0^{-1}\xi_z\Delta_y\hat{I}_y(s), \tag{12.81a}$$
$$\hat{H}^{s;\infty}_y(\boldsymbol{\xi}, s) \simeq -sc_0^{-1}\xi_z\Delta_x\hat{I}_x(s), \tag{12.81b}$$
$$\hat{H}^{s;\infty}_z(\boldsymbol{\xi}, s) \simeq sc_0^{-1}\xi_y\Delta_x\hat{I}_x(s) - sc_0^{-1}\xi_x\Delta_y\hat{I}_y(s). \tag{12.81c}$$

With the far-field amplitudes at our disposal, it is possible to evaluate the EM power scattered by the rectangular sheet. For that purpose, we shall define the following quantity (see [2, Eq. (29.4-19)] and [3, Eqs. (3.10) with (3.17)]):

$$\hat{P}^{\mathrm{s}}(s) = \frac{\zeta_0}{32\pi^2} \int_{\boldsymbol{\xi}\in\Omega} \hat{\boldsymbol{H}}^{\mathrm{s};\infty}(\boldsymbol{\xi}, s) \cdot \hat{\boldsymbol{H}}^{\mathrm{s};\infty}(\boldsymbol{\xi}, -s)\mathrm{d}\Omega, \tag{12.82}$$

where $\Omega = \{\boldsymbol{\xi} \cdot \boldsymbol{\xi} = 1\}$ denotes a unit sphere. The thus defined quantity can be for sinusoidally in time-varying EM fields associated with the time-averaged EM power carried by the scattered wave. Upon substituting (12.81a)–(12.81c) in (12.82) and carrying out the integration over the unit sphere, we end up with

$$\hat{P}^{\mathrm{s}}(s) = -\frac{\zeta_0}{12\pi}\frac{s^2}{c_0^2}\left[\Delta_x^2 \hat{I}_x(s)\hat{I}_x(-s) + \Delta_y^2 \hat{I}_y(s)\hat{I}_y(-s)\right]. \tag{12.83}$$

In the (limiting) real-frequency domain (FD) pertaining to the (imaginary-valued) parameter $\{s = \delta + \mathrm{i}\omega, \delta \downarrow 0, \omega \in \mathbb{R}\}$, the time-averaged scattered power can be expressed as

$$\hat{P}^{\mathrm{s}}(\mathrm{i}\omega) = \frac{\zeta_0}{12\pi}k^2\left[\Delta_x^2|\hat{I}_x(\mathrm{i}\omega)|^2 + \Delta_y^2|\hat{I}_y(\mathrm{i}\omega)|^2\right], \tag{12.84}$$

where we used the wave number, $k = \omega/c_0$, and the property $\hat{I}(\mathrm{i}\omega)\hat{I}(-\mathrm{i}\omega) = |\hat{I}(\mathrm{i}\omega)|^2$ applying to a real-valued signal $I(t)$.

By virtue of the forward scattering theorem [3, Chapter 3], the EM power scattered from the loss-free sheet can also be expressed as [3, Eq. (3.12)]

$$\hat{P}^{\mathrm{s}}(s) = \frac{1}{4s\mu_0}\left[\hat{e}^{\mathrm{i}}(s)\,\boldsymbol{\alpha} \cdot \hat{\boldsymbol{E}}^{\mathrm{s};\infty}(\boldsymbol{\beta}, -s) - \hat{e}^{\mathrm{i}}(-s)\,\boldsymbol{\alpha} \cdot \hat{\boldsymbol{E}}^{\mathrm{s};\infty}(\boldsymbol{\beta}, s)\right], \tag{12.85}$$

where $\boldsymbol{\alpha}$ and $\boldsymbol{\beta}$ denote the unit vectors in the direction of polarization and propagation, respectively, and $\hat{e}^{\mathrm{i}}(s)$ is the plane-wave amplitude (see (12.1)). Consequently, considering the E-polarized plane-wave excitation, the use of (12.80a) and (12.80b) in (12.85) yields

$$\begin{aligned}\hat{P}^{\mathrm{s}}(s) = \tfrac{1}{4}\Big\{&\hat{e}^{\mathrm{i}}(s)\left[\alpha_x\Delta_x\hat{I}_x(-s) + \alpha_y\Delta_y\hat{I}_y(-s)\right] \\ &+ \hat{e}^{\mathrm{i}}(-s)\left[\alpha_x\Delta_x\hat{I}_x(s) + \alpha_y\Delta_y\hat{I}_y(s)\right]\Big\}.\end{aligned} \tag{12.86}$$

As a matter of fact, (12.86) is equivalent to (12.83). Apparently, the equality of (12.83) and (12.86) is achieved if

$$-\frac{\zeta_0}{6\pi}\frac{s^2}{c_0^2}\Delta_x^2\hat{I}_x(s) = \alpha_x\Delta_x\hat{e}^{\mathrm{i}}(s) \text{ and } -\frac{\zeta_0}{6\pi}\frac{s^2}{c_0^2}\Delta_y^2\hat{I}_y(s) = \alpha_y\Delta_y\hat{e}^{\mathrm{i}}(s). \tag{12.87}$$

These relations become transparent in the real-FD, where they have the following form:

$$\frac{\zeta_0}{6\pi}k^2\Delta_x^2\hat{I}_x(\mathrm{i}\omega) = \alpha_x\Delta_x\hat{e}^{\mathrm{i}}(\mathrm{i}\omega) \text{ and } \frac{\zeta_0}{6\pi}k^2\Delta_y^2\hat{I}_y(\mathrm{i}\omega) = \alpha_y\Delta_y\hat{e}^{\mathrm{i}}(\mathrm{i}\omega). \tag{12.88}$$

Here, the quantity $(\zeta_0/6\pi)k^2\Delta^2$ can be associated with the radiation resistance of a short-wire antenna of length $\Delta > 0$ (see [4, Section 4.4a, Eq. (10)]). Consequently, our computational model of a small loss-free sheet can be represented by two (mutually orthogonal) short-wire antennas oriented along the x- and y-directions. In accordance with (12.88), the corresponding components of the (vectorial) induced electric current and the incident electric-field strength are interrelated via the radiation resistances of the cross-polarized dipoles.

References

[1] M. Štumpf. *Time-Domain Electromagnetic Reciprocity in Antenna Modeling*. Hoboken, NJ: John Wiley & Sons, Inc., 2019.

[2] A. T. de Hoop, *Handbook of Radiation and Scattering of Waves*. London, UK: Academic Press, 1995.

[3] M. Štumpf. *Electromagnetic Reciprocity in Antenna Theory*. Hoboken, NJ: John Wiley & Sons, Inc., 2018.

[4] J. A. Kong. *Electromagnetic Wave Theory*. New York, NY: John Wiley & Sons, Inc., 1986.

Chapter 13
Pulsed EM-field scattering by apertures

Electromagnetic (EM)-field penetration through a slot in a planar sheet is a fundamental problem of EM theory that has been the subject of in-depth research in the real-frequency domain (FD) (see [1, 2] and the references therein). In this chapter, it is demonstrated how the problem of EM scattering by apertures in a perfectly electrically conducting (PEC) planar sheet can be solved in the time domain (TD). To that end, we shall first employ the Cagniard–DeHoop method of moments (CdH-MoM) to tackle the two-dimensional (2-D) problem consisting of a slot in a PEC screen. Second, under the assumption that the slot is relatively narrow, we shall pursue the approach previously applied to narrow strips (see Chapter 7) and describe its EM scattering properties in terms of Kirchoff-circuit equivalent parameters such as the "aperture admittance" [3]. With an emphasis on the proper problem formulation, the analysis is further generalized to incorporate the presence of a dielectric half-space, a rectangular filled groove and a homogeneous slot's filling. Finally, it is shown that the CdH-MoM computational methodology can be readily extended to analyzing the pulsed EM scattering by a bounded aperture of arbitrary shape.

Throughout the chapter, the EM reciprocity theorem of the time-convolution type is taken as the point of departure (see [4, Section 28.4] and [5, Section 1.4.1]). For details regarding the EM scattering by slots in the real-FD, the reader is referred to thorough texts on the subject (see [6, Section 5.2] and [7, Section 14.9]).

13.1 Pulsed EM scattering by a bounded slot

To illustrate the computational methodology, we shall first analyze the pulsed EM scattering by an infinite PEC sheet with a slot extending over a domain $\{a < x < b, -\infty < y < \infty, -\delta/2 < z < \delta/2\}$, where $\delta > 0$ is the (relatively small) thickness of the sheet and $w = b - a > 0$ represents the width of the slot (see Figure 13.1). The slotted conducting screen is located in the homogeneous, isotropic and loss-free medium that is characterized by two (scalar, real-valued and positive) parameters, $\{\epsilon_0, \mu_0\}$, again. The screen is supposed to be irradiated by a uniform, H-polarized plane wave that can be in the s-domain represented via

$$\hat{H}_y^{\rm i}(x, z, s) = \eta_0 \hat{e}^{\rm i}(s) \exp\{-s[\kappa_0 x - p(\kappa_0) z]\}, \qquad \text{(4.18 revisited)}$$

where $\kappa_0 = \sin(\beta)/c_0$, $p(\kappa_0) = \cos(\beta)/c_0$ and recall that $\hat{e}^{\rm i}(s)$ denotes the Laplace-transform image of the (causal) plane-wave signature $e^{\rm i}(t)$.

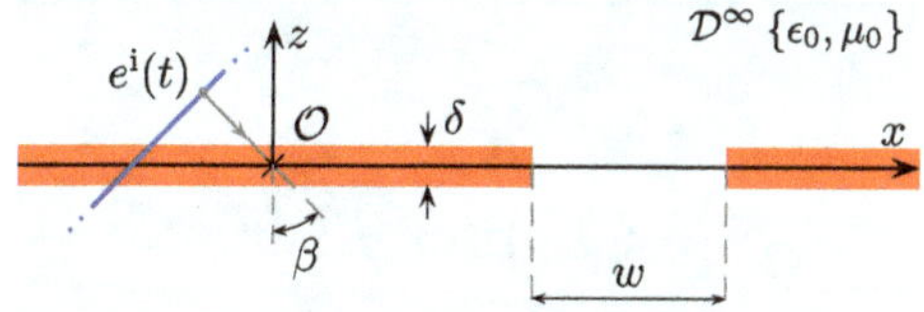

Figure 13.1 Pulsed EM plane wave impinging upon a plane screen with a slot

The problem under consideration could be, in principle, analyzed with the aid of the CdH-MoM-based solution methodology similar to the one introduced in Section 10.2.1. If that approach were chosen, the unknown quantity would be the electric-current distribution induced on the conductive screens. Owing to their unbounded support, however, a truncation of the spatial computational domain would be necessary. But, since we are always interested in the response to a temporally bounded, causal excitation pulse only, the omission of sufficiently distant parts of the screen (the effect of which occurs at the field point beyond the time window of observation) does not affect the result. Alternatively, the presence of the slot can be accounted for via the concept of compensation in which the "unperturbed" state corresponds to the configuration with the infinite PEC screen [8]. Accordingly, the scattered EM field (denoted by superscript ^{s}) is defined as the difference between the total EM field and the excitation one (denoted by superscript ^{e}), that is (cf. (9.1)),

$$\{\hat{H}_y^{\mathrm{s}}, \hat{E}_x^{\mathrm{s}}, \hat{E}_z^{\mathrm{s}}\}(x, z, s) = \{\hat{H}_y - \hat{H}_y^{\mathrm{e}}, \hat{E}_x - \hat{E}_x^{\mathrm{e}}, \hat{E}_z - \hat{E}_z^{\mathrm{e}}\}(x, z, s), \tag{13.1}$$

where the excitation field describes the total, H-polarized EM field in the presence of the infinite PEC screen. As a consequence, the solution domain extends over the bounded slot only, where (the tangential component of) the electric-field strength is the unknown (space–time) EM quantity to be sought. This problem formulation is also adopted in the ensuing sections.

13.1.1 Problem formulation

The EM scattering problem is formulated with the aid of the EM reciprocity theorem of the time-convolution type [5, Section 1.4.1] that is applied to interrelate the scattered EM field (see (13.1)) with the testing field (denoted by superscript ^{T}). The testing field is taken to be a causal EM wave field whose non-vanishing field components, $\{H_y^{\mathrm{T}}, E_x^{\mathrm{T}}, E_z^{\mathrm{T}}\}$, are excited by an electric-field testing-source distribution along the slot's domain. Hence, applying the EM reciprocity theorem to the unbounded domain above the screen, we arrive at

$$\int_{x=a}^{b} \hat{H}_y^{\mathrm{T}}(x, \delta, s)\hat{E}_x^{\mathrm{s}}(x, \delta, s)\mathrm{d}x = \int_{x=a}^{b} \hat{H}_y^{\mathrm{s}}(x, \delta, s)\hat{E}_x^{\mathrm{T}}(x, \delta, s)\mathrm{d}x, \tag{13.2}$$

as $\delta \downarrow 0$, where we employed the property of causality of the interrelated EM fields (see [5, Section 1.4.1]). In a similar manner, upon applying the reciprocity theorem to the lower half-space, it is found that

$$\int_{x=a}^{b} \hat{H}_y^{\mathrm{T}}(x,-\delta,s)\hat{E}_x^{\mathrm{s}}(x,-\delta,s)\mathrm{d}x = \int_{x=a}^{b} \hat{H}_y^{\mathrm{s}}(x,-\delta,s)\hat{E}_x^{\mathrm{T}}(x,-\delta,s)\mathrm{d}x, \tag{13.3}$$

as $\delta \downarrow 0$. Now, making use of (see (13.1))

$$\hat{H}_y^{\mathrm{s}}(x,\delta,s) = \hat{H}_y(x,\delta,s) - 2\hat{H}_y^{\mathrm{i}}(x,0,s), \tag{13.4a}$$

$$\hat{H}_y^{\mathrm{s}}(x,-\delta,s) = \hat{H}_y(x,-\delta,s), \tag{13.4b}$$

along with the continuity conditions in the slot, that is,

$$\lim_{z\downarrow\delta/2}\{\hat{H}_y,\hat{E}_x^{\mathrm{s}},\hat{E}_x^{\mathrm{T}}\}(x,z,s) = \lim_{z\uparrow-\delta/2}\{\hat{H}_y,\hat{E}_x^{\mathrm{s}},\hat{E}_x^{\mathrm{T}}\}(x,z,s)$$
$$\text{for all } \{-w/2 < x < w/2\}, \tag{13.5}$$

the reciprocity relations (13.2) and (13.3) lead to

$$\int_{x=a}^{b} \left[\hat{H}_y^{\mathrm{T}}(x,\delta,s) - \hat{H}_y^{\mathrm{T}}(x,-\delta,s)\right]\hat{E}_x^{\mathrm{s}}(x,0,s)\mathrm{d}x$$
$$= -2\int_{x=a}^{b} \hat{H}_y^{\mathrm{i}}(x,0,s)\hat{E}_x^{\mathrm{T}}(x,0,s)\mathrm{d}x, \tag{13.6}$$

as $\delta \downarrow 0$. It is emphasized here that $\hat{E}_x^{\mathrm{s}}(x,0,s) = \hat{E}_x(x,0,s)$ since $\hat{E}_x^{\mathrm{e}}(x,0,s) = 0$ for all $x \in \mathbb{R}$, and that the testing magnetic field in (13.6), $\hat{H}_y^{\mathrm{T}}(x,\pm\delta,s)$, corresponds to the prescribed testing-source distributions $\lim_{z\downarrow 0}\hat{E}_x^{\mathrm{T}}(x,z,s) = \partial\hat{K}_y^{\mathrm{T}}(x,s)$ and $\lim_{z\uparrow 0}\hat{E}_x^{\mathrm{T}}(x,z,s) = \partial\hat{K}_y^{\mathrm{T}}(x,s)$, respectively. Hence, for the (symmetric) problem configuration shown in Figure 13.1, $\hat{H}_y^{\mathrm{T}}(x,-\delta,s) = -\hat{H}_y^{\mathrm{T}}(x,\delta,s)$, and we end up with (cf. (8.1))

$$\int_{x=a}^{b} \hat{H}_y^{\mathrm{T}}(x,\delta,s)\hat{E}_x^{\mathrm{s}}(x,0,s)\mathrm{d}x = -\int_{x=a}^{b} \hat{H}_y^{\mathrm{i}}(x,0,s)\hat{E}_x^{\mathrm{T}}(x,0,s)\mathrm{d}x, \tag{13.7}$$

as $\delta \downarrow 0$. Following the lines of reasoning applied in Section 8.1.1, the s-domain reciprocity relation (13.7) will next be solved for the unknown electric-field distribution in the slot. Therefore, the wave-slowness representation of the type (4.33) is used in (13.7) to represent the magnetic-field strengths. In the thus obtained relation, we change the order of integration and get (cf. (8.3))

$$\frac{s}{2\pi\mathrm{i}}\int_{\kappa=-\mathrm{i}\infty}^{\mathrm{i}\infty} \tilde{H}_y^{\mathrm{T}}(\kappa,\delta,s)\tilde{E}_x^{\mathrm{s}}(-\kappa,0,s)\mathrm{d}\kappa$$
$$= -\frac{s}{2\pi\mathrm{i}}\int_{\kappa=-\mathrm{i}\infty}^{\mathrm{i}\infty} \tilde{H}_y^{\mathrm{i}}(\kappa,0,s)\tilde{E}_x^{\mathrm{T}}(-\kappa,0,s)\mathrm{d}\kappa \qquad \text{for } \{s\in\mathbb{R}; s>0\}. \tag{13.8}$$

The testing magnetic-field strength in the transform domain is related to the testing source via

$$\tilde{H}_y^{\mathrm{T}}(\kappa,\delta,s) = [\epsilon_0/p(\kappa)]\tilde{E}_x^{\mathrm{T}}(\kappa,0,s), \tag{13.9}$$

as $\delta \downarrow 0$, where the vertical slowness parameter, $p(\kappa)$, was defined by (4.36). The transform-domain reciprocity relation (13.8) with (13.9) is next handled numerically via the CdH-MoM. The determination of space–time electric-field distribution in the slot is the result.

13.1.2 Problem solution

To solve the reciprocity relation (13.8) numerically, the space–time solution domain is first discretized. For the sake of simplicity, the spatial domain occupied by the slot is divided into $N-1$ segments of a constant length $\Delta = w/(N-1)$ in the uniform manner depicted in Figure 8.2. Likewise, the time coordinate, $\{t \in \mathbb{R}; t > 0\}$, is discretized uniformly as $\{t_k = k\Delta t; \Delta t > 0, k = 1,2,\ldots,M\}$, where Δt has the meaning of the (constant) time step. Consequently, the unknown electric-field distribution in the slot is expanded in a piecewise linear basis (cf. (8.6))

$$E_x^{\mathrm{s}}(x,0,t) \simeq \sum_{n=1}^{N}\sum_{k=1}^{M} e_k^{[n]}\Lambda^{[n]}(x)\Lambda_k(t), \tag{13.10}$$

where $e_k^{[n]}$ are unknown coefficients (in V/m) and the basis functions are given by (8.7a)–(8.7c) and (7.19). Following further the solution strategy from Section 8.1.2, the testing-source distribution is taken to be represented using the rectangular function of x (see (8.9a)–(8.9c)) and the Dirac-delta distribution in t, that is,

$$E_x^{\mathrm{T}}(x,0,t) = \Pi^{[S]}(x)\delta(t) \quad \text{for all } S = \{1,\ldots,N\}. \tag{13.11}$$

Upon substituting the transform-domain counterparts of (13.10) and (13.11) in the transform-domain reciprocity relation (13.8), we end up with a system of equations in the s-domain whose constituents can be transformed to the TD using the CdH technique (see Section 2.1.1). The inverse transform of the system leads to (cf. (7.42) and (8.40))

$$\sum_{k=1}^{m}\left(\ddot{\underline{\mathcal{Y}}}_{m-k+1} - 2\ddot{\underline{\mathcal{Y}}}_{m-k} + \ddot{\underline{\mathcal{Y}}}_{m-k-1}\right)\cdot \boldsymbol{E}_k = \boldsymbol{I}_m, \tag{13.12}$$

where $\ddot{\underline{\mathcal{Y}}}_k = \ddot{\underline{\mathcal{Y}}}(t_k)$ denotes a 2-D $[N \times N]$ "admittivity array" at $t = t_k = k\Delta t$ whose elements can be found directly from the impeditivity array pertaining to a PEC sheet. Thus, making use of the results from Appendix B, it is found that $\ddot{\mathcal{Y}}^{[S,n]}(t) = 2\eta_0^2\mathcal{Z}^{[S,n]}(t)$ for all $S = \{1,\ldots,N\}$, $n = \{1,\ldots,N\}$ and $t > 0$. Further, the elements of the excitation array follow from $I^{[S]}(t) = -\eta_0 V^{[S]}(t)$ using (8.13a)–(8.14) for all $S = \{1,\ldots,N\}$ and $t > 0$. Finally, the convolution-type system of equations

(13.12) can be solved for the unknown electric-field coefficients with the help of the marching-on-in-time technique (see Section 7.4). In this way, it is found that (cf. (7.44))

$$\boldsymbol{E}_m = \ddot{\boldsymbol{\mathcal{Y}}}_1^{-1} \cdot \left[\boldsymbol{I}_m - \sum_{k=1}^{m-1} \left(\ddot{\boldsymbol{\mathcal{Y}}}_{m-k+1} - 2\ddot{\boldsymbol{\mathcal{Y}}}_{m-k} + \ddot{\boldsymbol{\mathcal{Y}}}_{m-k-1} \right) \cdot \boldsymbol{E}_k \right]$$
$$\text{for all } m = \{1, \ldots, M\}, \tag{13.13}$$

where $\ddot{\boldsymbol{\mathcal{Y}}}_1^{-1}$ represents the inverse of the admittivity matrix at $t = \Delta t$.

The incorporation of a planarly layered background medium or/and additional slots can be effected using the methodology applied to thin screens in Chapters 9 and 10. Therefore, a detailed analysis of problem configurations further extending the one shown in Figure 13.1 is not given in the next sections. Instead, attention is paid to closed-form approximate solutions concerning the pulsed EM scattering by narrow slots.

13.2 Pulsed EM scattering by narrow slots

Throughout this section, it is assumed that slots are relatively narrow with respect to the spatial extent of the exciting EM plane-wave signature $e^{\mathrm{i}}(t)$. The analysis starts by analyzing the pulsed EM scattering by a narrow slot located on a dielectric half-space. Further, approximate closed-form expressions describing the EM scattering by a narrow rectangular groove are derived. Finally, in order to demonstrate how to handle accompanying mutual-coupling effects, we shall analyze problem configurations consisting of two narrow slots and rectangular grooves.

13.2.1 Slot on a dielectric half-space

The problem configuration under consideration is shown in Figure 13.2. It consists of a narrow slot in a PEC thin sheet that is located on a dielectric half-space. The narrow slot extends along $\{-w/2 < x < w/2, -\infty < y < \infty, -\delta/2 < z < \delta/2\}$, where both the width $w > 0$ and thickness $\delta > 0$ are assumed to be relatively small.

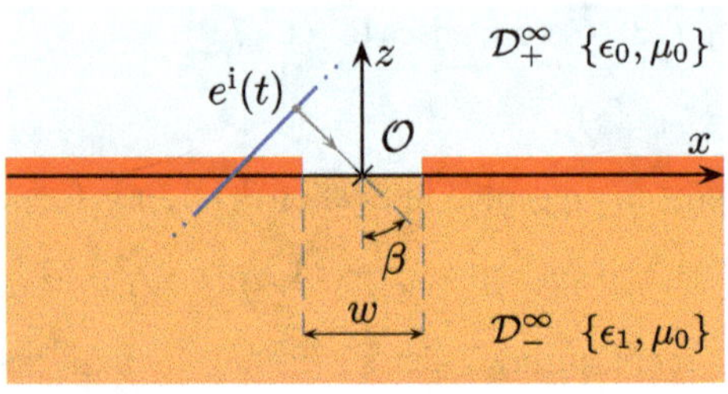

Figure 13.2 Pulsed EM plane wave impinging upon a narrow slot located on a dielectric half-space

The upper half-space, denoted by $\mathcal{D}^{\infty}_{+}$, is described by $\{\epsilon_0, \mu_0\}$, again, and the lower (dielectric) half-space is characterized by $\{\epsilon_1, \mu_0\}$. The corresponding EM wave speeds and admittances are then denoted by $c_{0,1} = (\epsilon_{0,1}\mu_0)^{-1/2} > 0$ and $\eta_{0,1} = 1/c_{0,1}\mu_0$, respectively. The structure is activated by the H-polarized plane wave that is defined by (4.18) in the s-domain.

The starting point for the analysis that follows is the reciprocity relation (13.6), where, in line with Figure 13.2, we set $a = -w/2$ and $b = w/2$. In the following step, the testing fields $\hat{H}^{\mathrm{T}}_y(x, \pm\delta, s)$ pertaining to the actual problem configuration are determined. To that end, we employ the wave-slowness representation along the x-direction (see (4.33)), again. Consequently, assuming the constant testing-source distribution over the slot, the testing EM fields can be represented via (cf. (7.5))

$$\hat{H}^{\mathrm{T}}_y(x, \delta, s) = \frac{s\epsilon_0}{2\pi\mathrm{i}} \int_{\kappa=-\mathrm{i}\infty}^{\mathrm{i}\infty} \exp(-s\kappa x)\, \mathrm{i}_0(s\kappa w/2) \frac{\mathrm{d}\kappa}{p_0(\kappa)}, \tag{13.14a}$$

$$\hat{H}^{\mathrm{T}}_y(x, -\delta, s) = -\frac{s\epsilon_1}{2\pi\mathrm{i}} \int_{\kappa=-\mathrm{i}\infty}^{\mathrm{i}\infty} \exp(-s\kappa x)\, \mathrm{i}_0(s\kappa w/2) \frac{\mathrm{d}\kappa}{p_1(\kappa)}, \tag{13.14b}$$

as $\delta \downarrow 0$, where the slowness parameters, $p_{0,1}(\kappa)$, pertaining to $\mathcal{D}^{\infty}_{\pm}$, respectively, are defined by (9.25). The slowness integral representations are next substituted in the starting reciprocity relation, where we change the order of the integrations with respect to κ and x. Assuming further the uniform electric-field distribution along the slot, i.e. (cf. (7.2))

$$\hat{E}^{\mathrm{s}}_x(x, 0, s) = \frac{\hat{V}^{\mathrm{s}}(s)}{w} \left[\mathrm{H}(x + w/2) - \mathrm{H}(x - w/2)\right], \tag{13.15}$$

the reciprocity relation can be readily cast into the following form (cf. (7.7a)):

$$\left[\hat{Y}^{+}_{\mathrm{ext}}(s) + \hat{Y}^{-}_{\mathrm{ext}}(s)\right] \hat{V}^{\mathrm{s}}(s) = -2\eta_0 \hat{e}^{\mathrm{i}}(s)\, \mathrm{i}_0(s\kappa_0 w/2), \tag{13.16}$$

where

$$\hat{Y}^{\pm}_{\mathrm{ext}}(s) = \frac{s\epsilon_{0,1}}{2\pi\mathrm{i}} \int_{\kappa=-\mathrm{i}\infty}^{\mathrm{i}\infty} \mathrm{i}_0^2(s\kappa w/2) \frac{\mathrm{d}\kappa}{p_{0,1}(\kappa)}, \tag{13.17}$$

is interpreted as the aperture admittance pertaining to the medium in $\mathcal{D}^{\infty}_{\pm}$, respectively. On the right-hand side of (13.16), we may, on account of the narrow-slot assumption, use $\mathrm{i}_0(s\kappa_0 w/2) = 1 + O(s^2\kappa_0^2 w^2)$ as $s\kappa_0 w \downarrow 0$. Then, in the homogeneous case, if $\epsilon_1 = \epsilon_0$, (13.16) condenses to (cf. (7.9))

$$\hat{Y}^{+}_{\mathrm{ext}}(s)\, \hat{V}^{\mathrm{s}}(s) = -\eta_0 \hat{e}^{\mathrm{i}}(s). \tag{13.18}$$

Pursuing now the approach presented in Section 7.1.1, it is found that (cf. (7.16a))

$$\hat{Y}^{+}_{\mathrm{ext}}(s) \simeq -(s\epsilon_0/\pi)[\log(sw/c_0) - 3/2 - \log(2) + \gamma] \quad \text{as } sw/c_0 \ll 1, \tag{13.19}$$

with $\gamma = 0.57721\ldots$ representing Euler's constant. Its limit pertaining to the real-FD, $s \to \mathrm{i}\omega$, then immediately follows as

$$\hat{Y}^{+}_{\mathrm{ext}}(\mathrm{i}\omega) \simeq \omega\epsilon_0/2 + (\mathrm{i}\omega\epsilon_0/\pi) \left[\log(\lambda/w) + 3/2 - \gamma - \log(\pi)\right], \tag{13.20}$$

provided that $w/\lambda \ll 1$ with $\lambda = 2\pi c_0/\omega$. Upon introducing the wave number $k = 2\pi/\lambda$ in the real-FD expression (13.20), it can be easily shown that the result agrees with the aperture admittance as known from the literature [3, Eq. (24) with (25)].

To solve (13.16) in the TD, we may follow the method described in Section 7.1.2. Hence, the voltage TD response is first expanded as

$$V^{\mathrm{s}}(t) \simeq \sum_{k=1}^{M} v_k \Lambda_k(t), \tag{13.21}$$

where v_k are the coefficients to be found and $\Lambda_k(t)$ denotes the triangular function (7.19). The s-domain image of (13.21) is then substituted in (13.16) and the thus obtained relation is transformed to the TD. Consequently, applying the marching-on-in-time technique, the voltage coefficients can be computed from (cf. (7.20))

$$v_m = -\bar{\Phi}_1^{-1}\left[c_0\Delta t e^{\mathrm{i}}(t_m)/2 + \sum_{k=1}^{m-1} v_k\left(\bar{\Phi}_{m-k+1} - 2\bar{\Phi}_{m-k} + \bar{\Phi}_{m-k-1}\right)\right]$$
$$\text{for all } m = \{1, \ldots, M\}, \tag{13.22}$$

where we used $\bar{\Phi}_k = \bar{\Phi}(t_k)$ and

$$\bar{\Phi}(t) = [\Phi(t) + \Phi^-(t)]/2, \tag{13.23}$$
$$\Phi(t) = \left[\Upsilon^B(w,t) - 2\Upsilon^B(0,t) + \Upsilon^B(-w,t)\right]/w^2, \tag{7.21 revisited}$$
$$\Phi^-(t) = (\epsilon_1/\epsilon_0)\left[\Upsilon^B_-(w,t) - 2\Upsilon^B_-(0,t) + \Upsilon^B_-(-w,t)\right]/w^2. \tag{13.24}$$

Furthermore, we used (B.9) to define the TD function pertaining to the dielectric half-space, viz

$$\Upsilon^B_-(x,t) = \frac{1}{2\pi}\left\{\frac{x^2}{2}\cosh^{-1}\left(\frac{c_1 t}{|x|}\right) + \frac{c_1 t x}{2}\left(\frac{c_1^2t^2}{x^2} - 1\right)^{1/2}\right.$$
$$\left. - c_1 t|x|\tan^{-1}\left[\left(\frac{c_1^2t^2}{x^2} - 1\right)^{1/2}\right]\right\}\mathrm{H}(t - |x|/c_1)$$
$$+ (c_1 t x/2)\mathrm{H}(x)\mathrm{H}(t). \tag{13.25}$$

In the homogeneous case, $\epsilon_1 = \epsilon_0$, which leads to $\bar{\Phi}(t) = \Phi(t)$. In such a case, solution (13.22) can be viewed as a limit of the CdH-MoM-based solution (13.13).

13.2.2 *Filled rectangular groove*

In this subsection, we shall analyze the pulsed EM scattering by a narrow rectangular groove in a PEC ground that occupies domain $\{-w/2 < x < w/2, -\infty < y < \infty, 0 < z < -d\}$, where $w > 0$ is its width, and $d > 0$ denotes its depth (see Figure 13.3). The groove is filled by a homogeneous and isotropic medium of electric permittivity ϵ_1 and magnetic permeability μ_1. The corresponding EM wave speed and wave admittance are $c_1 = (\epsilon_1\mu_1)^{-1/2} > 0$ and $\eta_1 = 1/c_1\mu_1$, respectively.

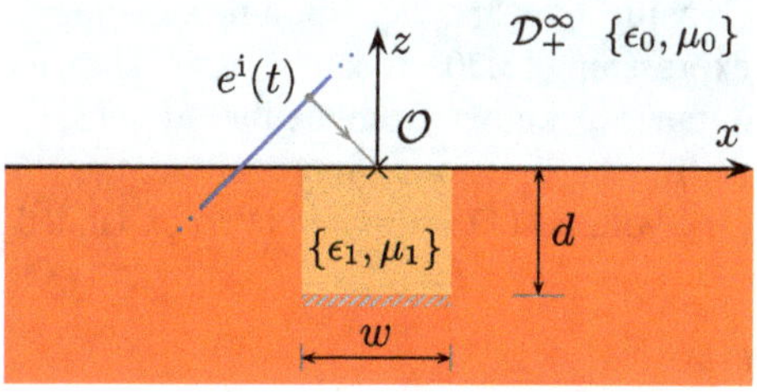

Figure 13.3 Pulsed EM plane wave impinging upon a narrow, rectangular filled groove

It is further assumed that the width of the groove is relatively small, such that the dominant EM-field components inside the gap are the transverse electromagnetic (TEM) x-independent ones, $\{\hat{H}_y, \hat{E}_x\}$. Therefore, inside the groove, they are then governed by

$$\partial_z \hat{H}_y + s\epsilon_1 \hat{E}_x = 0, \tag{13.26a}$$

$$\partial_z \hat{E}_x + s\mu_1 \hat{H}_y = 0. \tag{13.26b}$$

Further, it is assumed that the groove is on its lower terminating surface loaded such that the following condition applies:

$$\lim_{z \downarrow -d} \hat{H}_y(z, s) = Y^{\mathrm{g}} \lim_{z \downarrow -d} \hat{E}_x(z, s), \tag{13.27}$$

where Y^{g} denotes the corresponding surface admittance. Apparently, the limits $Y^{\mathrm{g}} \to 0$ and $Y^{\mathrm{g}} \to \infty$ leads to the explicit-type boundary conditions on the perfectly magnetically conducting (PMC) and PEC terminating surfaces, respectively.

Assuming now the testing field (further denoted by ^{T}, again) that inside the gap satisfies (13.26a) and (13.26b) with (13.27), the application of the reciprocity theorem to the domain occupied by the groove leads to a reciprocity relation similar to (13.3) applying to the unbounded lower half-space. Consequently, the reciprocity relation (13.6) (with $a = -w/2$ and $b = w/2$) can be used as the point of departure, again. Since the testing magnetic field pertaining to the uniform testing source $\hat{E}_x^{\mathrm{T}}(x, \delta, s) = 1/w$, for $\{-w/2 < x < w/2\}$ as $\delta \downarrow 0$, has been obtained before (see (13.14a)), it remains to determine the testing-field distribution pertaining to the gap's domain. Accordingly, the testing field excited by

$$\lim_{z \uparrow 0} \hat{E}_x^{\mathrm{T}}(z, s) = 1/w \tag{13.28}$$

follows upon solving the one-dimensional (1-D) system of equations (see (13.26a)–(13.27)). This way leads to (cf. (13.14b))

$$\hat{H}_y^{\mathrm{T}}(x, -\delta, s) = -\frac{\eta_1}{w} \frac{R + \exp(-2sd/c_1)}{R - \exp(-2sd/c_1)} \big[\mathrm{H}(x + w/2) - \mathrm{H}(x - w/2)\big], \tag{13.29}$$

where

$$R = (Y^{\mathrm{g}} - \eta_1)/(Y^{\mathrm{g}} + \eta_1) \tag{13.30}$$

is the corresponding reflection coefficient. Finally, making use of (13.14a) and (13.29) in the starting reciprocity relation (13.6), while assuming the uniform distribution of the electric field in the slot (see (13.15)), we arrive at (cf. (13.16))

$$\left[\hat{Y}^{+}_{\mathrm{ext}}(s) + \hat{Y}_{\mathrm{gap}}(s)\right]\hat{V}^{\mathrm{s}}(s) = -2\eta_0 \hat{e}^{\mathrm{i}}(s)\,\mathrm{i}_0(s\kappa_0 w/2), \tag{13.31}$$

where $\hat{Y}^{+}_{\mathrm{ext}}(s)$ is defined by (13.17) and the "gap admittance" follows as

$$\hat{Y}_{\mathrm{gap}}(s) = \frac{\eta_1}{w}\frac{R + \exp(-2sd/c_1)}{R - \exp(-2sd/c_1)}. \tag{13.32}$$

It is noted that if the pulse travel time across the gap is large such that the effect of reflections can be neglected, (13.32) comes down to

$$\hat{Y}_{\mathrm{gap}}(s) = \eta_1/w \text{ as } d/c_1 \to \infty. \tag{13.33}$$

Next, if the terminating surface at $z = -d$ is a PEC surface, then (13.32) can be expressed as

$$\hat{Y}_{\mathrm{gap}}(s) = (\eta_1/w)\coth(sd/c_1) \text{ as } Y^{\mathrm{g}} \to \infty, \tag{13.34}$$

which for short grooves, as $d/c_1 \downarrow 0$, has the form $\hat{Y}_{\mathrm{gap}}(s) \simeq 1/wsL^{\mathrm{g}}$ with $L^{\mathrm{g}} = \mu_1 d$. In a similar manner, the response of a narrow groove terminated by a PMC surface can be described through

$$\hat{Y}_{\mathrm{gap}}(s) = (\eta_1/w)\tanh(sd/c_1) \text{ as } Y^{\mathrm{g}} \to 0, \tag{13.35}$$

which takes the form $\hat{Y}_{\mathrm{gap}}(s) \simeq sC^{\mathrm{g}}/w$ as $d/c_1 \downarrow 0$ with $C^{\mathrm{g}} = \epsilon_1 d$.

The derived s-domain (13.31) can be transformed to and solved in the TD. Upon expanding the induced voltage pulse in terms of the triangular functions (see (13.21)), the desired coefficients can be evaluated using the step-by-step updating procedure described by (13.22), again. In the latter equation, however, $\bar{\Phi}(t)$ must be modified by using the appropriate $\Phi^{-}(t)$ pertaining to the actual configuration in the lower half-space. For a relatively long, narrow filled groove, whose gap admittance is given by (13.33), we use

$$\Phi^{-}(t) = (\epsilon_1/\epsilon_0)(c_1 t/2w)\mathrm{H}(t) \text{ as } d/c_1 \to \infty. \tag{13.36}$$

If reflections against the bottom surface cannot be neglected anymore, we use

$$\Phi^{-}(t) = \frac{\epsilon_1}{\epsilon_0}\left[\frac{c_1 t}{2w}\mathrm{H}(t) + \frac{1}{w}\sum_{n=1}^{\infty}(c_1 t - 2nd)\mathrm{H}(c_1 t - 2nd)\right] \text{ as } Y^{\mathrm{g}} \to \infty, \tag{13.37}$$

applying to the PEC termination and, finally,

$$\Phi^-(t) = \frac{\epsilon_1}{\epsilon_0}\left[\frac{c_1 t}{2w}\mathrm{H}(t) + \frac{1}{w}\sum_{n=1}^{\infty}(-1)^n(c_1 t - 2nd)\mathrm{H}(c_1 t - 2nd)\right] \quad \text{as } Y^{\mathrm{g}} \to 0, \tag{13.38}$$

applying to the PMC termination. Note that the first (early-time) term in (13.37) and (13.38) is identical with (13.36), as expected. Finally note that in any bounded time window of observation, the number of terms in the summations in (13.37) and (13.38) is always finite.

13.2.3 Slot with a homogeneous filling

If the domain occupied by a slot in a PEC screen is filled by a homogeneous medium that differs from the one of the background medium (see Figure 13.4), its presence must be properly accounted for. EM reciprocity-based formulations meeting this goal are hence discussed in this subsection. The analysis is provided for a narrow slot filled by a highly contrasting medium with dielectric and conductive properties, represented by $\{\epsilon, \sigma\}$, and for a metamaterial filling whose EM properties are described by parameters $\{\epsilon, \mu, \sigma, \Gamma\}$ (see Section 3.1).

13.2.3.1 Conductive and dielectric filling

Across a thin layer with conductive and dielectric properties, the tangential component of the magnetic-field strength is not continuous (see (3.4b)), a consequence of which the reciprocity relation (13.7) cannot be applied. Instead, we combine (13.2) and (13.3) to find (cf. (13.6))

$$\begin{aligned} 2\int_{x=-w/2}^{w/2} & \hat{H}_y^{\mathrm{T}}(x,\delta,s)\hat{E}_x^{\mathrm{s}}(x,0,s)\mathrm{d}x \\ & - \int_{x=-w/2}^{w/2}\left[\hat{H}_y(x,\delta,s) - \hat{H}_y(x,-\delta,s)\right]\hat{E}_x^{\mathrm{T}}(x,0,s)\mathrm{d}x \\ & \qquad = -2\int_{x=-w/2}^{w/2}\hat{H}_y^{\mathrm{i}}(x,0,s)\hat{E}_x^{\mathrm{T}}(x,0,s)\mathrm{d}x, \end{aligned} \tag{13.39}$$

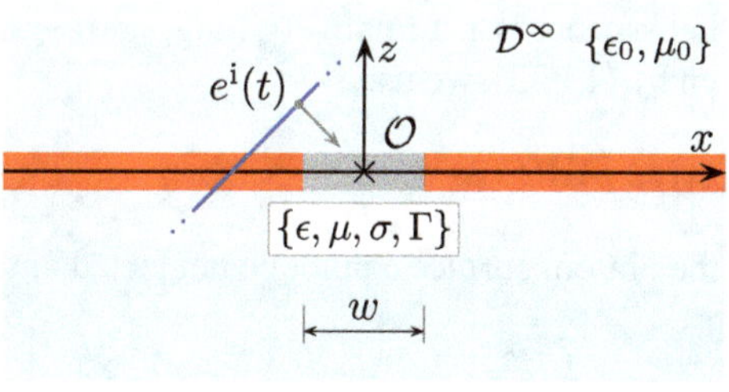

Figure 13.4 Pulsed EM plane wave impinging upon a narrow slot filled by a homogeneous medium

as $\delta \downarrow 0$, where we used the continuity of $\hat{E}_x^{\mathrm{s,T}}$ across the layer and the (odd) symmetry property applying to the testing magnetic-field strength, $\hat{H}_y^{\mathrm{T}}(x, -\delta, s) = -\hat{H}_y^{\mathrm{T}}(x, \delta, s)$. The jump of the magnetic-field strength across the thin filling is related to the tangential electric-field strength via the equivalent layer's admittance (see (3.4b) and (4.19a))

$$\hat{H}_y(x, \delta, s) - \hat{H}_y(x, -\delta, s) = -\hat{Y}^{\mathrm{E}}(s)\hat{E}_x(x, 0, s) \quad \text{for } \{-w/2 < x < w/2\}. \tag{13.40}$$

Substituting the cross-layer condition (13.40) in the reciprocity relation (13.39), we arrive at

$$\int_{x=-w/2}^{w/2} \hat{H}_y^{\mathrm{T}}(x, 0^+, s)\hat{E}_x^{\mathrm{s}}(x, 0, s)\mathrm{d}x + \tfrac{1}{2}\hat{Y}^{\mathrm{E}}(s)\int_{x=-w/2}^{w/2} \hat{E}_x^{\mathrm{s}}(x, 0, s)\hat{E}_x^{\mathrm{T}}(x, 0, s)\mathrm{d}x = -\int_{x=-w/2}^{w/2} \hat{H}_y^{\mathrm{i}}(x, 0, s)\hat{E}_x^{\mathrm{T}}(x, 0, s)\mathrm{d}x, \tag{13.41}$$

that for the vanishing EM contrast of the filling with respect to the background medium, represented by $\hat{Y}^{\mathrm{E}}(s) \to 0$, comes down to (13.7). Assuming, again, the uniform testing-field distribution across the slot, the testing field can be represented via (13.14a). If, in addition, the electric field is assumed to be uniformly distributed across the slot according to (13.15), the reciprocity relation (13.41) can be cast into the following form (cf. (13.18)):

$$\left[\hat{Y}_{\mathrm{ext}}^{+}(s) + \hat{Y}^{\mathrm{E}}(s)/2w\right]\hat{V}^{\mathrm{s}}(s) = -\eta_0\hat{e}^{\mathrm{i}}(s). \tag{13.42}$$

Upon expanding the voltage pulse in terms of the triangular functions according to (13.21), (13.42) can be transformed to the TD and solved in a step-by-step manner with the aid of the marching-on-in-time scheme. In this way, we arrive at (13.22), again, where we replace (13.23) with (cf. (7.26))

$$\bar{\Phi}(t) = \Phi(t) + (c_0/4w\alpha^{\mathrm{E}})(G^{\mathrm{E}}/\eta_0)(\alpha^{\mathrm{E}}t + 1)\mathrm{H}(t), \tag{13.43}$$

where $\alpha^{\mathrm{E}} = G^{\mathrm{E}}/C^{\mathrm{E}} > 0$ (see (3.19)) and recall that $\Phi(t)$ is defined via (7.21). For a vanishing equivalent layer capacitance, the parameter α^{E} tends to infinity, and we get

$$\lim_{C^{\mathrm{E}}\downarrow 0} \bar{\Phi}(t) = \Phi(t) + (c_0 t/4w)(G^{\mathrm{E}}/\eta_0)\mathrm{H}(t). \tag{13.44}$$

On the other hand, if the layer's conductance, G^{E}, is vanishingly small, while $C^{\mathrm{E}} = O(1)$, the parameter α^{E} approaches the zero value, which yields

$$\lim_{G^{\mathrm{E}}\downarrow 0} \bar{\Phi}(t) = \Phi(t) + (C^{\mathrm{E}}/4\epsilon_0 w)\mathrm{H}(t), \tag{13.45}$$

thus representing the effect of a loss-free dielectric filling. It can be seen from (13.44) and (13.45) that if both $G^{\mathrm{E}} \downarrow 0$ and $C^{\mathrm{E}} \downarrow 0$, then $\bar{\Phi}(t) = \Phi(t)$, which corresponds to the homogeneous case, $\epsilon_1 = \epsilon_0$, the problem configuration analyzed in Section 13.2.1.

13.2.3.2 Metasurface filling

When crossing a metasurface with high-contrast, combined magneto-dielectric characteristics, the tangential components of both electric- and magnetic-field strengths exhibit the jumps that are proportional to the equivalent layer's impedance and admittance, $\hat{Z}^{\mathrm{M}}$ and $\hat{Y}^{\mathrm{E}}$, respectively (see (3.10a) and (3.10b)). Consequently, the starting point for analyzing the EM scattering by a narrow slot filled by a metasurface are reciprocity relations (13.2) and (13.3), again, that are adapted to the actual problem configuration. Hence, upon recalling the meaning of the scattered field (see (13.1)), and employing the properties of the testing field, that is, $\hat{E}_x^{\mathrm{T}}(x, \pm\delta, s) = \hat{E}_x^{\mathrm{T}}(x, 0, s)$ and $\hat{H}_y^{\mathrm{T}}(x, -\delta, s) = -\hat{H}_y^{\mathrm{T}}(x, \delta, s)$, (13.2) and (13.3) are cast into the following form:

$$\int_{x=-w/2}^{w/2} \hat{H}_y^{\mathrm{T}}(x, \delta, s)\hat{E}_x(x, \delta, s)\mathrm{d}x = \int_{x=-w/2}^{w/2} \left[\hat{H}_y(x, \delta, s) - 2\hat{H}_y^{\mathrm{i}}(x, 0, s)\right]\hat{E}_x^{\mathrm{T}}(x, 0, s)\mathrm{d}x, \tag{13.46a}$$

and

$$\int_{x=-w/2}^{w/2} \hat{H}_y^{\mathrm{T}}(x, \delta, s)\hat{E}_x(x, -\delta, s)\mathrm{d}x = -\int_{x=-w/2}^{w/2} \hat{H}_y(x, -\delta, s)\hat{E}_x^{\mathrm{T}}(x, 0, s)\mathrm{d}x, \tag{13.46b}$$

as $\delta \downarrow 0$, respectively. In the reciprocity relations (13.46a) and (13.46b), the limits of the actual tangential EM-field components, $\{\hat{H}_y, \hat{E}_x\}$, are taken from above and from below of the metasurface filling, respectively. These limiting values are interrelated via the corresponding cross-layer transition conditions (4.19a) and (4.19b) that are further employed in the reciprocity relations to eliminate the magnetic component. Pursuing this approach, it can be found that (cf. (13.41))

$$\begin{aligned}
&\int_{x=-w/2}^{w/2} \hat{H}_y^{\mathrm{T}}(x, 0^+, s)\hat{E}_x(x, 0^+, s)\mathrm{d}x \\
&+ \left[\tfrac{1}{4}\hat{Y}^{\mathrm{E}}(s) + \hat{Y}^{\mathrm{M}}(s)\right]\int_{x=-w/2}^{w/2} \hat{E}_x(x, 0^+, s)\hat{E}_x^{\mathrm{T}}(x, 0, s)\mathrm{d}x \\
&+ \left[\tfrac{1}{4}\hat{Y}^{\mathrm{E}}(s) - \hat{Y}^{\mathrm{M}}(s)\right]\int_{x=-w/2}^{w/2} \hat{E}_x(x, 0^-, s)\hat{E}_x^{\mathrm{T}}(x, 0, s)\mathrm{d}x \\
&= -2\int_{x=-w/2}^{w/2} \hat{H}_y^{\mathrm{i}}(x, 0, s)\hat{E}_x^{\mathrm{T}}(x, 0, s)\mathrm{d}x,
\end{aligned} \tag{13.47a}$$

and

$$\begin{aligned}&\int_{x=-w/2}^{w/2} \hat{H}_y^{\mathrm{T}}(x,0^+,s)\hat{E}_x(x,0^-,s)\mathrm{d}x \\ &+ \left[\tfrac{1}{4}\hat{Y}^{\mathrm{E}}(s)+\hat{Y}^{\mathrm{M}}(s)\right]\int_{x=-w/2}^{w/2} \hat{E}_x(x,0^-,s)\hat{E}_x^{\mathrm{T}}(x,0,s)\mathrm{d}x \\ &+ \left[\tfrac{1}{4}\hat{Y}^{\mathrm{E}}(s)-\hat{Y}^{\mathrm{M}}(s)\right]\int_{x=-w/2}^{w/2} \hat{E}_x(x,0^+,s)\hat{E}_x^{\mathrm{T}}(x,0,s)\mathrm{d}x = 0. \qquad (13.47\mathrm{b})\end{aligned}$$

Assuming, again, the uniform spatial distribution for the testing source, $\hat{E}_x^{\mathrm{T}}(x,0,s)$, as well as for the actual electric fields on both sides of the metasurface planar filling, i.e. (cf. (13.15))

$$\hat{E}_x(x,0^\pm,s) = \frac{\hat{V}^\pm(s)}{w}\big[\mathrm{H}(x+w/2)-\mathrm{H}(x-w/2)\big], \qquad (13.48)$$

respectively, the reciprocity relations (13.47a) and (13.47b) can be cast into the matrix form

$$\begin{bmatrix} \hat{Y}_{++}(s) & \hat{Y}_{+-}(s) \\ \hat{Y}_{-+}(s) & \hat{Y}_{--}(s) \end{bmatrix} \cdot \begin{bmatrix} \hat{V}^+(s) \\ \hat{V}^-(s) \end{bmatrix} = \begin{bmatrix} -2\hat{H}_y^{\mathrm{i}}(0,0,s) \\ 0 \end{bmatrix}, \qquad (13.49)$$

where

$$\hat{Y}_{++}(s) = \hat{Y}_{--}(s) = \hat{Y}_{\mathrm{ext}}^{+}(s) + \left[\tfrac{1}{4}\hat{Y}^{\mathrm{E}}(s)+\hat{Y}^{\mathrm{M}}(s)\right]/w, \qquad (13.50\mathrm{a})$$

$$\hat{Y}_{+-}(s) = \hat{Y}_{-+}(s) = \left[\tfrac{1}{4}\hat{Y}^{\mathrm{E}}(s)-\hat{Y}^{\mathrm{M}}(s)\right]/w. \qquad (13.50\mathrm{b})$$

The system of s-domain equations is next solved in the TD. To that end, the voltage pulses induced across the slot are expanded in terms of the triangular functions (cf. (13.21))

$$V^\pm(t) \simeq \sum_{k=1}^{M} v_k^\pm \Lambda_k(t), \qquad (13.51)$$

where $v_k^\pm$ are the coefficients representing the voltage pulses induced across the filled slot as $z \to 0^\pm$, respectively. Employing the piecewise-linear expansion of the (yet unknown) voltage pulses, the s-domain system of equations (13.49) can be transformed to the TD, and we get (cf. (10.24))

$$\sum_{k=1}^{m} \left(\underline{\boldsymbol{Y}}_{m-k+1} - 2\underline{\boldsymbol{Y}}_{m-k} + \underline{\boldsymbol{Y}}_{m-k-1}\right) \cdot \boldsymbol{V}_k = \boldsymbol{H}_m. \qquad (13.52)$$

Here, $\underline{\boldsymbol{Y}}_k = \underline{\boldsymbol{Y}}(t_k)$ denotes a $[2 \times 2]$ admittance array that is defined by

$$\underline{\boldsymbol{Y}}_k = \frac{2\eta_0}{c_0\Delta t}\begin{bmatrix} \Phi(t_k)+\Pi(t_k) & \Psi(t_k) \\ \Psi(t_k) & \Phi(t_k)+\Pi(t_k) \end{bmatrix}, \qquad (13.53)$$

where $\Phi(t)$ is given by (7.21), again. For the sake of simplicity, the remaining TD functions in (13.53), $\Pi(t)$ and $\Psi(t)$, are provided for a loss-free metasurface filling characterized by $\hat{Y}^{\mathrm{E}}(s) = sC^{\mathrm{E}}$ and $\hat{Z}^{\mathrm{M}}(s) = sL^{\mathrm{M}}$. Under this assumption, it is found that

$$\Pi(t) = \frac{1}{4w}\left[\frac{C^{\mathrm{E}}/\epsilon_0}{2} + \frac{c_0^2 t^2}{L^{\mathrm{M}}/\mu_0}\right]\mathrm{H}(t), \tag{13.54}$$

$$\Psi(t) = \frac{1}{4w}\left[\frac{C^{\mathrm{E}}/\epsilon_0}{2} - \frac{c_0^2 t^2}{L^{\mathrm{M}}/\mu_0}\right]\mathrm{H}(t). \tag{13.55}$$

Further, $\boldsymbol{V}_k$ represents a $[2 \times 1]$ voltage array of unknown coefficients

$$\boldsymbol{V}_k = \begin{bmatrix} v_k^+ \\ v_k^- \end{bmatrix}, \tag{13.56}$$

and, finally, the right-hand side of the system of equations is given by

$$\boldsymbol{H}_m = \begin{bmatrix} -2\eta_0 e^{\mathrm{i}}(t_m) \\ 0 \end{bmatrix}. \tag{13.57}$$

The convolution-type system of equations (13.52) is now fully specified and can be hence solved for the voltage coefficients. Using the marching-on-in-time technique, again, we get (cf. (10.29))

$$\boldsymbol{V}_m = \underline{\mathbf{Y}}_1^{-1} \cdot \left[\boldsymbol{H}_m - \sum_{k=1}^{m-1}\left(\underline{\mathbf{Y}}_{m-k+1} - 2\underline{\mathbf{Y}}_{m-k} + \underline{\mathbf{Y}}_{m-k-1}\right) \cdot \boldsymbol{V}_k\right]$$
$$\text{for all } m = \{1, \ldots, M\}, \tag{13.58}$$

where $\underline{\mathbf{Y}}_1^{-1}$ denotes the inverse of the admittance matrix at $t = \Delta t$. The stable updating procedure (13.58) yields the coefficients that can be subsequently substituted in (13.51) to determine the corresponding electric-field distributions via (the TD counterpart of) (13.48).

13.2.4 Two slots in a PEC screen

An efficient analysis of the pulsed EM scattering by slotted screens calls for a closed-form description of the accompanying EM coupling phenomena. This is exactly the main purpose of this subsection, where the pulsed EM plane-wave interaction with two relatively narrow slots in a PEC sheet is analyzed.

Figure 13.5 shows the analyzed problem configuration. Without loss of generality, the first slot has its center at the origin and extends over $\mathcal{G}_A = \{-w/2 < x < w/2, -\infty < y < \infty, -\delta/2 < z < \delta/2\}$, where $w > 0$ is its (relatively) small width and $\delta \downarrow 0$. The second narrow slot is shifted by a distance $|x_B|$ and occupies domain $\mathcal{G}_B = \{x_B - w/2 < x < x_B + w/2, -\infty < y < \infty, -\delta/2 < z < \delta/2\}$. The slotted screen is supposed to be located in the homogeneous, isotropic and loss-free medium that is described by parameters $\{\epsilon_0, \mu_0\}$, again. The structure is irradiated by the uniform, H-polarized EM plane wave (see (4.18)).

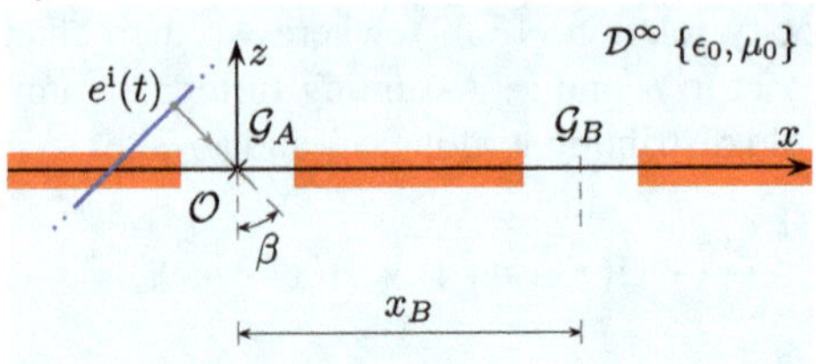

Figure 13.5 Pulsed EM plane wave impinging upon two narrow slots in a PEC screen

The problem at hand can be approached by generalizing the reciprocity relation (13.7) pertaining to a single slot in a narrow PEC screen. Following this way, we may arrive at (cf. (10.14))

$$\begin{aligned}
&\int_{x=-w/2}^{w/2} \left[\hat{H}_y^{\mathrm{T};A}(x,0^+,s) + \hat{H}_y^{\mathrm{T};B}(x,0^+,s)\right]\hat{E}_x^{\mathrm{s};A}(x,0,s)\mathrm{d}x \\
&+ \int_{x=x_B-w/2}^{x_B+w/2} \left[\hat{H}_y^{\mathrm{T};B}(x,0^+,s) + \hat{H}_y^{\mathrm{T};A}(x,0^+,s)\right]\hat{E}_x^{\mathrm{s};B}(x,0,s)\mathrm{d}x \\
&= -\int_{x=-w/2}^{w/2} \hat{H}_y^{\mathrm{i}}(x,0,s)\hat{E}_x^{\mathrm{T};A}(x,0,s)\mathrm{d}x \\
&- \int_{x=x_B-w/2}^{x_B+w/2} \hat{H}_y^{\mathrm{i}}(x,0,s)\hat{E}_x^{\mathrm{T};B}(x,0,s)\mathrm{d}x, \qquad (13.59)
\end{aligned}$$

where $\hat{E}_x^{\mathrm{s};A,B}$ denotes the unknown electric-field distributions across slots $\mathcal{G}_{A,B}$, respectively, and a similar meaning applies to the testing-source fields $\hat{E}_x^{\mathrm{T};A,B}$. The testing magnetic-field strengths, $\hat{H}_y^{\mathrm{T};A}(x,0^+,s)$ and $\hat{H}_y^{\mathrm{T};B}(x,0^+,s)$, generated by the (uniform) testing-source distributions on slots $\mathcal{G}_A$ and $\mathcal{G}_B$, respectively, are determined using the corresponding excitation conditions, that is,

$$\lim_{z\downarrow 0} \hat{E}_x^{\mathrm{T};A}(x,z,s) = \left[\mathrm{H}(x+w/2) - \mathrm{H}(x-w/2)\right]/w, \qquad (13.60a)$$

$$\lim_{z\downarrow 0} \hat{E}_x^{\mathrm{T};B}(x,z,s) = \left[\mathrm{H}(x-x_B+w/2) - \mathrm{H}(x-x_B-w/2)\right]/w,$$

$$\text{for all } x \in \mathbb{R}, \qquad (13.60b)$$

and using the wave-slowness representation in the x-direction (see (4.33)). Along these lines, it is found at once (cf. (13.14a))

$$\hat{H}_y^{\mathrm{T};A}(x,0^+,s) = \frac{s\epsilon_0}{2\pi\mathrm{i}}\int_{\kappa=-\mathrm{i}\infty}^{\mathrm{i}\infty} \exp(-s\kappa x)\,\mathrm{i}_0(s\kappa w/2)\frac{\mathrm{d}\kappa}{p(\kappa)}, \qquad (13.61a)$$

$$\hat{H}_y^{\mathrm{T};B}(x,0^+,s) = \frac{s\epsilon_0}{2\pi\mathrm{i}}\int_{\kappa=-\mathrm{i}\infty}^{\mathrm{i}\infty} \exp[-s\kappa(x-x_B)]\,\mathrm{i}_0(s\kappa w/2)\frac{\mathrm{d}\kappa}{p(\kappa)}, \qquad (13.61b)$$

and recall that the vertical slowness parameter, $p(\kappa)$, was defined previously by (4.36). The thus represented testing-field distributions are further substituted in

the s-domain reciprocity relation (13.59), where we then change the order of the integrations with respect to κ and x. Assuming further the uniform distribution of the desired electric-field distributions along $\mathcal{G}_A$ and $\mathcal{G}_B$, i.e.

$$\hat{E}_x^{\mathrm{s};A}(x,0,s) = \frac{\hat{V}^{\mathrm{s};A}(s)}{w}\left[\mathrm{H}(x+w/2) - \mathrm{H}(x-w/2)\right], \tag{13.62a}$$

$$\hat{E}_x^{\mathrm{s};B}(x,0,s) = \frac{\hat{V}^{\mathrm{s};B}(s)}{w}\left[\mathrm{H}(x-x_B+w/2) - \mathrm{H}(x-x_B-w/2)\right], \tag{13.62b}$$

the reciprocity relation (13.59) can be cast into the following matrix form (cf. (10.18)):

$$\begin{bmatrix} \hat{Y}_{AA}(s) & \hat{Y}_{AB}(s) \\ \hat{Y}_{BA}(s) & \hat{Y}_{BB}(s) \end{bmatrix} \cdot \begin{bmatrix} \hat{V}^{\mathrm{s};A}(s) \\ \hat{V}^{\mathrm{s};B}(s) \end{bmatrix} = \begin{bmatrix} -\hat{H}_y^{\mathrm{i}}(0,0,s) \\ -\hat{H}_y^{\mathrm{i}}(x_B,0,s) \end{bmatrix}. \tag{13.63}$$

In the obtained system of equations (13.63), the diagonal admittance terms, in fact, correspond to the aperture admittance, i.e. (cf. (13.17))

$$\hat{Y}_{AA}(s) = \hat{Y}_{BB}(s) = \frac{s\epsilon_0}{2\pi\mathrm{i}} \int_{\kappa=-\mathrm{i}\infty}^{\mathrm{i}\infty} \mathrm{i}_0^2(s\kappa w/2)\frac{\mathrm{d}\kappa}{p(\kappa)}, \tag{13.64}$$

while the off-diagonal components, $\hat{Y}_{AB}(s)$ and $\hat{Y}_{BA}(s)$, describe the mutual coupling between the slots. They can be represented via

$$\hat{Y}_{AB}(s) = \hat{Y}_{BA}(s) = \frac{s\epsilon_0}{2\pi\mathrm{i}} \int_{\kappa=-\mathrm{i}\infty}^{\mathrm{i}\infty} \exp(-s\kappa x_B)\,\mathrm{i}_0^2(s\kappa w/2)\frac{\mathrm{d}\kappa}{p(\kappa)}. \tag{13.65}$$

Assuming now that the distance between the slots, $|x_B|$, is large compared to the relatively small slots' width, w, we may apply the small-argument expansion for the modified spherical Bessel functions, $\mathrm{i}_0^2(x) = 1 + O(x^2)$ as $x \downarrow 0$, and get

$$\hat{Y}_{AB}(s) \simeq \frac{s\epsilon_0}{2\pi\mathrm{i}} \int_{\kappa=-\mathrm{i}\infty}^{\mathrm{i}\infty} \exp(-s\kappa x_B)\frac{\mathrm{d}\kappa}{p(\kappa)} = \frac{s\epsilon_0}{\pi}\mathrm{K}_0(s|x_B|/c_0). \tag{13.66}$$

To get the last analytical expression, we first deformed the integration contour into the CdH path, $\tau = \kappa x_B$ for $\{\tau \in \mathbb{R}; \tau > T_B = |x_B|/c_0\}$, and then employed the integral representation of the modified Bessel function of the second kind [9, Eq. (9.6.23)].

With the aid of expressions (13.64) and (13.66), the system of equations (13.63) is next solved in the TD. To that end, the unknown voltage pulse are expanded in a piecewise linear manner via the triangular functions (7.19). In the s-domain, this can be represented via (cf. (13.21) and (10.23a) with (10.23b))

$$\hat{V}^{\mathrm{s};A}(s) \simeq \sum_{k=1}^{M} v_k^A \hat{\Lambda}_k(s), \tag{13.67a}$$

$$\hat{V}^{\mathrm{s};B}(s) \simeq \sum_{k=1}^{M} v_k^B \hat{\Lambda}_k(s), \tag{13.67b}$$

where v_k^A and v_k^B denote the unknown coefficients (in V) that through (13.62a) and (13.62b) represent the desired electric-field strengths induced across slots $\mathcal{G}_A$ and $\mathcal{G}_B$, respectively. Next, substituting of (13.64) with (13.66)–(13.67b) and transforming the thus obtained system to the TD, we, again, end up with (cf. (10.24))

$$\sum_{k=1}^{m} \left(\underline{\boldsymbol{Y}}_{m-k+1} - 2\underline{\boldsymbol{Y}}_{m-k} + \underline{\boldsymbol{Y}}_{m-k-1}\right) \cdot \boldsymbol{V}_k = \boldsymbol{H}_m, \qquad \text{(13.52 revisited)}$$

where $\underline{\boldsymbol{Y}}_k = \underline{\boldsymbol{Y}}(t_k)$ denotes a $[2 \times 2]$ admittance array, whose elements can be related to the impedance array (10.25) via $\underline{\boldsymbol{Y}}_k = 2\eta_0^2 \underline{\boldsymbol{Z}}_k$. Therefore, we may write

$$\underline{\boldsymbol{Y}}_k = \frac{2\eta_0}{c_0 \Delta t} \begin{bmatrix} \Phi(t_k) & \Psi(t_k) \\ \Psi(t_k) & \Phi(t_k) \end{bmatrix}, \qquad (13.68)$$

where the TD functions, $\Phi(t)$ and $\Psi(t)$, are repeated here for convenience

$$\Phi(t) = [\Upsilon^B(w,t) - 2\Upsilon^B(0,t) + \Upsilon^B(-w,t)]/w^2, \qquad \text{(7.21 revisited)}$$

$$\Psi(t) = \cosh^{-1}(t/T_B)\mathrm{H}(t - T_B)/2\pi, \qquad \text{(10.26 revisited)}$$

and recall that $T_B = |x_B|/c_0$ and $\Upsilon^B(x,t)$ is defined by (B.9). Next, $\boldsymbol{V}_k$ represents a $[2 \times 1]$ voltage array of unknown coefficients

$$\boldsymbol{V}_k = \begin{bmatrix} v_k^A \\ v_k^B \end{bmatrix}, \qquad (13.69)$$

and, finally, the right-hand side of the solved system of equations is specified via

$$\boldsymbol{H}_m = \begin{bmatrix} -\eta_0 e^{\mathrm{i}}(t_m) \\ -\eta_0 e^{\mathrm{i}}(t_m - \kappa_0 x_B) \end{bmatrix}. \qquad (13.70)$$

As the system of equations (13.52) is now fully determined, it can be solved with the aid of the marching-on-in-time technique (13.58), again. The step-by-step updating procedure yields the voltage coefficients that represent the induced voltage pulses via (13.67a) and (13.67b). Finally, the corresponding electric-field strengths in the interacting slots immediately follow from (13.62a) and (13.62b).

13.2.5 Two filled rectangular grooves

As the final problem of Section 13.2, we shall investigate the pulsed EM scattering by a pair of relatively narrow rectangular grooves embedded in a PEC ground (see Figure 13.6). The grooves occupy domains $\mathcal{G}_A = \{-w/2 < x < w/2, -\infty < y < \infty, 0 < z < -d_1\}$ and $\mathcal{G}_B = \{x_B - w/2 < x < x_B + w/2, -\infty < y < \infty, 0 < z < -d_2\}$, where $w > 0$ is their width and $d_{1,2} > 0$ denotes their depths (not necessarily small). The gaps $\mathcal{G}_{A,B}$ are filled by homogeneous, isotropic and loss-free media whose EM constitutive properties are described by $\{\epsilon_{1,2}, \mu_{1,2}\}$, respectively. The corresponding EM wave speeds and wave admittances are $c_{1,2} = (\epsilon_{1,2}\mu_{1,2})^{-1/2} > 0$ and $\eta_{1,2} = 1/c_{1,2}\mu_{1,2}$, respectively. The width of the grooves is

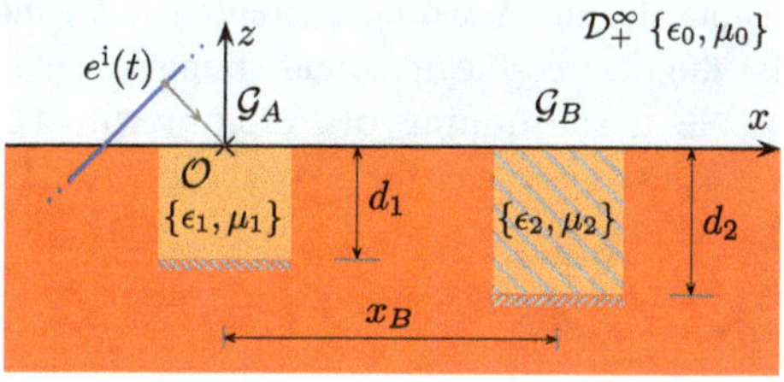

Figure 13.6 Pulsed EM plane wave impinging upon two narrow, rectangular filled grooves

supposed to be relatively small, again, such that the EM wave propagation in the cavities is governed by the 1-D x-independent EM field equations (see (13.26a) and (13.26b)). The non-vanishing TEM-field components are on the bottom surface of $\mathcal{G}_A$ and $\mathcal{G}_B$ interrelated via the admittance Y_1^{g} and Y_2^{g}, respectively (see (13.27)). Again, the structure is irradiated by the uniform H-polarized EM plane wave (see (4.18)). Consequently, the analyzed problem configuration as well as its excitation is y-independent.

The actual problem under consideration is tackled via the EM reciprocity theorem, again. To this end, the reciprocity relation (13.6) is adapted to the configuration at hand, which leads to

$$\int_{x=-w/2}^{w/2} \left[\hat{H}_y^{\mathrm{T};A}(x,0^+,s) - \hat{H}_y^{\mathrm{T};A}(x,0^-,s)\right] \hat{E}_x^{\mathrm{s};A}(x,0,s)\mathrm{d}x + \int_{x=x_B-w/2}^{x_B+w/2} \left[\hat{H}_y^{\mathrm{T};A}(x,0^+,s) - \hat{H}_y^{\mathrm{T};A}(x,0^-,s)\right] \hat{E}_x^{\mathrm{s};B}(x,0,s)\mathrm{d}x = -2\int_{x=-w/2}^{w/2} \hat{H}_y^{\mathrm{i}}(x,0,s)\hat{E}_x^{\mathrm{T};A}(x,0,s)\mathrm{d}x, \tag{13.71a}$$

in which the testing magnetic-field strengths, $\hat{H}_y^{\mathrm{T};A}(x,0^+,s)$ and $\hat{H}_y^{\mathrm{T};A}(x,0^-,s)$, are generated by the uniform testing-source distributions $\lim_{z\downarrow 0} \hat{E}_x^{\mathrm{T};A}(x,z,s) = 1/w$ and $\lim_{z\uparrow 0} \hat{E}_x^{\mathrm{T};A}(x,z,s) = 1/w$, respectively, as prescribed over the aperture of $\mathcal{G}_A$. To have the same number of equations as the number of unknown quantities, $\hat{E}_x^{\mathrm{s};A}$ and $\hat{E}_x^{\mathrm{s};B}$, the second reciprocity relation is arrived at by applying the testing-source distribution over the aperture of $\mathcal{G}_B$. In this way, we obtain

$$\int_{x=-w/2}^{w/2} \left[\hat{H}_y^{\mathrm{T};B}(x,0^+,s) - \hat{H}_y^{\mathrm{T};B}(x,0^-,s)\right] \hat{E}_x^{\mathrm{s};A}(x,0,s)\mathrm{d}x + \int_{x=x_B-w/2}^{x_B+w/2} \left[\hat{H}_y^{\mathrm{T};B}(x,0^+,s) - \hat{H}_y^{\mathrm{T};B}(x,0^-,s)\right] \hat{E}_x^{\mathrm{s};B}(x,0,s)\mathrm{d}x = -2\int_{x=x_B-w/2}^{x_B+w/2} \hat{H}_y^{\mathrm{i}}(x,0,s)\hat{E}_x^{\mathrm{T};B}(x,0,s)\mathrm{d}x, \tag{13.71b}$$

where the meaning of the EM field quantities is similar to the ones in (13.71a). To cast the reciprocity relations (13.71a) and (13.71b) into the matrix form, we shall first specify the testing magnetic-field strengths over the apertures. First, the uniform testing source is applied to the aperture of $\mathcal{G}_{A,B}$ from above $z = 0$, which results in the testing-field distributions $\hat{H}_y^{\mathrm{T};A,B}(x, 0^+, s)$, respectively, as represented by (13.61a) and (13.61b). Second, under the assumption of the TEM-field configuration in $\mathcal{G}_A$ and $\mathcal{G}_B$, the testing source applied from the domain occupied by the grooves leads to (cf. (13.29))

$$\begin{aligned}\hat{H}_y^{\mathrm{T};A}(x, 0^-, s) = &-(\eta_1/w)[R_1 + \exp(-2sd_1/c_1)]/[R_1 - \exp(-2sd_1/c_1)] \\ &\times \left[\mathrm{H}(x + w/2) - \mathrm{H}(x - w/2)\right], \end{aligned} \tag{13.72a}$$

$$\begin{aligned}\hat{H}_y^{\mathrm{T};B}(x, 0^-, s) = &-(\eta_2/w)[R_2 + \exp(-2sd_2/c_2)]/[R_2 - \exp(-2sd_2/c_2)] \\ &\times \left[\mathrm{H}(x - x_B + w/2) - \mathrm{H}(x - x_B - w/2)\right], \end{aligned} \tag{13.72b}$$

where the reflection coefficients corresponding to $\mathcal{G}_{A,B}$ are given by (cf. (13.30))

$$R_{1,2} = (Y_{1,2}^{\mathrm{g}} - \eta_{1,2})/(Y_{1,2}^{\mathrm{g}} + \eta_{1,2}), \tag{13.73}$$

respectively. Consequently, making use of (13.61a) with (13.61b) and (13.72a) with (13.72b) in (13.71a) and (13.71b), while assuming the uniform distribution of the electric-field strength induced across the narrow slots (see (13.62a) and (13.62b)), the reciprocity relations (13.71a) and (13.71b) can be cast into the matrix form (13.63), where the elements of the admittance array are given by

$$\hat{Y}_{AA}(s) = \left[\hat{Y}_{\mathrm{ext}}^{+}(s) + \hat{Y}_{\mathrm{gap}}^{A}(s)\right]/2, \tag{13.74a}$$

$$\hat{Y}_{BB}(s) = \left[\hat{Y}_{\mathrm{ext}}^{+}(s) + \hat{Y}_{\mathrm{gap}}^{B}(s)\right]/2, \tag{13.74b}$$

$$\hat{Y}_{AB}(s) = \hat{Y}_{BA} = \frac{s\epsilon_0}{2\pi\mathrm{i}} \int_{\kappa=-\mathrm{i}\infty}^{\mathrm{i}\infty} \exp(-s\kappa x_B)\, \mathrm{i}_0^2(s\kappa w/2) \frac{\mathrm{d}\kappa}{2p(\kappa)}, \tag{13.74c}$$

where $\hat{Y}_{\mathrm{ext}}^{+}(s)$ is given by (13.17) and gap admittances pertaining to $\mathcal{G}_A$ and $\mathcal{G}_B$ can be expressed as (cf. (13.32))

$$\hat{Y}_{\mathrm{gap}}^{A}(s) = \frac{\eta_1}{w} \frac{R_1 + \exp(-2sd_1/c_1)}{R_1 - \exp(-2sd_1/c_1)}, \tag{13.75a}$$

$$\hat{Y}_{\mathrm{gap}}^{B}(s) = \frac{\eta_2}{w} \frac{R_2 + \exp(-2sd_2/c_2)}{R_2 - \exp(-2sd_2/c_2)}, \tag{13.75b}$$

respectively. Under the assumption that the horizontal offset, $|x_B|$, is relatively large compared to the slots' width, w, the slowness integral representation of the mutual-coupling admittance can be expressed in terms of the modified Bessel function of the second kind [9, Eq. (9.6.23)], again. Hence, we get the closed-form approximation

$$\hat{Y}_{AB}(s) \simeq (s\epsilon_0/2\pi)\mathrm{K}_0(s|x_B|/c_0), \tag{13.76}$$

which is half of the admittance characterizing the EM coupling between two narrow slots in a thin a PEC sheet (see (13.66)).

Since the s-domain system of equations (13.63) is now fully determined, it may be solved in the TD. To achieve this goal, the voltage pulses induced across the apertures are expanded according to (13.67a) and (13.67b) and the resulting equations are transformed analytically to the TD. In this way, we arrive at (13.52), again, where we use the following TD admittance array:

$$\underline{\boldsymbol{Y}}_k = \frac{2\eta_0}{c_0\Delta t}\begin{bmatrix} \bar{\Phi}_A(t_k) & \Psi(t_k)/2 \\ \Psi(t_k)/2 & \bar{\Phi}_B(t_k) \end{bmatrix}, \tag{13.77}$$

where $\Psi(t)$ is given by (10.26) and (cf. (13.23))

$$\bar{\Phi}_{A,B}(t) = [\Phi(t) + \Phi^-_{A,B}(t)]/2. \tag{13.78}$$

The TD function $\Phi(t)$ is given by (7.21), again, and $\Phi^-_{A,B}(t)$ immediately follows from (13.36)–(13.38). For instance, if the gaps are at $z = -d_{1,2}$ terminated by PEC surfaces, we may use (cf. (13.37))

$$\Phi^-_A(t) = \frac{\epsilon_1}{\epsilon_0}\left[\frac{c_1 t}{2w}\mathrm{H}(t) + \frac{1}{w}\sum_{n=1}^{\infty}(c_1 t - 2nd_1)\mathrm{H}(c_1 t - 2nd_1)\right] \text{ as } Y_1^{\mathrm{g}} \to \infty, \tag{13.79a}$$

$$\Phi^-_B(t) = \frac{\epsilon_2}{\epsilon_0}\left[\frac{c_2 t}{2w}\mathrm{H}(t) + \frac{1}{w}\sum_{n=1}^{\infty}(c_2 t - 2nd_2)\mathrm{H}(c_2 t - 2nd_2)\right] \text{ as } Y_2^{\mathrm{g}} \to \infty, \tag{13.79b}$$

to evaluate the diagonal components of the admittance array. Other special cases can be handled in a similar way with the aid of (13.36) and (13.38). In the final step, the marching-on-in-time scheme (13.58) can be used to evaluate the desired voltage coefficients.

13.3 Pulsed EM scattering by a bounded aperture

If the physical extent of an aperture is predominant in one direction such that it can be approximately modeled by a slot of a constant width, the results from the previous sections are applicable. In general, however, the surface occupied by an aperture is bounded, which calls for a dedicated three-dimensional (3-D) computational methodology. A brief description of such a methodology is the main objective of this section.

We shall analyze the problem of EM penetration through an aperture in an infinite PEC screen (see Figure 13.7). The aperture occupies the bounded domain $\mathcal{V} = \{(x,y) \in \Sigma \subset \mathbb{R}^2, -\delta/2 < z < \delta/2\}$, where $\delta > 0$ denotes the (vanishing) thickness of the PEC screen. As in Section 12.1, the incident EM field is represented by an E-polarized plane wave via

$$\hat{\boldsymbol{E}}^{\mathrm{i}}(x,y,z,s) = \boldsymbol{\alpha}\,\hat{e}^{\mathrm{i}}(s)\exp\{-s[\kappa_0 x + \sigma_0 y - p(\kappa_0,\sigma_0)z]\}, \tag{12.1 revisited}$$

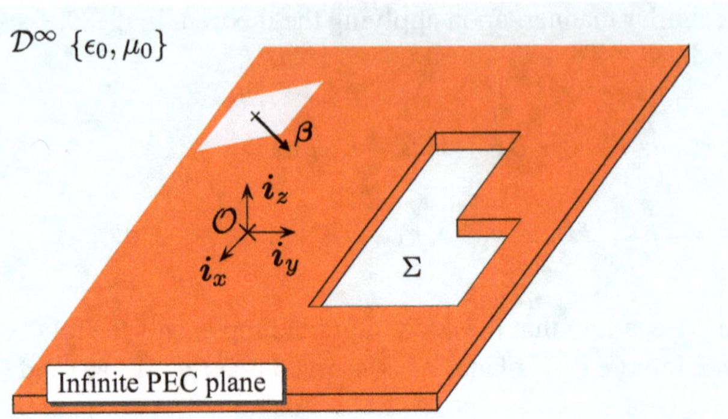

Figure 13.7 Pulsed EM plane wave impinging upon a bounded aperture in an infinite PEC screen

with $\boldsymbol{\alpha} = [\sin(\phi), -\cos(\phi), 0]$ being a unit vector in the direction of polarization and recall that $\kappa_0 = \cos(\phi)\sin(\theta)/c_0$ and $\sigma_0 = \sin(\phi)\sin(\theta)/c_0$ denote the slowness parameters in the x- and y-directions, respectively. Pursuing again the approach based on the concept of compensation (see Section 13.1), the scattered EM field (denoted by superscript $^{\rm s}$) is defined with respect to the excitation EM field (denoted by superscript $^{\rm e}$) pertaining to the infinite PEC screen, that is, in the absence of the hole. Accordingly, we may write (cf. (13.1))

$$\{\hat{\boldsymbol{E}}^{\rm s}, \hat{\boldsymbol{H}}^{\rm s}\}(x, y, z, s) = \{\hat{\boldsymbol{E}}, \hat{\boldsymbol{H}}\}(x, y, z, s) - \{\hat{\boldsymbol{E}}^{\rm e}, \hat{\boldsymbol{H}}^{\rm e}\}(x, y, z, s). \tag{13.80}$$

Owing to $\boldsymbol{i}_z \times \hat{\boldsymbol{E}} = \boldsymbol{0}$ and $\boldsymbol{i}_z \times \hat{\boldsymbol{E}}^{\rm e} = \boldsymbol{0}$ over the PEC screen with and without the aperture, respectively, also $\boldsymbol{i}_z \times \hat{\boldsymbol{E}}^{\rm s} = \boldsymbol{0}$ over the PEC screen with the aperture. Consequently, the calculation of (the space-time distribution of) the tangential component of the scattered electric-field strength over the surface of an aperture, Σ, is the goal of this section.

13.3.1 Problem formulation

The problem will be formulated with the help of the EM reciprocity theorem of the time-convolution type [5, Section 1.4.1], again, by following the lines of reasoning presented in Section 13.1.1. Hence, the theorem is first applied to the half-space above the screen and to the causal scattered and testing EM-field states. The testing field, denoted by superscript $^{\rm T}$, is generated by the equivalent magnetic-current surface density, $\partial\hat{\boldsymbol{K}}^{\rm T}$, distributed over the aperture. Consequently, we arrive at (cf. (12.3) and (13.2))

$$\begin{aligned}\int_{\boldsymbol{r}\in\Sigma} & \hat{\boldsymbol{H}}^{\rm T}(x, y, \delta, s) \cdot \partial\hat{\boldsymbol{K}}^{\rm s}(x, y, \delta, s){\rm d}A \\ &= \int_{\boldsymbol{r}\in\Sigma} \hat{\boldsymbol{H}}^{\rm s}(x, y, \delta, s) \cdot \partial\hat{\boldsymbol{K}}^{\rm T}(x, y, \delta, s){\rm d}A,\end{aligned} \tag{13.81a}$$

as $\delta \downarrow 0$. In a similar manner, upon applying the theorem in the half-space below the screen, we find

$$\int_{\boldsymbol{r}\in\Sigma} \hat{\boldsymbol{H}}^{\mathrm{T}}(x,y,-\delta,s) \cdot \partial\hat{\boldsymbol{K}}^{\mathrm{s}}(x,y,-\delta,s)\mathrm{d}A = \int_{\boldsymbol{r}\in\Sigma} \hat{\boldsymbol{H}}^{\mathrm{s}}(x,y,-\delta,s) \cdot \partial\hat{\boldsymbol{K}}^{\mathrm{T}}(x,y,-\delta,s)\mathrm{d}A, \tag{13.81b}$$

as $\delta \downarrow 0$, again. Recalling that the excitation field represents the total EM field in the presence of the infinite PEC plane, it is observed that (cf. (13.4a) and (13.4b))

$$\boldsymbol{i}_z \times \hat{\boldsymbol{H}}^{\mathrm{s}}(x,y,\delta,s) = \boldsymbol{i}_z \times \hat{\boldsymbol{H}}(x,y,\delta,s) - 2\,\boldsymbol{i}_z \times \hat{\boldsymbol{H}}^{\mathrm{i}}(x,y,0,s), \tag{13.82a}$$

$$\boldsymbol{i}_z \times \hat{\boldsymbol{H}}^{\mathrm{s}}(x,y,-\delta,s) = \boldsymbol{i}_z \times \hat{\boldsymbol{H}}(x,y,-\delta,s). \tag{13.82b}$$

Taking further $\partial\hat{\boldsymbol{K}}^{\mathrm{T}}(x,y,\delta,s) = \partial\hat{\boldsymbol{K}}^{\mathrm{T}}(x,y,-\delta,s) = \partial\hat{\boldsymbol{K}}^{\mathrm{T}}(x,y,s)$, which implies that $\boldsymbol{i}_z \times \hat{\boldsymbol{H}}^{\mathrm{T}}(x,y,\delta,s) = -\boldsymbol{i}_z \times \hat{\boldsymbol{H}}^{\mathrm{T}}(x,y,-\delta,s)$, and noting that both $\partial\hat{\boldsymbol{K}}^{\mathrm{s}}$ and $\partial\hat{\boldsymbol{K}}^{\mathrm{T}}$ have only nonzero components parallel to the PEC plane, we may write the following pair of reciprocity relations (cf. (13.46a) and (13.46b)):

$$\int_{\boldsymbol{r}\in\Sigma} \hat{\boldsymbol{H}}^{\mathrm{T}}(x,y,\delta,s) \cdot \partial\hat{\boldsymbol{K}}^{\mathrm{s}}(x,y,\delta,s)\mathrm{d}A = \int_{\boldsymbol{r}\in\Sigma} \left[\hat{\boldsymbol{H}}(x,y,\delta,s) - 2\hat{\boldsymbol{H}}^{\mathrm{i}}(x,y,0,s)\right] \cdot \partial\hat{\boldsymbol{K}}^{\mathrm{T}}(x,y,s)\mathrm{d}A, \tag{13.83a}$$

and

$$\int_{\boldsymbol{r}\in\Sigma} \hat{\boldsymbol{H}}^{\mathrm{T}}(x,y,\delta,s) \cdot \partial\hat{\boldsymbol{K}}^{\mathrm{s}}(x,y,-\delta,s)\mathrm{d}A = -\int_{\boldsymbol{r}\in\Sigma} \hat{\boldsymbol{H}}(x,y,-\delta,s) \cdot \partial\hat{\boldsymbol{K}}^{\mathrm{T}}(x,y,s)\mathrm{d}A \tag{13.83b}$$

as $\delta \downarrow 0$. Note that (13.83a) and (13.83b) are applicable to the analysis of a bounded aperture with a metasurface filling. Now, if the equivalent magnetic-current surface density is continuous when crossing the aperture, that is, $\partial\hat{\boldsymbol{K}}^{\mathrm{s}}(x,y,\delta,s) = \partial\hat{\boldsymbol{K}}^{\mathrm{s}}(x,y,-\delta,s) = \partial\hat{\boldsymbol{K}}^{\mathrm{s}}(x,y,s)$, which applies to both apertures with both vacuum and dielectric filling, for instance, the reciprocity relations (13.83a) and (13.83b) can be combined to get (cf. (13.39))

$$2\int_{\boldsymbol{r}\in\Sigma} \hat{\boldsymbol{H}}^{\mathrm{T}}(x,y,\delta,s) \cdot \partial\hat{\boldsymbol{K}}^{\mathrm{s}}(x,y,s)\mathrm{d}A - \int_{\boldsymbol{r}\in\Sigma} \left[\hat{\boldsymbol{H}}(x,y,\delta,s) - \hat{\boldsymbol{H}}(x,y,-\delta,s)\right] \cdot \partial\hat{\boldsymbol{K}}^{\mathrm{T}}(x,y,s)\mathrm{d}A = -2\int_{\boldsymbol{r}\in\Sigma} \hat{\boldsymbol{H}}^{\mathrm{i}}(x,y,0,s) \cdot \partial\hat{\boldsymbol{K}}^{\mathrm{T}}(x,y,s)\mathrm{d}A. \tag{13.84}$$

Across the actual aperture as shown in Figure 13.7, however, the tangential components of the total magnetic-field strengths are continuous, which finally leads to (cf. (13.7))

$$\int_{\boldsymbol{r}\in\Sigma} \hat{\boldsymbol{H}}^{\mathrm{T}}(x,y,\delta,s)\cdot\partial\hat{\boldsymbol{K}}^{\mathrm{s}}(x,y,s)\mathrm{d}A = -\int_{\boldsymbol{r}\in\Sigma} \hat{\boldsymbol{H}}^{\mathrm{i}}(x,y,0,s)\cdot\partial\hat{\boldsymbol{K}}^{\mathrm{T}}(x,y,s)\mathrm{d}A. \tag{13.85}$$

It is now observed that (13.85) bears similarity with the reciprocity relation pertaining to the (complementary) problem consisting of a PEC screen having the shape of the aperture (see (12.5) with $\hat{Z}^{\mathrm{E}}(s)\downarrow 0$). As a matter of fact, this equation can be established via the duality principle [7, Section 9.6.2]. The latter reciprocity relation (13.85) will be solved for the induced magnetic-current density using the CdH-MoM, again. Therefore, the 2-D wave-slowness representation in the plane parallel to the screen is employed (see (4.75)) to cast the reciprocity relation into its equivalent form (cf. (12.6))

$$\begin{aligned}
&\left(\frac{s}{2\mathrm{i}\pi}\right)^2\int_{\kappa=-\mathrm{i}\infty}^{\mathrm{i}\infty}\mathrm{d}\kappa\int_{\sigma=-\mathrm{i}\infty}^{\mathrm{i}\infty}\Big[\tilde{H}_x^{\mathrm{T}}(\kappa,\sigma,\delta,s)\partial\tilde{K}_x^{\mathrm{s}}(-\kappa,-\sigma,s)\\
&\qquad\qquad+\tilde{H}_y^{\mathrm{T}}(\kappa,\sigma,\delta,s)\partial\tilde{K}_y^{\mathrm{s}}(-\kappa,-\sigma,s)\Big]\mathrm{d}\sigma\\
&=-\left(\frac{s}{2\mathrm{i}\pi}\right)^2\int_{\kappa=-\mathrm{i}\infty}^{\mathrm{i}\infty}\mathrm{d}\kappa\int_{\sigma=-\mathrm{i}\infty}^{\mathrm{i}\infty}\Big[\tilde{H}_x^{\mathrm{i}}(\kappa,\sigma,0,s)\partial\tilde{K}_x^{\mathrm{T}}(-\kappa,-\sigma,s)\\
&\qquad\qquad+\tilde{H}_y^{\mathrm{i}}(\kappa,\sigma,0,s)\partial\tilde{K}_y^{\mathrm{T}}(-\kappa,-\sigma,s)\Big]\mathrm{d}\sigma\\
&\qquad\qquad\text{for }\{s\in\mathbb{R};s>0\}.
\end{aligned}\tag{13.86}$$

Recalling that the (transform-domain) testing fields, $\tilde{H}_x^{\mathrm{T}}(\kappa,\sigma,\delta,s)$ and $\tilde{H}_y^{\mathrm{T}}(\kappa,\sigma,\delta,s)$, are generated by the testing-source distribution on Σ as $z\downarrow 0$, we find (cf. (12.31c) and (12.31d))

$$\begin{aligned}
\tilde{H}_x^{\mathrm{T}}(\kappa,\sigma,\delta,s)=&-(s/\mu_0)\Omega_0^2(\kappa)\partial\tilde{K}_x^{\mathrm{T}}(\kappa,\sigma,s)/sp(\kappa,\sigma)\\
&+(s/\mu_0)\kappa\sigma\,\partial\tilde{K}_y^{\mathrm{T}}(\kappa,\sigma,s)/sp(\kappa,\sigma),
\end{aligned}\tag{13.87a}$$

$$\begin{aligned}
\tilde{H}_y^{\mathrm{T}}(\kappa,\sigma,\delta,s)=&-(s/\mu_0)\Omega_0^2(\sigma)\partial\tilde{K}_y^{\mathrm{T}}(\kappa,\sigma,s)/sp(\kappa,\sigma)\\
&+(s/\mu_0)\sigma\kappa\,\partial\tilde{K}_x^{\mathrm{T}}(\kappa,\sigma,s)/sp(\kappa,\sigma),
\end{aligned}\tag{13.87b}$$

where $\Omega_0(\kappa)=(1/c_0^2-\kappa^2)^{1/2}$ and $p(\kappa,\sigma)$ is the slowness parameter in the z-direction (see (4.77)). The reciprocity relation (13.86) with (13.87a) and (13.87b) will be solved numerically in the following section.

13.3.2 Problem solution

Owing to its similarity with (12.32b), we may adopt the solution approach introduced in Section 12.2.1 to solve the actual reciprocity relation (13.86). Accordingly,

the unknown magnetic-current surface densities are represented as piecewise linear functions over the uniformly discretized space–time solution domain, that is,

$$\partial K_x^{\mathrm{s}}(x,y,t) \simeq \frac{1}{\Delta_y}\sum_{u=1}^{U}\sum_{k=1}^{M} v_k^{[u]}\Lambda^{[u]}(x)\Pi^{[u]}(y)\Lambda_k(t), \qquad \text{(12.33a revisited)}$$

$$\partial K_y^{\mathrm{s}}(x,y,t) \simeq \frac{1}{\Delta_x}\sum_{v=1}^{V}\sum_{k=1}^{M} v_k^{[v]}\Lambda^{[v]}(y)\Pi^{[v]}(x)\Lambda_k(t), \qquad \text{(12.33b revisited)}$$

where $v_k^{[u]}$ and $v_k^{[v]}$ are magnetic-current coefficients (in V) pertaining to the uth and vth nodes of the spatial grid, respectively (see Figure 12.2). Furthermore, as for the testing-current densities, we take

$$\partial K_x^{\mathrm{T}}(x,y,t) = \Pi^{[P]}(x)\delta(y-y_P)\delta(t), \qquad \text{(12.34a revisited)}$$

$$\partial K_y^{\mathrm{T}}(x,y,t) = \Pi^{[Q]}(y)\delta(x-x_Q)\delta(t),$$

$$\text{for all } P=\{1,\ldots,U\} \text{ and } Q=\{1,\ldots,V\}, \qquad \text{(12.34b revisited)}$$

again. Having specified the induced and testing current densities, their transform-domain counterparts can be used to cast the reciprocity relation (13.86) into its matrix form. Symbolically, it can be represented in the form shown in Figure 13.8. Here, the 1-D arrays $\hat{V}_x^{[u]}$ and $\hat{V}_y^{[v]}$ of dimensions $[U\times 1]$ and $[V\times 1]$ represent, via coefficients $v_k^{[u]}$ and $v_k^{[v]}$, the induced magnetic-current surface densities $\partial\hat{K}_x^{\mathrm{s}}(x,y,s)$

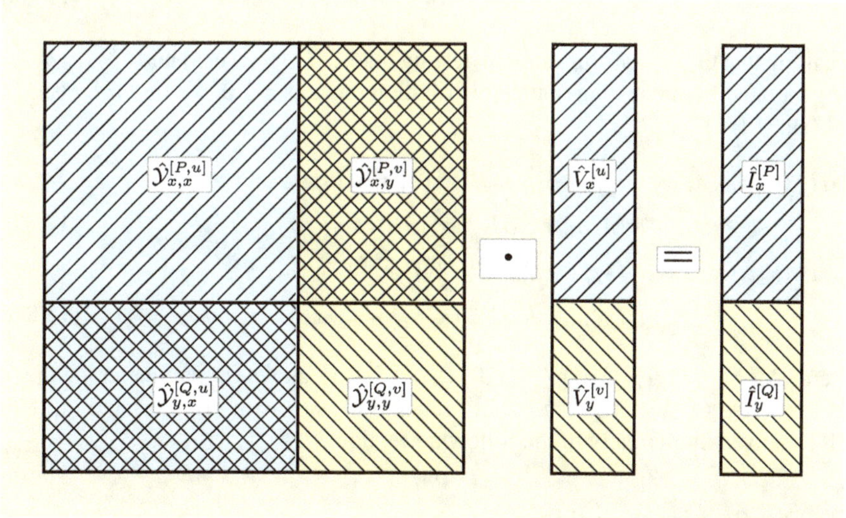

Figure 13.8 Symbolical partitioned matrix description of the EM scattering problem consisting a bounded aperture in a PEC screen

and $\partial\hat{K}_y^{\mathrm{s}}(x,y,s)$, respectively. The blocks of the 2-D $[(U+V)\times(U+V)]$ admittance array $\hat{\boldsymbol{\mathcal{Y}}}$ can be, again, expressed through the generic integrals $\hat{\Upsilon}(x,y,s)$ and $\hat{\Pi}(x,y,s)$ (see Sections K.1 and K.2). As a matter of fact, the admittance-array elements can be readily evaluated using the impedance array pertaining to the (complementary) PEC screen (see (12.12a)–(12.12d)). Hence, adopting this approach, we write (cf. (12.35))

$$\boldsymbol{\mathcal{Y}}(t) = 2\eta_0^2\boldsymbol{\mathcal{Z}}(t) \quad \text{for all } t > 0. \tag{13.88}$$

The additional factor 2 has its origin in the definition of the scattered EM field with respect to the short-circuited aperture (see (13.80) and [6, Section 5.4.2]). Finally, the right-hand side of the discretized reciprocity relation is composed of 1-D arrays, $\hat{I}_x^{[P]}$ and $\hat{I}_y^{[Q]}$, of dimensions $[U\times 1]$ and $[V\times 1]$, respectively. Their elements have been specified via (12.41)–(12.44) in the TD for the E-polarized plane-wave excitation (see (12.1)). This solution approach leads to a system of equations of the time-convolution type (cf. (12.45))

$$\sum_{k=1}^{m}\left(\underline{\boldsymbol{\mathcal{Y}}}_{m-k+1} - 2\underline{\boldsymbol{\mathcal{Y}}}_{m-k} + \underline{\boldsymbol{\mathcal{Y}}}_{m-k-1}\right)\cdot\boldsymbol{V}_k = \boldsymbol{I}_m, \tag{13.89}$$

where $\underline{\boldsymbol{\mathcal{Y}}}_k$ is a shorthand notation for the admittance array evaluated at $t=t_k$, that is, $\underline{\boldsymbol{\mathcal{Y}}}_k = \boldsymbol{\mathcal{Y}}(t_k)$. Furthermore, $\boldsymbol{V}_k$ denotes the 1-D $[(U+V)\times 1]$ array of the magnetic-current coefficients, and $\boldsymbol{I}_m$ is the 1-D $[(U+V)\times 1]$ excitation array at $t=t_m$. Recall that the TD excitation array can be readily filled using (12.41)–(12.44). The marching-on-in-time technique finally yields the desired array of magnetic-current coefficients (cf. (12.47))

$$\boldsymbol{V}_m = \underline{\boldsymbol{\mathcal{Y}}}_1^{-1}\cdot\left[\boldsymbol{I}_m - \sum_{k=1}^{m-1}\left(\underline{\boldsymbol{\mathcal{Y}}}_{m-k+1} - 2\underline{\boldsymbol{\mathcal{Y}}}_{m-k} + \underline{\boldsymbol{\mathcal{Y}}}_{m-k-1}\right)\cdot\boldsymbol{V}_k\right]$$
$$\text{for all } m=\{1,\ldots,M\}, \tag{13.90}$$

where $\underline{\boldsymbol{\mathcal{Y}}}_1^{-1}$ denotes the inverse of the admittance matrix at $t=\Delta t$. In conclusion, it is noted that the use of the calculated coefficients in (12.33a) and (12.33b) leads to (the space–time distribution of) the magnetic-current surface density on the aperture. These surface currents subsequently enable to evaluate the pertaining scattered EM fields. Since the excitation EM field is supposed to be known, the total EM field immediately follows. Therefore, in such calculations, the definition of the actual scattered EM field (with respect to the short-circuited aperture) must be kept in mind (see (13.80)).

13.4 Pulsed EM scattering by a small rectangular aperture

A special form of the TD solution introduced in the previous section can be arrived at for a (relatively) small rectangular aperture in an infinite PEC screen (see

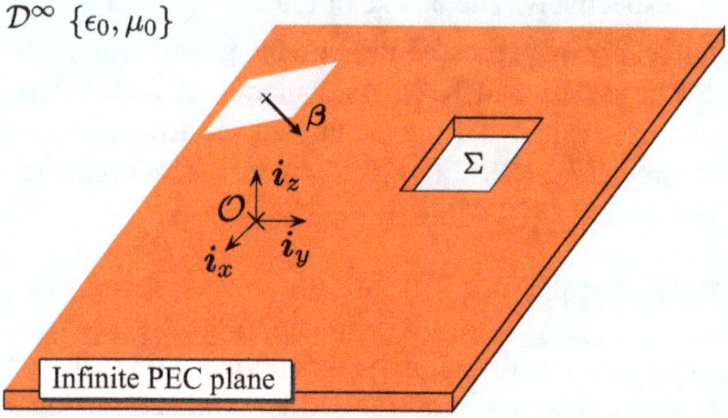

Figure 13.9 Pulsed EM plane wave impinging upon a small rectangular aperture in an infinite PEC screen

Figure 13.9). Without loss of generality, the center of the aperture is located at the origin such that its surface extends over $\Sigma = \{-\Delta_x < x < \Delta_x, \Delta_y < y < \Delta_y\}$, where $2\Delta_x > 0$ and $2\Delta_y > 0$ denote the (relatively small) lengths of its sides along the x- and y-directions, respectively. The medium surrounding the screen with the aperture is supposed to be homogeneous, isotropic and loss-free in its EM properties, again. As in the previous section, it is assumed that the aperture is irradiated by an E-polarized uniform plane wave (see (12.1)).

The EM scattering by the small aperture will be analyzed with the help of the reciprocity relation (13.86) with (13.87a). Its numerical solution calls for the definition of induced- and testing-current distributions. Therefore, the magnetic-current densities, satisfying $\partial K_x^{\mathrm{s}}(\pm\Delta_x, y, t) = 0$ and $\partial K_y^{\mathrm{s}}(x, \pm\Delta_y, t) = 0$ for all $t > 0$, are further postulated in the form used before as (see Figure 12.6)

$$\partial K_x^{\mathrm{s}}(x, y, t) \simeq \frac{K_x(t)}{2\Delta_y}\Lambda(x)\Pi(y) = \frac{\Lambda(x)\Pi(y)}{2\Delta_y}\sum_{k=1}^{M} v_{k;x}\Lambda_k(t), \tag{12.61a revisited}$$

$$\partial K_y^{\mathrm{s}}(x, y, t) \simeq \frac{K_y(t)}{2\Delta_x}\Lambda(y)\Pi(x) = \frac{\Lambda(y)\Pi(x)}{2\Delta_x}\sum_{k=1}^{M} v_{k;y}\Lambda_k(t), \tag{12.61b revisited}$$

where $v_{k;x}$ and $v_{k;y}$ denote the magnetic-current coefficients to be found. Furthermore, the testing currents are chosen to be represented by the “razor-type” functions, again, viz

$$\partial K_x^{\mathrm{T}}(x, y, t) = \Pi(2x)\delta(y)\delta(t), \tag{12.62a revisited}$$

$$\partial K_y^{\mathrm{T}}(x, y, t) = \Pi(2y)\delta(x)\delta(t). \tag{12.62b revisited}$$

In the subsequent step, the transform-domain images of the induced and testing magnetic-current densities are substituted in the reciprocity relation (13.86). This yields two mutually independent equations in the s-domain that interrelate the induced magnetic currents with the corresponding components of the incident magnetic-field strength. Their transform back to the TD leads to the convolution-type system of (uncoupled) equations (13.89), the admittance array in which follows from (cf. (13.88))

$$\underline{\mathcal{Y}}_k = 2\eta_0^2 \underline{\mathcal{Z}}_k \quad \text{for all } k = \{1, \dots, M\}, \tag{13.91}$$

where the impedance array, $\underline{\mathcal{Z}}_k$, is defined via (12.53). The additional factor 2 in (13.91) with respect to admittance array pertaining to a metasurface (see (12.63)) can be justified by comparing the reciprocity relations (12.32b) (supplemented with (12.31c), (12.31d)) and (13.86) (supplemented with (13.87a) and (13.87b)). The excitation array on the right-hand side of the solved equation (13.89) has the form described by (12.66). Its components for the E-polarized incident plane wave (12.1) are given by (12.67a)–(12.68b), again. Finally, the desired array of the unknown coefficients, $\boldsymbol{V}_m$ (see (12.69)), then immediately follows upon carrying out the step-by-step procedure (13.90) for all $m = \{1, \dots, M\}$.

13.5 Illustrative examples

With the aid of the techniques presented in the previous sections, we shall next analyze the TD electric-field distribution induced across a slot in a PEC sheet. The analyzed problem configurations shall be irradiated by the H-polarized EM field as represented through (4.18) in the s-domain. The electric-field plane-wave signature has the shape of a bipolar triangle, again, with the unit amplitude $e_{\rm m} = 1.0\,{\rm V/m}$ (see (8.15) and Figure 8.3). Next, we take $\beta = 0$.

We start by calculating the electric-field response at the center of a slot in a PEC thin sheet located in the unbounded, homogeneous, isotropic and loss-free medium (see Figure 13.1). In the first case, the CdH-MoM solution (13.13) is compared with the narrow-slot approximation introduced in Section 13.2.1, where we take $\epsilon_1 = \epsilon_0$. The corresponding pulse shapes for $c_0 t_{\rm w}/w = 10$ are shown in Figure 13.10(a). Since the lines are virtually on the top of each other, we may conclude that the TD solutions represented by (13.13) and (13.22) are consistent. In the second example, we choose $c_0 t_{\rm w}/w = 1.0$, so that the narrow-slot approximation is no longer applicable. To validate the result obtained via the CdH-MoM solution (13.13) with $N = 11$ nodes, the problem has also been analyzed with the aid of the finite-integration technique (FIT) as implemented in CST Microwave Studio®. Figure 13.10(b) shows that the numerical solutions correlate very well.

Further, we shall employ the results of Section 13.2.2 to evaluate the electric-field strength induced across the aperture of narrow rectangular grooves terminated at $z = -d$ by a PEC surface. The width of the grooves under analysis is taken to be related to the excitation pulse-time width via $c_0 t_{\rm w}/w = 10$. In the first case, we analyze the response of a rectangular groove that is filled by vacuum. Its depth is equal to the spatial extent of the excitation pulse via $d = c_0 t_{\rm w}$, which implies that $d = 10w$.

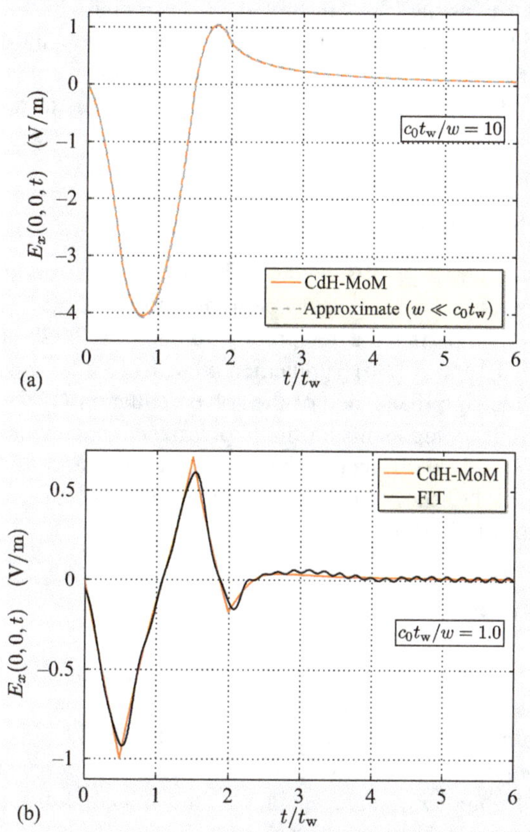

Figure 13.10 Plane-wave induced electric-field strength at the center of the slot in a PEC sheet. (a) The CdH-MoM solution (13.13) and the narrow-slot approximation (13.22) with $\epsilon_l = \epsilon_0$ for $c_0 t_w / w = 10$. (b) The CdH-MoM solution (13.13) and the FIT solution for $c_0 t_w / w = 1.0$.

Figure 13.11(a) demonstrates that the introduced approximate solution (13.22) with (13.37) yields the field response that agrees with the FIT-based numerical solution well. The second groove under consideration is filled by a homogeneous material specified by $\epsilon_1 = 4\epsilon_0$ and $\mu_1 = 2\mu_0$, which implies $c_1 = c_0/2\sqrt{2} \simeq 0.3536c_0$. Its depth is now $d = c_0 t_w/2 = 5w$, which is one half of the groove's depth from the first example. On the other hand, since $d = c_1 t_w \sqrt{2}$, the actual depth is *relatively* greater with respect to the spatial extent of the pulse inside the gap. The resulting pulse shapes as calculated approximately using the model from Section 13.2.2 and numerically via the FIT are shown in Figure 13.11(b). Again, a good correlation between the results is achieved.

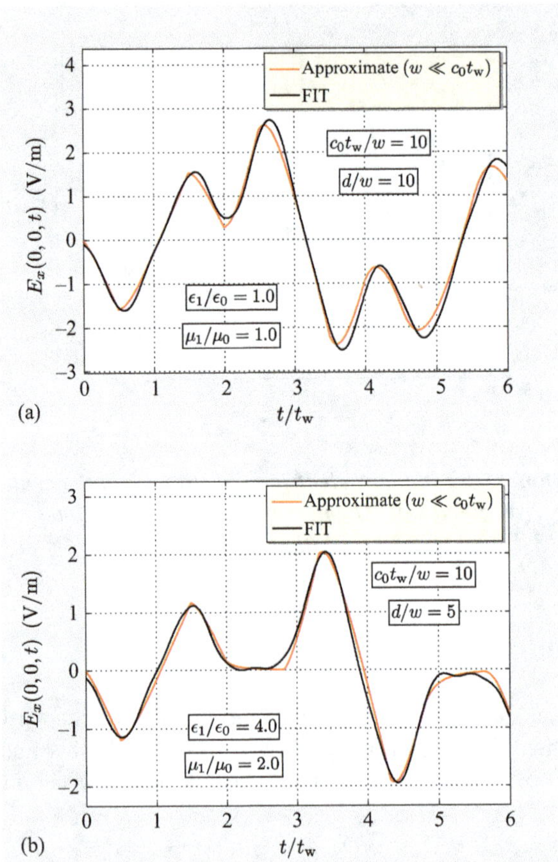

Figure 13.11 Plane-wave induced electric-field strength across the aperture of narrow rectangular grooves. (a) The narrow-slot approximation (13.22) with (13.37) and the FIT solution for $d/w = 10$, $\epsilon_1 = \epsilon_0$ and $\mu_1 = \mu_0$. (b) The narrow-slot approximation (13.22) with (13.37) and the FIT solution for $d/w = 5$, $\epsilon_1 = 4\epsilon_0$ and $\mu_1 = 2\mu_0$.

In the final example, the solution procedure proposed in Section 13.2.3.1 is validated with the help of FIT, again. To that end, we shall calculate the electric-field pulse induced across a slot filled by a homogeneous material with conductive and dielectric properties. The width of the slot under consideration is a tenth of the spatial support of the excitation pulse, that is, $c_0 t_w / w = 10$, again. The EM properties of the filling are described by the layer conductance ratio, $G^E/\eta_0 = 1.0$, and by $C^E/\eta_0 t_w = 1.0$, thus implying the layer admittance time constant $C^E/G^E = t_w$. Figure 13.12 clearly demonstrates the validity of the narrow-slot approximate solution.

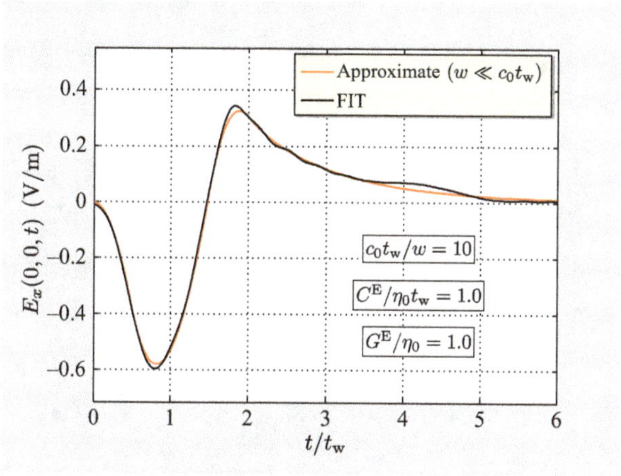

Figure 13.12 Plane-wave induced electric-field strength at the center of the narrow filled slot. The narrow-slot approximation (13.22) with (13.43) and the FIT solution for $c_0 t_w/w = 10$, $G^E/\eta_0 = 1.0$ and $C^E/\eta_0 t_w = 1.0$

13.6 Concluding remarks

The computational methodologies described throughout this chapter lead to the tangential electric-field distribution induced over apertures. In this final section, it is briefly indicated how this result can be used to calculate the pertaining scattered EM fields.

13.6.1 EM scattering by a slot

In this subsection, we shall briefly analyze the y-independent EM field scattered by a slot (see (13.1)). In particular, we start by expressing the scattered EM field at (x_0, z_0) located in the homogeneous upper half-space $\mathcal{D}_+^\infty$. The relevant expression can be found with the help of the EM reciprocity theorem of the time-convolution type [5, Section 1.4.1] by interrelating the desired scattered field (denoted by superscript ^{s}) with the testing (Green's) state (denoted by superscript ^{T}) that is activated by the point source located at (x_0, z_0). Taking $\hat{E}_x^T(x, 0^+, s) = 0$ for all $x \in \mathbb{R}$, the expression follows as (cf. (7.36))

$$\hat{H}_y^s(x_0, z_0, s) = \frac{s\epsilon_0}{\pi} \int_{x=-w/2}^{w/2} \mathrm{K}_0[s\rho_0(x)/c_0] \hat{E}_x^s(x, 0^+, s) \mathrm{d}x$$
$$\text{for } x_0 \in \mathbb{R} \text{ and } z_0 > 0, \tag{13.92}$$

with $\rho_0(x) = [(x - x_0)^2 + z_0^2]^{1/2} > 0$. Sufficiently far away from a relatively narrow slot, the integral in (13.92) can be approximated by (cf. (7.37))

$$\hat{H}_y^s(x_0, z_0, s) \simeq \frac{s\epsilon_0 \hat{V}^s(s)}{\pi} \mathrm{K}_0(sr_0/c_0), \tag{13.93}$$

where we used $r_0 = \rho_0(0) = (x_0^2 + z_0^2)^{1/2} > 0$, again. The TD counterpart of (13.93) can be then readily obtained with the aid of [9, Eq. (29.3.119)] in the following form:

$$H_y^{\mathrm{s}}(x_0, z_0, t) \simeq \epsilon_0 \partial_t V^{\mathrm{s}}(t) *_t \frac{\mathrm{H}(t - r_0/c_0)}{\pi(t^2 - r_0^2/c_0^2)^{1/2}}, \tag{13.94}$$

where $V^{\mathrm{s}}(t)$ is supposed to be known through the voltage coefficients (e.g. (13.22)) via the temporal expansion (13.21). If the field point is now located in the lower dielectric half-space $\mathcal{D}_-^\infty$ (see Figure 13.2), the scattered magnetic-field strength can be evaluated in a similar manner as

$$\hat{H}_y^{\mathrm{s}}(x_0, -z_0, s) = -\frac{s\epsilon_1}{\pi} \int_{x=-w/2}^{w/2} \mathrm{K}_0[s\rho_0(x)/c_1] \hat{E}_x^{\mathrm{s}}(x, 0^-, s) \mathrm{d}x$$

$$\text{for } x_0 \in \mathbb{R} \text{ and } z_0 > 0. \tag{13.95}$$

Again, for a relatively narrow slot, we may find a straightforward approximation of the TD scattered field in $\mathcal{D}_-^\infty$

$$H_y^{\mathrm{s}}(x_0, -z_0, t) \simeq -\epsilon_1 \partial_t V^{\mathrm{s}}(t) *_t \frac{\mathrm{H}(t - r_0/c_1)}{\pi(t^2 - r_0^2/c_1^2)^{1/2}}, \tag{13.96}$$

applying relatively far away from the slot, again. If the slot is filled by a metasurface with combined magneto-dielectric properties, E_x^{s} field is no longer continuous across the slot and we have to carefully distinguish between the voltage pulses on the top surface, where $V^{\mathrm{s}}(t) = V^+(t)$, and on the lower surface, where $V^{\mathrm{s}}(t) = V^-(t)$ (see (13.48)). The expression for the scattered field inside a narrow rectangular groove (see Figure 13.3) can be derived by following similar lines of reasoning. Hence, applying the reciprocity theorem to interrelate the scattered field with the pertaining Green's state in the domain occupied by the groove, we get

$$\begin{aligned} H_y^{\mathrm{s}}(-z_0, t) = &-(\eta_1/w) V^{\mathrm{s}}(t - z_0/c_1) \\ &- (\eta_1/w) \sum_{n=0}^{\infty} V^{\mathrm{s}}[t - 2(n+1)d/c_1 + z_0/c_1] \\ &- (\eta_1/w) \sum_{n=0}^{\infty} V^{\mathrm{s}}[t - 2(n+1)d/c_1 - z_0/c_1], \end{aligned} \tag{13.97}$$

where we assumed the PEC terminating surface along $\{-w/2 < x < w/2, z = -d\}$ and recall that $c_1 = (\epsilon_1\mu_1)^{-1/2} > 0$ and $\eta_1 = (\epsilon_1/\mu_1)^{1/2} > 0$. Owing to the property of causality, only a finite number of terms in the summations of (13.97) is needed in a bounded time window of observation. Finally, it is noted that since the excitation field is in $z < 0$ identically zero, (13.95) with (13.96) and (13.97), in fact, do represent the *total field* in the lower region (see (13.1)).

13.6.2 EM scattering by a small rectangular aperture

The reciprocity-based approach using the testing Green's state can also be applied to express the far-field scattering amplitudes (see (12.71)) pertaining to a relatively

small rectangular aperture in an infinite PEC screen. Pursuing this approach, we find (cf. (12.81a)–(12.81c))

$$\hat{E}_x^{\mathrm{s};\infty}(\boldsymbol{\xi}, s) \simeq 2sc_0^{-1}\xi_z\Delta_y\hat{K}_y(s), \tag{13.98a}$$

$$\hat{E}_y^{\mathrm{s};\infty}(\boldsymbol{\xi}, s) \simeq -2sc_0^{-1}\xi_z\Delta_x\hat{K}_x(s), \tag{13.98b}$$

$$\hat{E}_z^{\mathrm{s};\infty}(\boldsymbol{\xi}, s) \simeq 2sc_0^{-1}\xi_y\Delta_x\hat{K}_x(s) - 2sc_0^{-1}\xi_x\Delta_y\hat{K}_y(s). \tag{13.98c}$$

Via the spherical-wave amplitudes (13.98a)–(13.98c), we may then evaluate the EM power scattered by the rectangular aperture. Hence, using $\zeta_0\hat{\boldsymbol{H}}^{\mathrm{s};\infty} = \boldsymbol{\xi} \times \hat{\boldsymbol{E}}^{\mathrm{s};\infty}$ in (12.82), we may write

$$\hat{P}^{\mathrm{s}}(s) = \frac{\eta_0}{32\pi^2}\int_{\boldsymbol{\xi}\in\Omega} \hat{\boldsymbol{E}}^{\mathrm{s};\infty}(\boldsymbol{\xi}, s) \cdot \hat{\boldsymbol{E}}^{\mathrm{s};\infty}(\boldsymbol{\xi}, -s)\mathrm{d}\Omega, \tag{13.99}$$

where $\Omega = \{\boldsymbol{\xi} \cdot \boldsymbol{\xi} = 1\}$ denotes a unit sphere and recall that $\hat{P}^{\mathrm{s}}$ can be for sinusoidally in time-varying EM fields associated with the time-averaged EM power carried by the scattered wave. Next, upon substituting (13.98a)–(13.98c) in (13.99) and carrying out the integration over the unit sphere, we end up with (cf. (12.83))

$$\hat{P}^{\mathrm{s}}(s) = -\frac{\eta_0}{3\pi}\frac{s^2}{c_0^2}\left[\Delta_x^2\hat{K}_x(s)\hat{K}_x(-s) + \Delta_y^2\hat{K}_y(s)\hat{K}_y(-s)\right]. \tag{13.100}$$

The time-averaged scattered power then follows from (13.100) by taking the limit $\{s = \delta + \mathrm{i}\omega, \delta \downarrow 0, \omega \in \mathbb{R}\}$ as (cf. (12.84))

$$\hat{P}^{\mathrm{s}}(\mathrm{i}\omega) = \frac{\eta_0}{3\pi}k^2\left[\Delta_x^2|\hat{K}_x(\mathrm{i}\omega)|^2 + \Delta_y^2|\hat{K}_y(\mathrm{i}\omega)|^2\right], \tag{13.101}$$

where we used the wave number, $k = \omega/c_0$, again.

It will be further demonstrated that the results given above comply with Babinet's principle [10, Section 13.4]. In particular, we shall analyze the EM scattering by a rectangular PEC sheet and its complementary structure represented by a rectangular aperture in an infinite PEC screen. The aperture has the same dimensions as the sheet, but it is rotated through $\pi/2$ in the horizontal plane. Both structures are assumed to be irradiated by a uniform E-polarized plane wave propagating normally with respect to the xy-plane (see Figure 13.13). Its polarization and propagation vectors can be then described by $\boldsymbol{\alpha} = [\sin(\phi), -\cos(\phi), 0]$ and $\boldsymbol{\beta} = [0, 0, -1]$. On account of (13.91) and upon comparison of (12.59a), (12.59b) with (12.68a), (12.68b), respectively, it is straightforward to show that the replacement $\Delta_{x,y}$ by $\Delta_{y,x}$, respectively, allows to interrelate the electric currents induced on a small conducting sheet, $I_x(t)$ and $I_y(t)$ (see (12.51a) and (12.51b)), in terms of the magnetic currents induced on the corresponding aperture, $K_x(t)$ and $K_y(t)$ (see (12.61a) and (12.61b)), in the following manner:

$$I_x(t) = -2\eta_0 K_y(t) \text{ and } I_y(t) = 2\eta_0 K_x(t). \tag{13.102}$$

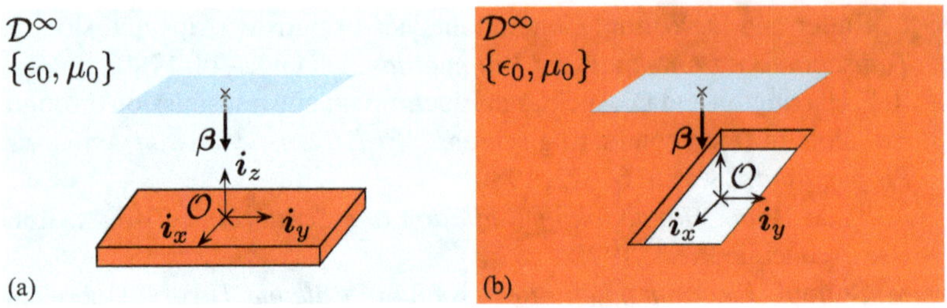

Figure 13.13 Babinet's principle applied to (a) a rectangular sheet and (b) a (complementary) rectangular aperture in an infinite screen

Relations (13.102) can be used in (12.80a)–(12.80c) to express the far-field scattered wave amplitudes pertaining to the small sheet in the forward direction, $\boldsymbol{\xi} = \boldsymbol{\beta} = -\boldsymbol{i}_z$, as

$$\hat{E}_x^{\mathrm{s};\infty}(\boldsymbol{\beta}, s)|_{\mathrm{sheet}} \simeq -s\mu_0\Delta_y\hat{I}_x(s) = 2sc_0^{-1}\Delta_y\hat{K}_y(s), \tag{13.103a}$$

$$\hat{E}_y^{\mathrm{s};\infty}(\boldsymbol{\beta}, s)|_{\mathrm{sheet}} \simeq -s\mu_0\Delta_x\hat{I}_y(s) = -2sc_0^{-1}\Delta_x\hat{K}_x(s), \tag{13.103b}$$

$$\hat{E}_z^{\mathrm{s};\infty}(\boldsymbol{\beta}, s)|_{\mathrm{sheet}} = 0. \tag{13.103c}$$

The corresponding far-field amplitudes pertaining to the complementary structure follow from (13.98a) to (13.98c) as

$$\hat{E}_x^{\mathrm{s};\infty}(\boldsymbol{\beta}, s)|_{\mathrm{aperture}} \simeq -2sc_0^{-1}\Delta_y\hat{K}_y(s), \tag{13.104a}$$

$$\hat{E}_y^{\mathrm{s};\infty}(\boldsymbol{\beta}, s)|_{\mathrm{aperture}} \simeq 2sc_0^{-1}\Delta_x\hat{K}_x(s), \tag{13.104b}$$

$$\hat{E}_z^{\mathrm{s};\infty}(\boldsymbol{\beta}, s)|_{\mathrm{aperture}} = 0. \tag{13.104c}$$

Upon inspection of (13.103a)–(13.103c) with (13.104a)–(13.104c), it is found that

$$\hat{\boldsymbol{E}}^{\mathrm{s};\infty}(\boldsymbol{\beta}, s)|_{\mathrm{sheet}} + \hat{\boldsymbol{E}}^{\mathrm{s};\infty}(\boldsymbol{\beta}, s)|_{\mathrm{aperture}} = \boldsymbol{0}, \tag{13.105}$$

which tells us that the sum of the scattered EM-field amplitudes behind the xy-plane is identically zero as in the absence of any screen. This result could also be deduced through the application of Babinet's principle [10, Section 13.4]. Finally note that the real-FD counterpart of (13.102) can be substituted in (12.84) to express the (time-averaged) power scattered from the small sheet in terms of the magnetic currents induced in the corresponding aperture. It is straightforward to show that the substitution yields (13.101), thus proving that the EM power scattered from the sheet and its complementary structure (see Figure 13.13) are equivalent.

References

[1] C. J. Bouwkamp. Diffraction theory: a critique of some recent developments. Research report no. EM–50. Technical report, Washington Square Coll. of Arts and Science, New York Univ., 1953.

[2] C. Butler and D. Wilton. General analysis of narrow strips and slots. *IEEE Transactions on Antennas and Propagation*, 28(1):42–48, 1980.
[3] R. F. Harrington and D. Auckland. Electromagnetic transmission through narrow slots in thick conducting screens. *IEEE Transactions on Antennas and Propagation*, 28(5):616–622, 1980.
[4] A. T. de Hoop. *Handbook of Radiation and Scattering of Waves*. London, UK: Academic Press, 1995.
[5] M. Štumpf. *Electromagnetic Reciprocity in Antenna Theory*. Hoboken, NJ: John Wiley & Sons, Inc., 2018.
[6] F. M. Tesche, M. V. Ianoz and T. Karlsson. *EMC Analysis Methods and Computational Models*. New York, NY: John Wiley & Sons, Inc., 1997.
[7] J. G. Van Bladel. *Electromagnetic Fields, 2nd edition*. Hoboken, NJ: John Wiley & Sons, 2007.
[8] R. Mittra. A vector form of compensation theorem and its application to boundary-value problems. *Applied Scientific Research*, B(11):26–42, 1964.
[9] M. Abramowitz and I. A. Stegun. *Handbook of Mathematical Functions*. New York, NY: Dover Publications, 1972.
[10] J. D. Kraus. *Antennas, 2nd edition*. New York, NY: McGraw Hill, 1988.

Chapter 14
Pulsed EM plane-wave scattering by time-varying metasurfaces

Space–time metasurfaces hold great promise for their virtually unlimited potentialities for the development of fundamentally new material-based devices [1, 2]. Transient electromagnetic (EM) scattering by structures with time-varying constitutive properties is, however, a little understood phenomenon that is analyzed either numerically via finite-difference approaches [3, 4], or analytically under the restriction of a periodically modulated medium [5, 6], or the constant wave impedance [7, 8].

In this chapter, we introduce a straightforward analytical procedure for analyzing the pulsed plane-wave EM scattering by simple time-varying metasurfaces. In particular, it is demonstrated that the approach is applicable to EM-impenetrable perfectly electrically/magnetically conducting (PEC/PMC) metasurfaces [9] and to thin highly conducting sheets with the time-dependent electric conductivity.

14.1 Pulsed EM plane-wave induced response of a time-varying PEC/PMC metasurface

For the sake of simplicity, we shall start by analyzing the pulsed EM scattering by a time-varying thin sheet whose EM reflection properties can be modeled by an EM-impenetrable wall that switches between PEC and PMC*. The time-varying PEC/PMC metasurface is located in a homogeneous, isotropic and loss-free medium that is described by $\{\epsilon_0, \mu_0\}$. The metasurface under consideration is supposed to be irradiated by a uniform E-polarized EM plane propagating along the positive x-direction and the negative z-direction. Such a plane wave can be described by (cf. (4.1))

$$E_y^{\mathrm{i}}(x, z, t) = e^{\mathrm{i}}[t - \kappa_0 x + p(\kappa_0) z], \tag{14.1}$$

where $e^{\mathrm{i}}(t)$ is the plane-wave signature satisfying $e^{\mathrm{i}}(t) = 0$ for $t < 0$, and the slowness parameters in the x- and z-directions, $\kappa_0 = \sin(\beta)/c_0$ and $p(\kappa_0) = \cos(\beta)/c_0$, respectively, are defined through the angle of incidence β (see Figure 14.1). The excitation by a uniform H-polarized plane wave can be readily handled along

*The author wishes to thank Professor Christophe Caloz, KU Leuven, who drawn his attention to PEC/PMC time-varying metasurfaces. His insightful comments regarding their potentialities are gratefully acknowledged.

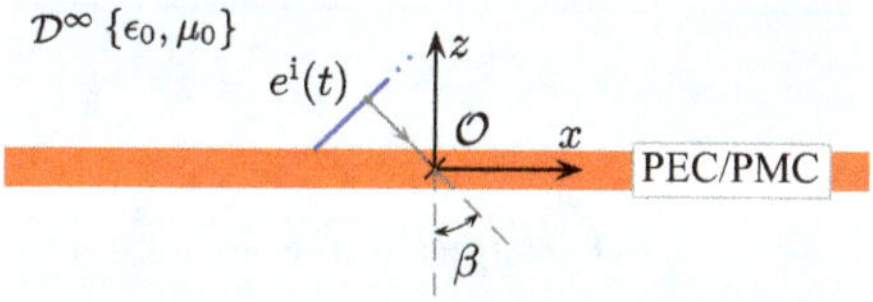

Figure 14.1 Pulsed EM plane wave impinging upon a time-varying PEC/PMC metasurface

similar lines. Since the time-varying metasurface is EM-impenetrable, we shall calculate the scattered EM field in the upper half-space, $z > 0$, only. Accordingly, the presence of the PEC/PMC metasurface is accounted for via a temporally mixed, surface boundary condition

$$m(t)E_y^+ + [1 - m(t)]p^{-1}(\kappa_0)\partial_z\partial_t^{-1}E_y^+ = 0 \quad \text{for all } x \in \mathbb{R} \text{ and } t > 0, \tag{14.2}$$

where the superscript $^+$ denotes the limiting value of the field on the upper surface as $z \downarrow \delta/2$. Furthermore, $m(t)$ is the modulation function that takes the value either 0 or 1. Then, if $m(t) = 1$, (14.2) boils down to the explicit-type boundary condition on the PEC surface, $E_y^+ = 0$. On the other hand, if $m(t) = 0$, the condition simplifies to $\partial_z\partial_t^{-1}E_y^+ = 0$ and hence $H_x^+ = \mu_0^{-1}\partial_z\partial_t^{-1}E_y^+ = 0$, which is met on the PMC surface.

It is next postulated that the total EM field above the metasurface can be written in the following form (cf. (4.16)):

$$\begin{aligned} E_y(x, z, t) &= E_y^i(x, z, t) + E_y^r(x, z, t) \\ &= e^i[t - \kappa_0 x + p(\kappa_0)z] + \mathcal{R}_\perp^0(x, t) *_{xt} e^i[t - \kappa_0 x - p(\kappa_0)z], \end{aligned} \tag{14.3}$$

which is the sum of the incident plane wave (14.1) and the reflected one, denoted by $E_y^r(x, z, t)$. The latter is represented through the space–time convolution (with respect to the x- and t-coordinates) of the reflection coefficient, $\mathcal{R}_\perp^0(x, t)$, and the plane-wave pulse shape traveling along the positive x- and z-directions. With reference to (14.2), (14.3) is next differentiated with respect to z. It can be easily verified that this leads to

$$\begin{aligned} \partial_z E_y(x, z, t) = {} & p(\kappa_0)\partial_t e^i[t - \kappa_0 x + p(\kappa_0)z] \\ & - p(\kappa_0)\mathcal{R}_\perp^0(x, t) *_{xt} \partial_t e^i[t - \kappa_0 x - p(\kappa_0)z]. \end{aligned} \tag{14.4}$$

Making use of the property $f(t) *_t \partial_t g(t) = \partial_t[f(t) *_t g(t)]$, (14.4) reveals that

$$\begin{aligned} p^{-1}(\kappa_0)\partial_z\partial_t^{-1}E_y(x, z, t) = {} & e^i[t - \kappa_0 x + p(\kappa_0)z] \\ & - \mathcal{R}_\perp^0(x, t) *_{xt} e^i[t - \kappa_0 x - p(\kappa_0)z]. \end{aligned} \tag{14.5}$$

Using now (14.3) with (14.5) in the boundary condition (14.2), we end up with

$$\mathcal{R}_{\perp}^{0}(x,t) *_{xt} E_y^{\mathrm{i}}(x,0,t) = \Gamma_0(t)\, E_y^{\mathrm{i}}(x,0,t)$$
$$\text{for all } x \in \mathbb{R} \text{ and } t > 0, \tag{14.6}$$

where

$$\Gamma_0(t) = -\frac{1}{2m(t) - 1} = 1 - 2m(t). \tag{14.7}$$

Equation (14.6) can be viewed as a convolution-type integral equation to be solved for the time-domain (TD) reflection coefficient $\mathcal{R}_{\perp}^{0}(x,t)$. Its (trivial) solution can be illustrated on the analysis of (time-invariant) PEC and PMC screens, for which the solution of (14.6) is available at once. Indeed, we may consider $m(t) = 1$ (the PEC case) or $m(t) = 0$ (the PMC case) for all $t > 0$ and get

$$\mathcal{R}_{\perp}^{0}(x,t) = \begin{cases} -\delta(x)\delta(t) & \text{for the PEC case} \\ +\delta(x)\delta(t) & \text{for the PMC case} \end{cases} \tag{14.8}$$

Apparently, the use of (14.8) in the general solution (14.3) then leads to

$$E_y(x,z,t) = e^{\mathrm{i}}[t - \kappa_0 x + p(\kappa_0)z] \mp e^{\mathrm{i}}[t - \kappa_0 x - p(\kappa_0)z]$$
$$\text{for all } x \in \mathbb{R},\ z > 0 \text{ and } t > 0, \tag{14.9}$$

respectively. A more general solution applying to an arbitrary binary sequence, $m(t)$, can be achieved as follows. First, we observe that the left-hand side of (14.6) represents, in fact, the reflected wave field along the surface of the screen, i.e.

$$E_y^{\mathrm{r}}(x,0,t) = \Gamma_0(t)\, E_y^{\mathrm{i}}(x,0,t). \tag{14.10}$$

Next, it is noted that the *plane-wave* reflected field in the upper half-space can be expressed using its (time-delayed) surface distribution according to

$$E_y^{\mathrm{r}}(x,z,t) = E_y^{\mathrm{r}}[x,0,t - p(\kappa_0)z]. \tag{14.11}$$

The latter property, applying to plane waves, can be easily proved from the postulated form of the reflected wave field (see (14.3)). Then, upon combining (14.10) and (14.11), the desired reflected field follows as

$$E_y^{\mathrm{r}}(x,z,t) = \Gamma_0[t - p(\kappa_0)z]\, e^{\mathrm{i}}[t - \kappa_0 x - p(\kappa_0)z]$$
$$\text{for all } x \in \mathbb{R},\ z > 0 \text{ and } t > 0. \tag{14.12}$$

The obtained TD solution (14.12) can be viewed as a plane wave propagating in the positive x- and z-directions, whose polarity is modulated by the actual value of $\Gamma_0[t - p(\kappa_0)z]$. Since $\Gamma_0(t)$ depends on the binary modulation sequence $m(t)$ via (14.7), it takes values ± 1 only. Consequently, $\Gamma_0[t - p(\kappa_0)z] = -1$ if $m[t - p(\kappa_0)z] = 1$ and $\Gamma_0[t - p(\kappa_0)z] = 1$ if $m[t - p(\kappa_0)z] = 0$, which implies that the polarity of the reflected wave signal as observed at (x,z,t) depends on the value of the modulation function at $t - T_0$, where $T_0 = p(\kappa_0)z$ represents the time to traverse the vertical distance from the metasurface to the field point at the plane-wave slowness $p(\kappa_0)$.

14.2 Pulsed EM plane-wave induced response of a time-varying conductive sheet

In this section, we demonstrate that the methodology applied to analyzing the pulsed EM scattering by an EM-impenetrable metasurface in Section 14.1 can also be applied to a EM-penetrable, thin highly contrasting layer with time-varying conductive properties. Accordingly, we shall next analyze the pulsed EM response of a homogeneous layer that occupies $\{-\infty < x < \infty, -\infty < y < \infty, -\delta/2 < z < \delta/2\}$, with $\delta \downarrow 0$, and whose EM properties are described by its time-dependent electric conductivity $\sigma(t)$ (see Figure 14.2). The pertaining TD saltus-type conditions have the following form (cf. (4.117a) and (4.117b)):

$$E_y^+ - E_y^- = 0, \tag{14.13a}$$

$$H_x^+ - H_x^- = G^{\mathrm{E}}(t)E_y(x,0,t), \quad \text{for all } x \in \mathbb{R} \text{ and } t > 0, \tag{14.13b}$$

where superscripts $^+$ and $^-$ denote the limiting values on the upper and lower sides of the sheet, respectively, and the time-dependent layer conductance is given by (cf. (3.5a))

$$G^{\mathrm{E}}(t) = \int_{\zeta=-\delta/2}^{\delta/2} \sigma(t)\mathrm{d}\zeta = \delta\sigma(t). \tag{14.14}$$

Recall that the layer's conductivity is assumed to be relatively high such that $G^{\mathrm{E}}(t)$ is of $O(1)$.

To solve the problem, it is postulated, again, that the total EM field above the sheet has the following form (cf. (14.3)):

$$\begin{aligned} E_y(x,z,t) &= E_y^{\mathrm{i}}(x,z,t) + E_y^{\mathrm{r}}(x,z,t) \\ &= e^{\mathrm{i}}[t - \kappa_0 x + p(\kappa_0)z] + \mathcal{R}_\perp(x,t) *_{xt} e^{\mathrm{i}}[t - \kappa_0 x - p(\kappa_0)z] \\ &\quad \text{for all } x \in \mathbb{R},\ z \geq 0 \text{ and } t > 0, \end{aligned} \tag{14.15}$$

where coefficient $\mathcal{R}_\perp(x,t)$ has the meaning of the space–time reflection coefficient. In a similar manner, the total EM field below the sheet is expressed as (cf. (4.4))

$$E_y(x,z,t) = E_y^{\mathrm{t}}(x,z,t) = \mathcal{T}_\perp(x,t) *_{xt} e^{\mathrm{i}}[t - \kappa_0 x + p(\kappa_0)z] \quad \text{for all } x \in \mathbb{R},\ z \leq 0 \text{ and } t > 0, \tag{14.16}$$

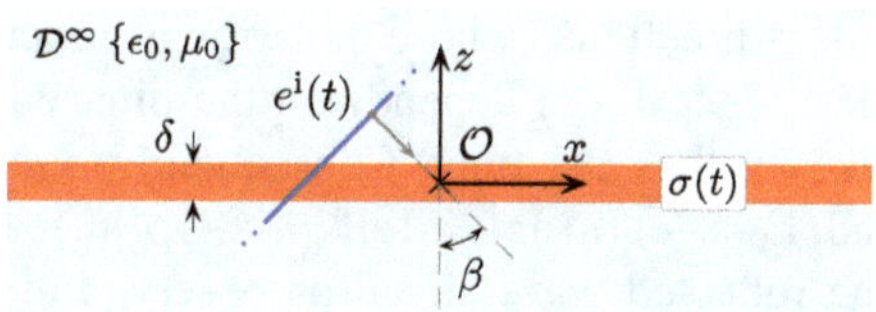

Figure 14.2 Pulsed EM plane wave impinging upon a time-varying conductive sheet

where $\mathcal{T}_\perp(x,t)$ denotes the pertaining (space–time) transmission coefficient. Since $H_x = \partial_z \partial_t^{-1} E_y/\mu_0$, (14.15) leads to

$$\begin{aligned} H_x(x,z,t) = {} & Y_\perp(\kappa_0)\mathrm{e}^{\mathrm{i}}[t - \kappa_0 x + p(\kappa_0)z] \\ & - Y_\perp(\kappa_0)\mathcal{R}_\perp(x,t) *_{xt} \mathrm{e}^{\mathrm{i}}[t - \kappa_0 x - p(\kappa_0)z] \\ & \text{for all } x \in \mathbb{R},\ z > 0 \text{ and } t > 0, \end{aligned} \tag{14.17}$$

and recall that $Y_\perp(\kappa_0) = \eta_0 \cos(\beta)$. Continuing in like manner with (14.16), it is found that

$$\begin{aligned} H_x(x,z,t) = {} & Y_\perp(\kappa_0)\mathcal{T}_\perp(x,t) *_{xt} \mathrm{e}^{\mathrm{i}}[t - \kappa_0 x + p(\kappa_0)z] \\ & \text{for all } x \in \mathbb{R},\ z < 0 \text{ and } t > 0. \end{aligned} \tag{14.18}$$

Substituting next (14.15)–(14.18) in the TD cross-layer conditions (14.13a) and (14.13b), we end up with the following system of equations:

$$\mathrm{e}^{\mathrm{i}}(t - \kappa_0 x) + \mathcal{R}_\perp(x,t) *_{xt} \mathrm{e}^{\mathrm{i}}(t - \kappa_0 x) - \mathcal{T}_\perp(x,t) *_{xt} \mathrm{e}^{\mathrm{i}}(t - \kappa_0 x) = 0, \tag{14.19a}$$

$$\begin{aligned} & Y_\perp(\kappa_0)\mathrm{e}^{\mathrm{i}}(t - \kappa_0 x) - Y_\perp(\kappa_0)\mathcal{R}_\perp(x,t) *_{xt} \mathrm{e}^{\mathrm{i}}(t - \kappa_0 x) \\ & - Y_\perp(\kappa_0)\mathcal{T}_\perp(x,t) *_{xt} \mathrm{e}^{\mathrm{i}}(t - \kappa_0 x) = G^{\mathrm{E}}(t)[\mathcal{T}_\perp(x,t) *_{xt} \mathrm{e}^{\mathrm{i}}(t - \kappa_0 x)]. \end{aligned} \tag{14.19b}$$

Upon eliminating the term containing the transmission coefficient, we arrive at

$$\begin{aligned} & \mathcal{R}_\perp(x,t) *_{xt} E_y^{\mathrm{i}}(x,0,t) = \Gamma_0^{\mathrm{E}}(t)\, E_y^{\mathrm{i}}(x,0,t) \\ & \text{for all } x \in \mathbb{R} \text{ and } t > 0, \end{aligned} \tag{14.20}$$

where (cf. (6.133))

$$\Gamma_0^{\mathrm{E}}(t) = -\frac{G^{\mathrm{E}}(t)}{G^{\mathrm{E}}(t) + 2Y_\perp(\kappa_0)}. \tag{14.21}$$

It is observed that the resulting equation (14.20) is similar to (14.6) applying to a PEC/PMC metasurface. As a matter of fact, under the limit $G^{\mathrm{E}}(t) \to \infty$ describing the PEC wall, $\Gamma_0^{\mathrm{E}}(t) = -1$ and (14.20) yields $\mathcal{R}_\perp(x,t) = -\delta(x)\delta(t)$, a limiting solution that was discussed before (see (14.8)). More generally, it is first observed we may use (14.15) to rewrite (14.20) as

$$E_y^{\mathrm{r}}(x,0,t) = \Gamma_0^{\mathrm{E}}(t)\, E_y^{\mathrm{i}}(x,0,t). \tag{14.22}$$

Owing to the property expressed via (14.11), the reflected EM-field in the upper half-space then follows as (cf. (14.12))

$$\begin{aligned} & E_y^{\mathrm{r}}(x,z,t) = \Gamma_0^{\mathrm{E}}[t - p(\kappa_0)z]\, \mathrm{e}^{\mathrm{i}}[t - \kappa_0 x - p(\kappa_0)z] \\ & \text{for all } x \in \mathbb{R},\ z \geq 0 \text{ and } t > 0. \end{aligned} \tag{14.23}$$

Equation (14.23) represents an EM plane wave propagating in the positive x- and z-directions. Its amplitude is modulated by the time-dependent function $\Gamma_0^{\mathrm{E}}(t)$ that is related to the equivalent layer's conductance, $G^{\mathrm{E}}(t)$, and the angle of incidence, β, via (14.21). In fact, it is seen that the amplitude of the reflected plane wave as observed at (x, z, t) is related to $\Gamma_0^{\mathrm{E}}(t - T_0)$, where $T_0 = p(\kappa_0)z$ denotes the travel time to traverse the vertical distance, $z > 0$, between the sheet and the point of observation. The transmitted field in the lower half-space can be obtained in a similar manner. First, we use (14.19a) to express the left-hand side of (14.20) through the difference of the transmitted and incident electric-field strength at $z = 0$, i.e.

$$E_y^{\mathrm{t}}(x, 0, t) - E_y^{\mathrm{i}}(x, 0, t) = \Gamma_0^{\mathrm{E}}(t)\, E_y^{\mathrm{i}}(x, 0, t), \tag{14.24}$$

which leads to

$$E_y^{\mathrm{t}}(x, 0, t) = [1 + \Gamma_0^{\mathrm{E}}(t)]\, E_y^{\mathrm{i}}(x, 0, t). \tag{14.25}$$

Consequently, the postulated form of the transmitted field (14.16) allows to establish the following property (cf. (14.11)):

$$E_y^{\mathrm{t}}(x, z, t) = E_y^{\mathrm{t}}[x, 0, t + p(\kappa_0)z], \tag{14.26}$$

which is finally combined with (14.25) to find the desired TD expression for the total EM field transmitted through the time-varying layer

$$E_y^{\mathrm{t}}(x, z, t) = \{1 + \Gamma_0^{\mathrm{E}}[t + p(\kappa_0)z]\}\, e^{\mathrm{i}}[t - \kappa_0 x + p(\kappa_0)z] \quad \text{for all } x \in \mathbb{R},\ z \leq 0 \text{ and } t > 0. \tag{14.27}$$

The resulting TD expression (14.27) can be interpreted as an EM plane wave propagating along the direction of propagation of the incident plane wave (see (14.1)). Its amplitude is modulated by the factor $\{1 + \Gamma_0^{\mathrm{E}}[t + p(\kappa_0)z]\}$, which depends on the equivalent layer's conductance, $G^{\mathrm{E}}(t)$, and the angle of incidence, β, via (14.21), again. Apparently, under the limit $G^{\mathrm{E}} \to \infty$, the modulation factor is identically zero, which yields the (expected) vanishing EM field behind the PEC wall.

14.2.1 Illustrative examples

Through the analytical results derived in Sections 14.1 and 14.2, we shall next calculate the electric-field space–time distribution in the presence of both time-varying PEC/PMC metasurface and conducting thin layer. In the examples that follow, the planar structures are irradiated by the E-polarized plane wave (see (14.1)), whose signature has the shape of the ringing rectangular pulse (cf. (6.13))

$$e^{\mathrm{i}}(t) = e_{\mathrm{m}} \sin(2\pi f_0 t)[\mathrm{H}(t) - \mathrm{H}(t - t_{\mathrm{w}})], \tag{14.28}$$

where we take the unit amplitude, $e_{\mathrm{m}} = 1.0\,\mathrm{V/m}$. Its frequency of oscillation, f_0, is related to the pulse-time width, t_{w}, via $f_0\, t_{\mathrm{w}} = 2.0$. As the angle of incidence, we take $\beta = \pi/4$, again.

In the first example, we shall analyze the electric-field response of a PEC/PMC metasurface whose reflection properties vary according to the following modulation function:

$$m(t) = \mathrm{H}(t) - \mathrm{H}(t - 2t_\mathrm{w}) + \mathrm{H}(t - 4t_\mathrm{w}). \tag{14.29}$$

The corresponding TD reflection coefficient, $\Gamma_0(t)$, as found via (14.7), is shown in Figure 14.3 along with the (normalized) excitation plane-wave signature. With reference to Figure 14.3, the metasurface under consideration behaves as a PEC wall in $\{0 < t/t_\mathrm{w} < 2\}$ and later in $\{4 < t/t_\mathrm{w} < \infty\}$. In the remaining finite interval, $\{2 < t/t_\mathrm{w} < 4\}$, the metasurface reflects as a PMC screen. The resulting spatial distributions in $\{0 < x/c_0 t_\mathrm{w} < 6, 0 < z/c_0 t_\mathrm{w} < 6\}$ at fixed instants $t/t_\mathrm{w} = \{1.5, 3.0, 4.5\}$ are shown in Figure 14.4(a), (b) and (c), respectively. In line with Figure 14.3 and the boundary condition (14.2) with (14.29), in Figure 14.4(a) we may see that the total electric-field strength just above the metasurface vanishes, which is characteristic for a PEC screen. Figure 14.4(b) then shows the corresponding field distribution at $t/t_\mathrm{w} = 3.0$, when the metasurface reflects as a PMC wall. Here, on the other hand, we may observe the constructive wave interference on the surface. The effect of transition from PEC to PMC at $t/t_\mathrm{w} = 2$ (see Figure 14.3) is clearly visible on the reflected wave field that suddenly changed its polarity at this instant. Finally, Figure 14.4(c) shows the total field distribution at $t/t_\mathrm{w} = 4.5$. Here, apparently, the metasurface acts as a PEC wall, again.

In the second example, we shall calculate the electric-field distribution in the presence of a time-varying, thin highly conducting sheet. For that purpose, we shall

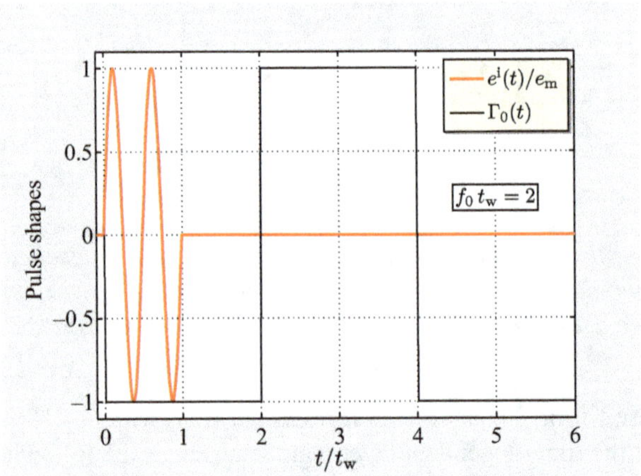

Figure 14.3 Incident plane-wave pulse shape and the TD reflection coefficient of the analyzed time-varying PEC/PMC metasurface

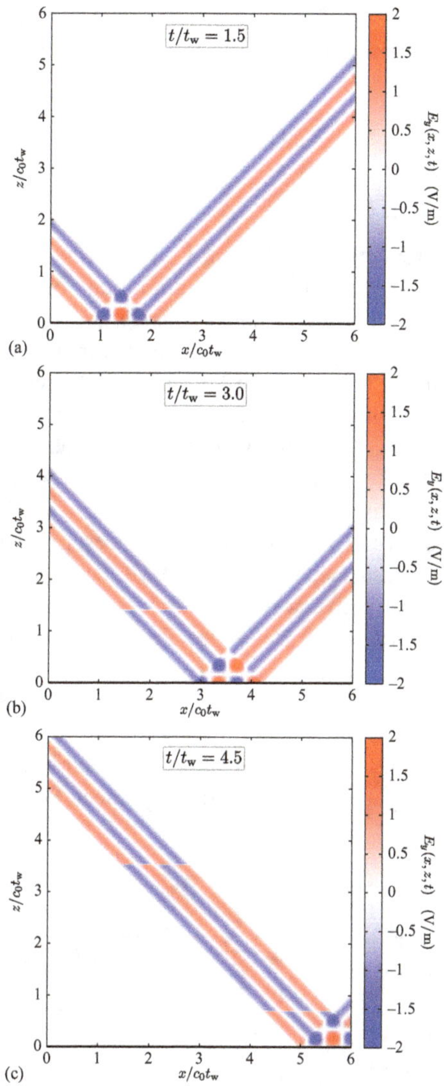

Figure 14.4 Spatial distributions of the electric-field strength excited by the impulsive plane wave in the presence of the time-varying PEC/PMC metasurface at (a) $t/t_W = 1.5$; (b) $t/t_W = 3.0$; (c) $t/t_W = 4.5$

employ the closed-form TD analytical formulas (14.23) with (14.15) and (14.27). It is assumed that the time-dependent layer conductance has the following form:

$$G^E(t) = 3\eta_0 \, [\mathrm{H}(t) + 19\,\mathrm{H}(t - 2t_W) - 19\,\mathrm{H}(t - 4t_W)]. \tag{14.30}$$

Consequently, the conductive layer has the equivalent conductance $G^E = 3\eta_0$ in $\{0 < t/t_W < 2\}$ and in $\{4 < t/t_W < \infty\}$ while $G^E = 60\eta_0$ in $\{2 < t/t_W < 4\}$.

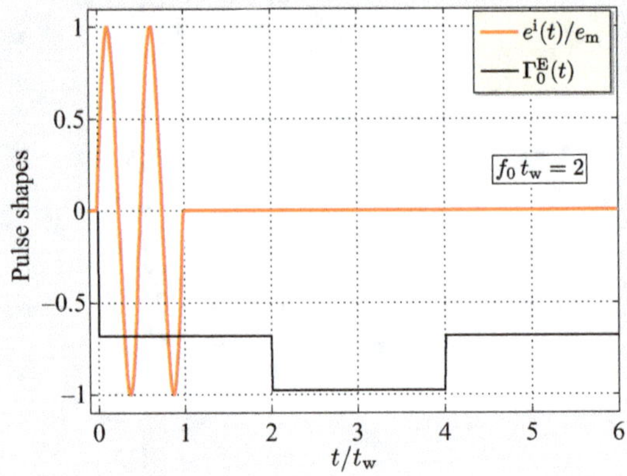

Figure 14.5 Incident plane-wave pulse shape and the TD reflection coefficient of the analyzed time-varying conducting layer

The corresponding TD reflection coefficient, $\Gamma_0^{\mathrm{E}}(t)$, as found via (14.21), is shown in Figure 14.5. It can be seen that $\Gamma_0^{\mathrm{E}}(t) \simeq -0.68$ in $\{0 < t/t_{\mathrm{w}} < 2\}$ and in $\{4 < t/t_{\mathrm{w}} < \infty\}$ while $\Gamma_0^{\mathrm{E}}(t) \simeq -0.98$ in $\{2 < t/t_{\mathrm{w}} < 4\}$. The pertaining spatial distributions in $\{0 < x/c_0 t_{\mathrm{w}} < 6, -3 < z/c_0 t_{\mathrm{w}} < 3\}$ at fixed instants $t/t_{\mathrm{w}} = \{1.5, 3.0, 4.5\}$ are shown in Figure 14.6(a), (b) and (c), respectively. Apparently, at $t/t_{\mathrm{w}} = 1.5$, when the equivalent layer conductance is relatively small, a non-negligible part of the field is transmitted through the sheet (see Figure 14.6(a)). On the other hand, at $t/t_{\mathrm{w}} = 3.0$, the layer behaves essentially as a PEC wall, thus eliminating the field in the lower half-space. The effect of transition can be observed at both reflected and transmitted wave field distributions. Finally, Figure 14.6(c) shows the total electric-field distribution at $t/t_{\mathrm{w}} = 4.5$. Here, owing to the lower value of the layer's conductance, the reflected field gets weaker and the field penetrates into the lower half-space, again.

14.3 Concluding remarks

In this chapter, we have limited ourselves to analyzing infinite and instantaneously reacting time-varying thin sheets. But, it is fairly straightforward to extend the analysis to bounded screens. To provide a simple yet illustrative example, we may briefly examine the pulsed EM response of a narrow strip with time-varying conductive properties. To that end, we start from the TD original of the reciprocity relation (7.1), where we use $E_y^{\mathrm{s}}(x, 0, t) = -E_y^{\mathrm{i}}(x, 0, t) + \partial J_y^{\mathrm{s}}(x, t)/G^{\mathrm{E}}(t)$ as obtained from the pertaining jump condition (see (14.13b)). Assuming further the uniform spatial distribution of both testing and induced electric-current surface densities (see (7.2)), and the piecewise linear temporal expansion of the induced current

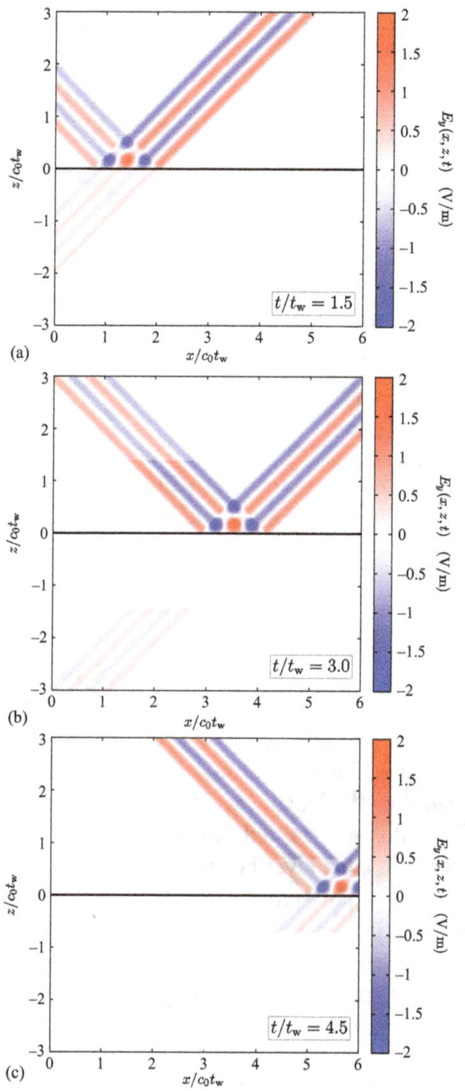

Figure 14.6 Spatial distributions of the electric-field strength excited by the impulsive plane wave in the presence of the time-varying conducting layer at (a) $t/t_w = 1.5$; (b) $t/t_w = 3.0$; (c) $t/t_w = 4.5$

(see (7.18)), the TD reciprocity relation can be readily cast into the following form (cf. (7.42)):

$$\frac{\zeta_0}{c_0 \Delta t} \sum_{k=1}^{m} \left[\Phi_{m-k+1} + \Delta\Phi(t_m|t_m - t_{k-1}) - 2\Phi_{m-k} - 2\Delta\Phi(t_m|t_m - t_k) \right.$$
$$\left. + \Phi_{m-k-1} + \Delta\Phi(t_m|t_m - t_{k+1}) \right] i_k = e^{\mathrm{i}}(t_m), \qquad (14.31)$$

where $\Phi_k = \Phi(t_k)$ corresponds to (7.21), and the correction term, accounting for the time-varying conductivity, is given by (cf. (7.25))

$$\Delta\Phi(t_m|t) = (c_0 t/w)[\eta_0/G^{\mathrm{E}}(t_m)]\mathrm{H}(t). \tag{14.32}$$

The obtained equation (14.31) can be solved iteratively, again (cf. (7.20)):

$$\begin{aligned} i_m &= [\Phi_1 + \Delta\Phi(t_m|\Delta t)]^{-1}\Big\{c_0\Delta t\eta_0 e^{\mathrm{i}}(t_m) \\ &\quad - \sum_{k=1}^{m-1} i_k\big[\Phi_{m-k+1} + \Delta\Phi(t_m|t_m - t_{k-1}) - 2\Phi_{m-k} - 2\Delta\Phi(t_m|t_m - t_k) \\ &\quad + \Phi_{m-k-1} + \Delta\Phi(t_m|t_m - t_{k+1})\big]\Big\} \quad \text{for all } m = \{1, \ldots, M\}. \end{aligned} \tag{14.33}$$

Owing to the fact that the factor $\Phi_1 + \Delta\Phi(t_m|\Delta t)$ depends on the actual value of the layer's conductivity, its inverse in the marching-on-in-time solution (14.33) must be evaluated for each time index m. The analysis of time-varying metasurfaces with combined magneto-dielectric properties via the Cagniard–DeHoop method of moments (CdH-MoM) seems to be feasible too. More details on this extensive research topic are left for a separate book.

References

[1] N. Engheta. Metamaterials with high degrees of freedom: space, time, and more. *Nanophotonics*, 1, 2020.

[2] C. Caloz and Z.-L. Deck-Léger. Spacetime metamaterials—part I: general concepts. *IEEE Transactions on Antennas and Propagation*, 68(3):1569–1582, 2019.

[3] Y. Vahabzadeh, N. Chamanara and C. Caloz. Generalized sheet transition condition FDTD simulation of metasurface. *IEEE Transactions on Antennas and Propagation*, 66(1):271–280, 2017.

[4] S. A. Stewart, T. J. Smy and S. Gupta. Finite-difference time-domain modeling of space–time-modulated metasurfaces. *IEEE Transactions on Antennas and Propagation*, 66(1):281–292, 2017.

[5] L. A. Pipes. Matrix analysis of linear time-varying circuits. *Journal of Applied Physics*, 25(9):1179–1185, 1954.

[6] T. T. Koutserimpas and R. Fleury. Electromagnetic waves in a time periodic medium with step-varying refractive index. *IEEE Transactions on Antennas and Propagation*, 66(10):5300–5307, 2018.

[7] N. V. Budko. Electromagnetic radiation in a time-varying background medium. *Physical Review A*, 80(5):053817, 2009.

[8] A. T. de Hoop and I. E. Lager. Closed-form analytic expressions for the pulsed-source radiated electromagnetic field in a class of media with time-varying wave speed. *Wave Motion*, 51(3):418–424, 2014.

[9] X. Wang and C. Caloz. Spread-spectrum selective camouflaging based on time-modulated metasurface. *IEEE Transactions on Antennas and Propagation*, 69(1):286–295, 2021.

Chapter 15
Pulsed EM response of transmission lines over a metasurface

Whenever the transverse dimension of a waveguide structure is relatively small such that its characteristic electromagnetic (EM) properties can be described via the concept of Kirchhoff's circuit lumped elements, the EM propagation along the structure's axis is governed by the transmission-line (TL) equations [1, Chapter VII]. To evaluate the pulse propagation along nonideal practical lines (e.g. an overhead wire over a lossy ground [2, Chapter 8]), one has to rely on approximate or/and numerical approaches. This strategy is also pursued in this chapter, where the Cagniard–DeHoop method of moments (CdH-MoM) is applied to analyze the pulsed EM scattering by a straight transmission line that is located above a thin highly contrasting layer. Accordingly, the computational approach introduced below bridges the gap between the EM reciprocity-based, EM-field-to-line coupling model [3, Chapters 11–13] and the CdH-MoM analysis of a narrow planar strip above the perfectly electrically conducting (PEC) ground [3, Chapter 17]. For its applications to the transient analysis of a transmission line in the presence of a highly conducting thin layer, the reader is referred to [4].

15.1 Transmission line above a PEC plane

We shall analyze the time-domain (TD) electric-current pulse propagation along a thin straight wire of a finite length, $\ell > 0$, and a (relatively small) radius, $a > 0$, with its axis extending along $\{-\ell/2 < x < \ell/2, y = 0, z = 0\}$. The wire is located above a thin highly contrasting layer of an infinite extent that occupies $\{-\infty < x < \infty, -\infty < y < \infty, -z_0 - \delta/2 < z < -z_0 + \delta/2\}$, where $\delta > 0$ is its thickness and $z_0 > 0$ denotes the wire's vertical distance from the layer. The transmission line can be at its ends loaded by passive circuit elements. The surrounding medium is described by scalar and real-valued constants, $\{\epsilon_0, \mu_0\}$, and is hence supposed to be homogeneous, isotropic and loss-free in its EM properties. In this section, it is further assumed that the layer exhibits infinite electric conductivity, $\sigma \to \infty$, or/and infinite permittivity, $\epsilon \to \infty$, which represents a PEC sheet. The transmission line under consideration is excited by a uniform EM plane wave or/and by a voltage pulse applied in a narrow gap (see Figure 15.1). Following the lines of reasoning pursued in Section 9.1, the scattered EM field is defined here as the difference between the total EM field in the actual problem configuration and the excitation field (denoted by

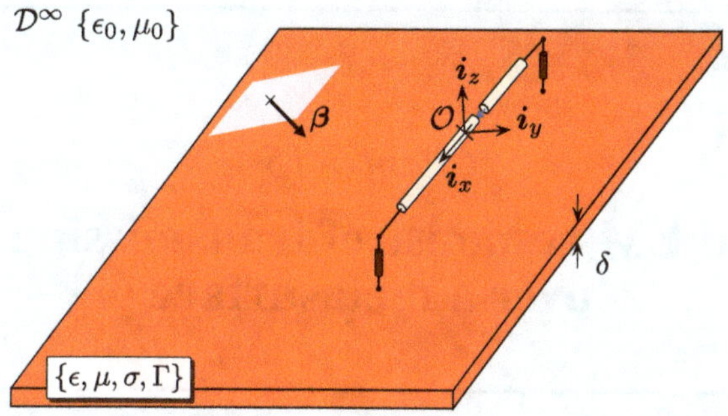

Figure 15.1 Transmission line above a metasurface

superscript e) that represents, as a matter of fact, the total EM field in the absence of the conducting wire. Consequently, the axial component of the scattered electric-field strength can be in the s-domain expressed as

$$\hat{E}^{\mathrm{s}}_x(x, \varrho, s) = \hat{E}_x(x, \varrho, s) - \hat{E}^{\mathrm{e}}_x(x, \varrho, s), \tag{15.1}$$

where $\varrho = \varrho(y, z) = (y^2 + z^2)^{1/2} > 0$. Since the axial component of the excitation electric-field strength satisfies the explicit-type boundary condition on the PEC surface, we have

$$\lim_{z \downarrow -z_0} \hat{E}^{\mathrm{s}}_x(x, \varrho, s) = 0 \quad \text{for all } x \in \mathbb{R} \text{ and } y \in \mathbb{R}. \tag{15.2}$$

15.1.1 Problem formulation

For the sake of simplicity, this subsection is concerned with an open-circuited, voltage-gap-excited transmission line above a PEC ground plane (see [3, Chapter 17]). The analyzed transmission-line problem is formulated via the EM reciprocity theorem of the time-convolution type [5, Section 28.4] that interrelates the actual scattered EM-field state (denoted by superscript s) with the (causal) testing one (denoted by superscript T) in the upper half-space above the PEC sheet. Accounting for the explicit-type boundary condition on the screen and assuming that the conductor is relatively thin such that the actual volume electric-current density in the wire can be replaced with the current flowing essentially along its axis, we arrive at (cf. (8.1))

$$\int_{x=-\ell/2}^{\ell/2} \hat{E}^{\mathrm{T}}_x(x, a, s)\hat{I}^{\mathrm{s}}_x(x, s)\mathrm{d}x = \int_{x=-\ell/2}^{\ell/2} \hat{E}^{\mathrm{s}}_x(x, a, s)\hat{I}^{\mathrm{T}}_x(x, s)\mathrm{d}x, \tag{15.3}$$

where $\hat{I}^{\mathrm{s}}_x(x, s)$ is the (unknown) axial current induced in the wire and $\hat{I}^{\mathrm{T}}_x(x, s)$ denotes the testing electric-current distribution. Here, the testing electric-field

strength, $\hat{E}_x^{\mathrm{T}}(x, \varrho, s)$, can be related to the testing current via the standard field expression (see [5, Eq. (26.9-2)], for example) provided that

$$\lim_{z \downarrow -z_0} \hat{E}_x^{\mathrm{T}}(x, \varrho, s) = 0 \quad \text{for all } x \in \mathbb{R} \text{ and } y \in \mathbb{R}, \tag{15.4}$$

is met. The reciprocity relation (15.3) is further solved via the CdH-MoM. With that goal in mind, the wave-slowness representation along the x-direction (see (4.33)) is employed. Under this representation, (the axial component of) the testing electric-field strength follows as (see (9.5) and (12.7a))

$$\tilde{E}_x^{\mathrm{T}}(\kappa, \varrho, s) = -(s/\epsilon_0)\Omega_0^2(\kappa)\tilde{K}(\kappa, \varrho, s)\tilde{I}_x^{\mathrm{T}}(\kappa, s), \tag{15.5}$$

where the transform-domain Green's function, $\tilde{K}(\kappa, \varrho, s)$, accounting for the reflection from the PEC sheet, is given by

$$\tilde{K}(\kappa, \varrho, s) = \mathrm{K}_0[s\Omega_0(\kappa)\varrho(y, z)]/2\pi - \mathrm{K}_0[s\Omega_0(\kappa)\varrho(y, z + 2z_0)]/2\pi. \tag{15.6}$$

In (15.6), $\mathrm{K}_0(x)$ denotes the modified Bessel function of the second kind [6, Section 9.6] and recall that $\Omega_0(\kappa) = (1/c_0^2 - \kappa^2)^{1/2}$. Along the surface of the wire, assuming that the wire is relatively close to the PEC plane and that $a \ll 2z_0$, the small-argument expansion of the modified Bessel function [6, Eq. (9.6.13)] leads to the following approximation:

$$\tilde{K}(\kappa, a, s) \simeq \log(2z_0/a)/2\pi. \tag{15.7}$$

Consequently, the axial component of the (transform-domain) testing field can be approximated via

$$\tilde{E}_x^{\mathrm{T}}(\kappa, a, s) \simeq -(s/c_0)Z^{\mathrm{c}}c_0^2\Omega_0^2(\kappa)\tilde{I}_x^{\mathrm{T}}(\kappa, s), \tag{15.8}$$

where $Z^{\mathrm{c}} = (\zeta_0/2\pi)\log(2z_0/a)$ represents the characteristic impedance of the line above the PEC plane (see [3, Eq. (11.9)]). It is worth noting that the approximate testing source-field relation (15.8) complies with the transmission-line equations (cf. (15.59a) and (15.59b)) interrelating the testing voltage and current along an equivalent, voltage-source-excited transmission line, i.e.

$$\partial_x \hat{V}^{\mathrm{T}}(x, s) + \hat{\zeta}(s)\hat{I}_x^{\mathrm{T}}(x, s) = -\hat{E}_x^{\mathrm{T}}(x, a, s), \tag{15.9a}$$

$$\partial_x \hat{I}_x^{\mathrm{T}}(x, s) + \hat{\eta}(s)\hat{V}^{\mathrm{T}}(x, s) = 0, \tag{15.9b}$$

where $\hat{\eta}(s) = 2\pi s\epsilon_0/\log(2z_0/a)$ and $\hat{\zeta}(s) = s\mu_0\log(2z_0/a)/2\pi$ denote the transverse admittance and the longitudinal impedance (per-unit-length) of the transmission line, respectively.

The s-domain testing field expressed via (15.8) is further substituted in the starting reciprocity relation (15.3), where we change the order of the integrations with respect to κ and x. This way then yields

$$-\frac{sZ^{\mathrm{c}}}{c_0}\frac{s}{2\pi\mathrm{i}}\int_{\kappa=-\mathrm{i}\infty}^{\mathrm{i}\infty} c_0^2\Omega_0^2(\kappa)\tilde{I}_x^{\mathrm{T}}(\kappa, s)\tilde{I}_x^{\mathrm{s}}(-\kappa, s)\mathrm{d}\kappa$$
$$= -\frac{s}{2\pi\mathrm{i}}\int_{\kappa=-\mathrm{i}\infty}^{\mathrm{i}\infty} \tilde{E}_x^{\mathrm{e}}(\kappa, a, s)\tilde{I}_x^{\mathrm{T}}(-\kappa, s)\mathrm{d}\kappa \quad \text{for } \{s \in \mathbb{R}; s > 0\}, \tag{15.10}$$

and we used, on the right-hand side, the explicit-type boundary condition along the horizontal PEC conductor, $\hat{E}_x^{\mathrm{s}}(x,a,s) = -\hat{E}_x^{\mathrm{e}}(x,a,s)$ for all $\{-\ell/2 < x < \ell/2\}$ (see (15.1)). Finally note that, in general, one has to also incorporate the contributions of the vertical component of the excitation electric-field strength along $\{x = \pm\ell/2, y = 0, -z_0 < z < 0\}$ (see (15.21)). In the following section, we shall present a numerical solution of the slowness-domain reciprocity relation (15.10) that yields the space–time distribution of the induced electric current.

15.1.2 Problem solution

In order to solve the transform-domain reciprocity relation (15.10) numerically, the space–time solution domain has to be discretized. Accordingly, the axis of the wire is divided into $N+1$ segments of a constant length $\Delta = \ell/(N+1)$, where N denotes the number of (inner) nodes along the uniform grid (see Figure 15.2 for $N = 5$). The position of a node is specified by

$$x_n = -\ell/2 + n\Delta \quad \text{for } n = \{1, \ldots, N\}. \tag{15.11}$$

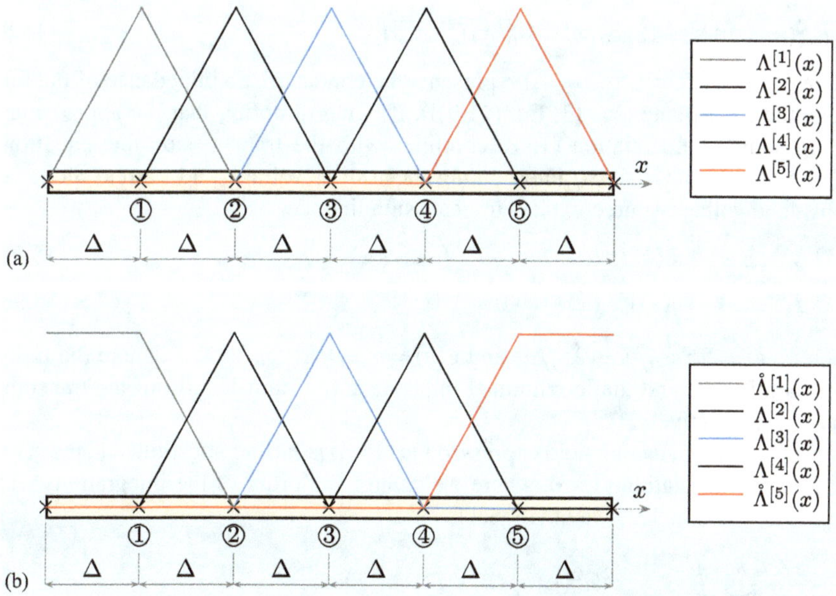

Figure 15.2 Uniform discretization of the spatial solution domain. (a) Open-circuited transmission line. (b) Loaded transmission line.

Furthermore, the time axis $\{t \in \mathbb{R}; t > 0\}$ is discretized uniformly with the constant time step $\{t_k = k\Delta t; \Delta t > 0, k = 1, 2, \ldots, M\}$. Consequently, the unknown axial current can be expanded in a series of triangular functions (cf. (8.6))

$$I_x^{\mathrm{s}}(x,t) \simeq \sum_{n=1}^{N}\sum_{k=1}^{M} i_k^{[n]} \Lambda^{[n]}(x)\Lambda_k(t), \tag{15.12}$$

where $i_k^{[n]}$ are unknown coefficients (in A) and the temporal triangular function is defined by (7.19). For an open-circuited transmission line, the spatial basis functions, $\Lambda^{[n]}(x)$, are given by (8.7a) for all $n = \{1, \ldots, N\}$. If the transmission line is not open-circuited at its terminals, non-vanishing axial currents can be incorporated by replacing the triangular functions $\Lambda^{[1]}(x)$ and $\Lambda^{[N]}(x)$ with $\mathring{\Lambda}^{[1]}(x)$ and $\mathring{\Lambda}^{[N]}(x)$, respectively (see Figure 15.2). These "edge spatial basis functions" can be described by (cf. (8.7a))

$$\mathring{\Lambda}^{[1]}(x) = \begin{cases} 1 & \text{for } x \in [-\ell/2, x_1], \\ 1 - (x - x_1)/\Delta & \text{for } x \in [x_1, x_2], \end{cases} \tag{15.13a}$$

and

$$\mathring{\Lambda}^{[N]}(x) = \begin{cases} 1 + (x - x_N)/\Delta & \text{for } x \in [x_{N-1}, x_N], \\ 1 & \text{for } x \in [x_N, \ell/2]. \end{cases} \tag{15.13b}$$

As in Section 8.1.2, the testing current is chosen to be represented by the rectangular function of x and the Dirac-delta distribution with respect to the time coordinate. Hence, we may write

$$I_x^{\mathrm{T}}(x,t) = \Pi^{[S]}(x)\delta(t) \quad \text{for all } S = \{1, \ldots, N\}, \tag{15.14}$$

where the rectangular function is defined by (8.9a). Upon substituting the transform-domain counterparts of (15.12) and (15.14) in the reciprocity relation (15.10), we end up the system of equations in the s-domain, constituents of which can be transformed to the TD. This way leads to the following time-convolution-type system of equations, again (cf. (7.42)):

$$\sum_{k=1}^{m} \left(\underline{\boldsymbol{Z}}_{m-k+1} - 2\underline{\boldsymbol{Z}}_{m-k} + \underline{\boldsymbol{Z}}_{m-k-1}\right) \cdot \boldsymbol{I}_k = \boldsymbol{V}_m, \tag{12.25 revisited}$$

where $\underline{\boldsymbol{Z}}_k = \underline{\boldsymbol{Z}}(t_k)$ denotes a 2-D $[N \times N]$ TD impedance array evaluated at $t = t_k = k\Delta t$, $\boldsymbol{I}_k$ is a 1-D $[N \times 1]$ array of the electric-current coefficients, $i_k^{[n]}$, and, finally, $\boldsymbol{V}_m$ denotes a 1-D $[N \times 1]$ excitation voltage array at $t = t_m$. The elements

of the impedance array can be readily found from the TD originals of the following s-domain expressions (cf. (B.2a)), viz

$$\hat{Z}^{[S,n]}(s) = \frac{Z^{\mathrm{c}}}{c_0 \Delta t} \frac{1}{\Delta} \Big[\hat{\Upsilon}^A(x_S - x_n + 3\Delta/2, s) - 3\hat{\Upsilon}^A(x_S - x_n + \Delta/2, s) + 3\hat{\Upsilon}^A(x_S - x_n - \Delta/2, s) - \hat{\Upsilon}^A(x_S - x_n - 3\Delta/2, s) \Big]$$
$$\text{for all } S = \{1, \ldots, N\} \text{ and } n = \{1, \ldots, N\} \text{ if open-circuited, or for all } S = \{1, \ldots, N\} \text{ and } n = \{2, \ldots, N-1\} \text{ if loaded,} \tag{15.15a}$$

at its end points. In the latter case, the remaining impedance array elements follow as (cf. (B.2b) and (B.2c))

$$\hat{Z}^{[S,1]}(s) = -\frac{Z^{\mathrm{c}}}{c_0 \Delta t} \frac{1}{\Delta} \Big[\hat{\Upsilon}^A(x_S - x_1 + \Delta/2, s) - 2\hat{\Upsilon}^A(x_S - x_1 - \Delta/2, s) + \hat{\Upsilon}^A(x_S - x_1 - 3\Delta/2, s) \Big] + \frac{Z^{\mathrm{c}}}{c_0 \Delta t} \Big[\hat{\Upsilon}^B(x_S - x_1 + 3\Delta/2, s) - \hat{\Upsilon}^B(x_S - x_1 + \Delta/2, s) \Big] \quad \text{for all } S = \{1, \ldots, N\}, \tag{15.15b}$$

and, finally,

$$\hat{Z}^{[S,N]}(s) = \frac{Z^{\mathrm{c}}}{c_0 \Delta t} \frac{1}{\Delta} \Big[\hat{\Upsilon}^A(x_S - x_N + 3\Delta/2, s) - 2\hat{\Upsilon}^A(x_S - x_N + \Delta/2, s) + \hat{\Upsilon}^A(x_S - x_N - \Delta/2, s) \Big] - \frac{Z^{\mathrm{c}}}{c_0 \Delta t} \Big[\hat{\Upsilon}^B(x_S - x_N - \Delta/2, s) - \hat{\Upsilon}^B(x_S - x_N - 3\Delta/2, s) \Big] \quad \text{for all } S = \{1, \ldots, N\}. \tag{15.15c}$$

The TD impedance-array elements can be calculated through the TD originals of the following s-domain functions:

$$\hat{\Upsilon}^A(x, s) = -\frac{c_0^2}{2\pi\mathrm{i}} \int_{\kappa \in \mathbb{I}_0} \frac{\exp(s\kappa x)}{s^3 \kappa^3} \Omega_0^2(\kappa) \mathrm{d}\kappa, \tag{15.16a}$$

$$\hat{\Upsilon}^B(x, s) = -\frac{c_0^2}{2\pi\mathrm{i}} \int_{\kappa \in \mathbb{I}_0} \frac{\exp(s\kappa x)}{s^2 \kappa^2} \Omega_0^2(\kappa) \mathrm{d}\kappa, \quad \text{for } x \in \mathbb{R} \text{ and } \{s \in \mathbb{R}; s > 0\}, \tag{15.16b}$$

where the (indented) integration path in the complex κ-plane, $\mathbb{I}_0$, is depicted in Figure B.1. As the integrands in (15.16a) and (15.16b) are not multivalued functions of κ, their evaluation is straightforward. Indeed, upon employing Jordan's lemma and Cauchy's theorem to replace $\mathbb{I}_0$ with a closed contour encircling the simple pole singularity at the origin $\kappa = 0$, Cauchy's formula [7, Section 2.41] yields closed-form expressions for $\hat{\Upsilon}^{A,B}(x, s)$ at once. Their subsequent transform to the TD leads to

$$\Upsilon^A(x, t) = \big[(c_0^2 t^2 - x^2)/2\big] \mathrm{H}(x) \mathrm{H}(t), \tag{15.17a}$$

$$\Upsilon^B(x, t) = -x \mathrm{H}(x) \mathrm{H}(t), \quad \text{for } x \in \mathbb{R} \text{ and } t \in \mathbb{R}. \tag{15.17b}$$

Owing to the simplicity of TD functions (15.17a) and (15.17b), the calculation of TD impedance array elements, $Z^{[S,n]}(t)$, is computationally effortless and can be carried out via the TD originals of (15.15a)–(15.15c).

To fully define the TD system of equations (12.25), we shall further determine the elements of the excitation array $\boldsymbol{V}_m$. They will be worked out for both the delta-gap and EM plane-wave excitations. More general EM sources can be easily incorporated along the same lines.

15.1.2.1 Delta-gap excitation

If the transmission is excited by a causal voltage pulse, say $V^{\mathrm{T}}(t)$, applied in a relatively narrow gap with its center at $x = x_\delta \in (-\ell/2, \ell/2)$, the axial excitation field can be expressed as

$$\hat{E}_x^{\mathrm{e}}(x, a, s) = \hat{V}^{\mathrm{T}}(s)\delta(x - x_\delta). \tag{15.18}$$

Upon enforcing the explicit-type boundary condition along the wire, $\hat{E}_x^{\mathrm{s}}(x, a, s) = -\hat{E}_x^{\mathrm{e}}(x, a, s)$, and using (15.18) on the right-hand side of the reciprocity relation (15.3), the interaction integral can carried out at once. Transforming the result of integration to the TD, we end up with

$$\begin{aligned} V^{[S]}(t) &= -V^{\mathrm{T}}(t)\big[\mathrm{H}(x_\delta + \Delta/2 - x_S) - \mathrm{H}(x_\delta - \Delta/2 - x_S)\big] \\ &\text{for all } S = \{1, \ldots, N\}. \end{aligned} \tag{15.19}$$

Equation (15.19) specifies the elements of the excitation array $\boldsymbol{V}_m$ for the delta-gap excitation.

15.1.2.2 Plane-wave excitation

If the transmission line is irradiated by a uniform EM plane wave, say (cf. (12.1))

$$\hat{\boldsymbol{E}}^{\mathrm{i}}(x, y, z, s) = \boldsymbol{\alpha}\,\hat{e}^{\mathrm{i}}(s)\exp[-s(\kappa_0 x + \sigma_0 y - p_0 z)], \tag{15.20}$$

the contribution from the z-component of the excitation field, $\hat{E}_z^{\mathrm{e}} = \hat{E}_z^{\mathrm{e}}(x, y, z, s)$, may not be zero. Therefore, the right-hand side of the reciprocity relation (15.3) must be modified accordingly. This can be done by including additional interaction terms that account for the effect of vertical sections extending along $\{x = \pm\ell/2, -z_0 < z < 0\}$ (see [3, Eq. (11.17)]). The pertaining reciprocity relation then has the following form (cf. (15.3)):

$$\begin{aligned} \int_{x=-\ell/2}^{\ell/2} \hat{E}_x^{\mathrm{T}}(x, a, s)\hat{I}_x^{\mathrm{s}}(x, s)\mathrm{d}x &= -\int_{x=-\ell/2}^{\ell/2} \hat{E}_x^{\mathrm{e}}(x, a, s)\hat{I}_x^{\mathrm{T}}(x, s)\mathrm{d}x \\ &- \hat{I}_1^{\mathrm{T}}(s)\int_{z=-z_0}^{0} \hat{E}_z^{\mathrm{e}}(-\ell/2, 0, z, s)\mathrm{d}z + \hat{I}_2^{\mathrm{T}}(s)\int_{z=-z_0}^{0} \hat{E}_z^{\mathrm{e}}(\ell/2, 0, z, s)\mathrm{d}z, \end{aligned} \tag{15.21}$$

where $\hat{I}^{\mathrm{T}}_{1,2}(s)$ denotes the testing currents at the transmission line terminals at $x = \pm\ell/2$, respectively. With the meaning of the excitation field in mind, its x- and z-components follow as

$$\hat{E}^{\mathrm{e}}_x(x, a, s) = -\alpha_x\,\hat{e}^{\mathrm{i}}(s)\exp(-s\kappa_0 x)[1 - \exp(-2sp_0 z_0)], \tag{15.22a}$$

$$\hat{E}^{\mathrm{e}}_z(x, 0, z, s) = \alpha_z\,\hat{e}^{\mathrm{i}}(s)\exp(-s\kappa_0 x)\{\exp(sp_0 z) + \exp[-sp_0(2z_0 + z)]\}. \tag{15.22b}$$

The integrals on the right-hand side of the reciprocity relation (15.21) can be for the plane-wave excitation fields (15.22a) and (15.22b) evaluated in closed form, again. Consequently, taking the limit $sz_0/c_0 \downarrow 0$ and transforming the result to the TD, we get

$$\begin{aligned} V^{[S]}(t) &= 2z_0\alpha_x p_0\{e^{\mathrm{i}}[t - \kappa_0(x_S - \Delta/2)] - e^{\mathrm{i}}[t - \kappa_0(x_S + \Delta/2)]\}/\kappa_0 \\ &- z_0\alpha_z e^{\mathrm{i}}[t - \kappa_0(x_1 - \Delta/2)]\delta_{S,1} + z_0\alpha_z e^{\mathrm{i}}[t - \kappa_0(x_N + \Delta/2)]\delta_{S,N} \\ &\text{for all } S = \{1, \dots, N\}, \end{aligned} \tag{15.23}$$

where $\delta_{i,k}$ denotes the Kronecker delta ($\delta_{i,k} = 1$ if $i = k$ and $\delta_{i,k} = 0$ otherwise). Finally, if $\kappa_0 = 0$, the limit leads to

$$\begin{aligned} V^{[S]}(t) &= 2z_0\alpha_x p_0\Delta\partial_t e^{\mathrm{i}}(t) - z_0\alpha_z e^{\mathrm{i}}(t)\delta_{S,1} + z_0\alpha_z e^{\mathrm{i}}(t)\delta_{S,N} \\ &\text{for all } S = \{1, \dots, N\}. \end{aligned} \tag{15.24}$$

It is finally emphasized that (15.23) and (15.24) were derived from the excitation fields applying to the ideal PEC ground plane (see (15.22a) and (15.22b)). Their generalization pertaining to the transmission line in the presence of a metasurface with combined magnetic and dielectric properties can be readily found with the help of the results of Section 4.1.

In the last step, with the elements of the impedance and excitation arrays at our disposal, the TD system of equations can be solved. An efficient way to achieve the electric-current coefficients is the marching-on-in-time technique (see (12.27)).

15.1.3 Incorporating termination loads

In order to keep the computational model simple and efficient, passive loads connected to the terminations of the transmission line will be incorporated through a local perturbation of the boundary condition applying to the axial electric-field strength on the wire's surface. Hence, the boundary condition in the s-domain is expressed as

$$\lim_{\varrho \downarrow a} \hat{E}^{\mathrm{s}}_x(x, \varrho, s) = -\hat{E}^{\mathrm{e}}_x(x, a, s) + \hat{\zeta}(s)\delta(x - x_\zeta)\hat{I}^{\mathrm{s}}_x(x, s), \tag{15.25}$$

where $\hat{\zeta}(s)$ denotes the impedance of a lumped circuit element and x_ζ denotes the position of the connected load. With regard to the chosen testing-current distribution (see (15.14)), we take $x_\zeta = -\ell/2 + \Delta/2$ to model the load distributed around

$x = -\ell/2$ or/and $x_\zeta = \ell/2 - \Delta/2$ to model the load distributed around $x = \ell/2$. Substituting the boundary condition (15.25) in (the right-hand side of) the reciprocity relation (15.3), its slowness-domain form follows as (cf. (15.10))

$$\begin{aligned} &-\frac{sZ^{\mathrm{c}}}{c_0}\frac{s}{2\pi\mathrm{i}}\int_{\kappa=-\mathrm{i}\infty}^{\mathrm{i}\infty} c_0^2\Omega_0^2(\kappa)\tilde{I}_x^{\mathrm{T}}(\kappa,s)\tilde{I}_x^{\mathrm{s}}(-\kappa,s)\mathrm{d}\kappa \\ &\qquad -\frac{s\hat{\zeta}(s)}{2\pi\mathrm{i}}\hat{I}_x^{\mathrm{s}}(x_\zeta,s)\int_{\kappa=-\mathrm{i}\infty}^{\mathrm{i}\infty} \tilde{I}_x^{\mathrm{T}}(-\kappa,s)\exp(s\kappa x_\zeta)\mathrm{d}\kappa \\ &\qquad = -\frac{s}{2\pi\mathrm{i}}\int_{\kappa=-\mathrm{i}\infty}^{\mathrm{i}\infty} \tilde{E}_x^{\mathrm{e}}(\kappa,a,s)\tilde{I}_x^{\mathrm{T}}(-\kappa,s)\mathrm{d}\kappa \quad \text{for } \{s\in\mathbb{R}; s>0\}. \end{aligned} \tag{15.26}$$

In the transform-domain reciprocity relation, we use, assuming the spatial distribution shown in Figure 15.2(b), the following expansions:

$$\hat{I}_x^{\mathrm{s}}(x_\zeta,s) \simeq \sum_{k=1}^{M} i_k^{[1]}\hat{\Lambda}_k(s) \ \text{ for } x_\zeta = -\ell/2 + \Delta/2, \tag{15.27a}$$

$$\hat{I}_x^{\mathrm{s}}(x_\zeta,s) \simeq \sum_{k=1}^{M} i_k^{[N]}\hat{\Lambda}_k(s) \ \text{ for } x_\zeta = \ell/2 - \Delta/2. \tag{15.27b}$$

Upon evaluating the second integral on the left-hand side of (15.26), we obtain a 2-D $[N\times N]$ lumped-element impedance array, say $\underline{\hat{\boldsymbol{L}}}$, whose nonzero elements can be described by

$$\hat{L}^{[1,1]}(s) = \frac{\hat{\zeta}_A(s)}{2s^2\Delta t} \ \text{ or/and } \ \hat{L}^{[N,N]}(s) = \frac{\hat{\zeta}_B(s)}{2s^2\Delta t}, \tag{15.28}$$

where we used $\hat{\zeta}(s) = \hat{\zeta}_A(s)$ and $\hat{\zeta}(s) = \hat{\zeta}_B(s)$ to distinguish between the load impedances distributed around $x = -\ell/2$ and $x = \ell/2$, respectively. Thanks to the similarity of the expressions for $\hat{L}^{[1,1]}(s)$ and $\hat{L}^{[N,N]}(s)$, it is sufficient to provide the TD originals for the former one. Thus, the inverse of $\hat{L}^{[1,1]}(s)$ leads to

$$L^{[1,1]}(t) = (\Delta t/4C^{\mathrm{L}})(c_0 t/c_0\Delta t)^2\mathrm{H}(t), \tag{15.29}$$

for a capacitive impedance $\hat{\zeta}_A(s) = 1/sC^{\mathrm{L}}$ and

$$L^{[1,1]}(t) = (R^{\mathrm{L}}/2)(c_0 t/c_0\Delta t)\mathrm{H}(t), \tag{15.30}$$

for a resistive impedance $\hat{\zeta}_A(s) = R^{\mathrm{L}}$, and

$$L^{[1,1]}(t) = (L^{\mathrm{L}}/2\Delta t)\mathrm{H}(t), \tag{15.31}$$

for an inductive impedance $\hat{\zeta}_A(s) = sL^{\mathrm{L}}$. In the final step, the effect of terminal lumped loads is incorporated by replacing $\underline{\boldsymbol{Z}}_k$ with $\underline{\boldsymbol{Z}}_k - \underline{\boldsymbol{L}}_k$ for all $k = \{1,\ldots,M\}$ in the final system of equations (12.25).

15.2 Transmission line above a layer with dielectric and conductive properties

If the electric permittivity or/and conductivity of the thin layer are relatively high, yet not unbounded, the problem formulation as well as its solution presented in Sections 15.1.1 and 15.1.2, respectively, has to be modified. This is exactly the objective of this section, where the effect of bounded dielectric and conductive properties is incorporated in the CdH-MoM solution via the cross-layer conditions (3.4a) and (3.4b). The point of departure for our further analysis is the reciprocity relation (15.3), again, where we modify the expression for (the axial component of) the testing electric-field strength along the thin wire's surface. The testing field is derived under the 2-D wave-slowness representation in the plane parallel to the layer (see (4.75)). Under this representation, the non-vanishing tangential components of the testing fields can be expressed using [5, Eqs. (26.9-2) and (26.9-3)] as

$$\tilde{E}_x^{\mathrm{T}}(\kappa,\sigma,z,s) = -(s/\epsilon_0)\Omega_0^2(\kappa)\tilde{G}(\kappa,\sigma,z,s)\tilde{I}_x^{\mathrm{T}}(\kappa,s), \tag{15.32a}$$

$$\tilde{E}_y^{\mathrm{T}}(\kappa,\sigma,z,s) = (s/\epsilon_0)\kappa\sigma\tilde{G}(\kappa,\sigma,z,s)\tilde{I}_x^{\mathrm{T}}(\kappa,s), \tag{15.32b}$$

$$\tilde{H}_y^{\mathrm{T}}(\kappa,\sigma,z,s) = \tilde{I}_x^{\mathrm{T}}(\kappa,s)\partial_z\tilde{G}(\kappa,\sigma,z,s). \tag{15.32c}$$

The (transform-domain) testing fields expressed via (15.32a)–(15.32c) are further determined with the aid of the pertaining excitation condition expressing the jump of the tangential magnetic-field strength across the testing electric current distributed along the wire

$$\lim_{z\downarrow 0}\tilde{H}_y^{\mathrm{T}}(\kappa,\sigma,z,s) - \lim_{z\uparrow 0}\tilde{H}_y^{\mathrm{T}}(\kappa,\sigma,z,s) = -\tilde{I}_x^{\mathrm{T}}(\kappa,s). \tag{15.33a}$$

Furthermore, across the layer, the testing field satisfies

$$\lim_{z\downarrow -z_0}\tilde{E}_x^{\mathrm{T}}(\kappa,\sigma,z,s) - \lim_{z\uparrow -z_0}\tilde{E}_x^{\mathrm{T}}(\kappa,\sigma,z,s) = 0, \tag{15.33b}$$

$$\lim_{z\downarrow -z_0}\tilde{H}_y^{\mathrm{T}}(\kappa,\sigma,z,s) - \lim_{z\uparrow -z_0}\tilde{H}_y^{\mathrm{T}}(\kappa,\sigma,z,s) = -\hat{Y}^{\mathrm{E}}(s)\tilde{E}_x^{\mathrm{T}}(\kappa,\sigma,-z_0,s), \tag{15.33c}$$

where $\hat{Y}^{\mathrm{E}}(s)$ denotes the equivalent layer's admittance (see (3.15b)). Now, combining (15.33a)–(15.33c) with (15.32a)–(15.32c), the conditions applying to the (transform-domain) Green's function immediately follow. Indeed, the excitation condition (15.33a) leads to

$$\lim_{z\downarrow 0}\partial_z\tilde{G}(\kappa,\sigma,z,s) - \lim_{z\uparrow 0}\partial_z\tilde{G}(\kappa,\sigma,z,s) = -1. \tag{15.34a}$$

Furthermore, the requirement for continuity of both $\tilde{E}_x^{\mathrm{T}}$ and $\tilde{E}_y^{\mathrm{T}}$ across the layer is met if the Green's function satisfies

$$\lim_{z\downarrow -z_0}\tilde{G}(\kappa,\sigma,z,s) - \lim_{z\uparrow -z_0}\tilde{G}(\kappa,\sigma,z,s) = 0. \tag{15.34b}$$

And, finally, the combination of jump condition (15.33c) with (15.32a) and (15.32c) gives

$$\lim_{z\downarrow -z_0} \partial_z \tilde{G}(\kappa,\sigma,z,s) - \lim_{z\uparrow -z_0} \partial_z \tilde{G}(\kappa,\sigma,z,s) = (s/c_0)\big[\hat{Y}^{\mathrm{E}}(s)/\eta_0\big] c_0^2\Omega_0^2(\kappa)\tilde{G}(\kappa,\sigma,-z_0,s). \qquad (15.34\mathrm{c})$$

After carrying out a few steps of algebra, the transform-domain Green's function satisfying conditions (15.34a)–(15.34c) can be at the level of the conductor, $z = 0$, written as

$$\tilde{G}(\kappa,\sigma,0,s) = \frac{1-\exp[-2sp(\kappa,\sigma)z_0]}{2sp(\kappa,\sigma)} + \frac{1}{2sp(\kappa,\sigma)} \frac{\exp[-2sp(\kappa,\sigma)z_0]}{1+\big[\hat{Y}^{\mathrm{E}}(s)/2\eta_0\big]\big[c_0^2\Omega_0^2(\kappa)/c_0 p(\kappa,\sigma)\big]}, \qquad (15.35)$$

where $p(\kappa,\sigma)$ has the meaning of slowness in the z-direction (see (4.77)). The first term on the right-hand side of (15.35) represents the Green's function pertaining to the conductor above a PEC plane. As a matter of fact, this term is the Fourier-transform image of (15.6) with respect to the y-coordinate. Consequently, the second term can be interpreted as a correction term (with respect to the ideal PEC plane), say $\Delta^{\mathrm{E}}\tilde{G}$, that accounts for the bounded value of the layer's admittance. Making use of (15.32a), we may then express the corresponding correction of the axial electric-field strength along the conductor

$$\Delta^{\mathrm{E}}\tilde{E}_x^{\mathrm{T}}(\kappa,\sigma,0,s) = -\frac{\tilde{I}_x^{\mathrm{T}}(\kappa,s)\exp[-2sp(\kappa,\sigma)z_0]}{2\eta_0\big[c_0 p(\kappa,\sigma)/c_0^2\Omega_0^2(\kappa)\big]+\hat{Y}^{\mathrm{E}}(s)}. \qquad (15.36)$$

As expected, the correction field, $\Delta^{\mathrm{E}}\tilde{E}_x^{\mathrm{T}}$, goes to zero as $\hat{Y}^{\mathrm{E}}(s)\to\infty$ that represents the PEC plane. In the following step, the wave-slowness representation (4.75) with (15.36) are combined to express the s-domain correction field along the wire. The thus obtained expression for $\Delta^{\mathrm{E}}\hat{E}_x^{\mathrm{T}}(x,0,0,s)$ is subsequently used to express the pertaining interaction integral (see (15.3)) in the form of complex slowness integrals. Upon interchanging the order of the integrations with respect to x and κ, we arrive at

$$\int_{x=-\ell/2}^{\ell/2} \Delta^{\mathrm{E}}\hat{E}_x^{\mathrm{T}}(x,0,0,s)\hat{I}_x^{\mathrm{s}}(x,s)\mathrm{d}x = -\left(\frac{s}{2\mathrm{i}\pi}\right)^2 \int_{\kappa=-\mathrm{i}\infty}^{\mathrm{i}\infty} \tilde{I}_x^{\mathrm{T}}(\kappa,s)\tilde{I}_x^{\mathrm{s}}(-\kappa,s)\mathrm{d}\kappa \times \int_{\sigma=-\mathrm{i}\infty}^{\mathrm{i}\infty} \exp[-2sp(\kappa,\sigma)z_0]\frac{\mathrm{d}\sigma}{2\eta_0\big[c_0 p(\kappa,\sigma)/c_0^2\Omega_0^2(\kappa)\big]+\hat{Y}^{\mathrm{E}}(s)}. \qquad (15.37)$$

In the ensuing steps, we shall evaluate, in an approximate manner, the inner integral with respect to σ. Hence, we first define

$$\hat{\zeta}^{\mathrm{E}} = \frac{s}{2\mathrm{i}\pi}\int_{\sigma=-\mathrm{i}\infty}^{\mathrm{i}\infty} \exp[-2sp(\kappa,\sigma)z_0]\frac{\mathrm{d}\sigma}{2\eta_0\big[c_0 p(\kappa,\sigma)/c_0^2\Omega_0^2(\kappa)\big]+\hat{Y}^{\mathrm{E}}(s)}. \qquad (15.38)$$

The integration in the complex σ-plane is further mapped into a Laplace-type integral with respect to a real-valued and positive parameter τ. For this reason, we substitute

$$2p(\kappa,\sigma)z_0 = \tau \quad \text{for } \{\tau \in \mathbb{R}; \tau > 0\}. \tag{15.39}$$

Under the mapping, upon combining the integrations in the lower and upper complex σ-half planes, we arrive at

$$\hat{\zeta}^{\mathrm{E}} = \frac{s\mu_0}{2\pi} c_0^2\Omega_0^2(\kappa) \int_{\tau=2z_0\Omega_0(\kappa)}^{\infty} \exp(-s\tau) \frac{\tau}{\tau + (z_0/c_0)\left[\hat{Y}^{\mathrm{E}}(s)/\eta_0\right] c_0^2\Omega_0^2(\kappa)} \times \frac{\mathrm{d}\tau}{\left[\tau^2 - 4z_0^2\Omega_0^2(\kappa)\right]^{1/2}}. \tag{15.40}$$

In order to carry out the integration with respect to τ in closed form, we take the limit $2z_0\Omega_0 \downarrow 0$ and make use of the integral formula [6, Eq. (5.1.28)]. Under this approximation, it is then found that

$$\hat{\zeta}^{\mathrm{E}} \simeq \frac{s\mu_0}{2\pi} c_0^2\Omega_0^2(\kappa) \mathrm{E}_1\left[\frac{sz_0}{c_0}\frac{\hat{Y}^{\mathrm{E}}(s)}{\eta_0} c_0^2\Omega_0^2(\kappa)\right] \exp\left[\frac{sz_0}{c_0}\frac{\hat{Y}^{\mathrm{E}}(s)}{\eta_0} c_0^2\Omega_0^2(\kappa)\right], \tag{15.41}$$

as $2z_0\Omega_0 \downarrow 0$, where $\mathrm{E}_1(x)$ denotes the exponential integral function [6, Eq. (5.1.1)]. If the equivalent layer's admittance is relatively large, the corresponding dominant term of (15.41) follows upon applying [6, Eq. (5.1.51)] as

$$\hat{\zeta}^{\mathrm{E}} \simeq \frac{1}{2\pi z_0}\frac{1}{\hat{Y}^{\mathrm{E}}(s)}, \tag{15.42}$$

as $2z_0\Omega_0 \downarrow 0$ and $\hat{Y}^{\mathrm{E}}(s)/\eta_0 \to \infty$. Substitution of this approximation in (15.37) leads to a straightforward expression for the correction interaction integral

$$\int_{x=-\ell/2}^{\ell/2} \Delta^{\mathrm{E}}\hat{E}_x^{\mathrm{T}}(x,0,0,s)\hat{I}_x^{\mathrm{s}}(x,s)\mathrm{d}x \simeq -\frac{s\hat{\zeta}^{\mathrm{E}}}{2\pi\mathrm{i}} \int_{\kappa=-\mathrm{i}\infty}^{\mathrm{i}\infty} \tilde{I}_x^{\mathrm{T}}(\kappa,s)\tilde{I}_x^{\mathrm{s}}(-\kappa,s)\mathrm{d}\kappa. \tag{15.43}$$

In the final step, the transform-domain counterparts of (15.12) and (15.14) are substituted in the right-hand side of (15.43) and the resulting integrals are evaluated analytically with the aid of Cauchy's formula [7, Section 2.41], again. Transforming the obtained s-domain expressions to the TD, we end up with the following elements of the TD correction impedance array (cf. (15.15a)):

$$\begin{aligned}\Delta^{\mathrm{E}}Z^{[S,n]}(t) &= -\Phi^A(x_S - x_n + 3\Delta/2, t) + 3\Phi^A(x_S - x_n + \Delta/2, t) \\ &\quad - 3\Phi^A(x_S - x_n - \Delta/2, t) + \Phi^A(x_S - x_n - 3\Delta/2, t) \\ &\quad \text{for all } S = \{1,\dots,N\} \text{ and } n = \{1,\dots,N\} \text{ if open-circuited, or} \\ &\quad \text{for all } S = \{1,\dots,N\} \text{ and } n = \{2,\dots,N-1\} \text{ if loaded,}\end{aligned} \tag{15.44a}$$

at its end points. In the latter circumstance, the TD correction impedance array elements follow from (cf. (15.15b) and (15.15c))

$$\begin{aligned}\Delta^{\mathrm{E}} Z^{[S,1]}(t) = {}& \Phi^A(x_S - x_1 + \Delta/2, t) - 2\Phi^A(x_S - x_1 - \Delta/2, t)\\ &+ \Phi^A(x_S - x_1 - 3\Delta/2, t) - \Phi^B(x_S - x_1 + 3\Delta/2, t)\\ &+ \Phi^B(x_S - x_1 + \Delta/2, t) \quad \text{for all } S = \{1, \dots, N\},\end{aligned} \tag{15.44b}$$

and, finally,

$$\begin{aligned}\Delta^{\mathrm{E}} Z^{[S,N]}(t) = {}& -\Phi^A(x_S - x_N + 3\Delta/2, t) + 2\Phi^A(x_S - x_N + \Delta/2, t)\\ &- \Phi^A(x_S - x_N - \Delta/2, t) + \Phi^B(x_S - x_N - \Delta/2, t)\\ &- \Phi^B(x_S - x_N - 3\Delta/2, t) \quad \text{for all } S = \{1, \dots, N\}.\end{aligned} \tag{15.44c}$$

The TD functions in (15.44a)–(15.44c) are given simply by

$$\Phi^A(x, t) = \mathcal{R}^{\mathrm{E}}(t)\,(x^2/2\Delta c_0 \Delta t)\mathrm{H}(x), \tag{15.45a}$$

$$\Phi^B(x, t) = \mathcal{R}^{\mathrm{E}}(t)\,(x/c_0 \Delta t)\mathrm{H}(x), \tag{15.45b}$$

and $\mathcal{R}^{\mathrm{E}}(t)$ corresponds, in fact, to the original of $c_0\hat{\zeta}^{\mathrm{E}}/s^2$. Hence, it has the following form:

$$\mathcal{R}^{\mathrm{E}}(t) = \frac{c_0}{2\pi z_0}\frac{\alpha^{\mathrm{E}} t - 1 + \exp(-\alpha^{\mathrm{E}} t)}{\alpha^{\mathrm{E}}\, G^{\mathrm{E}}}\mathrm{H}(t) \quad \text{for all } t \in \mathbb{R}, \tag{15.46}$$

with $\alpha^{\mathrm{E}} = G^{\mathrm{E}}/C^{\mathrm{E}}$ (see (3.19)). If $\alpha^{\mathrm{E}} \downarrow 0$, which is a characteristic of a loss-free dielectric sheet, we get $\mathcal{R}^{\mathrm{E}}(t) = (1/4\pi z_0)(c_0^2 t^2/c_0 C^{\mathrm{E}})\mathrm{H}(t)$. On the other hand, taking the dominant term in the limit $t \to \infty$, we find $\mathcal{R}^{\mathrm{E}}(t) = (1/2\pi z_0)(c_0 t/G^{\mathrm{E}})\mathrm{H}(t)$, which corresponds to a purely conductive layer. In the last step, the effect of a finite layer's admittance is incorporated by replacing $\underline{\boldsymbol{Z}}_k$ with $\underline{\boldsymbol{Z}}_k + \Delta^{\mathrm{E}}\underline{\boldsymbol{Z}}_k$ for all $k = \{1, \dots, M\}$ in the final system of equations (12.25).

15.3 Transmission line above a metasurface

The problem under consideration is next extended by analyzing the pulse-excited transmission line above a metasurface with combined magnetic and dielectric properties (see Figure 15.1). The starting point is the s-domain reciprocity relation (15.3), again, the testing field in which is modified accordingly. Hence, adopting further the transform-domain testing-field expressions (15.32a)–(15.32c) with the excitation condition (15.33a), the cross-layer conditions (15.33b) and (15.33c) are generalized to account for the effect of both electric and magnetic currents in the layer, that is (cf. (15.33b) and (15.33c)),

$$\lim_{z \downarrow -z_0} \tilde{E}_x^{\mathrm{T}}(\kappa, \sigma, z, s) - \lim_{z \uparrow -z_0} \tilde{E}_x^{\mathrm{T}}(\kappa, \sigma, z, s) = -\hat{Z}^{\mathrm{M}}(s)\langle \tilde{H}_y^{\mathrm{T}} \rangle_-^+(\kappa, \sigma, s), \tag{15.47a}$$

$$\lim_{z \downarrow -z_0} \tilde{H}_y^{\mathrm{T}}(\kappa, \sigma, z, s) - \lim_{z \uparrow -z_0} \tilde{H}_y^{\mathrm{T}}(\kappa, \sigma, z, s) = -\hat{Y}^{\mathrm{E}}(s)\langle \tilde{E}_x^{\mathrm{T}} \rangle_-^+(\kappa, \sigma, s), \tag{15.47b}$$

where the equivalent layer impedance and admittance, $\hat{Z}^{\rm M}(s)$ and $\hat{Y}^{\rm E}(s)$, are defined by (3.15a) and (3.15b), respectively, and recall that the meaning of $\langle\rangle^{+}_{-}$ is specified by (3.11). Consequently, the transform-domain Green's function satisfies the excitation condition (15.34a) and the following saltus-type conditions (cf. (15.34b) and (15.34c)):

$$\lim_{z\downarrow -z_0}\tilde{G}(\kappa,\sigma,z,s)-\lim_{z\uparrow -z_0}\tilde{G}(\kappa,\sigma,z,s)=\frac{c_0}{s}\frac{\hat{Z}^{\rm M}(s)/\zeta_0}{c_0^2\Omega_0^2(\kappa)}\langle\partial_z\tilde{G}\rangle^{+}_{-}(\kappa,\sigma,s),\tag{15.48a}$$

$$\lim_{z\downarrow -z_0}\partial_z\tilde{G}(\kappa,\sigma,z,s)-\lim_{z\uparrow -z_0}\partial_z\tilde{G}(\kappa,\sigma,z,s)=\frac{s}{c_0}\frac{\hat{Y}^{\rm E}(s)/\eta_0}{c_0^{-2}\Omega_0^{-2}(\kappa)}\langle\tilde{G}\rangle^{+}_{-}(\kappa,\sigma,s).\tag{15.48b}$$

Through straightforward algebra, the pertinent (transform-domain) Green's function at the level of conductor, $z=0$, can be expressed in the following form (cf. (15.35)):

$$\begin{aligned}\tilde{G}(\kappa,\sigma,0,s)=&\frac{1-\exp[-2sp(\kappa,\sigma)z_0]}{2sp(\kappa,\sigma)}\\&+\frac{1}{2sp(\kappa,\sigma)}\frac{\exp[-2sp(\kappa,\sigma)z_0]}{1+\left[\hat{Y}^{\rm E}(s)/2\eta_0\right]\left[c_0^2\Omega_0^2(\kappa)/c_0p(\kappa,\sigma)\right]}\\&+\frac{1}{2sp(\kappa,\sigma)}\frac{\left[\hat{Z}^{\rm M}(s)/2\zeta_0\right]\left[c_0p(\kappa,\sigma)/c_0^2\Omega_0^2(\kappa)\right]}{1+\left[\hat{Z}^{\rm M}(s)/2\zeta_0\right]\left[c_0p(\kappa,\sigma)/c_0^2\Omega_0^2(\kappa)\right]}\exp[-2sp(\kappa,\sigma)z_0].\end{aligned}\tag{15.49}$$

Likewise in (15.35), the first term in (15.49) corresponds to the Green's function pertaining to the conductor above a PEC plane. The second term has been, in Section 15.2, interpreted as the correction Green's function (with respect to the ideal PEC plane), $\Delta^{\rm E}\tilde{G}$, that incorporates the effect of the (bounded) equivalent layer's admittance $\hat{Y}^{\rm E}(s)$. Following the same lines of reasoning, the last term can be then viewed as correction, say $\Delta^{\rm M}\tilde{G}$, with respect to the layer with dielectric and conductive properties. It should be noted that as both $\hat{Y}^{\rm E}(s)\to\infty$ and $\hat{Z}^{\rm M}(s)\to\infty$, $\tilde{G}(\kappa,\sigma,0,s)=1/2sp(\kappa,\sigma)$, which represents, as a matter of fact, the transform-domain (free-space) Green's function in the absence of layer. This property is in congruence with the effect of EM transparency as discussed and demonstrated in Chapter 4. Via (15.32a), the corresponding correction of the axial field component along the conductor then follows as (cf. (15.36))

$$\Delta^{\rm M}\tilde{E}_x^{\rm T}(\kappa,\sigma,0,s)=-\frac{[\hat{Z}^{\rm M}(s)/4]\tilde{I}_x^{\rm T}(\kappa,s)\exp[-2sp(\kappa,\sigma)z_0]}{1+[\hat{Z}^{\rm M}(s)/2\zeta_0][c_0p(\kappa,\sigma)/c_0^2\Omega_0^2(\kappa)]}.\tag{15.50}$$

Note that $\Delta^{\rm M}\tilde{E}_x^{\rm T}$ goes to zero as $\hat{Z}^{\rm M}(s)\to 0$, as was expected. In the next step, (15.50) is combined with the slowness representation to get the field correction

along the line, $\Delta^{\mathrm{M}}\hat{E}_x^{\mathrm{T}}(x,0,0,s)$. The resulting slowness representation is subsequently used to express the pertaining correction interaction integral, which leads to (cf. (15.37))

$$\int_{x=-\ell/2}^{\ell/2} \Delta^{\mathrm{M}}\hat{E}_x^{\mathrm{T}}(x,0,0,s)\hat{I}_x^{\mathrm{s}}(x,s)\mathrm{d}x = -\left(\frac{s}{2\mathrm{i}\pi}\right)^2 \int_{\kappa=-\mathrm{i}\infty}^{\mathrm{i}\infty} \tilde{I}_x^{\mathrm{T}}(\kappa,s)\tilde{I}_x^{\mathrm{s}}(-\kappa,s)\mathrm{d}\kappa$$
$$\times \int_{\sigma=-\mathrm{i}\infty}^{\mathrm{i}\infty} \exp[-2sp(\kappa,\sigma)z_0]\frac{[\hat{Z}^{\mathrm{M}}(s)/4]\mathrm{d}\sigma}{1+[\hat{Z}^{\mathrm{M}}(s)/2\zeta_0][c_0 p(\kappa,\sigma)/c_0^2\Omega_0^2(\kappa)]}. \tag{15.51}$$

The inner integral with respect to σ will be carried out approximately, again. To that end, we first define

$$\hat{\zeta}^{\mathrm{M}} = \frac{s}{2\mathrm{i}\pi}\int_{\sigma=-\mathrm{i}\infty}^{\mathrm{i}\infty} \exp[-2sp(\kappa,\sigma)z_0]\frac{[\hat{Z}^{\mathrm{M}}(s)/4]\mathrm{d}\sigma}{1+[\hat{Z}^{\mathrm{M}}(s)/2\zeta_0][c_0 p(\kappa,\sigma)/c_0^2\Omega_0^2(\kappa)]} \tag{15.52}$$

that has the form similar to (15.38). Accordingly, we may adopt the procedure that led to the approximate closed-form (15.41) and get

$$\hat{\zeta}^{\mathrm{M}} \simeq \frac{s\mu_0}{2\pi}c_0^2\Omega_0^2(\kappa)\mathrm{E}_1\left[\frac{sz_0}{c_0}\frac{4\zeta_0}{\hat{Z}^{\mathrm{M}}(s)}c_0^2\Omega_0^2(\kappa)\right]\exp\left[\frac{sz_0}{c_0}\frac{4\zeta_0}{\hat{Z}^{\mathrm{M}}(s)}c_0^2\Omega_0^2(\kappa)\right], \tag{15.53}$$

as $2z_0\Omega_0 \downarrow 0$. For a low-permeability layer, the relative layer's impedance, $\hat{Z}^{\mathrm{M}}(s)/\zeta_0$, is relatively small, so that the large-argument asymptotic expansion [6, Eq. (5.1.51)] applies. In such a case, we arrive at (cf. (15.42))

$$\hat{\zeta}^{\mathrm{M}} \simeq \frac{1}{2\pi z_0}\frac{\hat{Z}^{\mathrm{M}}(s)}{4}, \tag{15.54}$$

as $2z_0\Omega_0 \downarrow 0$ and $\hat{Z}^{\mathrm{M}}(s)/\zeta_0 \to 0$. On the other hand, if the equivalent layer's impedance is relatively large, the small-argument series expansion [6, Eq. (5.1.11)] reveals the logarithmic singularity of $\hat{\zeta}^{\mathrm{M}}$. For the sake of generality, the pertaining correction impedance array is derived using the closed-form approximation (15.53). Consequently, assuming vanishing magnetic losses, and hence $\hat{Z}^{\mathrm{M}}(s) = sL^{\mathrm{M}}$, the correction interaction integral reads (cf. (15.43))

$$\int_{x=-\ell/2}^{\ell/2} \Delta^{\mathrm{M}}\hat{E}_x^{\mathrm{T}}(x,0,0,s)\hat{I}_x^{\mathrm{s}}(x,s)\mathrm{d}x \simeq -\frac{s\mu_0}{2\pi}\frac{s}{2\pi\mathrm{i}}\int_{\kappa=-\mathrm{i}\infty}^{\mathrm{i}\infty} c_0^2\Omega_0^2(\kappa)$$
$$\times \mathrm{E}_1\left[\beta_0 c_0^2\Omega_0^2(\kappa)\right]\exp\left[\beta_0 c_0^2\Omega_0^2(\kappa)\right]\tilde{I}_x^{\mathrm{T}}(\kappa,s)\tilde{I}_x^{\mathrm{s}}(-\kappa,s)\mathrm{d}\kappa, \tag{15.55}$$

where we used the parameter $\beta_0 = (4z_0/c_0)(\zeta_0/L^{\mathrm{M}}) = 4z_0\mu_0/L^{\mathrm{M}}$. With the transform-domain counterparts of (15.12) and (15.14), the integral with respect to κ can be readily carried out analytically using Cauchy's formula [7, Section 2.41].

This way leads to an s-domain expression, the inverse of which yields the following elements of the TD correction impedance array (cf. (15.15a)):

$$\Delta^{\mathrm{M}} Z^{[S,n]}(t) = \frac{Z^{\mathrm{r}}}{c_0 \Delta t} \frac{1}{\Delta}\Big[\bar{\Upsilon}^A(x_S - x_n + 3\Delta/2, t) - 3\bar{\Upsilon}^A(x_S - x_n + \Delta/2, t) + 3\bar{\Upsilon}^A(x_S - x_n - \Delta/2, t) - \bar{\Upsilon}^A(x_S - x_n - 3\Delta/2, t)\Big]$$

$$\text{for all } S = \{1, \ldots, N\} \text{ and } n = \{1, \ldots, N\} \text{ if open-circuited, or for all } S = \{1, \ldots, N\} \text{ and } n = \{2, \ldots, N-1\} \text{ if loaded,} \tag{15.56a}$$

at its end points with $Z^{\mathrm{r}} = (\zeta_0/2\pi)\mathrm{E}_1(\beta_0)\exp(\beta_0)$. In the latter case, we further use (cf. (15.15b) and (15.15c))

$$\Delta^{\mathrm{M}} Z^{[S,1]}(t) = -\frac{Z^{\mathrm{r}}}{c_0 \Delta t} \frac{1}{\Delta}\Big[\bar{\Upsilon}^A(x_S - x_1 + \Delta/2, t) - 2\bar{\Upsilon}^A(x_S - x_1 - \Delta/2, t) + \bar{\Upsilon}^A(x_S - x_1 - 3\Delta/2, t)\Big] + \frac{Z^{\mathrm{r}}}{c_0 \Delta t}\Big[\Upsilon^B(x_S - x_1 + 3\Delta/2, t) - \Upsilon^B(x_S - x_1 + \Delta/2, t)\Big] \quad \text{for all } S = \{1, \ldots, N\}, \tag{15.56b}$$

and, finally,

$$\Delta^{\mathrm{M}} Z^{[S,N]}(t) = \frac{Z^{\mathrm{r}}}{c_0 \Delta t} \frac{1}{\Delta}\Big[\bar{\Upsilon}^A(x_S - x_N + 3\Delta/2, t) - 2\bar{\Upsilon}^A(x_S - x_N + \Delta/2, t) + \bar{\Upsilon}^A(x_S - x_N - \Delta/2, t)\Big] - \frac{Z^{\mathrm{r}}}{c_0 \Delta t}\Big[\Upsilon^B(x_S - x_N - \Delta/2, t) - \Upsilon^B(x_S - x_N - 3\Delta/2, t)\Big] \quad \text{for all } S = \{1, \ldots, N\}. \tag{15.56c}$$

The TD function that occurs in (15.56a)–(15.56c) has the following form:

$$\bar{\Upsilon}^A(x, t) = \frac{1}{2}\left[c_0^2 t^2 \frac{(1+\beta_0)\mathrm{E}_1(\beta_0)\exp(\beta_0) - 1}{\mathrm{E}_1(\beta_0)\exp(\beta_0)} - x^2\right]\mathrm{H}(x)\mathrm{H}(t), \tag{15.57}$$

and $\Upsilon^B(x, t)$ is given by (15.17b). It is worth noting that $\bar{\Upsilon}^A(x, t) \to \Upsilon^A(x, t)$ (see (15.17b)) as $\beta_0 \downarrow 0$. This applies to a high-permeability layer for which $Z^{\mathrm{r}} \to -(\zeta_0/2\pi)[\log(\beta_0) + \gamma]$ with $\gamma = 0.57721\ldots$ [6, Eq. (5.1.11)]. In the last step, again, the effect of a finite layer's impedance is incorporated by replacing $\underline{\boldsymbol{Z}}_k$ with $\underline{\boldsymbol{Z}}_k + \Delta^{\mathrm{M}}\underline{\boldsymbol{Z}}_k$ for all $k = \{1, \ldots, M\}$ in the final system of equations (12.25).

15.3.1 Illustrative examples

First, to validate the computational approach presented in Section 15.1, we shall calculate the space–time distribution along a gap-excited, short-circuited transmission line above the PEC plane. The transmission line of length $\ell > 0$ and radius $a = \ell/200$ is located at $z_0 = \ell/50$ so that $Z^{\mathrm{c}} = (\zeta_0/2\pi)\log(8) \simeq 124.7\,\Omega$.

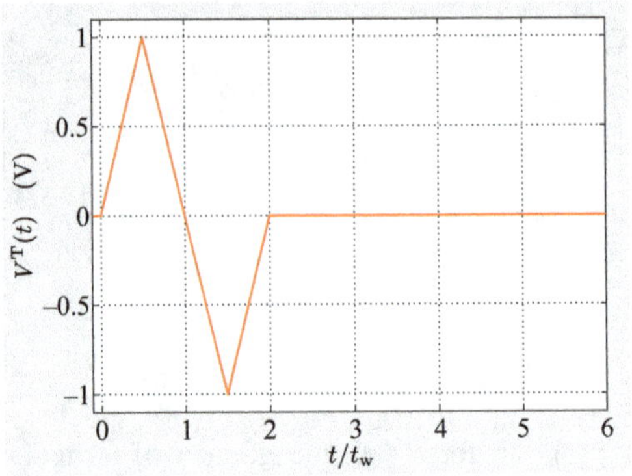

Figure 15.3 Exciting voltage pulse shape

The transmission line is supposed to be activated at its center, $x = x_\delta = 0$, by the voltage pulse having the shape of a bipolar triangle, i.e. (cf. (8.15))

$$V^{\mathrm{T}}(t) = \frac{2V_{\mathrm{m}}}{t_{\mathrm{w}}}\left[t\,\mathrm{H}(t) - 2\left(t - \frac{t_{\mathrm{w}}}{2}\right)\mathrm{H}\left(t - \frac{t_{\mathrm{w}}}{2}\right)\right.$$
$$\left. + 2\left(t - \frac{3t_{\mathrm{w}}}{2}\right)\mathrm{H}\left(t - \frac{3t_{\mathrm{w}}}{2}\right) - (t - 2t_{\mathrm{w}})\,\mathrm{H}(t - 2t_{\mathrm{w}})\right], \tag{15.58}$$

with $c_0 t_{\mathrm{w}} = 2.0\,\ell$ and $V_{\mathrm{m}} = 1.0\,\mathrm{V}$ (see Figure 15.3).

The corresponding CdH-MoM solution (see Section 15.1.2) will be compared with an analytical one, achieved by solving the following transmission-line equations:

$$\partial_x \hat{V}^{\mathrm{s}}(x,s) + \hat{\zeta}(s)\hat{I}^{\mathrm{s}}_x(x,s) = \hat{V}^{\mathrm{T}}(s)\delta(x - x_\delta), \tag{15.59a}$$
$$\partial_x \hat{I}^{\mathrm{s}}_x(x,s) + \hat{\eta}(s)\hat{V}^{\mathrm{s}}(x,s) = 0, \tag{15.59b}$$

where $\hat{\eta}(s) = 2\pi s\epsilon_0/\log(2z_0/a)$ and $\hat{\zeta}(s) = s\mu_0\log(2z_0/a)/2\pi$, with the short-circuit end conditions represented by

$$\hat{V}^{\mathrm{s}}(x = \pm\ell/2, s) = 0. \tag{15.60}$$

Adopting the solution methodology described in [3, p. 134], the pertaining TD solution can be expressed as a sum of causal TD constituents

$$I^{\mathrm{s}}_x(x,t) = \sum_{n=0}^{\infty} I^{[n]}(x,t), \tag{15.61}$$

where

$$I^{[n]}(x,t) = I^{\mathrm{T};[n]}[t - (x - x_\delta)/c_0] + I^{\mathrm{T};[n]}[t - (\ell - x - x_\delta)/c_0]$$
$$+ I^{\mathrm{T};[n]}[t - (2\ell - x + x_\delta)/c_0] + I^{\mathrm{T};[n]}[t - (\ell + x + x_\delta)/c_0]$$
$$\text{for } \{x_\delta \le x \le \ell/2\}, \tag{15.62a}$$

and

$$I^{[n]}(x,t) = I^{\mathrm{T};[n]}[t-(x_\delta - x)/c_0] + I^{\mathrm{T};[n]}[t-(\ell + x + x_\delta)/c_0] + I^{\mathrm{T};[n]}[t-(2\ell + x - x_\delta)/c_0] + I^{\mathrm{T};[n]}[t-(\ell - x - x_\delta)/c_0] \quad \text{for } \{-\ell/2 \le x \le x_\delta\}, \tag{15.62b}$$

with

$$I^{\mathrm{T};[n]}(t) = V^{\mathrm{T}}(t - 2n\ell/c_0)/2Z^{\mathrm{c}}. \tag{15.63}$$

Apparently, since $V^{\mathrm{T}}(t) = 0$ for $t \le 0$, the number of terms in (15.61) is always finite. Figure 15.4 shows the resulting electric-current pulse shape observed at the short-circuited end at $x = -\ell/2$. To calculate the TD response via the CdH-MoM, we used $N = 19$ discretization points (see Figure 15.2(b)). The apparent discrepancy in the pulses' amplitude can be easily decreased by using a higher grid density.

In the second example, we analyze the TD response of the gap-excited transmission line, again, but now over a non-perfectly conductive thin layer. In addition, the conductor is in this case terminated by the characteristic impedance Z^{c} at its both ends. The EM properties of the layer are described by the relative layer's conductance $G^{\mathrm{E}}/\eta_0 = 10$. The calculated electric-current pulses at the center of the conductor are shown in Figure 15.5. It is seen that the pulse shapes calculated via the CdH-MoM and the finite-integration technique (FIT) as implemented in CST Microwave Studio® agree very well. To demonstrate the impact of the finite layer's conductivity on the induced electric current, we have also plotted the corresponding response above the PEC layer, that is, under the limit $G^{\mathrm{E}} \to \infty$.

In the final example, we shall investigate the TD response of the gap-excited transmission line that is located above a thin, highly contrasting metasurface with combined magneto-dielectric properties. In this case, the conductor is assumed to be open-circuited at its terminals. The metasurface is characterized by parameters

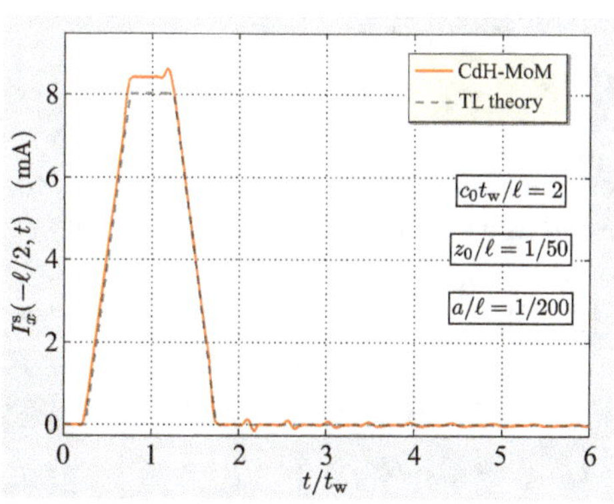

Figure 15.4 Electric current flowing across the short-circuited end ($x = -\ell/2$) of the transmission line above a PEC ground

$C^{\mathrm{E}}/\eta_0 t_{\mathrm{w}} = 10$ and $L^{\mathrm{M}}\eta_0/t_{\mathrm{w}} = 2.0$. Figure 15.6 shows the pertaining electric-current TD response at $x = 0$ as calculated via the CdH-MoM with $N = 19$. It is interesting to observe that (the initial part of) the pulse closely resembles the self-response of the gap-excited wire antenna in the absence of the layer (void) as analyzed via the FIT. This result clearly illustrates, in agreement with the pertaining conclusions drawn in Chapter 4, that the layer exhibiting a relatively high contrast in both electric permittivity and magnetic permeability (with respect to the surrounding medium) becomes electromagnetically transparent. Finally, to illustrate the effect of bounded constitutive parameters, the electric-current self-response of a gap-excited, open-circuited transmission line above a PEC ground plane has also been evaluated using the analytical result based on the transmission-line theory [3, Eqs. (17.27) and (17.28)]. The corresponding result is presented in Figure 15.6.

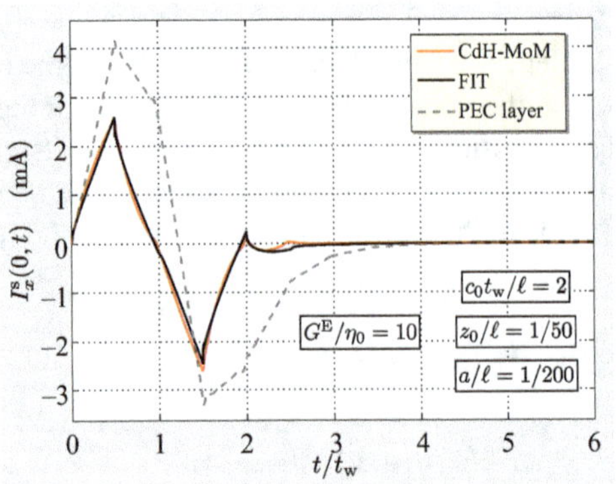

Figure 15.5 Electric current at the center ($x = 0$) of the matched transmission line above a conductive layer with $G^{\mathrm{E}}/\eta_0 = 10$

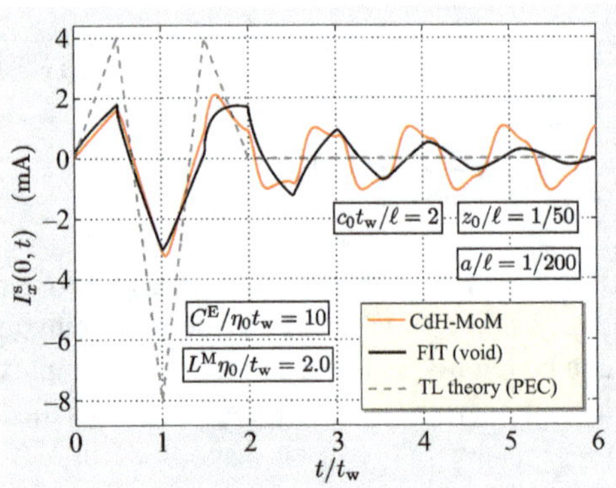

Figure 15.6 Electric current at the center ($x = 0$) of the open-circuited transmission line above a metasurface with $C^{\mathrm{E}}/\eta_0 t_{\mathrm{w}} = 10$ and $L^{\mathrm{M}}\eta_0/t_{\mathrm{w}} = 2.0$

15.4 Concluding remarks

In this concluding section, it is demonstrated that the CdH-MoM can be readily applied to analyzing the TD response of a multiconductor transmission line. Moreover, we shall briefly discuss how the effect of a lossy ground can be incorporated in the computational methodology. Finally, it will be shown that the introduced CdH-MoM solutions can be further generalized by removing some assumptions of the conventional transmission-line theory.

15.4.1 Multiconductor transmission line

The CdH-MoM computational approach can be easily generalized to calculate the induced electric-current space–time distribution along a multiconductor transmission line above an infinite PEC ground. Without loss of generality, we shall describe the EM interaction between two parallel, thin conductors extending along $\mathcal{W}_{\mathrm{A}} = \{-\ell_A/2 < x < \ell_A/2, y = 0, z = 0\}$ and $\mathcal{W}_{\mathrm{B}} = \{-\ell_B/2 < x - x_0 < \ell_B/2, y = y_0, z = 0\}$. Hence, the conductors are located at the height $z_0 > 0$ above the reference plane and at the distance $|y_0| > 0$ far apart. For the sake of simplicity, the transmission lines are assumed to be open-circuited at its ends. Denoting their (relatively small) radius by $a > 0$, again, application of the reciprocity theorem of the time-convolution leads to the following pair of relations (cf. (10.1) and (15.3)):

$$\int_{x=-\ell_A/2}^{\ell_A/2} \hat{E}_x^{\mathrm{T};A}(x,a,s)\hat{I}_x^{\mathrm{s};A}(x,s)\mathrm{d}x + \int_{x=x_0-\ell_B/2}^{x_0+\ell_B/2} \hat{E}_x^{\mathrm{T};A}(x,|y_0|,s)\hat{I}_x^{\mathrm{s};B}(x,s)\mathrm{d}x$$
$$= \int_{x=-\ell_A/2}^{\ell_A/2} \hat{E}_x^{\mathrm{s}}(x,a,s)\hat{I}_x^{\mathrm{T};A}(x,s)\mathrm{d}x, \tag{15.64a}$$

and

$$\int_{x=x_0-\ell_B/2}^{x_0+\ell_B/2} \hat{E}_x^{\mathrm{T};B}(x,a,s)\hat{I}_x^{\mathrm{s};B}(x,s)\mathrm{d}x + \int_{x=-\ell_A/2}^{\ell_A/2} \hat{E}_x^{\mathrm{T};B}(x,|y_0|,s)\hat{I}_x^{\mathrm{s};A}(x,s)\mathrm{d}x$$
$$= \int_{x=x_0-\ell_B/2}^{x_0+\ell_B/2} \hat{E}_x^{\mathrm{s}}(x,a,s)\hat{I}_x^{\mathrm{T};B}(x,s)\mathrm{d}x, \tag{15.64b}$$

where $\hat{I}_x^{\mathrm{s};A}(x,s)$ and $\hat{I}_x^{\mathrm{s};B}(x,s)$ denote the (unknown) electric currents induced in $\mathcal{W}_{\mathrm{A}}$ and $\mathcal{W}_{\mathrm{B}}$, respectively, and $\hat{I}_x^{\mathrm{T};A,B}(x,s)$ are the pertaining testing currents. The reciprocity relations can be readily expressed in terms of complex slowness integrals and subsequently solved numerically via the approach specified in Section 15.1.2. Hence, the conductors $\mathcal{W}_A$ and $\mathcal{W}_B$ can be discretized into $N_A + 1$ and $N_B + 1$ segments, so that their partitions follow as

$$\Delta\mathcal{W}_A = \{x_n = -\ell_A/2 + n\Delta_A, y = 0, z = 0\} \quad \text{for } n = \{1, \ldots, N_A\}, \tag{15.65a}$$

with $\Delta_A = \ell_A/(N_A+1)$, and

$$\Delta\mathcal{W}_B = \{x_n - x_0 = -\ell_B/2 + (n - N_A)\Delta_B, y = y_0, z = 0\}$$
$$\text{for } n = \{N_A+1,\ldots,N_A+N_B\}, \tag{15.65b}$$

where $\Delta_B = \ell_B/(N_B+1)$. In a similar manner, the time axis is discretized uniformly as $\{t_k = k\Delta t; k = 1,2,\ldots,M\}$. It is emphasized, again, that the uniform discretization is not mandatory, but it greatly simplifies the resulting algorithm and thus its code implementation. Applying the discretization to the reciprocity relation, while assuming the piecewise-linear space–time basis functions, we obtain a system of equations in the s-domain that has the form similar to the one depicted in Figure 10.2. The inversion of its constituents to the original domain leads to (12.25), again, where $\underline{\boldsymbol{Z}}_k = \underline{\boldsymbol{Z}}(t_k)$ denotes a 2-D $[(N_A+N_B)\times(N_A+N_B)]$ TD impedance array evaluated at $t = t_k = k\Delta t$, $\boldsymbol{I}_k$ is a 1-D $[(N_A+N_B)\times 1]$ array of the electric-current coefficients, $i_k^{[n]}$, and, finally, $\boldsymbol{V}_m$ denotes a 1-D $[(N_A+N_B)\times 1]$ excitation voltage array at $t = t_m$. With the help of (15.17a), the elements of the TD impedance array can be then expressed as (cf. (15.15a))

$$Z_{AA}^{[S,n]}(t) = \frac{Z^{\mathrm{c}}}{c_0\Delta t}\frac{1}{\Delta_A}\Big[\Upsilon^A(x_S - x_n + 3\Delta_A/2, t) - 3\Upsilon^A(x_S - x_n + \Delta_A/2, t)$$
$$+ 3\Upsilon^A(x_S - x_n - \Delta_A/2, t) - \Upsilon^A(x_S - x_n - 3\Delta_A/2, t)\Big]$$
$$\text{for all } S = \{1,\ldots,N_A\} \text{ and } n = \{1,\ldots,N_A\}, \tag{15.66}$$

describing the local interactions on conductor $\mathcal{W}_A$. In a similar fashion, the local interactions on $\mathcal{W}_B$ can be described through

$$Z_{BB}^{[S,n]}(t) = \frac{Z^{\mathrm{c}}}{c_0\Delta t}\frac{1}{\Delta_B}\Big[\Upsilon^A(x_S - x_n + 3\Delta_B/2, t) - 3\Upsilon^A(x_S - x_n + \Delta_B/2, t)$$
$$+ 3\Upsilon^A(x_S - x_n - \Delta_B/2, t) - \Upsilon^A(x_S - x_n - 3\Delta_B/2, t)\Big]$$
$$\text{for all } S = \{N_A+1,\ldots,N_A+N_B\}$$
$$\text{and } n = \{N_A+1,\ldots,N_A+N_B\}. \tag{15.67}$$

If the distance between the lines, $|y_0| > 0$, is relatively small, then the remote interaction from $\mathcal{W}_A$ to $\mathcal{W}_B$ can be accounted for via

$$Z_{AB}^{[S,n]}(t) = \frac{Z^{\mathrm{d}}}{c_0\Delta t}\frac{1}{\Delta_B}\Big[\Upsilon^A(x_S - x_n + \Delta_B + \Delta_A/2, t)$$
$$- \Upsilon^A(x_S - x_n + \Delta_B - \Delta_A/2, t) - 2\Upsilon^A(x_S - x_n + \Delta_A/2, t)$$
$$+ 2\Upsilon^A(x_S - x_n - \Delta_A/2, t) + \Upsilon^A(x_S - x_n - \Delta_B + \Delta_A/2, t)$$
$$- \Upsilon^A(x_S - x_n - \Delta_B - \Delta_A/2, t)\Big]$$
$$\text{for all } S = \{1,\ldots,N_A\} \text{ and } n = \{N_A+1,\ldots,N_A+N_B\}, \tag{15.68}$$

where

$$Z^{\rm d} = \frac{\zeta_0}{2\pi}\log\left[\frac{(y_0^2+4z_0^2)^{1/2}}{|y_0|}\right]. \tag{15.69}$$

Finally, the remote interaction from $\mathcal{W}_B$ to $\mathcal{W}_A$ can be described by

$$\begin{aligned} Z_{BA}^{[S,n]}(t) = {} & \frac{Z^{\rm d}}{c_0\Delta t}\frac{1}{\Delta_A}\Big[\Upsilon^A(x_S - x_n + \Delta_A + \Delta_B/2, t) \\ & - \Upsilon^A(x_S - x_n + \Delta_A - \Delta_B/2, t) - 2\Upsilon^A(x_S - x_n + \Delta_B/2, t) \\ & + 2\Upsilon^A(x_S - x_n - \Delta_B/2, t) + \Upsilon^A(x_S - x_n - \Delta_A + \Delta_B/2, t) \\ & - \Upsilon^A(x_S - x_n - \Delta_A - \Delta_B/2, t)\Big] \\ & \text{for all } S = \{N_A+1,\ldots,N_A+N_B\} \text{ and } n = \{1,\ldots,N_A\}. \end{aligned} \tag{15.70}$$

Alternatively, owing to the self-adjointness of the surrounding medium [8, Section 1.4.1], one may invoke the property of reciprocity represented by $Z_{BA}^{[S,n]}(t) = Z_{AB}^{[n,S]}(t)$ for all $S = \{N_A+1,\ldots,N_A+N_B\}$, $n = \{1,\ldots,N_A\}$ and $t > 0$.

In order to validate the coupling model, we have analyzed a gap-excited transmission line, $\mathcal{W}_{\rm A} = \{-\ell_A/2 < x < \ell_A/2, y = 0, z = 0\}$, in the presence of a loaded transmission, $\mathcal{W}_{\rm B} = \{-\ell_A/4 < x < \ell_A/4, y = \ell_A/20, z = 0\}$. The transmitting line, $\mathcal{W}_{\rm A}$, is excited at its center (i.e. $x_\delta = 0$) and the receiving line, $\mathcal{W}_{\rm B}$, is loaded at its center (i.e. $x_\zeta = 0$) by a resistor $R^{\rm L} = 100\,\Omega$. Again, we take $a = \ell_A/200$ and $z_0 = \ell_A/50$ with the triangular excitation pulse shape (see Figure 15.3) defined by $c_0 t_{\rm w} = 2.0\,\ell_A$ and $V_{\rm m} = 1.0\,$V. The calculated voltage pulse induced across the load resistance of the receiving transmission line is shown in Figure 15.7. Good correspondence with the pulse computed with the aid of the FIT is achieved.

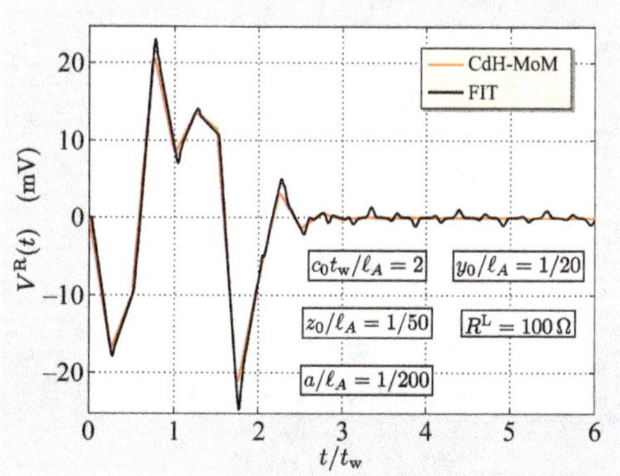

Figure 15.7 Voltage induced across the resistive load of the receiving transmission line

15.4.2 Transmission line above a lossy ground

In many EM compatibility-related applications, the transmission line is located over a lossy ground that is approximately represented by a homogeneous half-space, $\mathcal{D}_-^\infty = \{-\infty < x < \infty, -\infty < y < \infty, -z_0 < z < -\infty\}$, with dielectric and conductive properties (see [2, Chapter 8] and Figure 15.8). Therefore, it is next briefly indicated how to incorporate the influence of such a lossy ground in the CdH-MoM computational model. To that end, the cross-layer conditions (15.33b) with (15.33c) and (15.47a) with (15.47b) are replaced with the transform-domain surface-impedance (Leontovich) condition

$$\lim_{z\downarrow -z_0} \tilde{E}_x^{\mathrm{T}}(\kappa,\sigma,z,s) = -\hat{Z}(s) \lim_{z\downarrow -z_0} \tilde{H}_y^{\mathrm{T}}(\kappa,\sigma,z,s), \tag{15.71}$$

in which $\hat{Z}(s) = [s\mu_0/(s\epsilon_1 + \sigma)]^{1/2}$ is the pertaining s-domain surface impedance [8, Eq. (20)]. Following the solution methodology described in Sections 15.2 and 15.3, the correction of the axial component of the testing electric-field with respect to the ideal PEC plane follows as (cf. (15.36))

$$\Delta\tilde{E}_x^{\mathrm{T}}(\kappa,\sigma,0,s) = -\frac{\tilde{I}_x^{\mathrm{T}}(\kappa,s)\exp[-2sp(\kappa,\sigma)z_0]}{\eta_0[c_0 p(\kappa,\sigma)/c_0^2\Omega_0^2(\kappa)] + \hat{Y}(s)}, \tag{15.72}$$

where $\hat{Y}(s) = 1/\hat{Z}(s)$ denotes the surface admittance. Equation (15.72) is subsequently used to express the corresponding correction interaction term, the inner integral of which leads to (cf. (15.38))

$$\hat{\zeta}^{\mathrm{G}} = \frac{s}{2\mathrm{i}\pi}\int_{\sigma=-\mathrm{i}\infty}^{\mathrm{i}\infty} \exp[-2sp(\kappa,\sigma)z_0]\frac{\mathrm{d}\sigma}{\eta_0[c_0 p(\kappa,\sigma)/c_0^2\Omega_0^2(\kappa)] + \hat{Y}(s)}. \tag{15.73}$$

The integration with respect to σ can be carried out approximately, which yields (cf. (15.41))

$$\hat{\zeta}^{\mathrm{G}} \simeq \frac{s\mu_0}{2\pi}c_0^2\Omega_0^2(\kappa)\mathrm{E}_1\left[\frac{sz_0}{c_0}\frac{\zeta_0}{\hat{Z}(s)}c_0^2\Omega_0^2(\kappa)\right]\exp\left[\frac{sz_0}{c_0}\frac{\zeta_0}{\hat{Z}(s)}c_0^2\Omega_0^2(\kappa)\right], \tag{15.74}$$

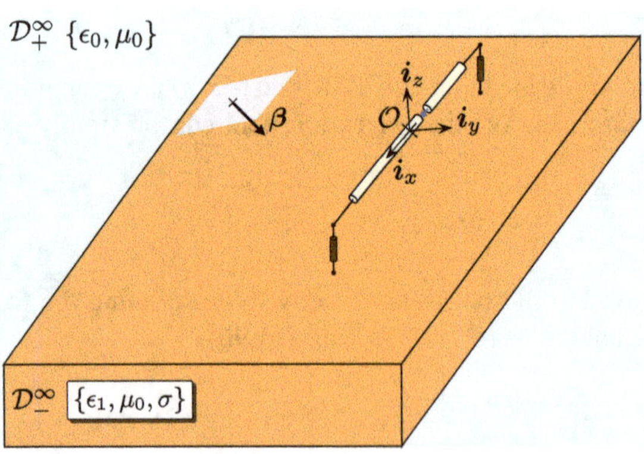

Figure 15.8 A straight transmission line above a lossy ground

as $2z_0\Omega_0 \downarrow 0$. If the surface impedance is relatively small, the expression for $\hat{\zeta}^{\mathrm{G}}$ can be further simplified. Indeed, making use of [6, Eq. (5.1.51)], we may find (cf. (15.42))

$$\hat{\zeta}^{\mathrm{G}} \simeq \hat{Z}(s)/2\pi z_0, \tag{15.75}$$

as $2z_0\Omega_0 \downarrow 0$ and $\hat{Z}(s)/\zeta_0 \downarrow 0$, which can be associated with the parameter termed as the *ground impedance* in the s-domain. As a matter of fact, (15.75) with the quasi-static approximation, $\hat{Z}(s) \simeq (s\mu_0/\sigma)^{1/2}$, represents the so-called simplified approximation of Vance [2, Section 8.3.3]. For the calculation of lightning-induced voltages, however, this approximation may not be accurate enough [2, Section 8.3.3]. To remedy the issue, we can use (15.74), where we let $c_0^2\Omega_0^2(\kappa) = 1$, or, equivalently, take $\kappa = 0$. Under this assumption, we get a κ-independent, closed-form expression for the s-domain ground impedance

$$\hat{\zeta}^{\mathrm{G}} \simeq \frac{s\mu_0}{2\pi}\mathrm{E}_1\left[\frac{sz_0}{c_0}\frac{\zeta_0}{\hat{Z}(s)}\right]\exp\left[\frac{sz_0}{c_0}\frac{\zeta_0}{\hat{Z}(s)}\right], \tag{15.76}$$

as $2z_0\Omega_0 \downarrow 0$ and $\kappa \to 0$. The closed-formula (15.76) has been introduced in [8], where its validity (with respect to the approximations due to Carson and Vance) is demonstrated. Approximation (15.76) of the slowness integral (15.73) can be finally used to cast the corresponding interaction integral into the following form (cf. (15.43)):

$$\int_{x=-\ell/2}^{\ell/2} \Delta\hat{E}_x^{\mathrm{T}}(x,0,0,s)\hat{I}_x^{\mathrm{s}}(x,s)\mathrm{d}x \simeq -\frac{s\hat{\zeta}^{\mathrm{G}}}{2\pi\mathrm{i}}\int_{\kappa=-\mathrm{i}\infty}^{\mathrm{i}\infty} \tilde{I}_x^{\mathrm{T}}(\kappa,s)\tilde{I}_x^{\mathrm{s}}(-\kappa,s)\mathrm{d}\kappa. \tag{15.77}$$

Through the use of Plancherel's identity [5, Eq. (B.2-19)], we may express the right-hand side of (15.77) as

$$\frac{s}{2\pi\mathrm{i}}\int_{\kappa=-\mathrm{i}\infty}^{\mathrm{i}\infty} \tilde{I}_x^{\mathrm{T}}(\kappa,s)\tilde{I}_x^{\mathrm{s}}(-\kappa,s)\mathrm{d}\kappa. = \int_{x=-\ell/2}^{\ell/2} \hat{I}_x^{\mathrm{T}}(x,s)\hat{I}_x^{\mathrm{s}}(x,s)\mathrm{d}x \tag{15.78}$$

Finally, upon combining the latter equation with (15.77), we can immediately deduce that (the axial component of) the correction field can be simply expressed via (cf. (15.72))

$$\Delta\hat{E}_x^{\mathrm{T}}(x,0,0,s) \simeq -\hat{\zeta}^{\mathrm{G}}\hat{I}_x^{\mathrm{T}}(x,s). \tag{15.79}$$

This result, however, could be found directly by postulating the following form of transmission-line equations (cf. (15.9a) and (15.9b)):

$$\partial_x\hat{V}^{\mathrm{T}}(x,s) + \left[\hat{\zeta}(s) + \hat{\zeta}^{\mathrm{G}}\right]\hat{I}_x^{\mathrm{T}}(x,s) = -\hat{E}_x^{\mathrm{T}}(x,a,s), \tag{15.80a}$$

$$\partial_x\hat{I}_x^{\mathrm{T}}(x,s) + \hat{\eta}(s)\hat{V}^{\mathrm{T}}(x,s) = 0. \tag{15.80b}$$

In this way, the effect of a lossy ground is (approximately) incorporated in the lumped per-unit-length parameter $\hat{\zeta}^{\mathrm{G}}$ [2, Section 8.3].

15.4.3 Extending the applicability of the numerical solution

In this final section, it is demonstrated that the CdH-MoM solutions discussed throughout this chapter can be readily generalized by removing some assumptions introduced by the use of TL theory. For the sake of simplicity, we shall begin with the extension of the results applying to an open-circuited, thin straight wire located above the infinite PEC plane (see Section 15.1.2). Here, upon removing the assumption that the wire is relatively close to the reference plane, the impedance array specified via (15.15a) is no longer valid and must be thus replaced with

$$\begin{aligned}\hat{Z}^{[S,n]}(s) = \frac{\zeta_0}{c_0\Delta t\Delta}\Big[&\hat{\Upsilon}(x_S - x_n + 3\Delta/2, 0, a, s) - \hat{\Upsilon}(x_S - x_n + 3\Delta/2, 0, 2z_0, s)\\ &- 3\hat{\Upsilon}(x_S - x_n + \Delta/2, 0, a, s) + 3\hat{\Upsilon}(x_S - x_n + \Delta/2, 0, 2z_0, s)\\ &+ 3\hat{\Upsilon}(x_S - x_n - \Delta/2, 0, a, s) - 3\hat{\Upsilon}(x_S - x_n - \Delta/2, 0, 2z_0, s)\\ &- \hat{\Upsilon}(x_S - x_n - 3\Delta/2, 0, a, s) + \hat{\Upsilon}(x_S - x_n - 3\Delta/2, 0, 2z_0, s)\Big]\\ &\text{for all } S = \{1, \ldots, N\} \text{ and } n = \{1, \ldots, N\},\end{aligned} \tag{15.81}$$

where we used (cf. (15.16a) and (K.1))

$$\begin{aligned}\hat{\Upsilon}(x, y, z, s) = &-\frac{c_0^2}{2\pi\mathrm{i}}\int_{\kappa\in\mathbb{I}_0}\frac{\exp(s\kappa x)}{s^3\kappa^3}\Omega_0^2(\kappa)\mathrm{d}\kappa\\ &\times\frac{1}{2\pi\mathrm{i}}\int_{\sigma=-\mathrm{i}\infty}^{\mathrm{i}\infty}\frac{\exp\{-s[-\sigma y + p(\kappa,\sigma)z]\}}{2\,p(\kappa,\sigma)}\mathrm{d}\sigma\\ &\text{for } x \in \mathbb{R}, y \in \mathbb{R}, z \geq 0 \text{ and } \{s \in \mathbb{R}; s > 0\},\end{aligned} \tag{15.82}$$

where $p(\kappa, \sigma)$ is defined by (4.77) and the (indented) integration contour in the complex κ-plane is shown in Figure B.1. Following the inversion procedure as used in [3, Appendix G], for example, the TD original of $\hat{\Upsilon}(x, y, z, s)$ is found as

$$\begin{aligned}\Upsilon(x, y, z, t) = \frac{\mathrm{H}(x)}{4\pi}&\left\{(c_0^2t^2 + \varrho^2 - x^2)\log\left[\frac{c_0t + (c_0^2t^2 - \varrho^2)^{1/2}}{\varrho}\right]\right.\\ &\left.- 2c_0t(c_0^2t^2 - \varrho^2)^{1/2}\right\}\mathrm{H}(c_0t - \varrho)\\ - \frac{\mathrm{sgn}(x)}{8\pi}&\left\{(c_0^2t^2 + \varrho^2 - x^2)\log\left[\frac{c_0t + (c_0^2t^2 - \varrho^2)^{1/2}}{R + |x|}\right]\right.\\ &\left.- 2c_0t(c_0^2t^2 - \varrho^2)^{1/2} + 4|x|(c_0t - R/2)\right\}\mathrm{H}(c_0t - R),\end{aligned} \tag{15.83}$$

where $\varrho = (y^2 + z^2)^{1/2}$ and $R = (x^2 + \varrho^2)^{1/2}$. The TD function, $\Upsilon(x, y, z, t)$, can be then applied to analyze the TD response of a wire above a reference plane via the TD counterpart of (15.81). To see the relation between (15.81) and its approximation (15.15a), we observe that

$$\Upsilon(x, 0, a, t) - \Upsilon(x, 0, 2z_0, t) \simeq \Upsilon^A(x, t) \log(2z_0/a)/2\pi, \tag{15.84}$$

as $z_0 \downarrow 0$, $a \ll z_0$ and recall that $\Upsilon^A(x, t)$ is given by (15.17a). An example where the latter TL approximation is not appropriate is demonstrated through Figure 15.9, where the electric-current response of a gap-excited thin wire over a PEC reference plane is shown. In this example, the cylindrical wire of a relatively small radius, $a = \ell/400$, was excited at its center by the unit triangular pulse (see Figure 15.3) with $c_0 t_w = \ell/2$. Since we now take $z_0 = \ell/5$, the relative height of the wire above the ground plane, $z_0/c_0 t_w = 2/5$, is not small enough to ensure the validity of the approximation described by (15.84). Indeed, it is clearly seen in Figure 15.9 that the current responses, denoted by CdH-MoM and CdH-MoM-TL, as obtained via the impedance array (15.81) and its TL approximation (15.15a), respectively, differ significantly from each other. For the sake of validation, the current response has also been computed via the FIT, thus confirming a good correlation between the CdH-MoM- and FIT-based results.

Following the same lines of reasoning, the TD function (15.83) may also serve for analyzing a system of mutually parallel conductors located over a PEC ground. Indeed, if the wires are close to the reference plane, (15.66) and (15.67) describe the pertaining local interactions on $\mathcal{W}_A$ and $\mathcal{W}_B$. If this assumption does not apply, (15.66) and (15.67) will have the form similar to the one of (15.81). Moreover, if the

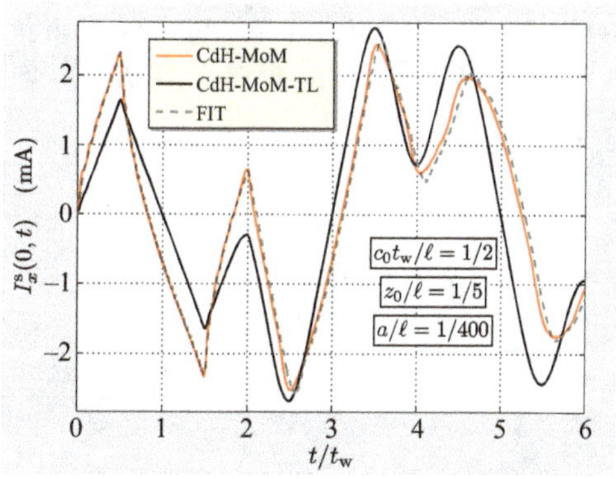

Figure 15.9 Electric current at the center ($x = 0$) of the open-circuited thin wire above a PEC ground

distance between the parallel wires is not relatively small, (15.68) and (15.70) must be modified accordingly. Therefore, we write (cf. (15.68))

$$
\begin{aligned}
Z_{AB}^{[S,n]}(t) = {} & \frac{\zeta_0}{c_0 \Delta t} \frac{1}{\Delta_B} \Big[\Upsilon(x_S - x_n + \Delta_B + \Delta_A/2, y_0, 0, t) \\
& - \Upsilon(x_S - x_n + \Delta_B + \Delta_A/2, y_0, 2z_0, t) \\
& - \Upsilon(x_S - x_n + \Delta_B - \Delta_A/2, y_0, 0, t) + \Upsilon(x_S - x_n + \Delta_B - \Delta_A/2, y_0, 2z_0, t) \\
& - 2\Upsilon(x_S - x_n + \Delta_A/2, y_0, 0, t) + 2\Upsilon(x_S - x_n + \Delta_A/2, y_0, 2z_0, t) \\
& + 2\Upsilon(x_S - x_n - \Delta_A/2, y_0, 0, t) - 2\Upsilon(x_S - x_n - \Delta_A/2, y_0, 2z_0, t) \\
& + \Upsilon(x_S - x_n - \Delta_B + \Delta_A/2, y_0, 0, t) - \Upsilon(x_S - x_n - \Delta_B + \Delta_A/2, y_0, 2z_0, t) \\
& - \Upsilon(x_S - x_n - \Delta_B - \Delta_A/2, y_0, 0, t) + \Upsilon(x_S - x_n - \Delta_B - \Delta_A/2, y_0, 2z_0, t) \Big] \\
& \text{for all } S = \{1, \ldots, N_A\} \text{ and } n = \{N_A + 1, \ldots, N_A + N_B\}.
\end{aligned} \tag{15.85}
$$

The remaining elements of the TD impedance array can be finally calculated by invoking the property of reciprocity via $Z_{BA}^{[S,n]}(t) = Z_{AB}^{[n,S]}(t)$ for all $S = \{N_A + 1, \ldots, N_A + N_B\}$, $n = \{1, \ldots, N_A\}$ and $t > 0$. As a matter of fact, the (approximate) coupling impedance array (15.68) leads to the instantaneous pulsed EM-field signal transfer between two coupled wires above a PEC plane. Such a close-range approximation is legitimate whenever the distance between the wires is negligible with respect to the spatial support of the excitation pulse.

Finally, it is noted that the TD function (15.83) can also be immediately applied to analyze (a system of) thin-wire antennas in the absence of the ground plane. Indeed, upon removing the "image-source contributions" due to the presence of the reference PEC plane, the TD array representing a thin-wire antenna in free space, for instance, has the following form (cf. (15.81)):

$$
\begin{aligned}
Z^{[S,n]}(t) = {} & \frac{\zeta_0}{c_0 \Delta t \Delta} \Big[\Upsilon(x_S - x_n + 3\Delta/2, 0, a, t) - 3\Upsilon(x_S - x_n + \Delta/2, 0, a, t) \\
& + 3\Upsilon(x_S - x_n - \Delta/2, 0, a, t) - \Upsilon(x_S - x_n - 3\Delta/2, 0, a, t) \Big] \\
& \text{for all } S = \{1, \ldots, N\} \text{ and } n = \{1, \ldots, N\}.
\end{aligned} \tag{15.86}
$$

In the same way, we can remove the influence of the ground plane from the coupling-impedance array (15.85). For more applications of the generic function (15.83) to thin-wire antenna problems, the reader is referred to [9].

References

[1] S. A. Schelkunoff. *Electromagnetic Waves*. New York, NY: D. Van Nostrand Company, Inc., 9th printing edition, 1957.

[2] F. M. Tesche, M. V. Ianoz and T. Karlsson. *EMC Analysis Methods and Computational Models*. New York, NY: John Wiley & Sons, Inc., 1997.

[3] M. Štumpf. *Time-Domain Electromagnetic Reciprocity in Antenna Modeling*. Hoboken, NJ: John Wiley & Sons, Inc., 2019.

[4] M. Štumpf. Transient response of a transmission line above a thin conducting sheet—a numerical model based on the Cagniard–DeHoop method of moments. *IEEE Antennas and Wireless Propagation Letters*, 20(9):1829–1833, 2021.
[5] A. T. de Hoop. *Handbook of Radiation and Scattering of Waves*. London, UK: Academic Press, 1995.
[6] M. Abramowitz and I. A. Stegun. *Handbook of Mathematical Functions*. New York, NY: Dover Publications, 1972.
[7] E. C. Titchmarsh. *The Theory of Functions, 2nd edition*. London, UK: Oxford University Press, 1939.
[8] M. Štumpf. Evaluating the ground impedance—a new methodology based on EM reciprocity. *IEEE Transactions on Electromagnetic Compatibility*, 61(3):928–934, 2018.
[9] M. Štumpf, I. E. Lager and G. Antonini. Time-domain analysis of thin-wire structures based on the Cagniard-DeHoop method of moments. *IEEE Transactions on Antennas and Propagation*, DOI: 10.1109/TAP.2022.3140782.

Chapter 16
Miscellaneous applications

In this chapter, we shall analyze selected illustrative problem configurations to demonstrate the use of some Cagniard–DeHoop (CdH)-based analytical tools for achieving closed-form time-domain (TD) solutions. First, to demonstrate the principle of a version of the CdH technique that is suitable for describing the *short-range* electromagnetic (EM) signal wireless transfer between antennas, we shall study the loop-to-loop signal transmission via a thin highly conductive layer. The introduced analytical methodology is directly applicable to the estimation of the (low-frequency) shielding effectiveness in standardized experimental scenarios (see [1, Appendix C], for instance), for instance, and can be further extended to more complex problem configurations consisting of fundamental EM radiators integrated in a layered structure. Consequently, we shall demonstrate how the analytical methodology described in Section 4.3.1 can be extended to incorporate plasmonic properties of the layer lying in between the transmitting and receiving loop antennas. Finally, a modified Kirchhoff approximation is applied to analyze the pulsed EM radiation from a two-dimensional (2-D) slot-excited Fabry–Pérot antenna. Again, the approximate analysis yields TD expressions that give some insights into the pulsed EM phenomenon.

16.1 Loop-to-loop pulsed EM close-range signal transfer via a conductive sheet

Section 4.3.1 is devoted to the description of the pulsed EM signal transfer via a metasurface. To demonstrate an analytical technique for achieving approximations applying to *close-range* signal transfers, a similar problem configuration consisting of two small loops and a thin highly conductive layer is considered in this section (see Figure 16.1). Namely, under the assumption that the distance between the interacting loop antennas is relatively small (with respect to the spatial support of the exciting electric-current pulse), we shall derive an approximate expression for the open-circuit voltage induced in the receiving loop. To that end, we shall employ a dedicated version of the CdH technique that has been previously applied to the short-range EM coupling between two loops in a simple inhomogeneous region [2]. The thus obtained approximate results will be validated using the exact TD solution that follows as a straightforward extension of (4.93) [3, Eq. (29)], viz

$$V^{\mathrm{R}}(t) = -\frac{\mathcal{A}^{\mathrm{T}}\mathcal{A}^{\mathrm{R}}}{(h^{\mathrm{T}}+h^{\mathrm{R}})^4}\frac{\partial_t^4 I^{\mathrm{T}}(t)}{2\pi\eta_0} *_t \frac{t(t^2-T^2)}{G^{\mathrm{E}}/\eta_0+2t/T}\mathrm{H}(t-T), \tag{16.1}$$

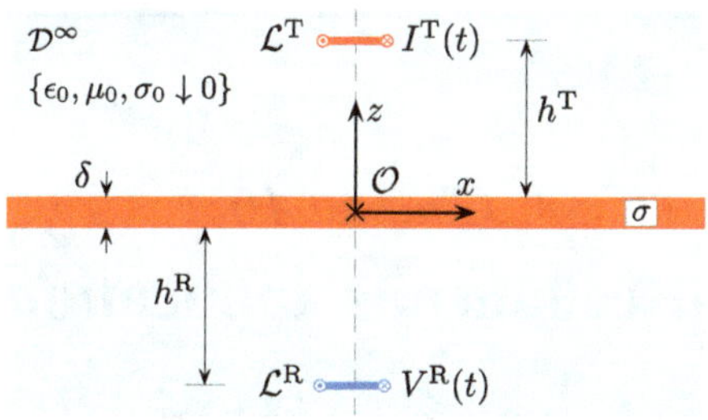

Figure 16.1 A thin highly conductive sheet in between two horizontal transmitting and receiving loop antennas

where $T = (h^{\rm R} + h^{\rm T})/c_0$ is the travel time of the pulse and the equivalent layer's conductance $G^{\rm E}$ follows from the layer's conductivity $\sigma > 0$ and thickness $\delta > 0$ via (3.5a).

16.1.1 Problem formulation

Closed-form expressions applying to the short-range signal transfer are further arrived at under the diffusive approximation. In this approximate model, it is assumed that the transfer between the loops is dominantly effected via the conduction current, while neglecting the contribution of displacement ones. For this purpose, it is assumed that the loops are located on a conductive (and homogeneous, isotropic) background with vanishingly small electric conductivity $\sigma_0 \downarrow 0$. The electric conductivity of the (relatively thin) layer, say $\sigma > 0$, is assumed to be high, such that the layer's conductance $G^{\rm E}$ is of $O(1)$.

The nonzero EM-field components parallel to the layer can be represented via [4, Section 26.3]

$$\hat{E}_\phi(r, z, s) = \partial_r \hat{\Phi}_z(r, z, s), \qquad \text{(4.71a revisited)}$$

$$\hat{H}_r(r, z, s) = \partial_r \partial_z \hat{\Phi}_z(r, z, s)/s\mu_0, \qquad \text{(4.71b revisited)}$$

with

$$\hat{\Phi}_z(r, z, s) = s\mu_0 \hat{I}^{\rm T}(s) \mathcal{A}^{\rm T} \hat{G}(r, z, s). \qquad \text{(4.72 revisited)}$$

Here, $\hat{I}^{\rm T}(s)$ represents the exciting electric current and $\hat{G}(r, z, s)$ denotes the pertaining (diffusive) Green's function in the s-domain. The tangential EM-field components do satisfy the cross-layer conditions (see (5.50a) and (5.50b)), i.e.

$$\hat{E}^+_\phi - \hat{E}^-_\phi = 0, \qquad (16.2a)$$

$$\hat{H}^+_r - \hat{H}^-_r = G^{\rm E} \hat{E}_\phi(r, 0, s), \quad \text{for all } r > 0, \qquad (16.2b)$$

as $\delta \downarrow 0$, from which the corresponding jump conditions applying to the Green's function can readily be deduced. Once the Green's function is determined, the voltage response induced in the receiving loop is found from

$$\hat{V}^{\mathrm{R}}(s) = -s\mu_0 \mathcal{A}^{\mathrm{R}} \hat{H}_z(0, -h^{\mathrm{R}}, s), \qquad \text{(4.73 revisited)}$$

again, where we replace (4.74) with

$$\hat{H}_z(r, z, s) = -\sigma_0 \hat{\Phi}_z(r, z, s) + \partial_z^2 \hat{\Phi}_z(r, z, s)/s\mu_0, \tag{16.3}$$

by virtue of the diffusive approximation.

16.1.2 Problem solution

The problem is next solved under a Fourier representation of the following type (cf. (4.75)):

$$\hat{E}_\phi(r, z, s) = \frac{s}{(2\pi \mathrm{i})^2} \int_{\lambda=-\mathrm{i}\infty}^{\mathrm{i}\infty} \mathrm{d}\lambda \int_{\nu=-\mathrm{i}\infty}^{\mathrm{i}\infty} \exp[-s^{1/2}(\lambda x + \nu y)] \tilde{E}_\phi(\lambda, \nu, z, s) \mathrm{d}\nu$$
$$\text{for } \{s \in \mathbb{R}; s > 0\}. \tag{16.4}$$

where $s^{1/2} > 0$ plays the role of scaling parameter. The thus defined representation entails $\partial_x \to -s^{1/2}\lambda$ and $\partial_y \to -s^{1/2}\nu$. Consequently, under the representation, the (transform-domain) Green's function is governed by (cf. (4.76))

$$[\partial_z^2 - s\pi_0^2(\lambda, \nu)]\tilde{G}(\lambda, \nu, z, s) = 0, \tag{16.5}$$

in which

$$\pi_0(\lambda, \nu) = (d_0^{-1} - \lambda^2 - \nu^2)^{1/2}, \tag{16.6}$$

with $\mathrm{Re}(\pi_0) \geq 0$ and $d_0 = 1/\sigma_0\mu_0$ has the meaning of the electric diffusive coefficient. Equation (16.5) will next be solved together with the (transform-domain) excitation condition

$$\lim_{z \downarrow h^{\mathrm{T}}} \partial_z \tilde{G}(\lambda, \nu, z, s) - \lim_{z \uparrow h^{\mathrm{T}}} \partial_z \tilde{G}(\lambda, \nu, z, s) = -1, \tag{16.7a}$$

and the corresponding cross-layer conditions (cf. (4.37b)–(4.37c))

$$\lim_{z \downarrow 0} \tilde{G} - \lim_{z \uparrow 0} \tilde{G} = 0, \tag{16.7b}$$

$$\lim_{z \downarrow 0} \partial_z \tilde{G} - \lim_{z \uparrow 0} \partial_z \tilde{G} = s\mu_0 G^{\mathrm{E}} \tilde{G}(\lambda, \nu, 0, s). \tag{16.7c}$$

The general (transform-domain) solution satisfying (16.5) can be next expressed in the following form:

$$\tilde{G} = \tilde{A} \exp[-s^{1/2}\pi_0(\lambda, \nu)|z - h^{\mathrm{T}}|] + \tilde{A}\tilde{R}_\perp \exp[-s^{1/2}\pi_0(\lambda, \nu)(z + h^{\mathrm{T}})]$$
$$\text{for } z > 0, \tag{16.8}$$

and

$$\tilde{G} = \tilde{A}\tilde{T}_{\perp} \exp[-s^{1/2}\pi_0(\lambda, \nu)(h^{\mathrm{T}} - z)] \quad \text{for } z < 0, \tag{16.9}$$

where $\tilde{A}$ is to be determined from the excitation condition (16.7a), while the reflection and transmission coefficients, $\tilde{R}_{\perp}$ and $\tilde{T}_{\perp}$, respectively, follow from the transition conditions (16.7b) and (16.7c). Pursuing this approach, it can be verified that

$$\tilde{T}_{\perp} = 1 + \tilde{R}_{\perp} = \frac{2d_0^{1/2}\pi_0}{(G^{\mathrm{E}}/\sigma_0)(s\mu_0\sigma_0)^{1/2} + 2d_0^{1/2}\pi_0}, \tag{16.10}$$

and $\tilde{A} = 1/2s^{1/2}\pi_0$. Hence, upon combining (16.9), (16.10) with (4.73) and (16.4), we arrive at the following representation of the voltage induced in $\mathcal{L}^{\mathrm{R}}$, that is (cf. (4.82)),

$$\hat{V}^{\mathrm{R}}(s) = -\frac{s^3\mu_0\hat{I}^{\mathrm{T}}(s)\mathcal{A}^{\mathrm{T}}\mathcal{A}^{\mathrm{R}}}{4\pi^2(s\mu_0\sigma_0)^{1/2}} \int_{\lambda=-\mathrm{i}\infty}^{\mathrm{i}\infty} \mathrm{d}\lambda \int_{\nu=-\mathrm{i}\infty}^{\mathrm{i}\infty} \exp[-s^{1/2}\pi_0(\lambda, \nu)(h^{\mathrm{T}} + h^{\mathrm{R}})]$$
$$\times \frac{\lambda^2 + \nu^2}{(G^{\mathrm{E}}/\sigma_0)(s\mu_0\sigma_0)^{1/2} + 2d_0^{1/2}\pi_0(\lambda, \nu)} \mathrm{d}\nu. \tag{16.11}$$

To cast representation (16.11) into the form that is amenable to its transformation back to the TD, we shall first introduce polar variables of integration, say ω and ψ, via

$$\lambda = \mathrm{i}\omega\cos(\psi), \tag{16.12a}$$
$$\nu = \mathrm{i}\omega\sin(\psi), \tag{16.12b}$$

where $\{0 \leq \psi < 2\pi\}$ and $\{\omega \in \mathbb{R}; \omega > 0\}$. The substitution yields $\mathrm{d}\lambda\mathrm{d}\nu = -\omega\mathrm{d}\omega\mathrm{d}\psi$, $\lambda^2 + \nu^2 = -\omega^2$, so that we get

$$\hat{V}^{\mathrm{R}}(s) = -\frac{s^3\mu_0\hat{I}^{\mathrm{T}}(s)\mathcal{A}^{\mathrm{T}}\mathcal{A}^{\mathrm{R}}}{2\pi(s\mu_0\sigma_0)^{1/2}} \int_{\omega=0}^{\infty} \exp[-s^{1/2}\bar{\pi}_0(\omega)(h^{\mathrm{T}} + h^{\mathrm{R}})]$$
$$\times \frac{\omega^3}{(G^{\mathrm{E}}/\sigma_0)(s\mu_0\sigma_0)^{1/2} + 2d_0^{1/2}\bar{\pi}_0(\omega)} \mathrm{d}\omega, \tag{16.13}$$

in which $\bar{\pi}_0(\omega) = (d_0^{-1} + \omega^2)^{1/2}$ (see (16.6)). In the next step, we substitute

$$\bar{\pi}_0(\omega)(h^{\mathrm{T}} + h^{\mathrm{R}}) = \tau \quad \text{for } \{\tau \in \mathbb{R}; \tau > 0\}, \tag{16.14}$$

which allows to write (cf. (4.88))

$$\hat{V}^{\mathrm{R}}(s) = -\frac{s^3\mu_0\hat{I}^{\mathrm{T}}(s)\mathcal{A}^{\mathrm{T}}\mathcal{A}^{\mathrm{R}}}{4\pi s^{1/2}(h^{\mathrm{T}} + h^{\mathrm{R}})^3} \int_{\tau=T_0}^{\infty} \exp(-s^{1/2}\tau)\frac{\tau(\tau^2 - T_0^2)}{\tau + s^{1/2}\beta} \mathrm{d}\tau, \tag{16.15}$$

where we used $\beta = \mu_0 G^{\mathrm{E}}(h^{\mathrm{T}} + h^{\mathrm{R}})/2$ and $T_0 = (h^{\mathrm{T}} + h^{\mathrm{R}})d_0^{-1/2} > 0$. The integration with respect to τ is trivial for $\beta = 0$. For our purposes, however, it is sufficient

to obtain the result of integration in the limit $\sigma_0 \downarrow 0$. To that end, we employ the following formula:

$$\lim_{T_0 \downarrow 0} \int_{\tau=T_0}^{\infty} \exp(-s^{1/2}\tau) \frac{\tau(\tau^2 - T_0^2)}{\tau + s^{1/2}\beta} \mathrm{d}\tau = 2s^{-3/2} - \beta s^{-1/2} + \beta^2 s^{1/2} - \beta^3 s^{3/2} \mathrm{E}_1(s\beta) \exp(s\beta), \tag{16.16}$$

where $\mathrm{E}_1(x)$ denotes the exponential integral [5, Eq. (5.1.4)] and obtain

$$\lim_{T_0 \downarrow 0} \hat{V}^{\mathrm{R}}(s) = -\frac{s\mu_0 \hat{I}^{\mathrm{T}}(s) \mathcal{A}^{\mathrm{T}} \mathcal{A}^{\mathrm{R}}}{2\pi(h^{\mathrm{T}} + h^{\mathrm{R}})^3} \left[1 - \frac{s\beta}{2} + \frac{s^2\beta^2}{2} - \frac{s^3\beta^3}{2} \mathrm{E}_1(s\beta) \exp(s\beta)\right]. \tag{16.17}$$

The latter result can be further used to derive useful asymptotic expressions. Accordingly, if $s\beta \downarrow 0$, which applies to a small layer's conductivity, or/and to the late-time behavior of the voltage response, we get

$$\lim_{T_0 \downarrow 0} \hat{V}^{\mathrm{R}}(s) = -\frac{s\mu_0 \hat{I}^{\mathrm{T}}(s) \mathcal{A}^{\mathrm{T}} \mathcal{A}^{\mathrm{R}}}{2\pi(h^{\mathrm{T}} + h^{\mathrm{R}})^3} + O(s\beta) = -sL^{\mathrm{C}} \hat{I}^{\mathrm{T}}(s) + O(s\beta), \tag{16.18}$$

as $s\beta \downarrow 0$, where $L^{\mathrm{C}} = (\mu_0/2\pi)\mathcal{A}^{\mathrm{T}}\mathcal{A}^{\mathrm{R}}/(h^{\mathrm{T}} + h^{\mathrm{R}})^3$ has the meaning of coupling inductance. On the other hand, as $s\beta \to \infty$, which applies to a high layer's conductivity, or/and to the early part of the response, we get

$$\lim_{T_0 \downarrow 0} \hat{V}^{\mathrm{R}}(s) = -\frac{3\hat{I}^{\mathrm{T}}(s) \mathcal{A}^{\mathrm{T}} \mathcal{A}^{\mathrm{R}}}{\pi G^{\mathrm{E}} (h^{\mathrm{T}} + h^{\mathrm{R}})^4} + O(1/s\beta) = -R^{\mathrm{C}} \hat{I}^{\mathrm{T}}(s) + O(1/s\beta), \tag{16.19}$$

as $s\beta \to \infty$, where $R^{\mathrm{C}} = (3/\pi G^{\mathrm{E}})\mathcal{A}^{\mathrm{T}}\mathcal{A}^{\mathrm{R}}/(h^{\mathrm{T}} + h^{\mathrm{R}})^4$ has the meaning of coupling resistance.

The obtained s-domain expressions can be readily transformed to the TD. Hence, with the help of [5, Eqs. (29.3.126)], we can write the original of (16.17) as (cf. (4.93))

$$\lim_{T_0 \downarrow 0} V^{\mathrm{R}}(t) = -\frac{1}{2\pi\eta_0} \frac{\mathcal{A}^{\mathrm{T}} \mathcal{A}^{\mathrm{R}}}{(h^{\mathrm{T}} + h^{\mathrm{R}})^4} \big[T\partial_t I^{\mathrm{T}}(t) - \tfrac{1}{4}(G^{\mathrm{E}}/\eta_0) T^2 \partial_t^2 I^{\mathrm{T}}(t) + \tfrac{1}{8}(G^{\mathrm{E}}/\eta_0)^2 T^3 \partial_t^3 I^{\mathrm{T}}(t) - \tfrac{1}{16}(G^{\mathrm{E}}/\eta_0)^3 T^4 \partial_t^4 I^{\mathrm{T}}(t) \overset{*}{,}_t \mathrm{H}(t)/(t+\beta)\big], \tag{16.20}$$

and recall that $T = (h^{\mathrm{R}} + h^{\mathrm{T}})/c_0$ denotes the pulse travel time. As the derived TD approximate expression (16.20) still involves the time-convolution operator, it does not introduce any significant computational savings with respect to the exact solution (16.1). As the short-range approximation consists of terms proportional to powers of the relative layers conductivity, G^{E}/η_0, it may, however, provide useful insights into the asymptotic behavior of the short-range signal transfer via a conductive layer. Indeed, the first term in (16.20) represents the inductive coupling in the absence of

the layer (cf. (16.18)), while the dominant term in the pulse transmission via a highly conductive screen can be described via (16.19) as

$$\lim_{T_0 \downarrow 0} V^{\mathrm{R}}(t) \simeq -(3/\pi)\left[\mathcal{A}^{\mathrm{T}}\mathcal{A}^{\mathrm{R}}/(h^{\mathrm{T}}+h^{\mathrm{R}})^4\right] I^{\mathrm{T}}(t)/G^{\mathrm{E}}, \tag{16.21}$$

as $G^{\mathrm{E}}/\eta_0 \gg 1$. Illustrative applications of the TD expressions are given in the following section.

16.1.3 Illustrative examples

The previously derived TD voltage responses are next evaluated numerically. To that end, it is assumed that the transmitting loop above the conductive screen is activated via the electric-current pulse, $I^{\mathrm{T}}(t)$, whose pulse shape is described by the right-hand side of (4.70), where we take $I_{\mathrm{m}} = 1.0\,\mathrm{mA}$ and $n = 6$, again (see Figure 4.7(a)). In virtue of the close-range approximation, the pulse-time width is equal to $10\,T$, so that $c_0 t_{\mathrm{w}} = 10\,(h^{\mathrm{T}} + h^{\mathrm{R}})$.

In the first example, the results of which are shown in Figure 16.2, it is demonstrated that the exact solution represented by (16.1) can be successfully estimated from its close-range approximation (16.20). Since the layer's conductance was chosen to be relatively small, $G^{\mathrm{E}} = 3.0\,\eta_0$, the voltage pulse is dominantly formed by the first (inductive-coupling) term of (16.20). As a matter of fact, it can be shown that the first two terms of the close-range approximation (16.20) approximates the exact solution very well.

In the second example, we have increased the layer's conductivity such that its relative conductance is high, $G^{\mathrm{E}}/\eta_0 = 300$. In such a case, the first two terms of (16.20) can be essentially neglected and the initial lobe of the response can roughly be approximated via (16.21). This observation is illustrated in Figure 16.3, where the exact voltage pulse as calculated using (16.1) is compared with the outcome of the early-time approximation (16.21).

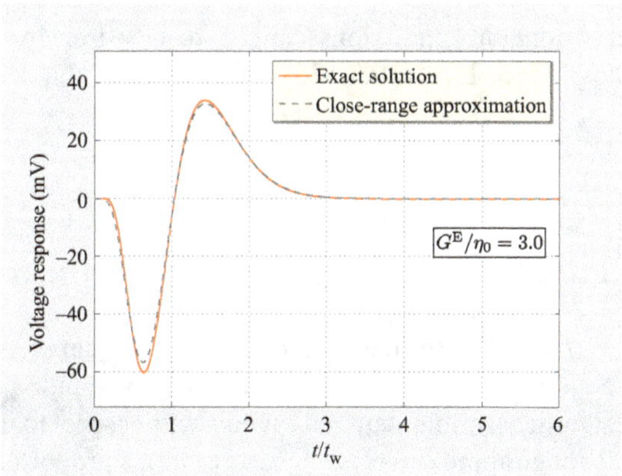

Figure 16.2 Induced voltage responses as calculated via (exact) (16.1) and its close-range approximation (16.20) with $G^{\mathrm{E}}/\eta_0 = 3.0$

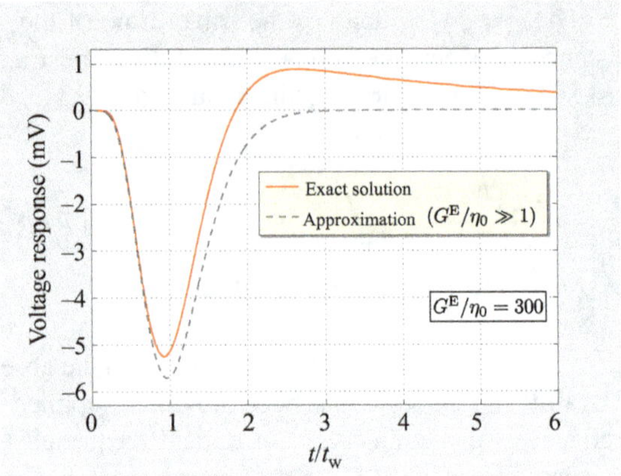

Figure 16.3 Induced voltage responses as calculated via (exact) (16.1) and its high-conductance approximation (16.21) with $G^E/\eta_0 = 300$

16.2 Loop-to-loop pulsed EM signal transfer via a plasmonic sheet

With the help of the theory presented in Section 4.3.1, we may evaluate the pulsed EM-field signal transfer across a plasmonic thin sheet. This is the subject matter of this section, where we analyze the signal transmission through a dispersive metal sheet that shows the EM contrast with respect to the surrounding medium in its electric conductivity described by the Drude-type conduction relaxation function in the s-domain [4, Section 19.5], viz

$$\hat{\sigma}(s) = \epsilon_0 \omega_{\text{pe}}^2/(s + \nu_\text{c}), \tag{5.21 revisited}$$

and recall that $\omega_{\text{pe}} = (n_\text{e}\text{e}^2/\epsilon_0 m_\text{e})^{1/2}$ is the electron plasma angular frequency, ν_c denotes the collision frequency, n_e denotes the number density of conduction electrons, $-\text{e}$ is the electron charge and m_e denotes the electron mass. Referring to the general form for the transmission coefficient as given by (4.81), we use $\hat{Z}^\text{M}(s) = 0$, $\hat{Y}^\text{E}(s) = \hat{G}^\text{E}(s) = \delta\hat{\sigma}(s)$ and obtain

$$\tilde{T}_\perp = \frac{c_0 p(\kappa, \sigma)}{c_0 p(\kappa, \sigma) + \hat{G}^\text{E}(s)/2\eta_0}. \tag{16.22}$$

The latter is next substituted in (4.82) and the result is transformed using the CdH approach applied in Section 4.3.1. In this way, employing the substitutions introduced by (4.83a), (4.83b) and (4.85), we get

$$\bar{T}_\perp[\gamma(\tau), s] = \frac{\tau/T}{\tau/T + \hat{G}^\text{E}(s)/2\eta_0}, \tag{16.23}$$

and recall that $T = (h^{\rm T} + h^{\rm R})/c_0$ denotes the travel time of the wave front taken to traverse the distance between the loops. Substituting the transmission coefficient (16.23) in (4.88) and transforming the result to the TD, we end up with (cf. (4.92))

$$V^{\rm R}(t) = V_0^{\rm R}(t) + \frac{\mu_0 \mathcal{A}^{\rm T}\mathcal{A}^{\rm R}\partial_t^4 I^{\rm T}(t)}{4\pi(h^{\rm T}+h^{\rm R})^3} *_t \Psi_0 \int_{\tau=T}^{t} \exp[-(T/2\tau)\Psi_0(t-\tau)] \times \exp[-\nu_{\rm c}(t-\tau)](T/2\tau)(\tau^2 - T^2){\rm d}\tau, \tag{16.24}$$

where $\Psi_0 = \delta\omega_{\rm pe}^2/c_0$ and $V_0^{\rm R}(t)$ represents the response in the absence of the sheet, which is specified by (4.93). This result has been previously applied in [3] to evaluate the shielding effectiveness of metal screens at optical frequencies, where the concept of perfect electric conductor (PEC) is no longer applicable and the plasmonic behavior of conduction electrons in the metal has to be taken into account.

16.3 A Kirchhoff-approximation-based analysis of a slot-excited resonator antenna

The conventional Fabry–Pérot structure can be conceived as an open resonator antenna consisting of two reflecting screens in between of which the excited wave is reflected back and forth. Such a structure typically finds its applications in the antenna gain enhancement or in designing masers operating at infrared and optical frequencies. In the following sections, closed-form TD expressions describing the pulsed EM-field radiation from a slot-excited resonator antenna are derived. It is demonstrated that the CdH technique and the modified Kirchhoff approximation (see Sections 2.1.1 and 6.6) may serve the purpose very well. The applied solution methodology is largely based on [6] and can be readily generalized to more complex antenna structures.

16.3.1 *Problem formulation*

The problem configuration under consideration is shown in Figure 16.4. The analyzed antenna structure consists of a relatively narrow exciting slot and a PEC thin screen. The slot is excited via the prescribed impulsive electric-field distribution,

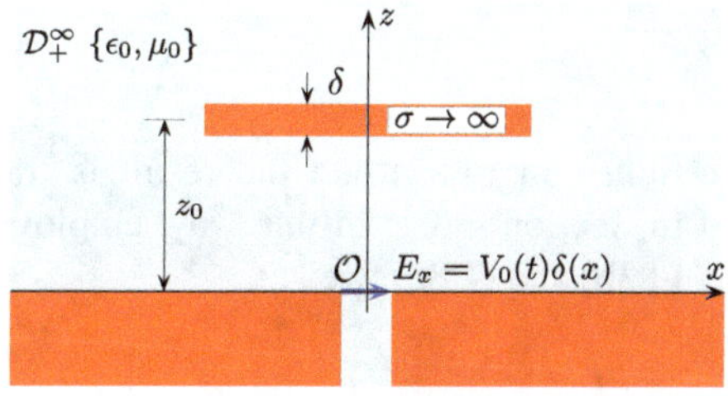

Figure 16.4 A slot-excited resonator antenna

which, as a matter of fact, represents a magnetic-line source. The antenna radiates pulsed EM waves into the unbounded half-space $\mathcal{D}_+^\infty$, whose EM properties are described by (scalar, real-valued) constants, $\{\epsilon_0, \mu_0\}$, implying the wave speed $c_0 = (\epsilon_0\mu_0)^{-1/2} > 0$. The EM fields radiated into the homogeneous and isotropic half-space are subsequently diffracted by the PEC screen occupying a domain $\{-w/2 < x < w/2, -\infty < y < \infty, z_0 - \delta/2 < z < z_0 + \delta/2\}$, where $\delta > 0$ denotes its relatively small thickness, $w > 0$ denotes its width and, finally, $z_0 > 0$ is its height above the slotted plane.

Since the antenna structure as well as its excitation are y-independent, the non-vanishing TM-polarized EM-field components, $\{E_x, E_z, H_y\}(x, z, t)$, are in $\mathcal{D}_+^\infty$ governed by an y-independent system of the source-free Maxwell's equations [4, Section 18.2]. Hence, throughout the half-space $\mathcal{D}_+^\infty$, the magnetic-field strength in the s-domain satisfies

$$(\partial_x^2 + \partial_z^2 - s^2/c_0^2)\hat{H}_y(x, z, s) = 0, \tag{16.25}$$

where, in view of the CdH solution methodology, the Laplace-transform parameter s is kept real-valued and positive. The 2-D (modified) Helmholtz equation is further supplemented with the excitation condition

$$\lim_{z \downarrow 0} \hat{E}_x(x, z, s) = \hat{V}_0(s)\delta(x) \quad \text{for all } x \in \mathbb{R}, \tag{16.26a}$$

where $\hat{V}_0(s)$ represents the s-domain exciting voltage applied in the slot and $\hat{E}_x = -\partial_z \hat{H}_y / s\epsilon_0$. The presence of the PEC screen above the slot is accounted for via the pertaining cross-layer conditions. Hence, with reference to (3.4a), for instance, the tangential component of the electric-field strength is continuous across the screen, that is,

$$\lim_{z \downarrow z_0 + \delta/2} \hat{E}_x(x, z, s) - \lim_{z \uparrow z_0 - \delta/2} \hat{E}_x(x, z, s) = 0 \quad \text{for all } x \in \mathbb{R}, \tag{16.26b}$$

as $\delta \downarrow 0$. On the other hand, the tangential component of the magnetic-field strength jumps across the screen by a certain amount that is not a priori known. To solve the difficulty, we employ a (modified) Kirchhoff-approximate jump condition

$$\begin{aligned}&\lim_{z \downarrow z_0 + \delta/2} \hat{H}_y(x, z, s) - \lim_{z \uparrow z_0 - \delta/2} \hat{H}_y(x, z, s) \\ &= -2\eta_0 \hat{V}_0(s) \sum_{n=1}^{\infty} \exp[-s z_0(2n - 1)/c_0]\left[\mathrm{H}(x - w/2) - \mathrm{H}(x + w/2)\right] \\ &\text{for all } x \in \mathbb{R},\end{aligned} \tag{16.26c}$$

as $\delta \downarrow 0$, resulting from an x-independent wave motion that consists of successively reflected uniform plane waves confined between the PEC planes in the finite interval $\{w/2 < x < w/2\}$ extending over the PEC screen. It can be anticipated that the approximate cross-layer condition (16.26c) yields radiated EM fields with acceptable accuracy for a relatively wide PEC screen. Their derivation is the subject of the following section.

16.3.2 Problem solution

Likewise in Section 4.2, the problem will be next solved under the wave-slowness representation taken along the direction parallel to the PEC screen. Consequently, the non-vanishing magnetic-field component in the s-domain can be represented via (cf. (4.33))

$$\hat{H}_y(x,z,s) = \frac{s}{2\pi\mathrm{i}}\int_{\kappa\in\mathbb{I}_0}\exp(-s\kappa x)\tilde{H}_y(\kappa,z,s)\mathrm{d}\kappa \quad \text{for } \{s\in\mathbb{R}; s>0\}, \tag{16.27}$$

where $\mathbb{I}_0$ is the (indented) integration path that runs along $\mathrm{Re}(\kappa)=0$ except for a vanishingly small semicircle with its center at the origin (see Figure B.1). Consequently, the general solution in the transform domain can be written as a sum of upgoing and downgoing wave constituents propagating below the screen, that is,

$$\tilde{H}_y(x,z,s) = \tilde{A}\exp[-sp(\kappa)z] + \tilde{B}\exp[sp(\kappa)z] \quad \text{for } \{0<z<z_0\}, \tag{16.28}$$

while

$$\tilde{H}_y(x,z,s) = \tilde{C}\exp[-sp(\kappa)z] \quad \text{for } z>z_0, \tag{16.29}$$

above the screen. Recall that $p(\kappa) = (1/c_0^2 - \kappa^2)^{1/2}$ has the meaning of the vertical slowness parameter for which we take $\mathrm{Re}(p)\geq 0$. The (unknown) coefficients $\tilde{A}$, $\tilde{B}$ and $\tilde{C}$ will next be determined with the aid of conditions (16.26a)–(16.26c). To that end, under the wave-slowness representation (16.27), the excitation condition (16.26a) is expressed as

$$\lim_{z\downarrow 0}\tilde{E}_x(\kappa,z,s) = \hat{V}_0(s), \tag{16.30a}$$

where we employed the integral representation of the Dirac-delta distribution. Furthermore, the cross-layer condition (16.26b) can be represented via

$$\lim_{z\downarrow z_0+\delta/2}\tilde{E}_x(\kappa,z,s) - \lim_{z\uparrow z_0-\delta/2}\tilde{E}_x(\kappa,z,s) = 0, \tag{16.30b}$$

and, finally, the transform-domain counterpart of (16.26b) reads

$$\lim_{z\downarrow z_0+\delta/2}\tilde{H}_y(\kappa,z,s) - \lim_{z\uparrow z_0-\delta/2}\tilde{H}_y(\kappa,z,s) = -2\eta_0\hat{V}_0(s)\mathrm{i}_0(s\kappa w/2)\frac{\exp(-sz_0/c_0)}{1-\exp(-2sz_0/c_0)}. \tag{16.30c}$$

In (16.30c), $\mathrm{i}_0(x)$ denotes the modified spherical Bessel function of the first kind. Consequently, making use of (16.28) and (16.29) in the transform-domain conditions (16.30a)–(16.30c), it is found that

$$\tilde{A} = [\epsilon_0/p(\kappa)]\hat{V}_0(s) + \tilde{B}, \tag{16.31a}$$

$$\tilde{C} = [\epsilon_0/p(\kappa)]\hat{V}_0(s) - \tilde{B}\{1-\exp[-2sp(\kappa)z_0]\}\exp[2sp(\kappa)z_0], \tag{16.31b}$$

$$\tilde{B} = \eta_0\hat{V}_0(s)\mathrm{i}_0(s\kappa w/2)\frac{\exp(-sz_0/c_0)}{1-\exp(-2sz_0/c_0)}\exp[-sp(\kappa)z_0]. \tag{16.31c}$$

Equations (16.31a)–(16.31c) determine the problem solution in the transform domain. The first terms on the right-hand side of (16.31a) and (16.31b) correspond to the incident wave (further denoted by superscript ^{i}) that can be, via the slowness representation (16.27), transformed to the TD at once. The result represents the cylindrical wave emanating from the slot that can be described as

$$H_y^{\mathrm{i}}(x,z,t) = \frac{\epsilon_0}{\pi}\partial_t V_0(t) *_t \frac{\mathrm{H}(t-T^{\mathrm{i}})}{(t^2 - T^{\mathrm{i}2})^{1/2}}, \tag{16.32}$$

where $T^{\mathrm{i}} = (x^2+z^2)^{1/2}/c_0$ is the travel time of the incident wave. Combining now (16.31a)–(16.31c) with (16.27)–(16.29), it is straightforward to demonstrate that the remaining Kirchhoff-diffracted wave constituents can be all represented via a generic constituent that can be expressed via

$$\hat{W}(X,Z,s) = \frac{1}{2\pi\mathrm{i}}\int_{\kappa\in\mathbb{I}_0} \exp\{-s[-\kappa X + p(\kappa)Z]\}\frac{\mathrm{d}\kappa}{\kappa}$$
$$\text{for } X\in\mathbb{R}, Z>0 \text{ and } \{s\in\mathbb{R}; s>0\}. \tag{16.33}$$

Relying on the CdH approach as pursued previously in Section 4.2, for instance, we can readily obtain the TD counterpart of $\hat{W}(X,Z,s)$ in closed form, viz

$$W(X,Z,t) = \delta(t - Z/c_0)\mathrm{H}(X) - \frac{1}{\pi}\frac{XZ}{c_0^2t^2 - Z^2}\frac{\mathrm{H}(t-T)}{(t^2-T^2)^{1/2}}, \tag{16.34}$$

in which $T = (X^2+Z^2)^{1/2}/c_0 > 0$. Then, employing the generic constituent to construct the final solution, we finally end up with

$$H_y(x,z,t) = H_y^{\mathrm{i}}(x,z,t) + \sum_{n=1}^{\infty} H_y^{\mathrm{d};n}(x,z,t), \tag{16.35}$$

in which H_y^{i} is given by (16.32) and

$$\begin{aligned} H_y^{\mathrm{d};n}(x,z,t) &= \frac{\eta_0}{w}\left\{V_0\left[t - \frac{z+2nz_0}{c_0}\right] + V_0\left[t - \frac{2nz_0 - z}{c_0}\right]\right\} \\ &\times [\mathrm{H}(x+w/2) - \mathrm{H}(x-w/2)] \\ &+ \frac{\eta_0}{w}V_0\left[t - \frac{(2n-1)z_0}{c_0}\right] *_t \frac{1}{\pi}\left\{\frac{(x-w/2)(z+z_0)}{c_0^2t^2-(z+z_0)^2}\frac{\mathrm{H}(t-T_{\mathrm{R}+})}{(t^2-T_{\mathrm{R}+}^2)^{1/2}}\right. \\ &- \frac{(x+w/2)(z+z_0)}{c_0^2t^2-(z+z_0)^2}\frac{\mathrm{H}(t-T_{\mathrm{L}+})}{(t^2-T_{\mathrm{L}+}^2)^{1/2}} + \frac{(x-w/2)(z_0-z)}{c_0^2t^2-(z_0-z)^2}\frac{\mathrm{H}(t-T_{\mathrm{R}-})}{(t^2-T_{\mathrm{R}-}^2)^{1/2}} \\ &\left. - \frac{(x+w/2)(z_0-z)}{c_0^2t^2-(z_0-z)^2}\frac{\mathrm{H}(t-T_{\mathrm{L}-})}{(t^2-T_{\mathrm{L}-}^2)^{1/2}}\right\} \end{aligned} \tag{16.36a}$$

represents the diffracted field below the PEC screen and

$$\begin{aligned} H_y^{\mathrm{d};n}(x,z,t) &= -\frac{\eta_0}{w} V_0\left(t-\frac{z}{c_0}\right)\delta_{n,1}[\mathrm{H}(x+w/2)-\mathrm{H}(x-w/2)] \\ &- \frac{\eta_0}{w} V_0\left[t-\frac{(2n-1)z_0}{c_0}\right] *_t \frac{1}{\pi}\left\{\frac{(x-w/2)(z-z_0)}{c_0^2t^2-(z-z_0)^2}\frac{\mathrm{H}(t-T_{\mathrm{R}-})}{(t^2-T_{\mathrm{R}-}^2)^{1/2}}\right. \\ &- \frac{(x+w/2)(z-z_0)}{c_0^2t^2-(z-z_0)^2}\frac{\mathrm{H}(t-T_{\mathrm{L}-})}{(t^2-T_{\mathrm{L}-}^2)^{1/2}} - \frac{(x-w/2)(z+z_0)}{c_0^2t^2-(z+z_0)^2}\frac{\mathrm{H}(t-T_{\mathrm{R}+})}{(t^2-T_{\mathrm{R}+}^2)^{1/2}} \\ &\left.+ \frac{(x+w/2)(z+z_0)}{c_0^2t^2-(z+z_0)^2}\frac{\mathrm{H}(t-T_{\mathrm{L}+})}{(t^2-T_{\mathrm{L}+}^2)^{1/2}}\right\} \end{aligned} \tag{16.36b}$$

represents the diffracted field above the screen. Here, $\delta_{n,1}$ denotes the Kronecker delta and

$$T_{\mathrm{R}+} = [(x-w/2)^2+(z+z_0)^2]^{1/2}/c_0, \tag{16.37a}$$

$$T_{\mathrm{L}+} = [(x+w/2)^2+(z+z_0)^2]^{1/2}/c_0, \tag{16.37b}$$

$$T_{\mathrm{R}-} = [(x-w/2)^2+(z_0-z)^2]^{1/2}/c_0, \tag{16.37c}$$

$$T_{\mathrm{L}-} = [(x+w/2)^2+(z_0-z)^2]^{1/2}/c_0 \tag{16.37d}$$

are the arrival times of the diffracted cylindrical wave fronts. Illustrative numerical examples of the derived space–time closed-form expressions are given in the following section.

16.3.3 Illustrative examples

In this section, the TD closed-form expressions describing the EM-field radiation from the slot-excited resonator antenna are evaluated numerically. For this purpose,

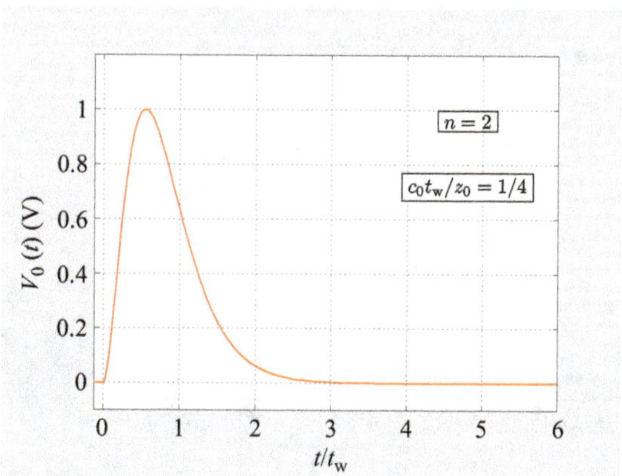

Figure 16.5 Excitation voltage pulse

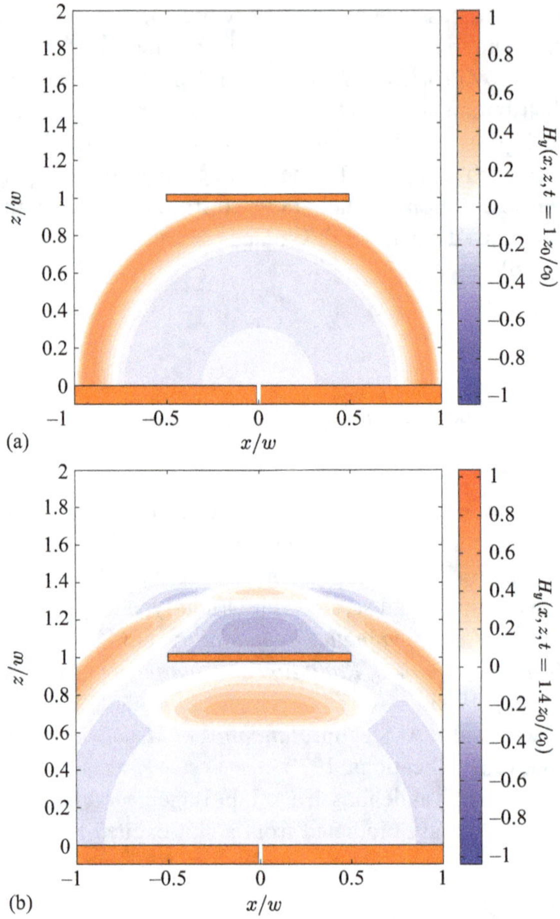

Figure 16.6 Spatial distribution of the magnetic-field strength radiated from the slot-excited resonator antenna as observed at (a) $c_0 t/z_0 = 1.0$*; (b)* $c_0 t/z_0 = 1.4$

a relatively narrow slot is excited by a voltage pulse whose shape is specified via (cf. (4.70))

$$V_0(t) = V_{\mathrm{m}}(t/t_{\mathrm{r}})^n \exp[-n(t/t_{\mathrm{r}} - 1)]\mathrm{H}(t), \tag{16.38}$$

where we take the unit amplitude, $V_{\mathrm{m}} = 1.0\,\mathrm{V}$, and $n = 2$ (see Figure 16.5). Then, the corresponding pulse-rise time approximately follows from $t_{\mathrm{r}} \simeq 0.5413\, t_{\mathrm{w}}$, again. The pulse-time width is related to the height of the PEC screen via $c_0 t_{\mathrm{w}} = z_0/4$, which allows to distinguish the individual wave constituents (see (16.35)) from each other. Finally, the width of the thin PEC screen is chosen to be $w = z_0$.

The spatial distribution of the magnetic-field strength is calculated via (16.35) in the spatial computational domain $\{-w \leq x \leq w, 0 \leq z \leq 2w\}$ and at two time points $c_0 t/z_0 = 1.0$ and $c_0 t/z_0 = 1.4$. Figure 16.6(a) shows the field distribution

at $c_0t/z_0 = 1.0$, that is, just when the slot-excited cylindrical wave front hits the PEC screen. Clearly, up to that instant, the total wave field is identical with the incident wave described by (16.32). Figure 16.6(b) then shows the relatively complex magnetic-field distribution at $c_0t/z_0 = 1.4$ that has shaped up upon hitting the PEC screen. In Figure 16.6(b), the downgoing and upgoing diffracted-wave constituents are clearly distinguishable. Later, after the leading wave front propagates out of the spatial window of observation, the field radiated from the slot is still partly trapped via its reflections against the parallel PEC planes. For more illustrative examples, we refer the reader to [6].

References

[1] S. Celozzi, R. Araneo and G. Lovat. *Electromagnetic Shielding*. Hoboken, NJ: John Wiley & Sons, Inc., 2008.

[2] M. Štumpf and G. Antonini. Transient close-range electromagnetic field coupling between loop antennas at the interface of dissipative half-spaces. *IEEE Transactions on Antennas and Propagation*, 69(3):1805–1808, 2021.

[3] M. Štumpf and A. T. de Hoop. Loop-to-loop pulsed electromagnetic signal transfer across a thin metal screen with Drude-type dispersive behavior. *IEEE Transactions on Electromagnetic Compatibility*, 60(4):885–889, 2018.

[4] A. T. de Hoop. *Handbook of Radiation and Scattering of Waves*. London, UK: Academic Press, 1995.

[5] M. Abramowitz and I. A. Stegun. *Handbook of Mathematical Functions*. New York, NY: Dover Publications, 1972.

[6] M. Štumpf, G. A. E. Vandenbosch and I. E. Lager. Modified Kirchhoff diffraction of pulsed EM waves radiated from a slot-excited Fabry–Pérot resonator antenna. In *12th European Conference on Antennas and Propagation*, pp. 1–5, 2018.

Appendix A
Diffracted-wave amplitude

The edge-diffracted wave on a semi-infinite perfectly electrically conducting (PEC) sheet is expressed in the form of a two-dimensional cylindrical wave propagating away from the edge (see (6.12)). To reveal the link between Sommerfeld's analytical solution [1] and the CdH-based TD solution represented by (6.12), we shall derive an explicit expression for the diffracted wave amplitude.* For readers' convenience, the lengthy derivation is presented in a step-by-step manner.

With reference to (6.7), the diffracted wave amplitude can be written as

$$\mathrm{Re}\{\tilde{A}(\kappa)\} = 2p^{+}(\kappa_0)\mathrm{Re}\left\{\frac{p^{-}(\kappa)}{\kappa-\kappa_0}\right\}. \tag{A.1}$$

In view of the (even) symmetry of the scattered field with respect to z, that is, $E_y^{\mathrm{s}}(x,z,t) = E_y^{\mathrm{s}}(x,-z,t)$ for all $x \in \mathbb{R}$ and $t > 0$ (see Figure 6.14(a)), we shall limit the following analysis to $z > 0$. Accordingly, the diffracted-wave amplitude is further evaluated along $\mathcal{C}$, whose parametric form is rewritten as (cf. (6.9))

$$\kappa(\tau) = \frac{\tau}{\rho}\cos(\theta) + \mathrm{i}\left(\frac{\tau^2}{\rho^2} - \frac{1}{c_0^2}\right)^{1/2}\sin(\theta), \tag{A.2}$$

where we used $\cos(\theta) = x/\rho$ and $\sin(\theta) = z/\rho$ with $\{0 \le \theta \le \pi\}$. In the first step, we introduce angle $\psi = \pi/2 - \beta$ (see Figure 6.3) and use $p^{+}(\kappa) = (1/c_0 + \kappa)^{1/2}$ with $\kappa_0 = \sin(\beta)/c_0 = \cos(\psi)/c_0$ to express $p^{+}(\kappa_0)$ as

$$p^{+}(\kappa_0) = \left[\frac{1}{c_0} + \frac{\cos(\psi)}{c_0}\right]^{1/2} = \left(\frac{2}{c_0}\right)^{1/2}\cos\left(\frac{\psi}{2}\right). \tag{A.3}$$

*The following derivation is in part based on notes [2] provided through the courtesy of Lorentz Chair Emeritus Professor A. T. de Hoop, TU Delft.

Next, using $p^-(\kappa) = (1/c_0 - \kappa)^{1/2}$ with (A.2), we get

$$
\begin{aligned}
p^-[\kappa(\tau)] &= \left[\frac{1}{c_0} - \frac{\tau}{\rho}\cos(\theta) - \mathrm{i}\left(\frac{\tau^2}{\rho^2} - \frac{1}{c_0^2}\right)^{1/2}\sin(\theta)\right]^{1/2} \\
&= \left\{\frac{1}{c_0}\left[\cos^2\left(\frac{\theta}{2}\right) + \sin^2\left(\frac{\theta}{2}\right)\right] - \frac{\tau}{\rho}\left[\cos^2\left(\frac{\theta}{2}\right) - \sin^2\left(\frac{\theta}{2}\right)\right]\right. \\
&\quad \left. - 2\mathrm{i}\left(\frac{\tau^2}{\rho^2} - \frac{1}{c_0^2}\right)^{1/2}\sin\left(\frac{\theta}{2}\right)\cos\left(\frac{\theta}{2}\right)\right\}^{1/2} \\
&= \left[\left(\frac{\tau}{\rho} + \frac{1}{c_0}\right)\sin^2\left(\frac{\theta}{2}\right) - 2\mathrm{i}\left(\frac{\tau^2}{\rho^2} - \frac{1}{c_0^2}\right)^{1/2}\sin\left(\frac{\theta}{2}\right)\cos\left(\frac{\theta}{2}\right)\right. \\
&\quad \left. - \left(\frac{\tau}{\rho} - \frac{1}{c_0}\right)\cos^2\left(\frac{\theta}{2}\right)\right]^{1/2} \\
&= \left(\frac{\tau}{\rho} + \frac{1}{c_0}\right)^{1/2}\sin\left(\frac{\theta}{2}\right) - \mathrm{i}\left(\frac{\tau}{\rho} - \frac{1}{c_0}\right)^{1/2}\cos\left(\frac{\theta}{2}\right),
\end{aligned}
\tag{A.4}
$$

where we employed trigonometric identities $1 = \cos^2(\theta/2) + \sin^2(\theta/2)$, $\cos(\theta) = \cos^2(\theta/2) - \sin^2(\theta/2)$, and $\sin(\theta) = 2\sin(\theta/2)\cos(\theta/2)$. Referring to (A.1), we further write

$$
\begin{aligned}
\frac{1}{\kappa(\tau) - \kappa_0} &= \frac{1}{\frac{\tau}{\rho}\cos(\theta) + \mathrm{i}\left(\frac{\tau^2}{\rho^2} - \frac{1}{c_0^2}\right)^{1/2}\sin(\theta) - \frac{1}{c_0}\cos(\psi)} \\
&= \frac{\frac{\tau}{\rho}\cos(\theta) - \mathrm{i}\left(\frac{\tau^2}{\rho^2} - \frac{1}{c_0^2}\right)^{1/2}\sin(\theta) - \frac{1}{c_0}\cos(\psi)}{\left[\frac{\tau}{\rho}\cos(\theta) - \frac{1}{c_0}\cos(\psi)\right]^2 + \left(\frac{\tau^2}{\rho^2} - \frac{1}{c_0^2}\right)\sin^2(\theta)} \\
&= \frac{\frac{\tau}{\rho}\cos(\theta) - \mathrm{i}\left(\frac{\tau^2}{\rho^2} - \frac{1}{c_0^2}\right)^{1/2}\sin(\theta) - \frac{1}{c_0}\cos(\psi)}{\frac{\tau^2}{\rho^2} + \frac{1}{c_0^2}\left[\cos^2(\psi) - \sin^2(\theta)\right] - \frac{2}{c_0}\frac{\tau}{\rho}\cos(\theta)\cos(\psi)}.
\end{aligned}
\tag{A.5}
$$

Employing next relations $\cos^2(\psi) - \sin^2(\theta) = \cos(\psi - \theta)\cos(\psi + \theta)$ and $2\cos(\psi)\cos(\theta) = \cos(\psi - \theta) + \cos(\psi + \theta)$, (A.5) is further cast into the following form:

$$
\begin{aligned}
\frac{1}{\kappa(\tau) - \kappa_0} &= \frac{\frac{\tau}{\rho}\cos(\theta) - \mathrm{i}\left(\frac{\tau^2}{\rho^2} - \frac{1}{c_0^2}\right)^{1/2}\sin(\theta) - \frac{1}{c_0}\cos(\psi)}{\frac{\tau^2}{\rho^2} + \frac{1}{c_0^2}\cos(\psi - \theta)\cos(\psi + \theta) - \frac{1}{c_0}\frac{\tau}{\rho}\left[\cos(\psi - \theta) + \cos(\psi + \theta)\right]} \\
&= \frac{\frac{\tau}{\rho}\cos(\theta) - \mathrm{i}\left(\frac{\tau^2}{\rho^2} - \frac{1}{c_0^2}\right)^{1/2}\sin(\theta) - \frac{1}{c_0}\cos(\psi)}{\left[\frac{\tau}{\rho} - \frac{1}{c_0}\cos(\psi - \theta)\right]\left[\frac{\tau}{\rho} - \frac{1}{c_0}\cos(\psi + \theta)\right]}.
\end{aligned}
\tag{A.6}
$$

Consequently, combination of (A.4) and (A.6) yields

$$\mathrm{Re}\left\{\frac{p^-[\kappa(\tau)]}{\kappa(\tau)-\kappa_0}\right\} = \frac{\left(\frac{\tau}{\rho}+\frac{1}{c_0}\right)^{1/2}\sin\left(\frac{\theta}{2}\right)\left[\frac{\tau}{\rho}\cos(\theta)-\frac{1}{c_0}\cos(\psi)\right]}{\left[\frac{\tau}{\rho}-\frac{1}{c_0}\cos(\psi-\theta)\right]\left[\frac{\tau}{\rho}-\frac{1}{c_0}\cos(\psi+\theta)\right]} - \frac{\left(\frac{\tau}{\rho}-\frac{1}{c_0}\right)^{1/2}\cos\left(\frac{\theta}{2}\right)\left(\frac{\tau^2}{\rho^2}-\frac{1}{c_0^2}\right)^{1/2}\sin(\theta)}{\left[\frac{\tau}{\rho}-\frac{1}{c_0}\cos(\psi-\theta)\right]\left[\frac{\tau}{\rho}-\frac{1}{c_0}\cos(\psi+\theta)\right]}, \tag{A.7}$$

which is further rewritten as

$$\begin{aligned}\mathrm{Re}\left\{\frac{p^-[\kappa(\tau)]}{\kappa(\tau)-\kappa_0}\right\} &= \frac{\left(\frac{\tau}{\rho}+\frac{1}{c_0}\right)^{1/2}}{\left[\frac{\tau}{\rho}-\frac{1}{c_0}\cos(\psi-\theta)\right]\left[\frac{\tau}{\rho}-\frac{1}{c_0}\cos(\psi+\theta)\right]}\\
&\times\left\{\sin\left(\frac{\theta}{2}\right)\left[\frac{\tau}{\rho}\cos(\theta)-\frac{1}{c_0}\cos(\psi)\right]-\left(\frac{\tau}{\rho}-\frac{1}{c_0}\right)\cos\left(\frac{\theta}{2}\right)\sin(\theta)\right\}\\
&= \frac{\left(\frac{\tau}{\rho}+\frac{1}{c_0}\right)^{1/2}}{\left[\frac{\tau}{\rho}-\frac{1}{c_0}\cos(\psi-\theta)\right]\left[\frac{\tau}{\rho}-\frac{1}{c_0}\cos(\psi+\theta)\right]}\\
&\left\{\frac{\tau}{\rho}\left[\sin\left(\frac{\theta}{2}\right)\cos(\theta)-\cos\left(\frac{\theta}{2}\right)\sin(\theta)\right]\right.\\
&\left.-\frac{1}{c_0}\left[\sin\left(\frac{\theta}{2}\right)\cos(\psi)-\cos\left(\frac{\theta}{2}\right)\sin(\theta)\right]\right\}.\end{aligned} \tag{A.8}$$

To arrive at an expression for the wave amplitude, we next combine (A.3) with (A.8) and get

$$\begin{aligned}\mathrm{Re}\{\tilde{A}[\kappa(\tau)]\} &= 2\left(\frac{2}{c_0}\right)^{1/2}\frac{\left(\frac{\tau}{\rho}+\frac{1}{c_0}\right)^{1/2}}{\left[\frac{\tau}{\rho}-\frac{1}{c_0}\cos(\psi-\theta)\right]\left[\frac{\tau}{\rho}-\frac{1}{c_0}\cos(\psi+\theta)\right]}\\
&\times\left\{\frac{1}{c_0}\cos\left(\frac{\psi}{2}\right)\sin\left(\frac{\theta}{2}\right)\left[2\sin\left(\frac{\psi-\theta}{2}\right)\sin\left(\frac{\psi+\theta}{2}\right)+1\right]\right.\\
&\left.-\frac{\tau}{\rho}\cos\left(\frac{\psi}{2}\right)\sin\left(\frac{\theta}{2}\right)\right\},\end{aligned} \tag{A.9}$$

where we used $\sin(\theta/2)\cos(\theta)-\cos(\theta/2)\sin(\theta)=-\sin(\theta/2)$ and $\sin(\theta/2)\cos(\psi)-\cos(\theta/2)\sin(\theta)=-\sin(\theta/2)\{2\sin[(\psi-\theta)/2]\sin[(\psi+\theta)/2]+1\}$. In the following step, we employ $2\cos(\psi/2)\sin(\theta/2)=\sin[(\psi+\theta)/2]-\sin[(\psi-\theta)/2]$ along with

$$\begin{aligned}&2\cos(\psi/2)\sin(\theta/2)\{2\sin[(\psi-\theta)/2]\sin[(\psi+\theta)/2]+1\}\\
&=\sin[(\psi+\theta)/2]\cos(\psi-\theta)-\sin[(\psi-\theta)/2]\cos(\psi+\theta),\end{aligned} \tag{A.10}$$

to find

$$\mathrm{Re}\{\tilde{A}[\kappa(\tau)]\} = \left(\frac{2}{c_0}\right)^{1/2} \frac{\left(\frac{\tau}{\rho}+\frac{1}{c_0}\right)^{1/2}}{\left[\frac{\tau}{\rho}-\frac{1}{c_0}\cos(\psi-\theta)\right]\left[\frac{\tau}{\rho}-\frac{1}{c_0}\cos(\psi+\theta)\right]}$$
$$\left\{\sin\left(\frac{\psi-\theta}{2}\right)\left[\frac{\tau}{\rho}-\frac{\cos(\psi+\theta)}{c_0}\right]-\sin\left(\frac{\psi+\theta}{2}\right)\left[\frac{\tau}{\rho}-\frac{\cos(\psi-\theta)}{c_0}\right]\right\}, \tag{A.11}$$

which finally gives

$$\mathrm{Re}\{\tilde{A}[\kappa(\tau)]\} = \left[2\left(\frac{c_0\tau}{\rho}+1\right)\right]^{1/2}\left\{\frac{\sin[(\psi-\theta)/2]}{c_0\tau/\rho-\cos(\psi-\theta)} - \frac{\sin[(\psi+\theta)/2]}{c_0\tau/\rho-\cos(\psi+\theta)}\right\}. \tag{A.12}$$

This expression can be used in (6.12) to evaluate the EM edge diffraction on a PEC screen.

References

[1] A. Sommerfeld. Mathematische Theorie der Diffraction. *Mathematische Annalen*, 47(2):317–374, 1896.

[2] A. T. de Hoop. *The 1956–1957 UCLA Notebook*. Notes on the pulsed plane-wave diffraction by a semi-infinite soft or semi-infinite hard planar scatterer of vanishing thickness. December 2013.

Appendix B
Impeditivity array of a PEC screen

The elements of the impeditivity array that occurs in the Cagniard-DeHoop method of moments (CdH-MoM) solution of electromagnetic (EM) scattering by a bounded perfectly electrically conducting (PEC) screen can be expressed through the following wave-slowness representation (cf. (E.1)):

$$\hat{\mathcal{Z}}^{[S,n]}(s) = \frac{\zeta_0}{sc_0\Delta t}\frac{1}{2\pi\mathrm{i}}\int_{\kappa=-\mathrm{i}\infty}^{\mathrm{i}\infty} \tilde{\Pi}^{[S]}(\kappa)\tilde{\Lambda}^{[n]}(-\kappa)\frac{\mathrm{d}\kappa}{2p(\kappa)}$$
$$\text{for all } S = \{1,\dots,N\}, n = \{1,\dots,N\} \text{ and } \{s \in \mathbb{R}; s > 0\}, \tag{B.1}$$

where $\zeta_0 = (\mu_0/\epsilon_0)^{1/2}$ is the free-space impedance and $p(\kappa) = (1/c_0^2 - \kappa^2)^{1/2}$ with $\mathrm{Re}(p) \geq 0$, thus entailing the branch cuts along $\{\mathrm{Im}(\kappa) = 0, 1/c_0 \leq |\mathrm{Re}(\kappa)| < \infty\}$. Furthermore, the integrand is bounded at $\kappa = 0$, which makes it possible to indent the integration path to the right with a vanishingly small semicircle without affecting the result of integration. The indented integration contour is denoted by $\mathbb{I}_0$ (see Figure B.1). Subsequently, employing the transform-domain counterparts of the basis and testing functions specified in Section 8.1.2, the array elements can be written

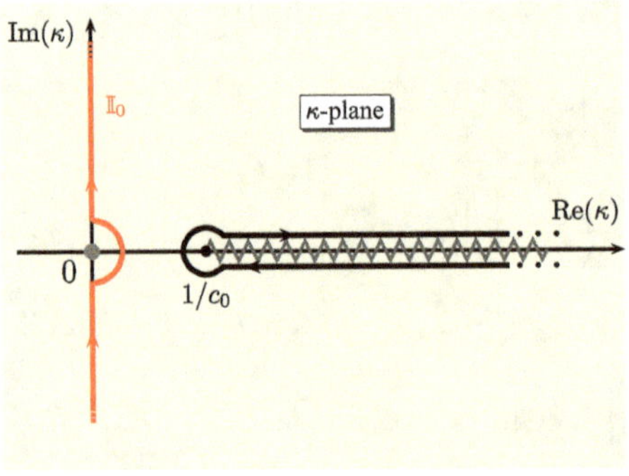

Figure B.1 Complex κ-plane with the CdH path encircling the branch cut for $x < 0$

out in terms of generic integral representations, $\hat{\Upsilon}^{A,B}(x,s)$, the TD counterparts of which are derived next in Sections B.1 and B.2. In this manner, we may write

$$\hat{\mathcal{Z}}^{[S,n]}(s) = \frac{\zeta_0}{c_0\Delta t}\frac{1}{\Delta}\Big[\hat{\Upsilon}^A(x_S - x_n + 3\Delta/2, s) - 3\hat{\Upsilon}^A(x_S - x_n + \Delta/2, s) + 3\hat{\Upsilon}^A(x_S - x_n - \Delta/2, s) - \hat{\Upsilon}^A(x_S - x_n - 3\Delta/2, s)\Big] \quad \text{for all } S = \{2,\ldots,N-1\} \text{ and } n = \{2,\ldots,N-1\}, \tag{B.2a}$$

$$\hat{\mathcal{Z}}^{[S,1]}(s) = \frac{\zeta_0}{c_0\Delta t}\frac{1}{\Delta}\Big[-\hat{\Upsilon}^A(x_S - x_1 + \Delta/2, s) + 2\hat{\Upsilon}^A(x_S - x_1 - \Delta/2, s) - \hat{\Upsilon}^A(x_S - x_1 - 3\Delta/2, s)\Big] + \frac{\zeta_0}{c_0\Delta t}\Big[\hat{\Upsilon}^B(x_S - x_1 + \Delta/2, s) - \hat{\Upsilon}^B(x_S - x_1 - \Delta/2, s)\Big] \quad \text{for all } S = \{2,\ldots,N-1\}, \tag{B.2b}$$

and

$$\hat{\mathcal{Z}}^{[S,N]}(s) = \frac{\zeta_0}{c_0\Delta t}\frac{1}{\Delta}\Big[\hat{\Upsilon}^A(x_S - x_N + 3\Delta/2, s) - 2\hat{\Upsilon}^A(x_S - x_N + \Delta/2, s) + \hat{\Upsilon}^A(x_S - x_N - \Delta/2, s)\Big] - \frac{\zeta_0}{c_0\Delta t}\Big[\hat{\Upsilon}^B(x_S - x_N + \Delta/2, s) - \hat{\Upsilon}^B(x_S - x_N - \Delta/2, s)\Big] \quad \text{for all } S = \{2,\ldots,N-1\}, \tag{B.2c}$$

with

$$\hat{\mathcal{Z}}^{[1,1]}(s) = \frac{\zeta_0}{c_0\Delta t}\frac{1}{\Delta}\Big[-\hat{\Upsilon}^A(\Delta/2, s) + \hat{\Upsilon}^A(0, s) + \hat{\Upsilon}^A(-\Delta/2, s) - \hat{\Upsilon}^A(-\Delta, s)\Big] + \frac{\zeta_0}{c_0\Delta t}\Big[\hat{\Upsilon}^B(\Delta/2, s) - \hat{\Upsilon}^B(0, s)\Big], \tag{B.2d}$$

and

$$\hat{\mathcal{Z}}^{[N,1]}(s) = \frac{\zeta_0}{c_0\Delta t}\frac{1}{\Delta}\Big[-\hat{\Upsilon}^A(x_N - x_1, s) + \hat{\Upsilon}^A(x_N - x_1 - \Delta/2, s) + \hat{\Upsilon}^A(x_N - x_1 - \Delta, s) - \hat{\Upsilon}^A(x_N - x_1 - 3\Delta/2, s)\Big] + \frac{\zeta_0}{c_0\Delta t}\Big[\hat{\Upsilon}^B(x_N - x_1, s) - \hat{\Upsilon}^B(x_N - x_1 - \Delta/2, s)\Big]. \tag{B.2e}$$

Finally, the remaining array elements can be found by invoking the property of symmetry, $\hat{\mathcal{Z}}^{[S,n]}(s) = \hat{\mathcal{Z}}^{[n,S]}(s)$.

B.1 Generic integral $\hat{\Upsilon}^A$

The first generic integral has the following form:

$$\hat{\Upsilon}^A(x,s) = \frac{1}{s}\frac{1}{2\pi\mathrm{i}}\int_{\kappa\in\mathbb{I}_0}\frac{\exp(s\kappa x)}{s^3\kappa^3}\frac{\mathrm{d}\kappa}{2p(\kappa)}$$
$$\text{for } x\in\mathbb{R} \text{ and } \{s\in\mathbb{R}; s>0\}, \tag{B.3}$$

where the integration path is depicted in Figure B.1. In order to derive the TD counterpart of (B.3), it is first observed that the integrand has a pole of order three at $\kappa = 0$. Moreover, in the complex κ-plane, we introduce branch cuts along $\{\mathrm{Im}(\kappa) = 0, 1/c_0 \leq |\mathrm{Re}(\kappa)| < \infty\}$ due to the square root in $p(\kappa)$. Pursuing the line of reasoning of the CdH technique (see Section 2.1.1), Cauchy's theorem and Jordan's lemma [1, p. 1054] are applied to replace the indented integration path, $\mathbb{I}_0$, with the CdH path encircling the branch cuts (see Figure B.1). Its parametric representation can be expressed via

$$\kappa(\tau) = -\tau/x \pm \mathrm{i}0 \quad \text{for all } \{\tau\in\mathbb{R}; \tau\geq |x|/c_0\}. \tag{B.4}$$

Introducing next the parameter τ as the new variable of integration and combining the contributions of the integrations just above and below the branch cut, (B.3) can be cast into the following form:

$$\hat{\Upsilon}^A(x,s) = -\frac{1}{2\pi s^4}\int_{\tau=|x|/c_0}^{\infty}\exp(-s\tau)\frac{x^3}{\tau^3}\frac{\mathrm{d}\tau}{(\tau^2 - x^2/c_0^2)^{1/2}}$$
$$+\frac{c_0^2}{4s^4}\left(\frac{s^2x^2}{c_0}+c_0\right)\mathrm{H}(x), \tag{B.5}$$

where the second term on the right-hand side represents the pole contribution. Owing to the indentation of $\mathbb{I}_0$ to the right half-plane, the pole yields a nonzero contribution for $x > 0$ only. Apart from the factor s^{-4}, the integral in (B.5) has the form of Laplace transformation. Consequently, Lerch's uniqueness theorem [2, Appendix] could be applied directly to find an expression for $\Upsilon^A(x,t)$. Such a TD expression would contain the time convolution, which may be computationally demanding. To avoid the difficulty, the integral with respect to τ is carried out analytically and the thus achieved s-domain expression is transformed back to the TD. In this way, after a few steps of algebra, we end up with

$$\Upsilon^A(x,t) = \frac{\mathrm{sgn}(x)}{4\pi}\left\{\frac{|x|^3}{3}\cosh^{-1}\left(\frac{c_0t}{|x|}\right)+\frac{5}{6}c_0tx^2\left(\frac{c_0^2t^2}{x^2}-1\right)^{1/2}\right.$$
$$\left.-c_0t\left(x^2+\frac{c_0^2t^2}{6}\right)\tan^{-1}\left[\left(\frac{c_0^2t^2}{x^2}-1\right)^{1/2}\right]\right\}\mathrm{H}(t-|x|/c_0)$$
$$+\frac{c_0t}{4}\left(x^2+\frac{c_0^2t^2}{6}\right)\mathrm{H}(x)\mathrm{H}(t) \quad \text{for all } x\in\mathbb{R} \text{ and } t>0, \tag{B.6}$$

where $\mathrm{sgn}(x)$ denotes the sign function. Finally, the limit for $x = 0$ follows as $\Upsilon^A(0,t) = (c_0^3t^3/48)\mathrm{H}(t)$.

B.2 Generic integral $\hat{\Upsilon}^B$

The second generic integral has the following form:

$$\hat{\Upsilon}^B(x,s) = \frac{1}{s}\frac{1}{2\pi\mathrm{i}}\int_{\kappa\in\mathbb{I}_0}\frac{\exp(s\kappa x)}{s^2\kappa^2}\frac{\mathrm{d}\kappa}{2p(\kappa)}$$
$$\text{for } x\in\mathbb{R} \text{ and } \{s\in\mathbb{R}; s>0\}. \tag{B.7}$$

To transform (B.7), we next adopt the strategy applied in Section B.1. Hence, the integrand is first continued analytically away from the imaginary axis, and under the application of Cauchy's theorem and Jordan's lemma [1, p. 1054], the indented integration path is replaced with the CdH path defined via (B.4). Subsequently, we change the variable of integration to τ, which leads to

$$\hat{\Upsilon}^B(x,s) = \frac{1}{2\pi s^3}\int_{\tau=|x|/c_0}^{\infty}\exp(-s\tau)\frac{x^2}{\tau^2}\frac{\mathrm{d}\tau}{(\tau^2 - x^2/c_0^2)^{1/2}}$$
$$+ (c_0 x/2s^2)\mathrm{H}(x), \tag{B.8}$$

in which we included the contribution of the double pole singularity at $\kappa = 0$. Carrying out the integration with respect to τ analytically and transforming the result to the TD, we end up with

$$\Upsilon^B(x,t) = \frac{1}{2\pi}\left\{\frac{x^2}{2}\cosh^{-1}\left(\frac{c_0 t}{|x|}\right) + \frac{c_0 t x}{2}\left(\frac{c_0^2 t^2}{x^2} - 1\right)^{1/2}\right.$$
$$\left. - c_0 t|x|\tan^{-1}\left[\left(\frac{c_0^2 t^2}{x^2} - 1\right)^{1/2}\right]\right\}\mathrm{H}(t - |x|/c_0)$$
$$+ (c_0 t x/2)\mathrm{H}(x)\mathrm{H}(t) \quad \text{for all } x\in\mathbb{R} \text{ and } t>0. \tag{B.9}$$

The corresponding limit for $x = 0$ then follows as $\Upsilon^B(0,t) = (c_0^2 t^2/4\pi)\mathrm{H}(t)$.

References

[1] A. T. de Hoop, *Handbook of Radiation and Scattering of Waves*, London, UK: Academic Press, 1995.

[2] M. Štumpf. *Electromagnetic Reciprocity in Antenna Theory*. Hoboken, NJ: John Wiley & Sons, Inc., 2018.

Appendix C

Plane-wave excitation array

The elements of the excitation array that occurs in the Cagniard-DeHoop method of moments (CdH-MoM) solution of electromagnetic (EM) scattering by a bounded screen can be expressed through the following wave-slowness representation:

$$\hat{V}^{[S]}(s) = \frac{s}{2\pi \mathrm{i}} \int_{\kappa=-\mathrm{i}\infty}^{\mathrm{i}\infty} \tilde{E}_y^{\mathrm{i}}(\kappa, 0, s) \tilde{\Pi}^{[S]}(-\kappa) \mathrm{d}\kappa$$
$$\text{for all } S = \{1, \ldots, N\} \text{ and } \{s \in \mathbb{R}; s > 0\}, \tag{C.1}$$

where we used the transform-domain counterpart of (8.8) and $\tilde{E}_y^{\mathrm{i}}(\kappa, 0, s)$ follows from the slowness representation of (4.1) at $z = 0$. The integrand of (C.1) is bounded at the origin $\kappa = 0$, including the normal incidence for which $\kappa_0 = 0$. Consequently, the integration path along $\mathrm{Re}(\kappa) = 0$ can be replaced with $\mathbb{I}_0$ (see Figure B.1) without changing the result of integration.

The elements of the excitation array can be further cast into the following form:

$$\begin{aligned} \hat{V}^{[S]}(s) = {} & \hat{e}^{\mathrm{i}}(s) \Big[\hat{\Theta}(x_N + \Delta/2 - x_S, s; \kappa_0) \\ & - \hat{\Theta}(x_N - \Delta/2 - x_S, s; \kappa_0) \Big] \exp(-s\kappa_0 x_N) \\ & - \hat{e}^{\mathrm{i}}(s) \Big[\hat{\Theta}(x_1 + \Delta/2 - x_S, s; \kappa_0) \\ & - \hat{\Theta}(x_1 - \Delta/2 - x_S, s; \kappa_0) \Big] \exp(-s\kappa_0 x_1) \\ & \text{for all } S = \{2, \ldots, N-1\}, \end{aligned} \tag{C.2a}$$

with

$$\begin{aligned} \hat{V}^{[1]}(s) = {} & \hat{e}^{\mathrm{i}}(s) \Big[\hat{\Theta}(x_N - x_1, s; \kappa_0) - \hat{\Theta}(x_N - \Delta/2 - x_1, s; \kappa_0) \Big] \exp(-s\kappa_0 x_N) \\ & - \hat{e}^{\mathrm{i}}(s) \Big[\hat{\Theta}(0, s; \kappa_0) - \hat{\Theta}(-\Delta/2, s; \kappa_0) \Big] \exp(-s\kappa_0 x_1), \end{aligned} \tag{C.2b}$$

and

$$\begin{aligned} \hat{V}^{[N]}(s) = {} & \hat{e}^{\mathrm{i}}(s) \Big[\hat{\Theta}(\Delta/2, s; \kappa_0) - \hat{\Theta}(0, s; \kappa_0) \Big] \exp(-s\kappa_0 x_N) \\ & - \hat{e}^{\mathrm{i}}(s) \Big[\hat{\Theta}(x_1 + \Delta/2 - x_N, s; \kappa_0) - \hat{\Theta}(x_1 - x_N, s; \kappa_0) \Big] \exp(-s\kappa_0 x_1), \end{aligned} \tag{C.2c}$$

where $\hat{\Theta}$ is the generic integral that is evaluated in closed form in the following section. Employing these results, we get s-domain expressions for $\hat{V}^{[S]}(s)$ that can be readily transformed to the TD. The thus obtained TD expressions are given in Section 8.1.2.

C.1 Generic integral $\hat{\Theta}$

The generic constituent can be expressed via the following complex slowness integral:

$$\hat{\Theta}(x, s; \kappa_0) = \frac{1}{s}\frac{1}{2\pi \mathrm{i}}\int_{\kappa\in\mathbb{I}_0} \exp(s\kappa x)\frac{\mathrm{d}\kappa}{\kappa(\kappa-\kappa_0)}$$
$$\text{for } x \in \mathbb{R} \text{ and } \{s \in \mathbb{R}; s > 0\}, \tag{C.3}$$

where the integration path is depicted in Figure B.1. The integrand has two simple pole singularities at $\kappa = 0$ and $\kappa = \kappa_0$ and can be hence evaluated with the aid of Cauchy's formula [1, Section 2.41]. In doing so, we shall distinguish two cases. At first, if $x < 0$, the indented integration path $\mathbb{I}_0$ can be in $\mathrm{Re}(\kappa) > 0$ supplemented with a semicircle of infinite radius with center at the origin. The integration along the resulting closed contour gives a non-vanishing contribution when the pole singularity lies inside the resulting closed contour only, that is,

$$\hat{\Theta}(x, s; \kappa_0) = -\frac{\exp(s\kappa_0 x)\mathrm{H}(\kappa_0)}{s\kappa_0} \quad \text{for } x < 0. \tag{C.4}$$

From (C.4), we may deduce the zero-offset limiting case, that is, $\hat{\Theta}(0, s; \kappa_0) = -\mathrm{H}(\kappa_0)/s\kappa_0$ with $\hat{\Theta}(0, s; 0) = 0$. A similar line of reasoning is pursued for $x > 0$. In this case, Jordan's lemma can be used to supplement the integration path with a large semicircle to the left. Owing to the chosen indentation, the contributions of both poles at $\kappa = 0$ and $\kappa = \kappa_0$ must also be accounted for, which yields

$$\hat{\Theta}(x, s; \kappa_0) = \frac{\exp(s\kappa_0 x)\mathrm{H}(-\kappa_0)}{s\kappa_0} - \frac{1}{s\kappa_0} \quad \text{for } x > 0. \tag{C.5}$$

The limiting case $\kappa_0 = 0$ pertaining to the normal incidence can be evaluated at once using Cauchy's formula, again. In this way, we find

$$\hat{\Theta}(x, s; 0) = \frac{1}{s}\frac{1}{2\pi \mathrm{i}}\int_{\kappa\in\mathbb{I}_0} \exp(s\kappa x)\frac{\mathrm{d}\kappa}{\kappa^2} = x\mathrm{H}(x). \tag{C.6}$$

These results are used to find closed-form expressions for the excitation-array elements $\hat{V}^{[N]}(s)$.

Reference

[1] E. C. Titchmarsh. *The Theory of Functions, 2nd edition.* London, UK: Oxford University Press, 1939.

Appendix D

Impeditivity array of a screen with conductive and dielectric properties

The elements of the impeditivity array that occurs in the Cagniard-DeHoop method of moments (CdH-MoM) solution of electromagnetic (EM) scattering by a bounded screen with conductive and dielectric properties can be expressed through the following wave-slowness representation:

$$\hat{\mathcal{R}}^{[S,n]}(s) = \frac{\hat{Z}^{\mathrm{E}}(s)}{c_0\Delta t}\frac{c_0}{s}\frac{1}{2\pi\mathrm{i}}\int_{\kappa=-\mathrm{i}\infty}^{\mathrm{i}\infty} \tilde{\Pi}^{[S]}(-\kappa)\tilde{\Lambda}^{[n]}(\kappa)\mathrm{d}\kappa$$
$$\text{for all } S = \{1,\ldots,N\}, n = \{1,\ldots,N\} \text{ and } \{s \in \mathbb{R}; s > 0\}. \tag{D.1}$$

Since the integrand in (D.1) is bounded at $\kappa = 0$, the integration contour along the imaginary axis can be indented to the right with a semicircular arc with center at the origin and vanishing small radius (see $\mathbb{I}_0$ in Figure B.1). Such a deformation does not change the result of integration. Consequently, the elements of $\underline{\mathcal{R}}$ can be expressed via generic integral representations, $\hat{\Phi}^{A,B}(x,s)$, the TD counterparts of which are derived next in Sections D.1 and D.2. In this manner, we may write

$$\hat{\mathcal{R}}^{[S,n]}(s) = \frac{\hat{Z}^{\mathrm{E}}(s)}{c_0\Delta t}\frac{1}{\Delta}\Big[\hat{\Phi}^{A}(x_n - x_S + 3\Delta/2, s) - 3\hat{\Phi}^{A}(x_n - x_S + \Delta/2, s)$$
$$+ 3\hat{\Phi}^{A}(x_n - x_S - \Delta/2, s) - \hat{\Phi}^{A}(x_n - x_S - 3\Delta/2, s)\Big]$$
$$\text{for all } S = \{2,\ldots,N-1\} \text{ and } n = \{2,\ldots,N-1\}, \tag{D.2a}$$

$$\hat{\mathcal{R}}^{[S,1]}(s) = \frac{\hat{Z}^{\mathrm{E}}(s)}{c_0\Delta t}\frac{1}{\Delta}\Big[\hat{\Phi}^{A}(x_1 - x_S - \Delta/2, s) - 2\hat{\Phi}^{A}(x_1 - x_S + \Delta/2, s)$$
$$+ \hat{\Phi}^{A}(x_1 - x_S + 3\Delta/2, s)\Big]$$
$$- \frac{\hat{Z}^{\mathrm{E}}(s)}{c_0\Delta t}\Big[\hat{\Phi}^{B}(x_1 - x_S + \Delta/2, s) - \hat{\Phi}^{B}(x_1 - x_S - \Delta/2, s)\Big]$$
$$\text{for all } S = \{2,\ldots,N-1\}, \tag{D.2b}$$

and

$$\hat{\mathcal{R}}^{[S,N]}(s) = \frac{\hat{Z}^{\mathrm{E}}(s)}{c_0 \Delta t} \frac{1}{\Delta} \Big[-\hat{\Phi}^A(x_N - x_S - 3\Delta/2, s) + 2\hat{\Phi}^A(x_N - x_S - \Delta/2, s) - \hat{\Phi}^A(x_N - x_S + \Delta/2, s) \Big] + \frac{\hat{Z}^{\mathrm{E}}(s)}{c_0 \Delta t} \Big[\hat{\Phi}^B(x_N - x_S + \Delta/2, s) - \hat{\Phi}^B(x_1 - x_S - \Delta/2, s) \Big] \quad \text{for all } S = \{2, \ldots, N-1\}, \tag{D.2c}$$

with

$$\hat{\mathcal{R}}^{[1,1]}(s) = \frac{\hat{Z}^{\mathrm{E}}(s)}{c_0 \Delta t} \frac{1}{\Delta} \Big[\hat{\Phi}^A(-\Delta/2, s) - \hat{\Phi}^A(0, s) - \hat{\Phi}^A(\Delta/2, s) + \hat{\Phi}^A(\Delta, s) \Big] - \frac{\hat{Z}^{\mathrm{E}}(s)}{c_0 \Delta t} \Big[\hat{\Phi}^B(0, s) - \hat{\Phi}^B(-\Delta/2, s) \Big], \tag{D.2d}$$

and

$$\hat{\mathcal{R}}^{[N,1]}(s) = \frac{\hat{Z}^{\mathrm{E}}(s)}{c_0 \Delta t} \frac{1}{\Delta} \Big[\hat{\Phi}^A(x_1 - x_N, s) - \hat{\Phi}^A(x_1 - x_N + \Delta/2, s) - \hat{\Phi}^A(x_1 - x_N + \Delta, s) + \hat{\Phi}^A(x_1 - x_N - 3\Delta/2, s) \Big] - \frac{\hat{Z}^{\mathrm{E}}(s)}{c_0 \Delta t} \Big[\hat{\Phi}^B(x_1 - x_N + \Delta/2, s) - \hat{\Phi}^B(x_1 - x_N, s) \Big]. \tag{D.2e}$$

Finally, the remaining array elements can be found by invoking the property of symmetry, $\hat{\mathcal{R}}^{[S,n]}(s) = \hat{\mathcal{R}}^{[n,S]}(s)$.

D.1 Generic integral $\hat{\Phi}^A$

The first generic integral has the following form:

$$\hat{\Phi}^A(x, s) = \frac{c_0}{s} \frac{1}{2\pi \mathrm{i}} \int_{\kappa \in \mathbb{I}_0} \frac{\exp(s\kappa x)}{s^3 \kappa^3} \mathrm{d}\kappa \quad \text{for } x \in \mathbb{R} \text{ and } \{s \in \mathbb{R}; s > 0\}, \tag{D.3}$$

where the integration path is depicted in Figure B.1. To evaluate the integral, Jordan's lemma is applied to supplement $\mathbb{I}_0$ with a semicircle of infinite radius to the left, if $x > 0$, or to the right, if $x < 0$. Owing to the indentation of $\mathbb{I}_0$, the integration along the resulting closed contour yields the zero value for $x < 0$. On the other hand, the closed contour for $x > 0$ can be contracted to a small circle around the pole singularity at $\kappa = 0$, the integration along which can be carried out with the aid of Cauchy's formula [1, Section 2.41]. This way yields

$$\hat{\Phi}^A(x, s) = (c_0 x^2 / 2s^2) \mathrm{H}(x) \quad \text{for all } x \in \mathbb{R}. \tag{D.4}$$

Finally, the TD counterpart of

$$\hat{\Gamma}^A(x,s) = [\hat{Z}^{\mathrm{E}}(s)/\Delta c_0 \Delta t]\hat{\Phi}^A(x,s) \tag{D.5}$$

is used in Section 8.2 to evaluate the elements of the impeditivity array. In doing so, the property $\hat{\Gamma}^A(x,s) = 0$ for $x \le 0$ is employed.

D.2 Generic integral $\hat{\Phi}^B$

The second generic integral has the following form:

$$\hat{\Phi}^B(x,s) = \frac{c_0}{s}\frac{1}{2\pi\mathrm{i}}\int_{\kappa\in\mathbb{I}_0} \frac{\exp(s\kappa x)}{s^2\kappa^2}\mathrm{d}\kappa$$
$$\text{for } x \in \mathbb{R} \text{ and } \{s \in \mathbb{R}; s > 0\}, \tag{D.6}$$

where the integration path is depicted in Figure B.1. To evaluate the integral, we follow the strategy described in Section D.1. Along these lines, we end up with

$$\hat{\Phi}^B(x,s) = (c_0 x/s^2)\mathrm{H}(x) \quad \text{for all } x \in \mathbb{R}. \tag{D.7}$$

Finally, the TD counterpart of

$$\hat{\Gamma}^B(x,s) = [\hat{Z}^{\mathrm{E}}(s)/c_0 \Delta t]\hat{\Phi}^B(x,s) \tag{D.8}$$

is used in Section 8.2 to evaluate the elements of the impeditivity array. It is noted that $\hat{\Gamma}^B(x,s) = 0$ for $x \le 0$.

Reference

[1] E. C. Titchmarsh. *The Theory of Functions, 2nd edition.* London, UK: Oxford University Press, 1939.

Appendix E
Admittivity array of a PMC screen

The elements of the admittivity array that occurs in the Cagniard-DeHoop method of moments (CdH-MoM) solution of electromagnetic (EM) scattering by a bounded perfectly magnetically conducting (PMC) screen can be expressed through the following wave-slowness representation (cf. (B.1)):

$$\hat{\mathcal{Y}}^{[S,n]}(s) = \frac{\eta_0}{sc_0\Delta t}\frac{c_0^2}{2\pi\mathrm{i}}\int_{\kappa=-\mathrm{i}\infty}^{\mathrm{i}\infty}\tilde{\Pi}^{[S]}(\kappa)\tilde{\Lambda}^{[n]}(-\kappa)\frac{p(\kappa)\mathrm{d}\kappa}{2}$$
$$\text{for all } S=\{2,\ldots,N-1\}, n=\{2,\ldots,N-1\} \text{ and } \{s\in\mathbb{R}; s>0\}, \quad \text{(E.1)}$$

where $\eta_0 = (\epsilon_0/\mu_0)^{1/2}$ is the free-space admittance and $p(\kappa) = (1/c_0^2 - \kappa^2)^{1/2}$ with $\mathrm{Re}(p) \geq 0$, which entails the branch cut along $\{\mathrm{Im}(\kappa) = 0, 1/c_0 \leq |\mathrm{Re}(\kappa)| < \infty\}$. As the integrand has not singularity at $\kappa = 0$, we may indent the integration path to the right with a vanishingly small semicircle without affecting the result of integration. The indented integration contour $\mathbb{I}_0$ is shown in Figure B.1. Subsequently, making use of the transform-domain counterparts of the basis and testing functions (see Section 8.1.2) in (E.1), the array elements can be written as

$$\hat{\mathcal{Y}}^{[S,n]}(s) = \frac{\eta_0}{c_0\Delta t}\frac{1}{\Delta}\Big[\hat{\Xi}(x_S - x_n + 3\Delta/2, s) - 3\hat{\Xi}(x_S - x_n + \Delta/2, s)$$
$$+ 3\hat{\Xi}(x_S - x_n - \Delta/2, s) - \hat{\Xi}(x_S - x_n - 3\Delta/2, s)\Big]$$
$$\text{for all } S=\{2,\ldots,N-1\} \text{ and } n=\{2,\ldots,N-1\}, \quad \text{(E.2)}$$

where $\hat{\Xi}(x,s)$ and its TD counterpart are specified in the ensuing text.

E.1 Generic integral $\hat{\Xi}$

The generic integral representation to be transformed to the TD has the following form (cf. (B.3)):

$$\hat{\Xi}(x,s) = \frac{c_0^2}{s}\frac{1}{2\pi\mathrm{i}}\int_{\kappa\in\mathbb{I}_0}\frac{\exp(s\kappa x)}{s^3\kappa^3}\frac{p(\kappa)\mathrm{d}\kappa}{2}$$
$$\text{for } x\in\mathbb{R} \text{ and } \{s\in\mathbb{R}; s>0\}. \quad \text{(E.3)}$$

To find the TD counterpart of $\hat{\Xi}$, we shall follow the CdH procedure as applied in Section B.1. Accordingly, upon deforming the integration contour into the CdH path

specified by (B.4), while introducing τ as the variable of integration and including the pole contribution, we arrive at

$$\hat{\Xi}(x,s) = \frac{1}{2\pi s^4}\int_{\tau=|x|/c_0}^{\infty} \exp(-s\tau)\frac{x^3}{\tau^3}(\tau^2 - x^2/c_0^2)^{1/2}\mathrm{d}\tau + \frac{c_0^2}{4s^4}\left(\frac{s^2x^2}{c_0} - c_0\right)\mathrm{H}(x). \tag{E.4}$$

The integral with respect to τ can be evaluated in closed form. The thus obtained s-domain expression is subsequently transformed to the TD and we end up with

$$\begin{aligned}\Xi(x,t) = &-\frac{\mathrm{sgn}(x)}{4\pi}\Bigg\{|x|\left(c_0^2t^2 - \frac{x^2}{6}\right)\cosh^{-1}\left(\frac{c_0t}{|x|}\right) - \frac{5}{3}c_0tx^2\left(\frac{c_0^2t^2}{x^2} - 1\right)^{1/2} \\ &+ \frac{\pi}{12}c_0^3t^3 - \frac{c_0^3t^3}{3}\tan^{-1}\left[\frac{c_0t}{|x|} + \left(\frac{c_0^2t^2}{x^2} - 1\right)^{1/2}\right] \\ &+ x^2c_0t\tan^{-1}\left[\left(\frac{c_0^2t^2}{x^2} - 1\right)^{1/2}\right]\Bigg\}\mathrm{H}(t - |x|/c_0) \\ &+ \frac{c_0t}{4}\left(x^2 - \frac{c_0^2t^2}{6}\right)\mathrm{H}(x)\mathrm{H}(t) \quad \text{for all } x \in \mathbb{R} \text{ and } t > 0.\end{aligned} \tag{E.5}$$

This result can be used to find the TD counterpart of $\hat{\mathfrak{Y}}^{[S,n]}(s)$ via (E.2).

Appendix F
Modified impeditivity array

The impeditivity elements that serve for analyzing the electromagnetic (EM) scattering by a grounded screen (see Section 9.1.2) and the mutual EM coupling between two electrically conducting screens (see (10.13)) can be expressed as follows (cf. (B.1)):

$$\hat{\mathcal{K}}^{[S,n]}(s) = \frac{\zeta_0}{sc_0\Delta t}\frac{1}{2\pi\mathrm{i}}\int_{\kappa=-\mathrm{i}\infty}^{\mathrm{i}\infty}\tilde{\Pi}^{[S]}(\kappa)\tilde{\Lambda}^{[n]}(-\kappa)\exp[-sp(\kappa)Z]\frac{\mathrm{d}\kappa}{2p(\kappa)}$$
$$\text{for all } S = \{1,\ldots,N\}, n = \{1,\ldots,N\} \text{ and } \{s \in \mathbb{R}; s > 0\}, \tag{F.1}$$

where $\zeta_0 = (\mu_0/\epsilon_0)^{1/2}$ is the free-space impedance, $Z > 0$ denotes the vertical propagation path and $p(\kappa) = (1/c_0^2 - \kappa^2)^{1/2}$ with $\mathrm{Re}(p) \geq 0$. Owing to the square root in $p(\kappa)$, the branch-cuts along $\{\mathrm{Im}(\kappa) = 0, 1/c_0 \leq |\mathrm{Re}(\kappa)| < \infty\}$ are introduced. Furthermore, it is noted that the integrand is bounded at the origin of the complex κ-plane, so that the integration path along $\mathrm{Re}(\kappa) = 0$ can be replaced with the indented path $\mathbb{I}_0$ as depicted in Figure F.1. Consequently, the array elements can

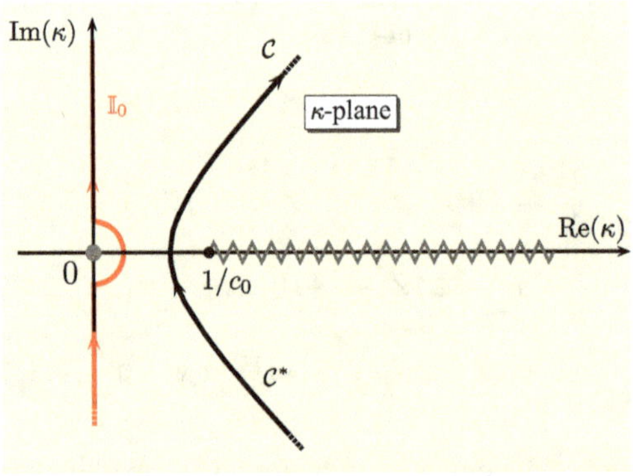

Figure F.1 Complex κ-plane with the hyperbolic CdH path for $x < 0$

be expressed in terms of generic integrals, say $\hat{\Pi}^{A,B}(x, z, s)$, the TD counterparts of which are derived next in Sections F.1 and F.2. In this way, we may write

$$\hat{\mathcal{K}}^{[S,n]}(s) = \frac{\zeta_0}{c_0\Delta t}\frac{1}{\Delta}\Big[\hat{\Pi}^A(x_S - x_n + 3\Delta/2, Z, s) - 3\hat{\Pi}^A(x_S - x_n + \Delta/2, Z, s) + 3\hat{\Pi}^A(x_S - x_n - \Delta/2, Z, s) - \hat{\Pi}^A(x_S - x_n - 3\Delta/2, Z, s)\Big]$$
$$\text{for all } S = \{2, \ldots, N-1\} \text{ and } n = \{2, \ldots, N-1\}, \tag{F.2a}$$

$$\hat{\mathcal{K}}^{[S,1]}(s) = \frac{\zeta_0}{c_0\Delta t}\frac{1}{\Delta}\Big[-\hat{\Pi}^A(x_S - x_1 + \Delta/2, Z, s) + 2\hat{\Pi}^A(x_S - x_1 - \Delta/2, Z, s) - \hat{\Pi}^A(x_S - x_1 - 3\Delta/2, Z, s)\Big] + \frac{\zeta_0}{c_0\Delta t}\Big[\hat{\Pi}^B(x_S - x_1 + \Delta/2, Z, s) - \hat{\Pi}^B(x_S - x_1 - \Delta/2, Z, s)\Big] \quad \text{for all } S = \{2, \ldots, N-1\}, \tag{F.2b}$$

and

$$\hat{\mathcal{K}}^{[S,N]}(s) = \frac{\zeta_0}{c_0\Delta t}\frac{1}{\Delta}\Big[\hat{\Pi}^A(x_S - x_N + 3\Delta/2, Z, s) - 2\hat{\Pi}^A(x_S - x_N + \Delta/2, Z, s) + \hat{\Pi}^A(x_S - x_N - \Delta/2, Z, s)\Big] - \frac{\zeta_0}{c_0\Delta t}\Big[\hat{\Pi}^B(x_S - x_N + \Delta/2, Z, s) - \hat{\Pi}^B(x_S - x_N - \Delta/2, Z, s)\Big] \quad \text{for all } S = \{2, \ldots, N-1\}, \tag{F.2c}$$

with

$$\hat{\mathcal{K}}^{[1,1]}(s) = \frac{\zeta_0}{c_0\Delta t}\frac{1}{\Delta}\Big[-\hat{\Pi}^A(\Delta/2, Z, s) + \hat{\Pi}^A(0, Z, s) + \hat{\Pi}^A(-\Delta/2, Z, s) - \hat{\Pi}^A(-\Delta, Z, s)\Big] + \frac{\zeta_0}{c_0\Delta t}\Big[\hat{\Pi}^B(\Delta/2, Z, s) - \hat{\Pi}^B(0, Z, s)\Big], \tag{F.2d}$$

and

$$\hat{\mathcal{K}}^{[N,1]}(s) = \frac{\zeta_0}{c_0\Delta t}\frac{1}{\Delta}\Big[-\hat{\Pi}^A(x_N - x_1, Z, s) + \hat{\Pi}^A(x_N - x_1 - \Delta/2, Z, s) + \hat{\Pi}^A(x_N - x_1 - \Delta, Z, s) - \hat{\Pi}^A(x_N - x_1 - 3\Delta/2, Z, s)\Big] + \frac{\zeta_0}{c_0\Delta t}\Big[\hat{\Pi}^B(x_N - x_1, Z, s) - \hat{\Pi}^B(x_N - x_1 - \Delta/2, Z, s)\Big]. \tag{F.2e}$$

Again, the remaining array elements can be found by invoking the property of symmetry, $\hat{\mathcal{K}}^{[S,n]}(s) = \hat{\mathcal{K}}^{[n,S]}(s)$. Finally, it is noted that the elements of $\mathcal{Z}^{[S,n]}$, as derived in Appendix B, correspond, in fact, to the limit of $\mathcal{K}^{[S,n]}$ as $Z \downarrow 0$.

F.1 Generic integral $\hat{\Pi}^A$

The first generic integral has the following form (cf. (B.3)):

$$\hat{\Pi}^A(x,z,s) = \frac{1}{s}\frac{1}{2\pi\mathrm{i}}\int_{\kappa\in\mathbb{I}_0}\frac{\exp\{-s[-\kappa x + p(\kappa)z]\}}{s^3\kappa^3}\frac{\mathrm{d}\kappa}{2p(\kappa)}$$
$$\text{for } x\in\mathbb{R},\ z>0 \text{ and } \{s\in\mathbb{R}; s>0\}. \tag{F.3}$$

To cast (F.3) into the form of Laplace transformation, we pursue the line of reasoning behind the CdH technique (see Section 2.1.1) and deform the integration contour, $\mathbb{I}_0$, to the CdH path that is defined via

$$-\kappa x + p(\kappa)z = \tau \quad \text{for } \{\tau\in\mathbb{R}; \tau>0\}. \tag{F.4}$$

Equation (F.4) holds true along a hyperbolic contour, say $\mathcal{C}\cup\mathcal{C}^*$ (here * denotes the complex conjugate), that is symmetrical with respect to $\mathrm{Im}(\kappa)=0$ (see Figure F.1). Upon solving (F.4) for κ, we get

$$\mathcal{C} = \left\{\kappa(\tau) = -(x/\rho^2)\tau + \mathrm{i}(z/\rho^2)(\tau^2-\rho^2/c_0^2)^{1/2}\right\}$$
$$\text{for } \{\rho/c_0 \le \tau < \infty\}, \tag{F.5}$$

where $\rho=(x^2+z^2)^{1/2}$. In the next step, the variable of integration in (F.3) is replaced with the time parameter, τ, which yields (cf. (B.5))

$$\hat{\Pi}^A(x,z,s) = -\frac{c_0^3}{2\pi s^4}\int_{\tau=\rho/c_0}^{\infty}\exp(-s\tau)\frac{xc_0\tau[x^2c_0^2\tau^2-3z^2(c_0^2\tau^2-\rho^2)]}{(c_0^2\tau^2-z^2)^3}$$
$$\times\frac{\mathrm{d}\tau}{(\tau^2-\rho^2/c_0^2)^{1/2}} + \frac{c_0^2}{4s^4}\left(\frac{s^2x^2}{c_0}+sz+c_0\right)\exp(-sz/c_0)\mathrm{H}(x), \tag{F.6}$$

where the second term on the right-hand side represents the pole contribution. With the aid of Lerch's uniqueness theorem [1, Appendix], we finally arrive at the desired TD expression

$$\Pi^A(x,z,t) = \frac{1}{12\pi}\int_{v=\rho}^{c_0t}(c_0t-v)^3\frac{xv[3z^2(v^2-\rho^2)-x^2v^2]}{(v^2-z^2)^3}\frac{\mathrm{d}v}{(v^2-\rho^2)^{1/2}}$$
$$+\frac{c_0t-z}{4}\left[x^2+\frac{z(c_0t-z)}{2}+\frac{(c_0t-z)^2}{6}\right]\mathrm{H}(x)\mathrm{H}(t-z/c_0)$$
$$\text{for all } x\in\mathbb{R},\ z>0 \text{ and } t>0. \tag{F.7}$$

The convolution-type integral in (F.7) can be evaluated via the recursive convolution technique (see Appendix L). A yet more efficient strategy is to carry out the integral analytically. In this way, after lengthy but straightforward algebra, we end up with

$$\begin{aligned}\Pi^A(x,z,t) = {} & \frac{\rho^3}{12\pi}\frac{x}{\rho}\left\{\left(1-\frac{4z^2}{\rho^2}\right)\cosh^{-1}\left(\frac{c_0t}{\rho}\right)+\frac{5}{2}\frac{c_0t}{\rho}\left(\frac{c_0^2t^2}{\rho^2}-1\right)^{1/2}\right.\\ & +\frac{c_0t}{2\rho}\left(\frac{9z^2}{\rho^2}-\frac{c_0^2t^2}{\rho^2}-6\right)\frac{1}{(1-z^2/\rho^2)^{1/2}}\tan^{-1}\left[\frac{(c_0^2t^2/\rho^2-1)^{1/2}}{(1-z^2/\rho^2)^{1/2}}\right]\\ & \left.+\frac{z}{\rho}\left(3-\frac{4z^2}{\rho^2}\right)\frac{1}{(1-z^2/\rho^2)^{1/2}}\tan^{-1}\left[\frac{z}{c_0t}\frac{(c_0^2t^2/\rho^2-1)^{1/2}}{(1-z^2/\rho^2)^{1/2}}\right]\right\}\mathrm{H}(t-\rho/c_0)\\ & +\frac{c_0t-z}{4}\left[x^2+\frac{z(c_0t-z)}{2}+\frac{(c_0t-z)^2}{6}\right]\mathrm{H}(x)\mathrm{H}(t-z/c_0). \qquad \text{(F.8)}\end{aligned}$$

Finally, the limiting case for $x = 0$ readily follows as $\Pi^A(0,z,t) = [(c_0t - z)^2/48](c_0t + 2z)\mathrm{H}(t - z/c_0)$, which is consistent with the results provided in Section B.1.

F.2 Generic integral $\hat{\Pi}^B$

The second generic integral to be transformed to the TD has the following form:

$$\hat{\Pi}^B(x,z,s) = \frac{1}{s}\frac{1}{2\pi\mathrm{i}}\int_{\kappa\in\mathbb{I}_0}\frac{\exp\{-s[-\kappa x+p(\kappa)z]\}}{s^2\kappa^2}\frac{\mathrm{d}\kappa}{2p(\kappa)}$$
$$\text{for } x\in\mathbb{R},\ z>0 \text{ and } \{s\in\mathbb{R}; s>0\}. \qquad \text{(F.9)}$$

Following the strategy from Section F.1, the integration contour is first deformed to the hyperbolic CdH path, $\mathcal{C}\cup\mathcal{C}^*$, the parametric form of which is described by (F.5). Consequently, upon mapping the variable of integration from κ to τ, we arrive at (cf. (B.8))

$$\begin{aligned}\hat{\Pi}^B(x,z,s) = {} & \frac{c_0^2}{2\pi s^3}\int_{\tau=\rho/c_0}^{\infty}\exp(-s\tau)\frac{x^2c_0^2\tau^2-z^2(c_0^2\tau^2-\rho^2)}{(c_0^2\tau^2-z^2)^2}\frac{\mathrm{d}\tau}{(\tau^2-\rho^2/c_0^2)^{1/2}}\\ & +(c_0x/2s^2)\exp(-sz/c_0)\mathrm{H}(x). \qquad \text{(F.10)}\end{aligned}$$

Recognizing the Laplace-type integral in (F.10), the TD counterpart of the s-domain expression immediately follows as

$$\begin{aligned}\Pi^B(x,z,t) = {} & \frac{1}{4\pi}\int_{v=\rho}^{c_0t}(c_0t-v)^2\frac{x^2v^2-z^2(v^2-\rho^2)}{(v^2-z^2)^2}\frac{\mathrm{d}v}{(v^2-\rho^2)^{1/2}}\\ & +[(c_0t-z)x/2]\mathrm{H}(x)\mathrm{H}(t-z/c_0)\\ & \text{for all } x\in\mathbb{R},\ z>0 \text{ and } t>0. \qquad \text{(F.11)}\end{aligned}$$

Again, it is noted that the convolution integral in (F.11) is amenable to handling via the recursive convolution technique (see Appendix L). Alternatively, the integral is carried out analytically, which yields

$$
\begin{aligned}
\Pi^B(x,z,t) = \frac{\rho^2}{4\pi}\Bigg\{ & \left(1-\frac{2z^2}{\rho^2}\right)\cosh^{-1}\left(\frac{c_0 t}{\rho}\right) + \frac{c_0 t}{\rho}\left(\frac{c_0^2 t^2}{\rho^2}-1\right)^{1/2} \\
& - \frac{2c_0 t}{\rho}\left(1-\frac{z^2}{\rho^2}\right)^{1/2}\tan^{-1}\left[\frac{(c_0^2t^2/\rho^2-1)^{1/2}}{(1-z^2/\rho^2)^{1/2}}\right] \\
& + \frac{2z}{\rho}\left(1-\frac{z^2}{\rho^2}\right)^{1/2}\tan^{-1}\left[\frac{z}{c_0 t}\frac{(c_0^2t^2/\rho^2-1)^{1/2}}{(1-z^2/\rho^2)^{1/2}}\right]\Bigg\}\mathrm{H}(t-\rho/c_0) \\
& + \frac{(c_0 t - z)x}{2}\mathrm{H}(x)\mathrm{H}(t-z/c_0).
\end{aligned} \tag{F.12}
$$

Finally, the limit for $x = 0$ can be expressed in the following way:

$$
\begin{aligned}
\Pi^B(0,z,t) &= (c_0^2t^2/4\pi)\tanh[\cosh^{-1}(c_0 t/z)]\mathrm{H}(t-z/c_0) \\
&\quad - (z^2/4\pi)\cosh^{-1}(c_0 t/z)\mathrm{H}(t-z/c_0),
\end{aligned} \tag{F.13}
$$

which boils down to $(c_0^2t^2/4\pi)\mathrm{H}(t)$ in the limit $z \downarrow 0$ (cf. Section B.2).

Reference

[1] M. Štumpf. *Electromagnetic Reciprocity in Antenna Theory*. Hoboken, NJ: John Wiley & Sons, Inc., 2018.

Appendix G

Interface impeditivity array

The elements of the impeditivity array that occurs in the Cagniard-DeHoop method of moments (CdH-MoM) solution of electromagnetic (EM) scattering by a bounded screen located on a dielectric half-space can be expressed as follows (cf. (B.1)):

$$\hat{\mathcal{X}}^{[S,n]}(s) = \frac{\zeta_0}{sc_0\Delta t}\frac{1}{2\pi\mathrm{i}}\int_{\kappa=-\mathrm{i}\infty}^{\mathrm{i}\infty}\tilde{\Pi}^{[S]}(\kappa)\tilde{\Lambda}^{[n]}(-\kappa)\frac{\mathrm{d}\kappa}{p_0(\kappa)+p_1(\kappa)}$$
$$\text{for all } S=\{1,\ldots,N\}, n=\{1,\ldots,N\} \text{ and } \{s\in\mathbb{R}; s>0\}, \tag{G.1}$$

where $\zeta_0 = (\mu_0/\epsilon_0)^{1/2}$ is the free-space impedance, and $p_{0,1}(\kappa) = (1/c_{0,1}^2 - \kappa^2)^{1/2}$ denotes the slowness parameters along the z-direction pertaining to $\mathcal{D}_{\pm}^{\infty}$, respectively (see Figure 9.3). We take $\mathrm{Re}(p_{0,1}) \geq 0$ throughout the complex κ-plane, which implies the branch cuts along $\{\mathrm{Im}(\kappa) = 0, 1/c_{0,1} \leq |\mathrm{Re}(\kappa)| < \infty\}$. Furthermore, it is noted that the integrand is bounded at the origin of the complex κ-plane, so that the original integration path along $\mathrm{Re}(\kappa) = 0$ can be replaced with the indented path $\mathbb{I}_0$ as depicted in Figure G.1. Subsequently, substituting the transform-domain counterparts of the basis and testing functions according to Section 8.1.2, the array elements

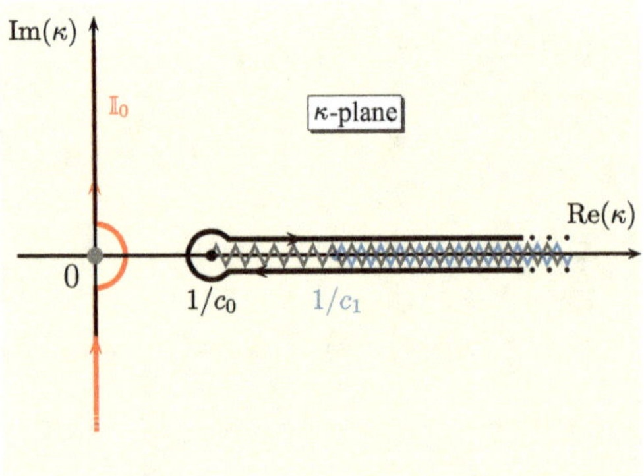

Figure G.1 Complex κ-plane with the CdH path encircling the branch cuts for $x < 0$

can be expressed in terms of generic integrals, $\hat{\Lambda}^{A,B}(x,s)$, the TD counterparts of which will be derived in Sections G.1 and G.2. In this fashion, we further write

$$\hat{\mathcal{X}}^{[S,n]}(s) = \frac{\zeta_0}{c_0\Delta t}\frac{1}{\Delta}\Big[\hat{\Lambda}^A(x_S - x_n + 3\Delta/2, s) - 3\hat{\Lambda}^A(x_S - x_n + \Delta/2, s) + 3\hat{\Lambda}^A(x_S - x_n - \Delta/2, s) - \hat{\Lambda}^A(x_S - x_n - 3\Delta/2, s)\Big] \quad \text{for all } S = \{2,\ldots,N-1\} \text{ and } n = \{2,\ldots,N-1\}, \tag{G.2a}$$

$$\hat{\mathcal{X}}^{[S,1]}(s) = \frac{\zeta_0}{c_0\Delta t}\frac{1}{\Delta}\Big[-\hat{\Lambda}^A(x_S - x_1 + \Delta/2, s) + 2\hat{\Lambda}^A(x_S - x_1 - \Delta/2, s) - \hat{\Lambda}^A(x_S - x_1 - 3\Delta/2, s)\Big] + \frac{\zeta_0}{c_0\Delta t}\Big[\hat{\Lambda}^B(x_S - x_1 + \Delta/2, s) - \hat{\Lambda}^B(x_S - x_1 - \Delta/2, s)\Big] \quad \text{for all } S = \{2,\ldots,N-1\}, \tag{G.2b}$$

and

$$\hat{\mathcal{X}}^{[S,N]}(s) = \frac{\zeta_0}{c_0\Delta t}\frac{1}{\Delta}\Big[\hat{\Lambda}^A(x_S - x_N + 3\Delta/2, s) - 2\hat{\Lambda}^A(x_S - x_N + \Delta/2, s) + \hat{\Lambda}^A(x_S - x_N - \Delta/2, s)\Big] - \frac{\zeta_0}{c_0\Delta t}\Big[\hat{\Lambda}^B(x_S - x_N + \Delta/2, s) - \hat{\Lambda}^B(x_S - x_N - \Delta/2, s)\Big] \quad \text{for all } S = \{2,\ldots,N-1\}, \tag{G.2c}$$

with

$$\hat{\mathcal{X}}^{[1,1]}(s) = \frac{\zeta_0}{c_0\Delta t}\frac{1}{\Delta}\Big[-\hat{\Lambda}^A(\Delta/2, s) + \hat{\Lambda}^A(0, s) + \hat{\Lambda}^A(-\Delta/2, s) - \hat{\Lambda}^A(-\Delta, s)\Big] + \frac{\zeta_0}{c_0\Delta t}\Big[\hat{\Lambda}^B(\Delta/2, s) - \hat{\Lambda}^B(0, s)\Big], \tag{G.2d}$$

and

$$\hat{\mathcal{X}}^{[N,1]}(s) = \frac{\zeta_0}{c_0\Delta t}\frac{1}{\Delta}\Big[-\hat{\Lambda}^A(x_N - x_1, s) + \hat{\Lambda}^A(x_N - x_1 - \Delta/2, s) + \hat{\Lambda}^A(x_N - x_1 - \Delta, s) - \hat{\Lambda}^A(x_N - x_1 - 3\Delta/2, s)\Big] + \frac{\zeta_0}{c_0\Delta t}\Big[\hat{\Lambda}^B(x_N - x_1, s) - \hat{\Lambda}^B(x_N - x_1 - \Delta/2, s)\Big]. \tag{G.2e}$$

Again, the remaining array elements are found by invoking the property of symmetry, $\hat{\mathcal{X}}^{[S,n]}(s) = \hat{\mathcal{X}}^{[n,S]}(s)$.

G.1 Generic integral $\hat{\Lambda}^A$

The first integral to be transformed to the TD has the following form (cf. (B.3)):

$$\hat{\Lambda}^A(x,s) = \frac{1}{s}\frac{1}{2\pi \mathrm{i}}\int_{\kappa\in\mathbb{I}_0}\frac{\exp(s\kappa x)}{s^3\kappa^3}\frac{\mathrm{d}\kappa}{p_0(\kappa)+p_1(\kappa)}$$
$$\text{for } x\in\mathbb{R} \text{ and } \{s\in\mathbb{R}; s>0\}, \tag{G.3}$$

where the integration path and the branch cuts due to $p_{0,1}(\kappa)$ are depicted in Figure G.1. To transform (G.3) to the TD, we next adopt the strategy applied in Section B.1. Accordingly, the integration contour, $\mathbb{I}_0$, is by virtue of Cauchy's theorem and Jordan's lemma deformed to the loop encircling the branch cuts. The loop is defined by the parametric equation (B.4), where τ has the meaning of the time parameter. Subsequently, the integration in (G.3) is next mapped from κ to τ, which yields (cf. (B.5))

$$\hat{\Lambda}^A(x,s) = -\frac{1}{\pi s^4}\int_{\tau=|x|/c_0}^{|x|/c_1}\exp(-s\tau)\frac{x^3}{\tau^3}\frac{\tau^2-x^2/c_0^2}{x^2/c_1^2-x^2/c_0^2}\frac{\mathrm{d}\tau}{(\tau^2-x^2/c_0^2)^{1/2}}$$
$$-\frac{1}{\pi s^4}\int_{\tau=|x|/c_1}^{\infty}\exp(-s\tau)\frac{x^3}{\tau^3}\frac{(\tau^2-x^2/c_0^2)^{1/2}}{(\tau^2-x^2/c_0^2)^{1/2}+(\tau^2-x^2/c_1^2)^{1/2}}$$
$$\times\frac{\mathrm{d}\tau}{(\tau^2-x^2/c_0^2)^{1/2}}+\frac{c_0^2}{2s^4}\frac{1}{1+c_0/c_1}\left(\frac{s^2x^2}{c_0}+c_1\right)\mathrm{H}(x), \tag{G.4}$$

where the last term on the right-hand side represents the pole contribution. Relying, again, on Lerch's uniqueness theorem [1, Appendix], (G.4) can be transformed to the TD and we get

$$\Lambda^A(x,t) = -\frac{1}{6\pi}\int_{v=|x|}^{\min[c_0t,(c_0/c_1)|x|]}(c_0t-v)^3\frac{x^3}{v^3}\frac{v^2-x^2}{(c_0/c_1)^2x^2-x^2}\frac{\mathrm{d}v}{(v^2-x^2)^{1/2}}$$
$$-\frac{1}{6\pi}\int_{v=|x|}^{c_0t}(c_0t-v)^3\frac{x^3}{v^3}\frac{(v^2-x^2)^{1/2}}{(v^2-x^2)^{1/2}+[v^2-(c_0/c_1)^2x^2]^{1/2}}$$
$$\times\frac{\mathrm{d}v}{(v^2-x^2)^{1/2}}+\frac{c_0t}{2}\frac{1}{1+c_0/c_1}\left(x^2+\frac{c_1}{c_0}\frac{c_0^2t^2}{6}\right)\mathrm{H}(x)\mathrm{H}(t). \tag{G.5}$$

The first integral in (G.5), say $\lambda^A(x,t;a)$, can be carried out analytically. After a few steps of algebra, we find

$$\lambda^A(x,t;a) = \frac{\mathrm{sgn}(x)}{2\pi}\frac{1}{(c_0/c_1)^2-1}\left\{|x|\left(c_0^2t^2-\frac{x^2}{6}\right)\cosh^{-1}(a)\right.$$
$$-c_0t\left(\frac{c_0^2t^2}{6}-x^2\right)\tan^{-1}\left[(a^2-1)^{1/2}\right]-a(a^2-1)^{1/2}$$
$$\left.\times\left[(c_0t/a)x^2+(c_0t/a)^2|x|-(c_0t/a)^3/6-|x|^3/6\right]\right\}, \tag{G.6}$$

where parameter a is defined by

$$a = \begin{cases} c_0t/|x| & \text{for } |x| \leq c_0t \leq (c_0/c_1)|x|, \\ c_0/c_1 & \text{for } c_0t > (c_0/c_1)|x|. \end{cases} \tag{G.7}$$

The zero-offset case follows as $\Lambda^A(0,t) = (c_0^3t^3/24)(c_1/c_0)/(1 + c_0/c_1)\mathrm{H}(t)$, which is consistent with the results provided in Section B.1. The second convolution-type integral in (G.5) can be efficiently evaluated via the recursive convolution technique (see Appendix L).

G.2 Generic integral $\hat{\Lambda}^B$

The second generic integral under consideration has the following form (cf. (B.7)):

$$\hat{\Lambda}^B(x,s) = \frac{1}{s}\frac{1}{2\pi\mathrm{i}}\int_{\kappa\in\mathbb{I}_0}\frac{\exp(s\kappa x)}{s^2\kappa^2}\frac{\mathrm{d}\kappa}{p_0(\kappa)+p_1(\kappa)}$$
$$\text{for } x \in \mathbb{R} \text{ and } \{s \in \mathbb{R}; s > 0\}. \tag{G.8}$$

To find its TD counterpart, we shall pursue the line of reasoning from the previous section. Hence, upon deforming the integration contour into the loop around the branch cuts and introducing the time parameter, τ, as the new variable of integration, we arrive at

$$\begin{aligned}\hat{\Lambda}^B(x,s) = {} & \frac{1}{\pi s^3}\int_{\tau=|x|/c_0}^{|x|/c_1}\exp(-s\tau)\frac{x^2}{\tau^2}\frac{\tau^2 - x^2/c_0^2}{x^2/c_1^2 - x^2/c_0^2}\frac{\mathrm{d}\tau}{(\tau^2 - x^2/c_0^2)^{1/2}} \\ & + \frac{1}{\pi s^3}\int_{\tau=|x|/c_1}^{\infty}\exp(-s\tau)\frac{x^2}{\tau^2}\frac{(\tau^2 - x^2/c_0^2)^{1/2}}{(\tau^2 - x^2/c_0^2)^{1/2} + (\tau^2 - x^2/c_1^2)^{1/2}} \\ & \times \frac{\mathrm{d}\tau}{(\tau^2 - x^2/c_0^2)^{1/2}} + \frac{c_0x}{s^2}\frac{1}{1 + c_0/c_1}\mathrm{H}(x).\end{aligned} \tag{G.9}$$

Transforming the latter expression to the TD, we get

$$\begin{aligned}\Lambda^B(x,t) = {} & \frac{1}{2\pi}\int_{v=|x|}^{\min[c_0t,(c_0/c_1)|x|]}(c_0t - v)^2\frac{x^2}{v^2}\frac{v^2 - x^2}{(c_0/c_1)^2x^2 - x^2}\frac{\mathrm{d}v}{(v^2 - x^2)^{1/2}} \\ & + \frac{1}{2\pi}\int_{v=|x|}^{c_0t}(c_0t - v)^2\frac{x^2}{v^2}\frac{(v^2 - x^2)^{1/2}}{(v^2 - x^2)^{1/2} + [v^2 - (c_0/c_1)^2x^2]^{1/2}} \\ & \times \frac{\mathrm{d}v}{(v^2 - x^2)^{1/2}} + \frac{c_0tx}{1 + c_0/c_1}\mathrm{H}(x)\mathrm{H}(t).\end{aligned} \tag{G.10}$$

The first integral in (G.10), say $\lambda^B(x,t;a)$, can be evaluated analytically. Upon carrying out a few steps of algebra, we find

$$\lambda^B(x,t;a) = \frac{1}{\pi}\frac{1}{(c_0/c_1)^2 - 1}\left\{\left(\frac{c_0^2t^2}{2} - \frac{x^2}{4}\right)\cosh^{-1}(a) \right.$$
$$+ c_0t|x|\tan^{-1}\left[(a^2-1)^{1/2}\right] - a(a^2-1)^{1/2}$$
$$\left. \times \left[(c_0t/a)|x| + (c_0t/a)^2/2 - x^2/4\right]\right\}, \tag{G.11}$$

where parameter a is defined by (G.7). The zero-offset limit follows as $\Lambda^B(0,t) = (c_0^2t^2/2\pi)\log(c_0/c_1)/(c_0^2/c_1^2 - 1)\mathrm{H}(t)$, which is compatible with the results presented in Section B.2. Finally, the second convolution-type integral in (G.10) can be carried out via the recursive convolution technique (see Appendix L), again.

Reference

[1] M. Štumpf. *Electromagnetic Reciprocity in Antenna Theory*. Hoboken, NJ: John Wiley & Sons, Inc., 2018.

Appendix H
Slab impeditivity array

The elements of the impeditivity array that occurs in the Cagniard-DeHoop method of moments (CdH-MoM) solution of electromagnetic (EM) scattering by a screen above a grounded slab can be expressed as follows (cf. (F.1) and (G.1)):

$$\hat{S}^{[S,n]}(s) = \frac{\zeta_0}{s c_0 \Delta t} \sum_{q=0}^{\infty} \frac{1}{2\pi \mathrm{i}} \int_{\kappa=-\mathrm{i}\infty}^{\mathrm{i}\infty} \tilde{\Pi}^{[S]}(\kappa) \tilde{\Lambda}^{[n]}(-\kappa) \tilde{G}^{[q]}(\kappa)$$
$$\times \exp\left[-s p_1(\kappa) Z^{[q]}\right] \frac{\mathrm{d}\kappa}{p_0(\kappa) + p_1(\kappa)}$$
$$\text{for all } S = \{1, \ldots, N\}, n = \{1, \ldots, N\} \text{ and } \{s \in \mathbb{R}; s > 0\}. \quad \text{(H.1)}$$

Here, $\zeta_0 = (\mu_0/\epsilon_0)^{1/2}$ denotes the free-space impedance, $p_{0,1}(\kappa) = (1/c_{0,1}^2 - \kappa^2)^{1/2}$ are the wave slowness parameters along the z-direction (see (9.25)), $Z^{[q]} = 2(q+1)z_0 > 0$ is the vertical propagation path and

$$\tilde{G}^{[q]}(\kappa) = \left[1 - \tilde{R}_{\perp}(\kappa)\right] \left[\tilde{R}_{\perp}(\kappa)\right]^q \quad \text{(H.2)}$$

is the (s-independent) factor describing reflections and transmissions associated with the grounded slab (see (9.42)). The integrand in (H.1) is bounded at the origin of the complex κ-plane, which allows to indent the integration contour to the right with a semicircular arc with center at $\kappa = 0$ and vanishingly small radius. In the complex κ-plane, we further introduce branch cuts along $\{\mathrm{Im}(\kappa) = 0, 1/c_{0,1} \leq |\mathrm{Re}(\kappa)| < \infty\}$ due to the square-root in $p_{0,1}(\kappa)$. The indented integration contour, denoted by $\mathbb{I}_0$, and the branch cuts are depicted in Figure H.1.

The elements of the impeditivity array are further expressed in terms of generic representations, say $\hat{\Psi}^{A,B}(x, z, s)$, the transformation of which back to the TD is thoroughly discussed in Sections H.1 and H.2. In this fashion, we may write

$$\hat{S}^{[S,n]}(s) = \frac{\zeta_0}{c_0 \Delta t} \frac{1}{\Delta} \sum_{q=0}^{\infty} \Big[\hat{\Psi}^A(x_S - x_n + 3\Delta/2, Z^{[q]}, s)$$
$$- 3\hat{\Psi}^A(x_S - x_n + \Delta/2, Z^{[q]}, s) + 3\hat{\Psi}^A(x_S - x_n - \Delta/2, Z^{[q]}, s)$$
$$- \hat{\Psi}^A(x_S - x_n - 3\Delta/2, Z^{[q]}, s)\Big]$$
$$\text{for all } S = \{2, \ldots, N-1\} \text{ and } n = \{2, \ldots, N-1\}, \quad \text{(H.3a)}$$

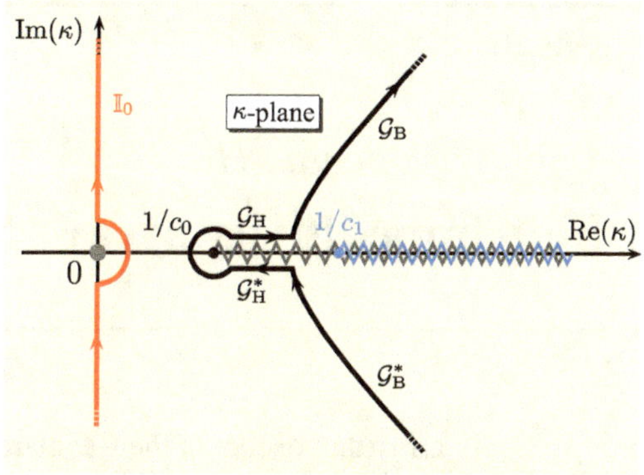

Figure H.1 Complex κ-plane with the head- and body-wave part of the CdH path for $x < 0$

$$\hat{\mathcal{S}}^{[S,1]}(s) = \frac{\zeta_0}{c_0\Delta t}\frac{1}{\Delta}\sum_{q=0}^{\infty}\Big[-\hat{\Psi}^A\big(x_S - x_1 + \Delta/2, Z^{[q]}, s\big)$$

$$+ 2\hat{\Psi}^A\big(x_S - x_1 - \Delta/2, Z^{[q]}, s\big) - \hat{\Psi}^A\big(x_S - x_1 - 3\Delta/2, Z^{[q]}, s\big)\Big]$$

$$+ \frac{\zeta_0}{c_0\Delta t}\Big[\hat{\Psi}^B\big(x_S - x_1 + \Delta/2, Z^{[q]}, s\big) - \hat{\Psi}^B\big(x_S - x_1 - \Delta/2, Z^{[q]}, s\big)\Big]$$

$$\text{for all } S = \{2, \ldots, N-1\}, \tag{H.3b}$$

and

$$\hat{\mathcal{S}}^{[S,N]}(s) = \frac{\zeta_0}{c_0\Delta t}\frac{1}{\Delta}\sum_{q=0}^{\infty}\Big[\hat{\Psi}^A\big(x_S - x_N + 3\Delta/2, Z^{[q]}, s\big)$$

$$- 2\hat{\Psi}^A\big(x_S - x_N + \Delta/2, Z^{[q]}, s\big) + \hat{\Psi}^A\big(x_S - x_N - \Delta/2, Z^{[q]}, s\big)\Big]$$

$$- \frac{\zeta_0}{c_0\Delta t}\Big[\hat{\Psi}^B\big(x_S - x_N + \Delta/2, Z^{[q]}, s\big) - \hat{\Psi}^B\big(x_S - x_N - \Delta/2, Z^{[q]}, s\big)\Big]$$

$$\text{for all } S = \{2, \ldots, N-1\}, \tag{H.3c}$$

with

$$\hat{\mathcal{S}}^{[1,1]}(s) = \frac{\zeta_0}{c_0\Delta t}\frac{1}{\Delta}\sum_{q=0}^{\infty}\Big[-\hat{\Psi}^A(\Delta/2, Z^{[q]}, s) + \hat{\Psi}^A(0, Z^{[q]}, s)$$

$$+ \hat{\Psi}^A(-\Delta/2, Z^{[q]}, s) - \hat{\Psi}^A(-\Delta, Z^{[q]}, s)\Big] + \frac{\zeta_0}{c_0\Delta t}\Big[\hat{\Psi}^B(\Delta/2, Z^{[q]}, s)$$

$$- \hat{\Psi}^B(0, Z^{[q]}, s)\Big], \tag{H.3d}$$

and

$$\hat{\mathcal{S}}^{[N,1]}(s) = \frac{\zeta_0}{c_0 \Delta t} \frac{1}{\Delta} \sum_{q=0}^{\infty} \Big[-\hat{\Psi}^A(x_N - x_1, Z^{[q]}, s)$$
$$+ \hat{\Psi}^A(x_N - x_1 - \Delta/2, Z^{[q]}, s) + \hat{\Psi}^A(x_N - x_1 - \Delta, Z^{[q]}, s)$$
$$- \hat{\Psi}^A(x_N - x_1 - 3\Delta/2, Z^{[q]}, s)\Big] + \frac{\zeta_0}{c_0 \Delta t} \Big[\hat{\Psi}^B(x_N - x_1, Z^{[q]}, s)$$
$$- \hat{\Psi}^B(x_N - x_1 - \Delta/2, Z^{[q]}, s)\Big]. \tag{H.3e}$$

Again, the remaining array elements can be found upon invoking the property of symmetry, $\hat{\mathcal{S}}^{[S,n]}(s) = \hat{\mathcal{S}}^{[n,S]}(s)$.

H.1 Generic integral $\hat{\Psi}^A$

The first generic integral to be transformed to the TD has the following form (see (B.3)):

$$\hat{\Psi}^A(x, z, s) = \frac{1}{s} \frac{1}{2\pi \mathrm{i}} \int_{\kappa \in \mathbb{I}_0} \frac{\exp\{-s[-\kappa x + p_1(\kappa) z]\}}{s^3 \kappa^3} \frac{\tilde{G}(\kappa) \mathrm{d}\kappa}{p_0(\kappa) + p_1(\kappa)}$$
$$\text{for } x \in \mathbb{R},\ z > 0 \text{ and } \{s \in \mathbb{R}; s > 0\}. \tag{H.4}$$

where $\tilde{G} = \tilde{G}^{[q]}$ (see (H.2)), for brevity. The TD counterpart of the generic integral representation (H.4) is arrived at with the aid of the CdH technique (see Section 2.1.1). To this end, the indented integration contour is deformed into the CdH path defined by (cf. (F.4))

$$-\kappa x + p_1(\kappa) z = \tau \quad \text{for } \{\tau \in \mathbb{R}; \tau > 0\}, \tag{H.5}$$

where τ denotes the time parameter. The deformation is permissible since the algebraic factor in the integrand of (H.4) is of $o(1)$ as $|\kappa| \to \infty$ and the contribution from the pole singularity at the origin is to be properly incorporated. Equation (H.5) is satisfied in $\{\mathrm{Im}(\kappa) = 0, -1/c_1 < \mathrm{Re}(\kappa) < 1/c_1\}$, where $p_1(\kappa)$ is real valued and positive. Furthermore, (H.5) holds true along a hyperbolic contour, say $\mathcal{G}_\mathrm{B} \cup \mathcal{G}_\mathrm{B}^*$, that is located symmetrically with respect to $\mathrm{Im}(\kappa) = 0$. The hyperbolic contour represents the *body-wave* part of the CdH path and can be parametrized via

$$\mathcal{G}_\mathrm{B} = \Big\{ \kappa(\tau) = -(x/\rho^2)\tau + \mathrm{i}(z/\rho^2)(\tau^2 - \rho^2/c_1^2)^{1/2} \Big\}$$
$$\text{for } \{\rho/c_1 \le \tau < \infty\}, \tag{H.6}$$

where $\rho = (x^2 + z^2)^{1/2}$. The hyperbolic path intersects the real κ-axis at $\tau = \rho/c_1$, where $\kappa(\tau) = \kappa_0 = -(x/\rho)c_1^{-1}$. Since $c_0 > c_1$, a provision must be made to avoid crossing the branch cut along $\{\mathrm{Im}(\kappa) = 0, 1/c_0 \le |\mathrm{Re}(\kappa)| < \infty\}$. Hence, for observation points where the condition

$$|\kappa_0| = \frac{|x|}{\rho} \frac{1}{c_1} > \frac{1}{c_0} \tag{H.7}$$

holds true, the body-wave CdH path is supplemented with a loop around the branch cut joining the points $\kappa_0 \pm \mathrm{i}0$ (see Figure H.1). This section of the CdH path, further denoted by $\mathcal{G}_\mathrm{H} \cup \mathcal{G}_\mathrm{H}^*$, is called as the *head-wave* part and can be parametrized via

$$\mathcal{G}_\mathrm{H} = \left\{\kappa(\tau) = -(x/\rho^2)\tau - (z/\rho^2)(\rho^2/c_1^2 - \tau^2)^{1/2} + \mathrm{i}0\right\}$$
$$\text{for } \{T_\mathrm{H} = |x|/c_0 + z(1/c_1^2 - 1/c_0^2)^{1/2} \le \tau \le \rho/c_1\}. \tag{H.8}$$

Introducing next the time parameter as the new variable of integration, the integral representation (H.4) can be cast into the form of integrals with the Laplace-transform kernel. In addition, if $x > 0$, the original path, $\mathbb{I}_0$, and the CdH path supplemented with joining arcs at $|\kappa| \to \infty$ surround the pole singularity at $\kappa = 0$, the contribution of which can be evaluated with the aid of Cauchy's formula [1, Section 2.41]. Following this way, the generic integral is finally expressed as a sum of three terms, say

$$\hat{\Psi}^A = \hat{\Psi}_\mathrm{H}^A + \hat{\Psi}_\mathrm{B}^A + \hat{\Psi}_\mathrm{P}^A, \tag{H.9}$$

where subscripts H, B and P denote the pertaining head-wave, body-wave and pole contributions, respectively. Employing the values of $\kappa(\tau)$ along $\mathcal{G}_\mathrm{B}$ as given by (H.6), the body-wave contribution reads

$$\hat{\Psi}_\mathrm{B}^A(x,z,s) = \frac{1}{\pi s^4}\int_{\tau=\rho/c_1}^{\infty} \exp(-s\tau)\mathrm{Re}\left\{\frac{\tilde{G}[\kappa(\tau)]}{\kappa^3(\tau)}\frac{p_1[\kappa(\tau)]}{p_0[\kappa(\tau)]+p_1[\kappa(\tau)]}\right\}_{\mathcal{G}_\mathrm{B}}$$
$$\times \frac{\mathrm{d}\tau}{(\tau^2 - \rho^2/c_1^2)^{1/2}}. \tag{H.10}$$

In a similar fashion, we may express the head-wave contribution

$$\hat{\Psi}_\mathrm{H}^A(x,z,s) = \frac{1}{\pi s^4}\int_{\tau=T_\mathrm{H}}^{\rho/c_1} \exp(-s\tau)\mathrm{Im}\left\{\frac{\tilde{G}[\kappa(\tau)]}{\kappa^3(\tau)}\frac{p_1[\kappa(\tau)]}{p_0[\kappa(\tau)]+p_1[\kappa(\tau)]}\right\}_{\mathcal{G}_\mathrm{H}}$$
$$\times \frac{\mathrm{d}\tau}{(\rho^2/c_1^2 - \tau^2)^{1/2}}\mathrm{H}[|x| - \rho(c_1/c_0)], \tag{H.11}$$

where $\kappa(\tau)$ along the head-wave CdH path is given by (H.8). Finally, the pole contribution follows as

$$\hat{\Psi}_\mathrm{P}^A(x,z,s) = \frac{\mathrm{d}^2}{\mathrm{d}\kappa^2}\left\{\tilde{G}(\kappa)\frac{\exp\{-s[-\kappa x + p_1(\kappa)z]\}}{p_0(\kappa)+p_1(\kappa)}\right\}_{\kappa=0}\frac{\mathrm{H}(x)}{2s^4}. \tag{H.12}$$

The TD counterpart of the body-wave contribution (H.10) can be found with the aid of Lerch's uniqueness theorem [2, Appendix] as

$$\Psi_\mathrm{B}^A(x,z,t) = \frac{1}{6\pi}\int_{v=\rho}^{c_1 t}(c_1 t - v)^3$$
$$\times \mathrm{Re}\left\{\frac{\tilde{G}[\bar{\kappa}(v)]}{c_1^3\bar{\kappa}^3(v)}\frac{c_1 p_1[\bar{\kappa}(v)]}{c_1 p_0[\bar{\kappa}(v)] + c_1 p_1[\bar{\kappa}(v)]}\right\}_{\mathcal{G}_\mathrm{B}}\frac{\mathrm{d}v}{(v^2-\rho^2)^{1/2}}, \tag{H.13}$$

where $c_1\bar{\kappa}(v)$ directly follows from (H.6), that is,

$$c_1\bar{\kappa}(v) = -(x/\rho^2)v + \mathrm{i}(z/\rho^2)(v^2-\rho^2)^{1/2}, \tag{H.14a}$$

$$c_1 p_1[\bar{\kappa}(v)] = (z/\rho^2)v + \mathrm{i}(x/\rho^2)(v^2-\rho^2)^{1/2}, \quad \text{for all } \{\rho \le v < \infty\}, \tag{H.14b}$$

along the body-wave section of the CdH path. The inverse-square root singularity in (H.13) can be handled via the substitution $v = \rho\cosh(u)$ in $\{0 \le u < \infty\}$, for example. In a similar way, we may derive the TD counterpart of the head-wave contribution (H.11) that reads

$$\Psi^A_{\mathrm{H}}(x,z,t) = \frac{1}{6\pi}\int_{v=c_1T_{\mathrm{H}}}^{\min(c_1t,\rho)} (c_1t-v)^3 \times \mathrm{Im}\left\{\frac{\tilde{G}[\bar{\kappa}(v)]}{c_1^3\bar{\kappa}^3(v)}\frac{c_1p_1[\bar{\kappa}(v)]}{c_1p_0[\bar{\kappa}(v)]+c_1p_1[\bar{\kappa}(v)]}\right\}_{\mathcal{G}_{\mathrm{H}}}\frac{\mathrm{d}v}{(\rho^2-v^2)^{1/2}}\mathrm{H}[|x|-\rho(c_1/c_0)], \tag{H.15}$$

where we use (H.8) to specify the corresponding values of $c_1\bar{\kappa}(v)$ along the head-wave section of the CdH path. Accordingly, we write

$$c_1\bar{\kappa}(v) = -(x/\rho^2)v - (z/\rho^2)(\rho^2-v^2)^{1/2} + \mathrm{i}0, \tag{H.16a}$$

$$c_1 p_1[\bar{\kappa}(v)] = (z/\rho^2)v - (x/\rho^2)(\rho^2-v^2)^{1/2}, \quad \text{for all } \{c_1T_{\mathrm{H}} \le v \le \rho\}. \tag{H.16b}$$

The inverse-square root singularity in (H.15) can be handled via the substitution $v = \rho\cos(w)$ in $\{0 \le w \le \pi/2\}$, for example. Further, in order to find the TD counterpart of the pole contribution, $\hat{\Psi}^A_{\mathrm{P}}$, it is necessary to carry out the differentiations with respect to κ as indicated in (H.12). Accordingly, after a few steps of straightforward algebra, it is found that

$$\frac{\mathrm{d}^2}{\mathrm{d}\kappa^2}\left\{\tilde{G}(\kappa)\frac{\exp\{-s[-\kappa x + p_1(\kappa)z]\}}{p_0(\kappa)+p_1(\kappa)}\right\}_{\kappa=0} = (-1)^q\left(\frac{c_0-c_1}{c_0+c_1}\right)^q \times \frac{2c_0c_1}{(c_0+c_1)^2}\left[s^2x^2c_0 + szc_0c_1 + 2c_0^2c_1(q+1) - c_0c_1^2\right]\exp(-sz/c_1). \tag{H.17}$$

Upon substituting (H.17) in (H.12) and transforming the thus obtained s-domain expression to the TD, we get

$$\Psi^A_{\mathrm{P}}(x,z,t) = (K_0/2)(c_1t-z)\{(c_1t-z)^2[2(c_0/c_1)(q+1)-1]/6 + z(c_1t-z)/2 + x^2\}\mathrm{H}(x)\mathrm{H}(t-z/c_1), \tag{H.18}$$

where

$$K_0 = (-1)^q\left(\frac{c_0-c_1}{c_0+c_1}\right)^q\frac{2c_0^2}{(c_0+c_1)^2}. \tag{H.19}$$

Note that for $q = 0$, the limit $c_1 \uparrow c_0$ of $\Psi^A_{\rm P}(x, z, t)$ yields the second term on the right-hand side of (F.7), as expected. Finally, the zero-offset case follows from the pole contribution as $\Psi^A(0, z, t) = \Psi^A_{\rm P}(0, z, t)$, that is,

$$\begin{aligned}\Psi^A(0, z, t) = (K_0/4)(c_1 t - z)\{(c_1 t - z)^2[2(c_0/c_1)(q + 1) - 1]/6 \\ + z(c_1 t - z)/2\}\mathrm{H}(t - z/c_1),\end{aligned} \tag{H.20}$$

which, in the limit $c_1 \uparrow c_0$ and for $q = 0$ yields $\Pi^A(0, z, t)$ as given below (F.8).

H.2 Generic integral $\hat{\Psi}^B$

The second generic integral under consideration can be expressed as

$$\begin{aligned}\hat{\Psi}^B(x, z, s) = \frac{1}{s}\frac{1}{2\pi\mathrm{i}}\int_{\kappa\in\mathbb{I}_0} \frac{\exp\{-s[-\kappa x + p_1(\kappa)z]\}}{s^2\kappa^2}\frac{\tilde{G}(\kappa)\mathrm{d}\kappa}{p_0(\kappa) + p_1(\kappa)} \\ \text{for } x \in \mathbb{R},\ z > 0 \text{ and } \{s \in \mathbb{R}; s > 0\},\end{aligned} \tag{H.21}$$

where $\tilde{G} = \tilde{G}^{[q]}$ (see (H.2)), for brevity. To find the TD counterpart of (H.21), we shall follow the CdH approach applied in the previous section. Hence, integrand is analytically continued away from the imaginary κ-axis and the indented contour $\mathbb{I}_0$ is deformed into the CdH path, while keeping $\mathrm{Re}(p_{0,1}) \geq 0$. Accounting for the contribution of the (double) pole singularity at $\kappa = 0$, the result of integration is written as a sum of head-wave, body-wave and pole contributions, i.e.

$$\hat{\Psi}^B = \hat{\Psi}^B_{\rm H} + \hat{\Psi}^B_{\rm B} + \hat{\Psi}^B_{\rm P}, \tag{H.22}$$

where (cf. (H.10))

$$\begin{aligned}\hat{\Psi}^B_{\rm B}(x, z, s) = \frac{1}{\pi s^3}\int_{\tau=\rho/c_1}^{\infty} \exp(-s\tau)\mathrm{Re}\left\{\frac{\tilde{G}[\kappa(\tau)]}{\kappa^2(\tau)}\frac{p_1[\kappa(\tau)]}{p_0[\kappa(\tau)] + p_1[\kappa(\tau)]}\right\}_{\mathcal{G}_{\rm B}} \\ \times \frac{\mathrm{d}\tau}{(\tau^2 - \rho^2/c_1^2)^{1/2}},\end{aligned} \tag{H.23}$$

and recall that the body-wave CdH path, $\mathcal{G}_{\rm B}$, is given by (H.6). In a similar manner, the head-wave contribution can be cast into the following form (cf. (H.11)):

$$\begin{aligned}\hat{\Psi}^B_{\rm H}(x, z, s) = \frac{1}{\pi s^3}\int_{\tau=T_{\rm H}}^{\rho/c_1} \exp(-s\tau)\mathrm{Im}\left\{\frac{\tilde{G}[\kappa(\tau)]}{\kappa^2(\tau)}\frac{p_1[\kappa(\tau)]}{p_0[\kappa(\tau)] + p_1[\kappa(\tau)]}\right\}_{\mathcal{G}_{\rm H}} \\ \times \frac{\mathrm{d}\tau}{(\rho^2/c_1^2 - \tau^2)^{1/2}}\mathrm{H}[|x| - \rho(c_1/c_0)].\end{aligned} \tag{H.24}$$

Moreover, with the help of Cauchy's formula [1, Section 2.41], we may write (cf. (H.12))

$$\hat{\Psi}^B_{\rm P}(x, z, s) = \frac{\mathrm{d}}{\mathrm{d}\kappa}\left\{\tilde{G}(\kappa)\frac{\exp\{-s[-\kappa x + p_1(\kappa)z]\}}{p_0(\kappa) + p_1(\kappa)}\right\}_{\kappa=0}\frac{\mathrm{H}(x)}{s^3}, \tag{H.25}$$

where

$$\frac{\mathrm{d}}{\mathrm{d}\kappa}\left\{\tilde{G}(\kappa)\frac{\exp\{-s[-\kappa x+p_1(\kappa)z]\}}{p_0(\kappa)+p_1(\kappa)}\right\}_{\kappa=0} = (-1)^q\left(\frac{c_0-c_1}{c_0+c_1}\right)^q \times\frac{2c_0c_1}{(c_0+c_1)^2}sxc_0\exp(-sz/c_1). \tag{H.26}$$

Upon noting the Laplace-transform kernel, $\exp(-s\tau)$, in (H.23) and (H.24), the TD results easily follow. Thus, the body-wave contribution can be written as (cf. (H.13))

$$\Psi^B_{\mathrm{B}}(x,z,t) = \frac{1}{2\pi}\int_{v=\rho}^{c_1t}(c_1t-v)^2 \times\mathrm{Re}\left\{\frac{\tilde{G}[\bar{\kappa}(v)]}{c_1^2\bar{\kappa}^2(v)}\frac{c_1p_1[\bar{\kappa}(v)]}{c_1p_0[\bar{\kappa}(v)]+c_1p_1[\bar{\kappa}(v)]}\right\}_{\mathcal{G}_{\mathrm{B}}}\frac{\mathrm{d}v}{(v^2-\rho^2)^{1/2}}, \tag{H.27}$$

with (H.14a) and (H.14b) specifying the values of $c_1\bar{\kappa}(v)$ and $c_1p_1[\bar{\kappa}(v)]$ along the body-wave path $\mathcal{G}_{\mathrm{B}}$. Next, the TD head-wave contribution reads (cf. (H.15))

$$\Psi^B_{\mathrm{H}}(x,z,t) = \frac{1}{2\pi}\int_{v=c_1T_{\mathrm{H}}}^{\min(c_1t,\rho)}(c_1t-v)^2 \times\mathrm{Im}\left\{\frac{\tilde{G}[\bar{\kappa}(v)]}{c_1^2\bar{\kappa}^2(v)}\frac{c_1p_1[\bar{\kappa}(v)]}{c_1p_0[\bar{\kappa}(v)]+c_1p_1[\bar{\kappa}(v)]}\right\}_{\mathcal{G}_{\mathrm{H}}}\frac{\mathrm{d}v}{(\rho^2-v^2)^{1/2}}\mathrm{H}[|x|-\rho(c_1/c_0)], \tag{H.28}$$

and, finally, the pole contribution follows upon transforming (H.25) with (H.26) to the TD, viz

$$\Psi^B_{\mathrm{P}}(x,z,t) = K_0(c_1t-z)x\mathrm{H}(x)\mathrm{H}(t-z/c_1), \tag{H.29}$$

with K_0 given by (H.19). Observe that the pole contribution under the limit $c_1\uparrow c_0$ yields the second term on the right-hand side of (F.11).

For the sake of completeness, we shall finally provide an expression concerning the zero-offset case. Since for $x=0$, the CdH path $\mathcal{G}_{\mathrm{B}}\cup\mathcal{G}^*_{\mathrm{B}}$ runs along the imaginary axis across the double-pole singularity at $\kappa=0$, integration by parts is employed to find

$$\Psi^B(0,z,t) = \frac{z}{\pi}\int_{v=z}^{c_1t}(c_1t-v)\tilde{H}[\bar{\kappa}(v)]\frac{\mathrm{d}v}{(v^2-z^2)^{1/2}} + \frac{1}{2\pi z}\int_{v=z}^{c_1t}(c_1t-v)^2\tilde{Q}[\bar{\kappa}(v)]\frac{v\,\mathrm{d}v}{(v^2-z^2)^{1/2}}, \tag{H.30}$$

where $c_1\bar{\kappa}(v)=\mathrm{i}(v^2-z^2)^{1/2}/z$, $c_1p_1[\bar{\kappa}(v)]=v/z$ and

$$c_1\tilde{H}(\kappa) = \tilde{G}(\kappa)/[p_0(\kappa)+p_1(\kappa)], \tag{H.31a}$$

$$c_1\tilde{Q}(\kappa) = [\mathrm{d}\tilde{H}(\kappa)/\mathrm{d}\kappa]/c_1\kappa. \tag{H.31b}$$

Recall that $\tilde{G} = \tilde{G}^{[q]}$ as given by (H.2). The inverse-square root singularities in (H.30) can be easily handled via a stretching the variable of integration according to $v = z\cosh(u)$ in $\{0 \leq u < \infty\}$, for instance.

References

[1] E. C. Titchmarsh. *The Theory of Functions, 2nd edition*. London, UK: Oxford University Press, 1939.

[2] M. Štumpf. *Electromagnetic Reciprocity in Antenna Theory*. Hoboken, NJ: John Wiley & Sons, Inc., 2018.

Appendix I

Modified admittivity array

The admittivity elements that serve for analyzing the mutual electromagnetic (EM) coupling between two magnetically conducting screens can be expressed as follows (cf. (11.24) and (11.25)):

$$\hat{\mathcal{L}}^{[S,n]}(s) = \frac{\eta_0}{sc_0\Delta t}\frac{c_0^2}{2\pi\mathrm{i}}\int_{\kappa=-\mathrm{i}\infty}^{\mathrm{i}\infty}\tilde{\Pi}_A^{[S]}(\kappa)\tilde{\Lambda}_B^{[n]}(-\kappa)\exp[-sp(\kappa)Z]\frac{p(\kappa)\mathrm{d}\kappa}{2}$$

$$\text{for all } S=\{2,\ldots,N_A-1\}, n=\{N_A+2,\ldots,N_A+N_B-1\}$$

$$\text{and } \{s\in\mathbb{R}; s>0\}, \tag{I.1}$$

where $\eta_0 = (\epsilon_0/\mu_0)^{1/2}$ is the free-space admittance, $Z > 0$ is the vertical propagation path and $p(\kappa) = (1/c_0^2 - \kappa^2)^{1/2}$ with $\mathrm{Re}(p) \geq 0$. Again, thanks to the square root in $p(\kappa)$, the branch cuts along $\{\mathrm{Im}(\kappa) = 0, 1/c_0 \leq |\mathrm{Re}(\kappa)| < \infty\}$ are introduced. Since the integrand is bounded at the origin of the complex κ-plane, the integration path along $\mathrm{Re}(\kappa) = 0$ can be replaced with the indented path $\mathbb{I}_0$ (see Figure F.1). Consequently, the admittivity-array entries can be expressed as

$$\begin{aligned}\hat{\mathcal{L}}^{[S,n]}(s) = \frac{\eta_0}{c_0\Delta t}\frac{1}{\Delta_B}\Big[&\hat{\Sigma}(x_S - x_n + \Delta_B + \Delta_A/2, Z, s)\\
&- \hat{\Sigma}(x_S - x_n + \Delta_B - \Delta_A/2, Z, s) - 2\hat{\Sigma}(x_S - x_n + \Delta_A/2, Z, s)\\
&+ 2\hat{\Sigma}(x_S - x_n - \Delta_A/2, Z, s) + \hat{\Sigma}(x_S - x_n - \Delta_B + \Delta_A/2, Z, s)\\
&- \hat{\Sigma}(x_S - x_n - \Delta_B - \Delta_A/2, Z, s)\Big]\end{aligned}$$

$$\text{for all } S=\{2,\ldots,N_A-1\} \text{ and } n=\{N_A+2,\ldots,N_A+N_B-1\}, \tag{I.2}$$

where $\hat{\Sigma}(x, z, s)$ and its TD counterpart are specified in the following section.

I.1 Generic integral $\hat{\Sigma}$

The generic integral to be transformed to the TD can be viewed as an extension of (E.3) applying to the self-coupling admittivity of a perfectly magnetically conducting

(PMC) screen. Incorporating now the nonzero vertical propagation path, the generic integral reads (cf. (F.3))

$$\hat{\Sigma}(x,z,s) = \frac{c_0^2}{s}\frac{1}{2\pi\mathrm{i}}\int_{\kappa\in\mathbb{I}_0}\frac{\exp\{-s[-\kappa x + p(\kappa)z]\}}{s^3\kappa^3}\frac{p(\kappa)\mathrm{d}\kappa}{2}$$
$$\text{for } x\in\mathbb{R},\ z>0 \text{ and } \{s\in\mathbb{R};\, s>0\}. \tag{I.3}$$

The TD original of (I.3) is arrived at using the CdH approach pursued in Section F.1. Carrying out this procedure, it is found that

$$\begin{aligned}\hat{\Sigma}(x,z,s) = {} & \frac{c_0^3}{2\pi s^4}\int_{\tau=\rho/c_0}^{\infty}\exp(-s\tau)\mathrm{Re}\left\{\frac{1}{c_0^3\kappa^3(\tau)}\right\}_{\mathcal{C}}\frac{\mathrm{d}\tau}{(\tau^2-\rho^2/c_0^2)^{1/2}}\\ & -\frac{c_0^3}{2\pi s^4}\int_{\tau=\rho/c_0}^{\infty}\exp(-s\tau)\mathrm{Re}\left\{\frac{1}{c_0\kappa(\tau)}\right\}_{\mathcal{C}}\frac{\mathrm{d}\tau}{(\tau^2-\rho^2/c_0^2)^{1/2}}\\ & +\frac{c_0^2}{4s^4}\left(\frac{s^2x^2}{c_0}+sz-c_0\right)\exp(-sz/c_0)\mathrm{H}(x),\end{aligned} \tag{I.4}$$

where the first integral on the left-hand side is, in fact, identical with the one in (F.6) and recall that the CdH path $\mathcal{C}$ is defined via (F.5). Since the second integral admits an analytical solution, say $\hat{\Sigma}^A$, that can be transformed to the TD, we end up with

$$\Sigma(x,z,t) = \Pi^A(x,z,t) + \Sigma^A(x,z,t) - \frac{(c_0t-z)^3}{12}\mathrm{H}(x)\mathrm{H}(t-z/c_0), \tag{I.5}$$

where Π^A is given by (F.8) and

$$\begin{aligned}\Sigma^A(x,z,t) = {} & -\frac{1}{4\pi}\frac{x}{\rho}\left\{\rho\left(c_0^2t^2+\frac{x^2}{6}+\frac{z^2}{2}\right)\cosh^{-1}\left(\frac{c_0t}{\rho}\right)\right.\\ & +z\left(c_0^2t^2+\frac{z^2}{3}\right)\frac{1}{(1-z^2/\rho^2)^{1/2}}\tan^{-1}\left[\frac{z}{c_0t}\frac{(c_0^2t^2/\rho^2-1)^{1/2}}{(1-z^2/\rho^2)^{1/2}}\right]\\ & -c_0t\left(\frac{c_0^2t^2}{3}+z^2\right)\frac{1}{(1-z^2/\rho^2)^{1/2}}\tan^{-1}\left[\frac{(c_0^2t^2/\rho^2-1)^{1/2}}{(1-z^2/\rho^2)^{1/2}}\right]\\ & \left.-\frac{5}{6}c_0t\rho^2\left(\frac{c_0^2t^2}{\rho^2}-1\right)^{1/2}\right\}\mathrm{H}(t-\rho/c_0).\end{aligned} \tag{I.6}$$

Equation (I.5) can be finally used to find the TD original of $\hat{\mathcal{L}}^{[S,n]}(s)$ via (I.2).

Appendix J
Coupling array of noncoplanar metasurfaces

The electromagnetic (EM) coupling between two noncoplanar metasurfaces can be analyzed through arrays, the inverse transformation of which is demonstrated on the elements defined by (11.26a) and (11.27a). Therefore, we shall look for the inverse of the following complex slowness integral:

$$\hat{\mathcal{F}}^{[S,n]}(s) = \frac{\mathrm{sign}(Z)}{c_0\Delta t}\frac{c_0}{2s}\frac{1}{2\pi\mathrm{i}}\int_{\kappa=-\mathrm{i}\infty}^{\mathrm{i}\infty} \tilde{\Pi}_A^{[S]}(\kappa)\tilde{\Lambda}_B^{[n]}(-\kappa)\exp[-sp(\kappa)|Z|]\mathrm{d}\kappa$$
$$\text{for all } S = \{1,\ldots,N_A\}, n = \{N_A+1,\ldots,N_A+N_B\},$$
$$\{s \in \mathbb{R}; s > 0\} \text{ and } Z \in \mathbb{R}. \tag{J.1}$$

Again, the slowness parameter in the z-direction is defined as $p(\kappa) = (1/c_0^2 - \kappa^2)^{1/2}$ with $\mathrm{Re}(p) \geq 0$. To find the TD original of (J.1), we introduce the branch cuts along $\{\mathrm{Im}(\kappa) = 0, 1/c_0 \leq |\mathrm{Re}(\kappa)| < \infty\}$. Furthermore, we note that the integrand is bounded at the origin of the complex κ-plane. Consequently, the original integration path can be replaced with the indented one, further denoted by $\mathbb{I}_0$ (see Figure F.1). The entries of the coupling array, $\hat{\mathcal{F}}^{[S,n]}$, can be then expressed in terms of generic integral representations, $\hat{\Omega}^{A,B}(x,z,s)$, the TD originals of which are derived in Sections J.1 and J.2. Accordingly, we write

$$\hat{\mathcal{F}}^{[S,n]}(s) = \frac{\mathrm{sign}(Z)}{2c_0\Delta t}\frac{1}{\Delta_B}\Big[\hat{\Omega}^A(x_S - x_n + \Delta_B + \Delta_A/2, |Z|, s)$$
$$- \hat{\Omega}^A(x_S - x_n + \Delta_B - \Delta_A/2, |Z|, s) - 2\hat{\Omega}^A(x_S - x_n + \Delta_A/2, |Z|, s)$$
$$+ 2\hat{\Omega}^A(x_S - x_n - \Delta_A/2, |Z|, s) + \hat{\Omega}^A(x_S - x_n - \Delta_B + \Delta_A/2, |Z|, s)$$
$$- \hat{\Omega}^A(x_S - x_n - \Delta_B - \Delta_A/2, |Z|, s)\Big]$$
$$\text{for all } S = \{2,\ldots,N_A-1\} \text{ and } n = \{N_A+2,\ldots,N_A+N_B-1\}, \tag{J.2a}$$

$$\hat{\mathcal{F}}^{[S,N_A+1]}(s) = \frac{\text{sign}(Z)}{2c_0\Delta t}\frac{1}{\Delta_B}\Big[-\hat{\Omega}^A(x_S - x_{N_A+1} + \Delta_A/2, |Z|, s)$$
$$+\hat{\Omega}^A(x_S - x_{N_A+1} - \Delta_A/2, |Z|, s) + \hat{\Omega}^A(x_S - x_{N_A+1} - \Delta_B + \Delta_A/2, |Z|, s)$$
$$-\hat{\Omega}^A(x_S - x_{N_A+1} - \Delta_B - \Delta_A/2, |Z|, s)\Big]$$
$$+\frac{\text{sign}(Z)}{2c_0\Delta t}\Big[\hat{\Omega}^B(x_S - x_{N_A+1} + \Delta_A/2, |Z|, s)$$
$$-\hat{\Omega}^B(x_S - x_{N_A+1} - \Delta_A/2, |Z|, s)\Big] \quad \text{for all } S = \{2, \ldots, N_A - 1\}, \tag{J.2b}$$

$$\hat{\mathcal{F}}^{[S,N_A+N_B]}(s) = \frac{\text{sign}(Z)}{2c_0\Delta t}\frac{1}{\Delta_B}\Big[\hat{\Omega}^A(x_S - x_{N_A+N_B} + \Delta_B + \Delta_A/2, |Z|, s)$$
$$-\hat{\Omega}^A(x_S - x_{N_A+N_B} + \Delta_A/2, |Z|, s) + \hat{\Omega}^A(x_S - x_{N_A+N_B} - \Delta_A/2, |Z|, s)$$
$$-\hat{\Omega}^A(x_S - x_{N_A+N_B} + \Delta_B - \Delta_A/2, |Z|, s)\Big]$$
$$-\frac{\text{sign}(Z)}{2c_0\Delta t}\Big[\hat{\Omega}^B(x_S - x_{N_A+N_B} + \Delta_A/2, |Z|, s)$$
$$-\hat{\Omega}^B(x_S - x_{N_A+N_B} - \Delta_A/2, |Z|, s)\Big] \quad \text{for all } S = \{2, \ldots, N_A - 1\}, \tag{J.2c}$$

$$\hat{\mathcal{F}}^{[1,n]}(s) = \frac{\text{sign}(Z)}{2c_0\Delta t}\frac{1}{\Delta_B}\Big[\hat{\Omega}^A(x_1 - x_n + \Delta_A/2 + \Delta_B, |Z|, s)$$
$$-\hat{\Omega}^A(x_1 - x_n + \Delta_B, |Z|, s) - 2\hat{\Omega}^A(x_1 - x_n + \Delta_A/2, |Z|, s)$$
$$+2\hat{\Omega}^A(x_1 - x_n, |Z|, s) + \hat{\Omega}^A(x_1 - x_n + \Delta_A/2 - \Delta_B, |Z|, s)$$
$$-\hat{\Omega}^A(x_1 - x_n - \Delta_B, |Z|, s)\Big]$$
$$\text{for all } n = \{N_A + 2, \ldots, N_A + N_B - 1\}, \tag{J.2d}$$

and

$$\hat{\mathcal{F}}^{[N_A,n]}(s) = \frac{\text{sign}(Z)}{2c_0\Delta t}\frac{1}{\Delta_B}\Big[\hat{\Omega}^A(x_{N_A} - x_n + \Delta_B, |Z|, s)$$
$$-\hat{\Omega}^A(x_{N_A} - x_n + \Delta_B - \Delta_A/2, |Z|, s) - 2\hat{\Omega}^A(x_{N_A} - x_n, |Z|, s)$$
$$+2\hat{\Omega}^A(x_{N_A} - x_n - \Delta_A/2, |Z|, s) + \hat{\Omega}^A(x_{N_A} - x_n - \Delta_B, |Z|, s)$$
$$-\hat{\Omega}^A(x_{N_A} - x_n - \Delta_B - \Delta_A/2, |Z|, s)\Big]$$
$$\text{for all } n = \{N_A + 2, \ldots, N_A + N_B - 1\}, \tag{J.2e}$$

with

$$\hat{\mathcal{F}}^{[1,N_A+1]}(s) = \frac{\mathrm{sign}(Z)}{2c_0\Delta t}\frac{1}{\Delta_B}\Big[-\hat{\Omega}^A(x_1 - x_{N_A+1} + \Delta_A/2, |Z|, s)$$
$$+\hat{\Omega}^A(x_1 - x_{N_A+1}, |Z|, s) + \hat{\Omega}^A(x_1 - x_{N_A+1} - \Delta_B + \Delta_A/2, |Z|, s)$$
$$-\hat{\Omega}^A(x_1 - x_{N_A+1} - \Delta_B, |Z|, s)\Big]$$
$$+\frac{\mathrm{sign}(Z)}{2c_0\Delta t}\Big[\hat{\Omega}^B(x_1 - x_{N_A+1} + \Delta_A/2, |Z|, s) - \hat{\Omega}^B(x_1 - x_{N_A+1}, |Z|, s)\Big], \tag{J.2f}$$

$$\hat{\mathcal{F}}^{[N_A,N_A+1]}(s) = \frac{\mathrm{sign}(Z)}{2c_0\Delta t}\frac{1}{\Delta_B}\Big[-\hat{\Omega}^A(x_{N_A} - x_{N_A+1}, |Z|, s)$$
$$+\hat{\Omega}^A(x_{N_A} - x_{N_A+1} - \Delta_A/2, |Z|, s) + \hat{\Omega}^A(x_{N_A} - x_{N_A+1} - \Delta_B, |Z|, s)$$
$$-\hat{\Omega}^A(x_{N_A} - x_{N_A+1} - \Delta_A/2 - \Delta_B, |Z|, s)\Big]$$
$$+\frac{\mathrm{sign}(Z)}{2c_0\Delta t}\Big[\hat{\Omega}^B(x_{N_A} - x_{N_A+1}, |Z|, s) - \hat{\Omega}^B(x_{N_A} - x_{N_A+1} - \Delta_A/2, |Z|, s)\Big], \tag{J.2g}$$

$$\hat{\mathcal{F}}^{[1,N_A+N_B]}(s) = \frac{\mathrm{sign}(Z)}{2c_0\Delta t}\frac{1}{\Delta_B}\Big[\hat{\Omega}^A(x_1 - x_{N_A+N_B} + \Delta_A/2 + \Delta_B, |Z|, s)$$
$$-\hat{\Omega}^A(x_1 - x_{N_A+N_B} + \Delta_B, |Z|, s) - \hat{\Omega}^A(x_1 - x_{N_A+N_B} + \Delta_A/2, |Z|, s)$$
$$+\hat{\Omega}^A(x_1 - x_{N_A+N_B}, |Z|, s)\Big]$$
$$-\frac{\mathrm{sign}(Z)}{2c_0\Delta t}\Big[\hat{\Omega}^B(x_1 - x_{N_A+N_B} + \Delta_A/2, |Z|, s) - \hat{\Omega}^B(x_1 - x_{N_A+N_B}, |Z|, s)\Big], \tag{J.2h}$$

and

$$\hat{\mathcal{F}}^{[N_A,N_A+N_B]}(s) = \frac{\mathrm{sign}(Z)}{2c_0\Delta t}\frac{1}{\Delta_B}\Big[\hat{\Omega}^A(x_{N_A} - x_{N_A+N_B} + \Delta_B, |Z|, s)$$
$$-\hat{\Omega}^A(x_{N_A} - x_{N_A+N_B} - \Delta_A/2 + \Delta_B, |Z|, s) - \hat{\Omega}^A(x_{N_A} - x_{N_A+N_B}, |Z|, s)$$
$$+\hat{\Omega}^A(x_{N_A} - x_{N_A+N_B} - \Delta_A/2, |Z|, s)\Big]$$
$$-\frac{\mathrm{sign}(Z)}{2c_0\Delta t}\Big[\hat{\Omega}^B(x_{N_A} - x_{N_A+N_B}, |Z|, s)$$
$$-\hat{\Omega}^B(x_{N_A} - x_{N_A+N_B} - \Delta_A/2, |Z|, s)\Big]. \tag{J.2i}$$

J.1 Generic integral $\hat{\Omega}^A$

The generic integral to be transformed to the TD is, as a matter of fact, an extension of (D.3). Incorporating now the nonzero vertical propagation path, the first generic integral reads

$$\hat{\Omega}^A(x,z,s) = \frac{c_0}{s}\frac{1}{2\pi \mathrm{i}}\int_{\kappa\in\mathbb{I}_0}\frac{\exp\{-s[-\kappa x + p(\kappa)z]\}}{s^3\kappa^3}\mathrm{d}\kappa$$
$$\text{for } x\in\mathbb{R},\ z>0 \text{ and } \{s\in\mathbb{R}; s>0\}. \tag{J.3}$$

To find the TD original of (J.3), we shall pursue the approach applied in Section F.1. Hence, upon mapping the variable of integration according to (F.4), we arrive at (cf. (F.6))

$$\hat{\Omega}^A(x,z,s) = \frac{c_0^3}{\pi s^4}\int_{\tau=\rho/c_0}^{\infty}\exp(-s\tau)\mathrm{Re}\left\{\frac{c_0 p[\kappa(\tau)]}{c_0^3\kappa^3(\tau)}\right\}_{\mathcal{C}}\frac{\mathrm{d}\tau}{(\tau^2-\rho^2/c_0^2)^{1/2}}$$
$$+\frac{c_0^2}{2s^4}\left(\frac{s^2x^2}{c_0}+sz\right)\exp(-sz/c_0)\mathrm{H}(x), \tag{J.4}$$

where the second term on the right-hand side represents the pole contribution and $\rho = (x^2+z^2)^{1/2}$. Relying upon Lerch's uniqueness theorem, the desired TD function is written as

$$\Omega^A(x,z,t) = \frac{1}{6\pi}\int_{v=\rho}^{c_0t}(c_0t-v)^3F(x,z,v)\frac{\mathrm{d}v}{(v^2-\rho^2)^{1/2}}$$
$$+\frac{c_0t-z}{2}\left[x^2+\frac{z(c_0t-z)}{2}\right]\mathrm{H}(x)\mathrm{H}(t-z/c_0), \tag{J.5}$$

where

$$F(x,z,v) = \frac{xz}{\rho^2}\frac{(v^2-\rho^2)[3v^2\rho^2-z^2(v^2-\rho^2)]-x^2v^4}{(v^2-z^2)^3}. \tag{J.6}$$

The convolution integral in (J.5) can be calculated with the aid of the recursive-convolution technique (see Appendix L). Alternatively, the integral can be carried out analytically. This way, after lengthy but straightforward algebra, leads to

$$\Omega^A(x,z,t) = \frac{\rho^3}{2\pi}\frac{xz}{\rho^2}\left\{\frac{2c_0t}{\rho}\cosh^{-1}\left(\frac{c_0t}{\rho}\right)-\frac{3}{2}\frac{c_0t}{\rho}\left(\frac{c_0^2t^2}{\rho^2}-1\right)^{1/2}\right.$$
$$-\frac{1}{2}\left(\frac{3z^2}{\rho^2}+\frac{c_0^2t^2}{\rho^2}-2\right)\frac{1}{(1-z^2/\rho^2)^{1/2}}\tan^{-1}\left[\frac{(c_0^2t^2/\rho^2-1)^{1/2}}{(1-z^2/\rho^2)^{1/2}}\right]$$
$$\left.+\frac{c_0t}{z}\left(\frac{2z^2}{\rho^2}-1\right)\frac{1}{(1-z^2/\rho^2)^{1/2}}\tan^{-1}\left[\frac{z}{c_0t}\frac{(c_0^2t^2/\rho^2-1)^{1/2}}{(1-z^2/\rho^2)^{1/2}}\right]\right\}\mathrm{H}(t-\rho/c_0)$$
$$+\frac{c_0t-z}{2}\left[x^2+\frac{z(c_0t-z)}{2}\right]\mathrm{H}(x)\mathrm{H}(t-z/c_0). \tag{J.7}$$

Finally, the limit for $x = 0$ readily follows from (J.7) as $\Omega^A(0, z, t) = [z(c_0t - z)^2/8]\mathrm{H}(t - z/c_0)$ and $\Omega^A(x, 0, t)$ can be found from (D.4) as $\Omega^A(x, 0, t) = (c_0tx^2/2)\mathrm{H}(x)\mathrm{H}(t)$.

J.2 Generic integral $\hat{\Omega}^B$

The second generic integral analyzed in this appendix has the following form (cf. (D.6)):

$$\hat{\Omega}^B(x, z, s) = \frac{c_0}{s}\frac{1}{2\pi\mathrm{i}}\int_{\kappa\in\mathbb{I}_0}\frac{\exp\{-s[-\kappa x + p(\kappa)z]\}}{s^2\kappa^2}\mathrm{d}\kappa$$
$$\text{for } x \in \mathbb{R},\ z > 0 \text{ and } \{s \in \mathbb{R}; s > 0\}. \tag{J.8}$$

Its TD counterpart is derived with the aid of the CdH technique as applied in Section F.1, again. Following this strategy, it is first found that (cf. (J.4))

$$\hat{\Omega}^B(x, z, s) = \frac{c_0^2}{\pi s^3}\int_{\tau=\rho/c_0}^{\infty}\exp(-s\tau)\mathrm{Re}\left\{\frac{c_0p[\kappa(\tau)]}{c_0^2\kappa^2(\tau)}\right\}_{\mathcal{C}}\frac{\mathrm{d}\tau}{(\tau^2 - \rho^2/c_0^2)^{1/2}}$$
$$+ (c_0x/s^2)\exp(-sz/c_0)\mathrm{H}(x), \tag{J.9}$$

which has the form that is amenable to its transform to the TD. Indeed, using some elementary rules of the Laplace transformation, we readily arrive at

$$\Omega^B(x, z, t) = \frac{1}{2\pi}\int_{v=\rho}^{c_0t}(c_0t - v)^2G(x, z, v)\frac{\mathrm{d}v}{(v^2 - \rho^2)^{1/2}}$$
$$+ (c_0t - z)x\mathrm{H}(x)\mathrm{H}(t - z/c_0), \tag{J.10}$$

in which

$$G(x, z, v) = \frac{vz}{\rho^2}\frac{x^2v^2 - (2x^2 + z^2)(v^2 - \rho^2)}{(v^2 - z^2)^2}. \tag{J.11}$$

Again, the convolution integral in (J.10) can be carried out using the recursive-convolution technique (see Appendix L). Alternatively, the integral is calculated analytically, which yields

$$\Omega^B(x, z, t) = \frac{\rho z}{\pi}\left\{\frac{c_0t}{\rho}\cosh^{-1}\left(\frac{c_0t}{\rho}\right) - \left(\frac{c_0^2t^2}{\rho^2} - 1\right)^{1/2}\right.$$
$$+ \left(1 - \frac{z^2}{\rho^2}\right)^{1/2}\tan^{-1}\left[\frac{(c_0^2t^2/\rho^2 - 1)^{1/2}}{(1 - z^2/\rho^2)^{1/2}}\right]$$
$$\left. - \frac{c_0t}{z}\left(1 - \frac{z^2}{\rho^2}\right)^{1/2}\tan^{-1}\left[\frac{z}{c_0t}\frac{(c_0^2t^2/\rho^2 - 1)^{1/2}}{(1 - z^2/\rho^2)^{1/2}}\right]\right\}\mathrm{H}(t - \rho/c_0)$$
$$+ (c_0t - z)x\mathrm{H}(x)\mathrm{H}(t - z/c_0). \tag{J.12}$$

It is finally noted that the limit of $\Omega^B(x, z, t)$ for $x = 0$ has the following form:

$$\Omega^B(0, z, t) = (z/\pi)\big[c_0 t \cosh^{-1}(c_0 t/z) - z(c_0^2 t^2/z^2 - 1)^{1/2}\big]\mathrm{H}(t - z/c_0), \tag{J.13}$$

and $\Omega^B(x, 0, t)$ immediately follows from the inverse of (D.7) as $\Omega^B(x, 0, t) = c_0 t x \mathrm{H}(x)\mathrm{H}(t)$.

Appendix K

Impedance array of a 3-D screen with dielectric and conductive properties

The elements of the impedance array that occurs in the Cagniard-DeHoop method of moments (CdH-MoM) solution of time-domain (TD) electromagnetic (EM) scattering by a 3-D bounded screen with dielectric and conductive properties can be expressed through (12.12a)–(12.13b). To transform these s-domain expressions to the TD, we further provide the TD counterparts of generic slowness integrals $\hat{\Upsilon}(x, y, s)$, $\hat{\Pi}(x, y, s)$ and $\hat{\Phi}(x, y, s)$.

K.1 Generic integral $\hat{\Upsilon}$

The elements of diagonal blocks given by (12.12a) and (12.12b) are expressed via the following generic integral [1, Eq. (G.3)]:

$$\hat{\Upsilon}(x, y, s) = \frac{c_0^2}{8\pi^2} \int_{\kappa \in \mathbb{I}_0} \frac{\exp(s\kappa x)}{s^3 \kappa^3} \Omega_0^2(\kappa) \mathrm{d}\kappa \int_{\sigma \in \mathbb{J}_0} \frac{\exp(s\sigma y)}{s\sigma} \frac{\mathrm{d}\sigma}{p(\kappa, \sigma)}$$
$$\text{for } \{x \in \mathbb{R}; x \neq 0\},\ \{y \in \mathbb{R}; y \neq 0\} \text{ and } \{s \in \mathbb{R}; s > 0\}, \tag{K.1}$$

where $\mathbb{I}_0$ and $\mathbb{J}_0$ are the integrations contours running along the imaginary axes in the complex κ- and σ-planes, respectively, with the indentation around the origin as depicted in Figure F.1. Further, recall that $\Omega_0(\kappa) = (1/c_0^2 - \kappa^2)^{1/2}$ and $p(\kappa, \sigma)$ has the meaning of slowness along the z-direction and is given by (4.77) with $\mathrm{Re}(p) \geq 0$.

The inverse transform of (K.1) has been carried out in [1, Section G.1] with the aid of the "Cartesian" CdH technique (see Section 2.1.2). Using this transformation technique, we arrive at

$$\Upsilon(x, y, t) = \Upsilon^A(x, y, t) + \Upsilon^B(x, y, t), \tag{K.2}$$

where

$$\Upsilon^A(x,y,t) = \frac{\mathrm{sgn}(x)\mathrm{sgn}(y)}{12\pi}\int_{v=r}^{c_0 t}(c_0 t - v)^3 f(x,y,v)\mathrm{d}v$$
$$+ \frac{\mathrm{sgn}(y)\mathrm{H}(x)}{4\pi}\left\{|y|\left(c_0^2t^2 - x^2 + \frac{y^2}{3}\right)\cosh^{-1}\left(\frac{c_0 t}{|y|}\right)\right.$$
$$- c_0 t\left(\frac{c_0^2t^2}{6} - x^2\right)\tan^{-1}\left[\left(\frac{c_0^2t^2}{y^2} - 1\right)^{1/2}\right]$$
$$\left. - \frac{7}{6}c_0 t y^2\left(\frac{c_0^2t^2}{y^2} - 1\right)^{1/2}\right\}\mathrm{H}(c_0 t - |y|), \tag{K.3}$$

with $r = (x^2 + y^2)^{1/2} > 0$ and

$$\Upsilon^B(x,y,t) = \frac{\mathrm{sgn}(x)\mathrm{H}(y)}{4\pi}\left\{|x|\left(c_0^2t^2 - \frac{x^2}{6}\right)\cosh^{-1}\left(\frac{c_0 t}{|x|}\right)\right.$$
$$- c_0 t\left(\frac{c_0^2t^2}{6} - x^2\right)\tan^{-1}\left[\left(\frac{c_0^2t^2}{x^2} - 1\right)^{1/2}\right]$$
$$\left. - \frac{5}{3}c_0 t x^2\left(\frac{c_0^2t^2}{x^2} - 1\right)^{1/2}\right\}\mathrm{H}(c_0 t - |x|)$$
$$+ \frac{c_0 t}{4}\left(\frac{c_0^2t^2}{6} - x^2\right)\mathrm{H}(x)\mathrm{H}(y)\mathrm{H}(t). \tag{K.4}$$

The function in the integrand on the right-hand side of (K.3) can be expressed as (cf. [1, eq. (G.20)])

$$f(x,y,v) = \frac{1}{2v}\left[\frac{1}{(v^2/x^2 - 1)^{1/2}} + \frac{1}{(v^2/y^2 - 1)^{1/2}}\right]$$
$$- \frac{1}{16}\frac{y^4}{x^2 v^3}\frac{1}{(v^2/y^2 - 1)^{5/2}}\left[3\frac{v^8}{y^8} + 6\frac{v^6}{y^6}\left(\frac{x^2}{y^2} - 1\right)\right.$$
$$\left. + \frac{v^4}{y^4}\left(15\frac{x^4}{y^4} - 10\frac{x^2}{y^2} + 3\right) + 4\frac{v^2}{y^2}\frac{x^2}{y^2}\left(1 - 5\frac{x^2}{y^2}\right) + 8\frac{x^4}{y^4}\right]$$
$$- \frac{1}{2}\frac{x^2}{v^3}\frac{1}{(v^2/x^2 - 1)^{1/2}} - \frac{3}{16}\frac{v}{x^2}\frac{1}{(v^2/y^2 - 1)^{5/2}}\left(3\frac{v^4}{y^4}\right.$$
$$\left. + 2\frac{v^2}{y^2}\left(\frac{x^2}{y^2} - 3\right) + \frac{x^2}{y^2}\left(3\frac{x^2}{y^2} - 2\right) + 3\right)$$
$$+ \frac{3}{4}\frac{v}{x^2}\frac{1}{(v^2/y^2 - 1)^{3/2}}\left(\frac{v^2}{y^2} + \frac{x^2}{y^2} - 1\right). \tag{K.5}$$

The convolution-type integral in (K.3), which is computationally the most expensive part, can be handled numerically either by a standard integration routine or, more efficiently, using the recursive convolution technique (see Appendix L). Yet alternatively, we may apply (multiple) integration by parts to rewrite the integral as

$$\int_{v=r}^{c_0 t}(c_0 t - v)^3 f(x,y,v)\mathrm{d}v = -(c_0 t - r)^3 \partial_v^{-1} f(x,y,r) - 3(c_0 t - r)^2 \partial_v^{-2} f(x,y,r) - 6\,(c_0 t - r)\partial_v^{-3} f(x,y,r) - 6\,\partial_v^{-4} f(x,y,r) + 6\,\partial_v^{-4} f(x,y,c_0 t), \tag{K.6}$$

where ∂_v^{-n} denotes the nth integration with respect to v. For the reader's convenience, the required integrations of (K.5) are further given in the form of explicit formulas. Hence, upon integrating $f(x,y,v)$ with respect to v, we obtain

$$\partial_v^{-1} f(x,y,v) = -\frac{1}{4}\left\{\tan^{-1}\left[\frac{1}{(v^2/x^2-1)^{1/2}}\right] + \tan^{-1}\left[\frac{1}{(v^2/y^2-1)^{1/2}}\right]\right\} - \frac{3}{(v^2/y^2-1)^{3/2}}\left(\frac{v^4}{4x^2y^2} - \frac{v^2}{2x^2} - \frac{v^2}{4y^2} + \frac{y^2}{4x^2} - \frac{x^2}{6y^2} + \frac{1}{4} + \frac{x^2}{12v^2}\right) - \frac{3}{(v^2/y^2-1)^{1/2}}\left(\frac{1}{3} + \frac{y^2}{4x^2} - \frac{v^2}{4x^2}\right) - \frac{x^2}{4v^2}\left(\frac{v^2}{x^2} - 1\right)^{1/2}. \tag{K.7a}$$

Integrating the function once more, we get

$$\partial_v^{-2} f(x,y,v) = -\frac{v}{4}\left\{\tan^{-1}\left[\frac{1}{(v^2/x^2-1)^{1/2}}\right] + \tan^{-1}\left[\frac{1}{(v^2/y^2-1)^{1/2}}\right]\right\} - \frac{|x|}{2}\log\left[v + (v^2-x^2)^{1/2}\right] - \frac{|y|}{2}\log\left[v + (v^2-y^2)^{1/2}\right] - \frac{3}{(v^2/y^2-1)^{1/2}}\left(\frac{v^3}{8x^2} - \frac{vy^2}{8x^2} + \frac{x^2}{12v}\right) + \frac{v}{4}\left(\frac{v^2}{x^2} - 1\right)^{1/2} + \frac{3y^2v}{8x^2}\left(\frac{v^2}{y^2} - 1\right)^{1/2} - \frac{x^2}{4v}\left(\frac{v^2}{x^2} - 1\right)^{3/2}. \tag{K.7b}$$

The third integral with respect to v can be then expressed as

$$\partial_v^{-3} f(x,y,v) = \frac{(2x^2 - v^2)}{8}\left\{\tan^{-1}\left[\frac{1}{(v^2/x^2-1)^{1/2}}\right] + \tan^{-1}\left[\frac{1}{(v^2/y^2-1)^{1/2}}\right]\right\} - \frac{v}{2}\left\{|x|\log\left[v + (v^2-x^2)^{1/2}\right] + |y|\log\left[v + (v^2-y^2)^{1/2}\right]\right\} + \frac{5x^2}{8}\left(\frac{v^2}{x^2} - 1\right)^{1/2} + \frac{y^2}{8}\left(\frac{v^2}{y^2} - 1\right)^{1/2}\left(3 - \frac{v^2}{x^2} + \frac{y^2}{x^2}\right) + \frac{y^4}{8x^2}\left(\frac{v^2}{y^2} - 1\right)^{3/2}, \tag{K.7c}$$

and, finally,

$$\partial_v^{-4} f(x,y,v) = \frac{v(6x^2 - v^2)}{24}\left\{\tan^{-1}\left[\frac{1}{(v^2/x^2-1)^{1/2}}\right]\right.$$
$$\left.+\tan^{-1}\left[\frac{1}{(v^2/y^2-1)^{1/2}}\right]\right\} + \frac{|x|}{4}\left(\frac{x^2}{6} - v^2\right)\cosh^{-1}\left(\frac{v}{|x|}\right)$$
$$-\frac{|y|}{4}\left(\frac{y^2}{3} + v^2 - x^2\right)\cosh^{-1}\left(\frac{v}{|y|}\right) - \frac{|x|}{4}\left(\frac{x^2}{3} + v^2\right)\log(|x|)$$
$$-\frac{|y|}{4}\left(\frac{5y^2}{6} + v^2 - x^2\right)\log(|y|) + \frac{5vx^2}{12}\left(\frac{v^2}{x^2} - 1\right)^{1/2} + \frac{7vy^2}{24}\left(\frac{v^2}{y^2} - 1\right)^{1/2}. \tag{K.7d}$$

With the integrals (K.7a)–(K.7d) at our disposal, the convolution integral in (K.3) can be evaluated analytically via (K.6).

K.2 Generic integral $\hat{\Pi}$

The elements of off-diagonal blocks given by (12.12c) and (12.12d) are expressed via the following generic integral [1, Eq. (14.44)]:

$$\hat{\Pi}(x,y,s) = -\frac{c_0^2}{8\pi^2 s^2}\int_{\kappa\in\mathbb{I}_0}\frac{\exp(s\kappa x)}{s\kappa}\mathrm{d}\kappa\int_{\sigma\in\mathbb{J}_0}\frac{\exp(s\sigma y)}{s\sigma}\frac{\mathrm{d}\sigma}{p(\kappa,\sigma)}$$
$$\text{for } \{x\in\mathbb{R}; x\neq 0\},\ \{y\in\mathbb{R}; y\neq 0\} \text{ and } \{s\in\mathbb{R}; s>0\}, \tag{K.8}$$

and recall that contours $\mathbb{I}_0$ and $\mathbb{J}_0$ run along $\mathrm{Re}(\kappa) = 0$ and $\mathrm{Re}(\sigma) = 0$, respectively, and are indented to the right at the origin (see Figure F.1, for instance). The inverse transform can be accomplished with the aid of the "Cartesian" CdH technique (see Section 2.1.2), again. After straightforward but lengthy steps, we arrive at

$$\Pi(x,y,t) = \frac{\mathrm{sgn}(x)\mathrm{sgn}(y)}{12\pi}\int_{v=r}^{c_0 t}(c_0 t - v)^3 g(x,y,v)\mathrm{d}v$$
$$+\frac{\mathrm{sgn}(y)\mathrm{H}(x)}{4\pi}\left\{|y|\left(c_0^2t^2 + \frac{y^2}{6}\right)\cosh^{-1}\left(\frac{c_0 t}{|y|}\right) - \frac{c_0^3t^3}{3}\tan^{-1}\left[\left(\frac{c_0^2t^2}{y^2} - 1\right)^{1/2}\right]\right.$$
$$\left.-\frac{5}{6}c_0 t y^2\left(\frac{c_0^2t^2}{y^2} - 1\right)^{1/2}\right\}\mathrm{H}(c_0 t - |y|)$$
$$+\frac{\mathrm{sgn}(x)\mathrm{H}(y)}{4\pi}\left\{|x|\left(c_0^2t^2 + \frac{x^2}{6}\right)\cosh^{-1}\left(\frac{c_0 t}{|x|}\right) - \frac{c_0^3t^3}{3}\tan^{-1}\left[\left(\frac{c_0^2t^2}{x^2} - 1\right)^{1/2}\right]\right.$$
$$\left.-\frac{5}{6}c_0 t x^2\left(\frac{c_0^2t^2}{x^2} - 1\right)^{1/2}\right\}\mathrm{H}(c_0 t - |x|) + \frac{c_0^3t^3}{12}\mathrm{H}(x)\mathrm{H}(y)\mathrm{H}(t), \tag{K.9}$$

where the function in the integrand can be expressed as (cf. [1, Eq. (14.46)])

$$g(x, y, v) = \frac{1}{2v}\left[\frac{1}{(v^2/x^2 - 1)^{1/2}} + \frac{1}{(v^2/y^2 - 1)^{1/2}}\right]. \tag{K.10}$$

The convolution-type integral can be, again, computed numerically via a standard integration routine or with the aid of the recursive convolution technique (see Appendix L). A yet more efficient approach is to carry out the integral analytically. This way results in the following formula:

$$\int_{v=r}^{c_0 t} (c_0 t - v)^3 g(x, y, v) \mathrm{d}v = h(x, y, c_0 t) + h(y, x, c_0 t), \tag{K.11}$$

in which

$$\begin{aligned} h(x, y, v) = {} & \frac{c_0^3 t^3}{2}\left\{\tan^{-1}\left[\frac{(v^2 - x^2)^{1/2}}{|x|}\right] - \tan^{-1}\left(\frac{|y|}{|x|}\right)\right\} \\ & - \frac{3|x|}{2}\left(v^2 + \frac{x^2}{6}\right)\log\left[\frac{v + (v^2 - x^2)^{1/2}}{r + |y|}\right] \\ & + \frac{5vx^2}{4}\left(\frac{v^2}{x^2} - 1\right)^{1/2} - \frac{3|x||y|}{2}\left(v - \frac{r}{6}\right). \end{aligned} \tag{K.12}$$

With the aid of (K.11) with (K.12), the convolution integral in (K.9) can be evaluated in a speedy manner.

K.3 Generic integral $\hat{\Phi}$

The effect of (bounded) electric permittivity and conductivity is incorporated via impedance elements $\hat{\mathcal{R}}_{x,x}^{[P,u]}(s)$ and $\hat{\mathcal{R}}_{y,y}^{[Q,v]}(s)$ that are expressed via the following generic integral [1, Eq. (15.4)]:

$$\hat{\Phi}(x, y, s) = -\frac{c_0}{4\pi^2}\int_{\kappa\in\mathbb{I}_0} \frac{\exp(s\kappa x)}{s^3\kappa^3}\mathrm{d}\kappa \int_{\sigma\in\mathbb{J}_0} \frac{\exp(s\sigma y)}{s\sigma}\mathrm{d}\sigma$$
$$\text{for } x \in \mathbb{R},\ y \in \mathbb{R} \text{ and } \{s \in \mathbb{R}; s > 0\}, \tag{K.13}$$

where $\mathbb{I}_0$ and $\mathbb{J}_0$ denote the indented integration paths, again. By virtue of Jordan's lemma [2, p. 1054], the integration paths can be for $x > 0$ and $y > 0$ supplemented by semicircular arcs of arbitrarily large radius in $\mathrm{Re}(\kappa) < 0$ and $\mathrm{Re}(\sigma) < 0$. Subsequently, Cauchy's theorem [3, Section 2.35] is employed to shrink the thus obtained closed contours in small circles around the origins. It is then straightforward to evaluate the integrals via Cauchy's formula [3, Section 2.41], which yields at once

$$\hat{\Phi}(x, y, s) = (c_0 x^2/2s^2)\mathrm{H}(x)\mathrm{H}(y). \tag{K.14}$$

This result in Section 12.1.2 is used to find the TD constituents of the pertaining impedance elements (see (12.15) and (12.16)).

References

[1] M. Štumpf. *Time-Domain Electromagnetic Reciprocity in Antenna Modeling*. Hoboken, NJ: John Wiley & Sons, Inc., 2019.
[2] A. T. de Hoop. *Handbook of Radiation and Scattering of Waves*. London, UK: Academic Press, 1995.
[3] E. C. Titchmarsh. *The Theory of Functions, 2nd edition*. London, UK: Oxford University Press, 1939.

Appendix L
Recursive convolution technique

The problem solution based on the Cagniard-DeHoop method of moments (CdH-MoM) leads to a system of equations of the convolution type (see (8.10)), the (time-dependent) coefficients of which are frequently expressed in the form of convolution-type integrals (e.g. (F.7)). To facilitate an efficient calculation of such integrals, a dedicated technique based on a recursive scheme (see [1]) was introduced in [2, Appendix H]. In this appendix, the recursive-convolution technique is briefly described.

L.1 Representation of the convolution-type integral

The time convolution of two causal signals, say $f(t)$ and $g(t)$, is defined as

$$[f *_t g](t) = \int_{\tau=0}^{t} f(t-\tau)g(\tau)\mathrm{d}\tau \quad \text{for all } t \geq 0. \tag{L.1}$$

The time-convolution integral (L.1) can be approximated along the discretized time axis, $\{t_k = k\Delta t; k = 1, 2, \ldots, M\}$, in several ways. Perhaps the most straightforward approach is to approximate the integral by a Gaussian quadrature. Since the error of a quadrature rule is proportional to the power of the time step $\Delta t > 0$, however, the computational burden grows with the linearly increasing domain of integration. To avoid the difficulty, the time convolution could be represented as the inverse Laplace transform of the product of image functions via the Bromwich inversion integral (see [3, p. 1053]). This way, however, requires that images of $f(t)$ and $g(t)$ are well defined and known, which is not always the case. As a matter of fact, it is assumed, for our purposes, that the Laplace transform of either $f(t)$ or $g(t)$ is available only. Then, supposing that the image of $f(t)$ is available, for example, we may use the Bromwich integral representation of $f(t-\tau)$ in (L.1) and get

$$[f *_t g](t) = \int_{\tau=0}^{t} \left[\frac{1}{2\pi\mathrm{i}} \int_{s\in\mathcal{B}} \exp[s(t-\tau)]\hat{f}(s)\mathrm{d}s\right] g(\tau)\mathrm{d}\tau, \tag{L.2}$$

where $\mathcal{B} = \{s \in \mathbb{C}; \mathrm{Re}(s) = \sigma_0, -\infty < \mathrm{Im}(s) < \infty\}$ denotes the Bromwich path that lies in the region of analyticity of $\hat{f}(s)$ (see Figure L.1). Limiting further ourselves to $\hat{f}(s) = s^{-n}$ with $n \in \{1, 2, \ldots\}$, this region extends over the right half of the complex s-plane. Since $\hat{f}(s) = o(1)$ as $|s| \to \infty$, the necessary condition for Jordan's lemma to apply is met, so that Cauchy's theorem [3, p. 1054] can be employed

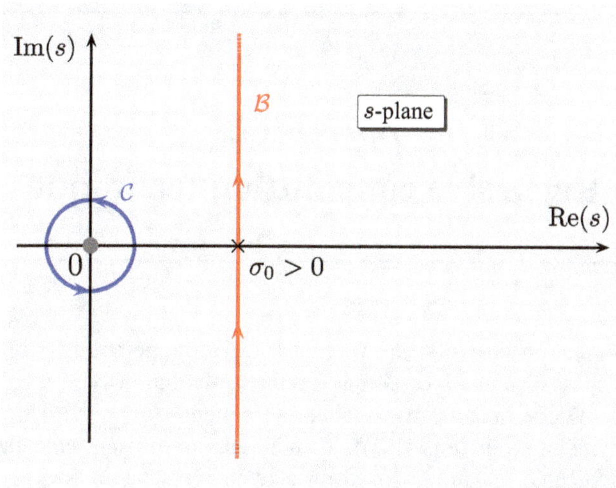

Figure L.1 Complex s-plane with the Bromwich path $\mathcal{B}$ and its replacement $\mathcal{C}$

to replace the Bromwich path with a circular path enclosing the origin of the complex s-plane, viz

$$\mathcal{C} = \left\{ s = \varrho \exp(\mathrm{i}\psi); \varrho > 0, 0 \leq \psi < 2\pi \right\}. \tag{L.3}$$

Subsequently, upon changing the order of the integrations with respect to τ and s, the time convolution can be cast into the following form:

$$[f *_t g](t) = \frac{1}{2\pi\mathrm{i}} \int_{s \in \mathcal{C}} \hat{G}(s,t) s^{-n} \mathrm{d}s, \tag{L.4}$$

in which

$$\hat{G}(s,t) = \int_{\tau=0}^{t} \exp[s(t-\tau)] g(\tau) \mathrm{d}\tau \tag{L.5}$$

is directly amenable to the recursive representation. Indeed, it is straightforward to show that

$$\hat{G}(s,t_k) = \hat{G}(s,t_{k-1}) \exp(s\Delta t) + \int_{\tau=t_{k-1}}^{t_k} \exp[s(t_k-\tau)] g(\tau) \mathrm{d}\tau, \tag{L.6}$$

thus enabling to evaluate the inner integral of (L.4) at $t = t_k$, for any parameter $s \in \mathcal{C}$, using its previous state at $t = t_{k-1}$ and an additional integral term with the constant domain of integration, $\Delta t > 0$. The latter can be approximated with the

aid of an appropriate integration rule. For instance, the 5-point Boole's rule [4, Eq. (25.4.14)] leads to

$$\int_{\tau=t_{k-1}}^{t_k} \exp[s(t_k-\tau)]g(\tau)\mathrm{d}\tau \simeq (14h/45)g(t_{k-1})\exp(s\Delta t)$$
$$+(64h/45)g(t_{k-1}+h)\exp[s(\Delta t-h)]$$
$$+(24h/45)g(t_{k-1}+2h)\exp[s(\Delta t-2h)]$$
$$+(64h/45)g(t_{k-1}+3h)\exp[s(\Delta t-3h)]+(14h/45)g(t_k), \tag{L.7}$$

with $h=\Delta t/4$. The thus approximated (L.6) is finally used in (L.4), where we replace the variable of integration, s, with ψ according to (L.3) and get

$$[f *_t g](t_k) = \frac{\varrho^{1-n}}{2\pi}\int_{\psi=0}^{2\pi} \hat{G}[\varrho\exp(\mathrm{i}\psi), t_k]\exp[\mathrm{i}(1-n)\psi]\mathrm{d}\psi. \tag{L.8}$$

The integral with respect to ψ can be finally handled numerically via the trapezoidal rule, for example, which yields

$$[f *_t g](t_k) \simeq \frac{\varrho^{1-n}}{P}\sum_{p=0}^{P-1} \hat{G}[\varrho\exp(\mathrm{i}p\Delta\psi), t_k]\exp[\mathrm{i}(1-n)p\Delta\psi], \tag{L.9}$$

with the constant step $\Delta\psi = 2\pi/P$. The accuracy of the numerical scheme will next be examined with the aid of a convolution integral that can be carried out analytically.

L.2 Illustrative example

We shall illustrate the numerical procedure described in the previous section on the evaluation of the following convolution-type integral, viz

$$[f *_t g](t) = \frac{1}{2}\int_{\tau=T}^{t} (t-\tau)^2\cosh^{-1}(\tau/T)\mathrm{d}\tau \quad \text{for all } t \geq T > 0. \tag{L.10}$$

Upon comparison of (L.10) with (L.1), it is found that

$$f(t) = (t^2/2)\mathrm{H}(t) \quad \Rightarrow \quad \hat{f}(s) = 1/s^3, \tag{L.11a}$$
$$g(t) = \cosh^{-1}(t/T)\mathrm{H}(t-T) \quad \Rightarrow \quad \hat{g}(s) = \mathrm{K}_0(sT)/s, \tag{L.11b}$$

where $\mathrm{K}_0(x)$ denotes the modified Bessel function of the second kind. In this case, the convolution integral can be carried out analytically, which yields

$$[f *_t g](t) = \frac{t}{2}\left(\frac{t^2}{3}+\frac{T^2}{2}\right)\log\left[\frac{t}{T}+\left(\frac{t^2}{T^2}-1\right)^{1/2}\right]\mathrm{H}(t-T)$$
$$-\left(\frac{11t^2}{36}+\frac{T^2}{9}\right)\left(t^2-T^2\right)^{1/2}\mathrm{H}(t-T). \tag{L.12}$$

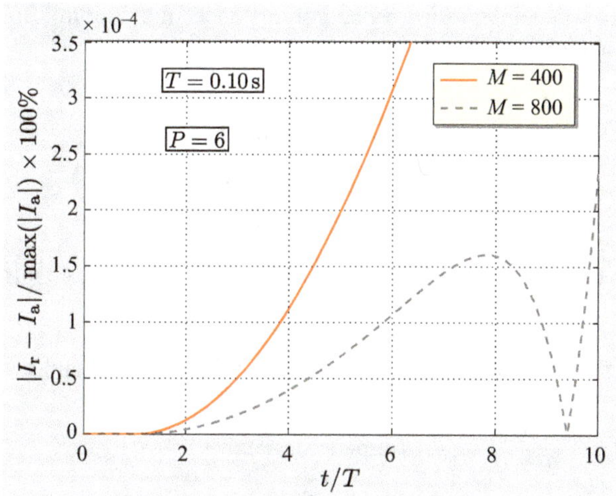

Figure L.2 Relative difference of the approximation evaluated according to (L.9) and the analytical reference (L.12)

We next use the analytical formula to assess the accuracy of the recursive-convolution numerical scheme.

In the illustrative example that follows, the time-convolution integral (L.10) is carried out along a partition of the bounded time window, $\{0 \leq t \leq 10\,T\}$, where we choose $T = 0.10\,\mathrm{s}$. To evaluate the approximation (L.9) for $n = 3$, the radius of $\mathcal{C}$ is taken to be inversely proportional to the upper bound of the time window, that is, $\varrho = 1/10\,T$. Furthermore, the number of division points along the (discretized) circle in the complex s-plane is set to $P = 6$. The results of integration as found using (L.9), further denoted by I_r, and via the analytical formula (L.12), denoted by I_a, were used to evaluate the relative (percentage) difference $|(I_\mathrm{r} - I_\mathrm{a})/\max(|I_\mathrm{a}|)| \times 100\%$. Figure L.2 shows that the recursive scheme converges to the analytical reference with the increasing number of discrete points along the time axis, or, equivalently, with reducing the time step Δt.

For readers' convenience, an illustrative MATLAB® implementation that can be used to approximate the convolution integral (L.10) via the recursive-convolution technique is given below:

```
% TIME WINDOW
T = 0.10; M = 400; t = linspace(0, 10*T, M+1);
% TIME STEP
dt = t(2)-t(1);
%
% ----------------- RECURSIVE ALGORITHM -----------------
g = @(t) acosh(t/T).*(t >= T);
%
% s-PARAMETERS
```

```
rho = 1/max(t); P = 6; psi = (0 : 2*pi/P: 2*pi-2*pi/P).';
s = rho*exp(1i*psi);
%
G = zeros(P, M+1); h = dt/4;
for k = 2 : M+1
    %
    % 5-point BOOLE'S APPROXIMATION
    G(:, k) = G(:, k-1).*exp(s*dt) ...
        + 14*h/45*g(t(k-1))*exp(s*dt) ...
        + 64*h/45*g(t(k-1) + h)*exp(s*(dt-h)) ...
        + 24*h/45*g(t(k-1) + 2*h)*exp(s*(dt-2*h)) ...
        + 64*h/45*g(t(k-1) + 3*h)*exp(s*(dt-3*h)) ...
        + 14*h/45*g(t(k));
    %
end
%
IN = sum(real(G.*repmat(1./s.^2, [1 M+1])))/P;
```

References

[1] A. Semlyen and A. Dabuleanu. Fast and accurate switching transient calculations on transmission lines with ground return using recursive convolutions. *IEEE Transactions on Power Apparatus and Systems*, 94(2):561–571, 1975.

[2] M. Štumpf. *Time-Domain Electromagnetic Reciprocity in Antenna Modeling*. Hoboken, NJ: John Wiley & Sons, Inc., 2019.

[3] A. T. de Hoop. *Handbook of Radiation and Scattering of Waves*. London, UK: Academic Press, 1995.

[4] M. Abramowitz and I. A. Stegun. *Handbook of Mathematical Functions*. New York, NY: Dover Publications, 1972.

Index

www.ingramcontent.com/pod-product-compliance
Lightning Source LLC
LaVergne TN
LVHW011828200926
848484LV00002B/337

* 9 7 8 1 8 3 9 5 3 6 1 3 7 *